Aquaculture Production Systems

Aquaculture Production Systems

Editor: Olando Martin

www.callistoreference.com

Callisto Reference,
118-35 Queens Blvd., Suite 400,
Forest Hills, NY 11375, USA

Visit us on the World Wide Web at:
www.callistoreference.com

ISBN: 978-1-64116-069-8 (Hardback)

Trademark Notice: Registered trademark of products or corporate names are used only for explanation and identification without intent to infringe.

Cataloging-in-Publication Data

Aquaculture production systems / edited by Olando Martin.
    p. cm.
Includes bibliographical references and index.
ISBN 978-1-64116-069-8
1. Aquaculture. 2. Fish culture. I. Martin, Olando.
SH135 .A68 2019
639.8--dc21

Table of Contents

Preface

Aquaculture is concerned with the breeding, rearing and harvesting of aquatic organisms as a source of food. It also includes interventions in the rearing process to enhance population of aquatic species. Higher consumption owing to rising world population has led to the disintegration of certain marine populations in aquatic ecosystems. A workable understanding of production systems is essential for sustaining a successful practice of aquaculture. Studies in aquaculture production systems explore techniques of mariculture and integrated multitrophic aquaculture practices. This book is a compilation of chapters that discuss the most vital concepts and emerging trends in the field of aquaculture production systems. It also attempts to present theories and studies in this field in a lucid and comprehensible manner. This book will be useful to marine biologists, ecologists and other professionals working in this field. It will also serve as a resource guide for students associated with this field.

This book is the end result of constructive efforts and intensive research done by experts in this field. The aim of this book is to enlighten the readers with recent information in this area of research. The information provided in this profound book would serve as a valuable reference to students and researchers in this field.

At the end, I would like to thank all the authors for devoting their precious time and providing their valuable contribution to this book. I would also like to express my gratitude to my fellow colleagues who encouraged me throughout the process.

Editor

13-*Cis*-Retinoic Acid-Induced Hyperglycemia in the Fresh Water Edible Crab, *Oziothelphusa Senex Senex* is mediated by Triggering Release of Hyperglycemic Hormone from Eyestalks

Sreenivasula Reddy P[1]*and Srilatha M[2]

[1]*Department of Zoology, Sri Venkateswara University, Tirupati-517 502, India*
[2]*Department of Biotechnology, Sri Venkateswara University, Tirupati-517 502, India*

Abstract

The present study was aimed to investigate the effect of 13-cis-retinoic acid (13-CRA) on hemolymph glucose levels in the fresh water edible crab, *Oziothelphusa senex* senex. Injection of 13-CRA significantly increased hemolymph glucose levels in a dose-dependent manner in intact crabs. Bilateral eyestalk ablation (ESX) resulted in significant decrease in hemolymph glucose levels. Injection of 13-CRA in to ESX crabs did not cause any significant changes in hemolymph glucose level as compared to ESX crabs suggesting that the effect of 13-CRA could be on the neuroendocrine system in the eyestalks increasing secretion of hyperglycemic hormone. To test this hypothesis, eyestalks were collected from control and 13-CRA injected crabs, and tested for hyperglycemic effect and also for the hyperglycemic hormone levels. The levels of hyperglycemic hormone and the hyperglycemic effect were significantly low in the eyestalks collected from 13-CRA injected crabs when compared with eyestalks from control crabs. From results, it is hypothesized that 13-CRA-induced hyperglycemia in the crab, O. senex, is mediated by triggering the release of hyperglycemic hormone from the eyestalk.

Keywords: 13-*cis*-retinoic acid; Eyestalk ablation; Hyperglycemia; Hemolymph glucose, *Oziothelphusa senex senex.*

Introduction

In crustaceans, glucose homeostasis is primarily under the control of an eyestalk hormone, namely crustacean hyperglycemic hormone (CHH). Crustacean hyperglycemic hormone as a diabetogenic factor was first reported by Abramowitz et al. [1], within eyestalks of decapod crustaceans. Since then, the chemical nature, mode of action, and the target tissues of CHH have been extensively studied in several crustaceans [2-5]. The action of CHH in inducing hyperglycemia is mainly through mobilization of glucose from the tissue carbohydrate pools [6]. It has been reported that CHH stimulates glycogenolysis by activating glycogen phosphorylase in both muscle and hepatopancreas [2,7].

Retinoic acids (RA) including 9-*cis*-retinoic acid (9-CRA) and 13-*cis*-retinoic acid (13-CRA) and all-*trans*-retinoic acid (ATRA) are the metabolites of the vitamin A. Retinoids render their biological activity by binding to nuclear receptors in vertebrates. *cis*-retinoic acids like 9-CRA and 13-CRA interacts with both retinoic acid receptors (RARα,β,γ) and retinoid X receptor (RXR), whereas ATRA mainly interacts with the RARs [8]. It was reported earlier that administration of *cis*-retinoic acid increases insulin release in cultured RINm5F cells [9, 10]. The antidiabetic effects of retinoids in human skeletal muscle [11] and diabetic rodents [12] are accepted to be the mediated through the RXR/RAR heterodimer, and RXR homodimer.

In several crustaceans endogenous retinoic acid has been discovered [13] though it is not recognized as a functional hormone. Although there are sporadic reports on the identification of retinoic acid in crustaceans [14], there is little information available on the role of retinoic acid in the regulation of physiology. Recently, we have reported hyperglycemia in the freshwater crab *Oziothelphusa senex senex* after 9-CRA administration [15]. In as much as (a) RA is identified in the crustacean eyestalks and in circulation [13], (b) RXR was discovered in the eyestalks of *O. senex*, and (c) the eyestalks are the major site of CHH secretion [2], the present study was undertaken to examine the possibility that 13-CRA has any role in regulating hemolymph glucose level in the crab *O. senex senex* and if so, to determine whether it has any effect on CHH in the eyestalks.

Materials and Methods

Collection and maintenance of animals

Intact, intermolt (Stage C_4) adult male crabs, *Oziothelphusa senex senex* Fabricius, with a body weight of 30 ± 3 g and carapace width of 36 ± 3 mm were collected from the rice fields and irrigation canals around Tirupati (13°36'N, 79°25'E), Andhra Pradesh, India. The animals were housed 6-8 per glass aquaria (length: width: height=60:30:30 cm) with 40 L sand-filtered tap water (Salinity: 0.5 ppt) and transferred to fresh water every day. They were acclimatized to the laboratory conditions (temperature 28 ± 1°C and12:12 h; light: dark cycle) for 7 days before use. The crabs were fed with sheep meat *ad libitum* once daily. Feeding was stopped one day before the commencement of the experiment to avoid changes due to prandial activity.

Test chemical

13-*cis*-retinoic acid (13-CRA; chemical purity ≥ 98% HPLC) was purchased from Sigma Chemical Company (St. Louis, MO, USA). 13-CRA was dissolved in acetone and then one aliquot of this solution was mixed with nine aliquots of crustacean saline (0.2 M NaCl, 5.4 mM

---

*Corresponding author: Sreenivasula Reddy, Department of Zoology, Sri Venkateswara University, Tirupati-517 502, India
E-mail: psreddy1955@gmail.com

KCl, 2.6 mM MgCl$_2$, 13.5 mM CaCl$_2$, 5.6 mM maleate, 10.8 mM Tris; pH 7.4) [16] to produce the final dose for injection.

Experimental Design

Three experiments were conducted. Experiment 1 was conducted to determine the dose of 13-CRA that induces maximum hyperglycemia in intact adult male crabs. Experiment 2 was performed to determine the time-course of action of 13-CRA in inducing hyperglycemia. Experiments 3, 4 and 5 were conducted to establish whether eyestalk hyperglycaemic hormone is involved in 13-CRA-induced hyperglycemia.

In experiment 1, intact adult male crabs were injected with different doses of 13-CRA through the base of the chelae with a micro-syringe (Hamilton make) in 10 μl volume. Control crabs were injected with 10 μl crustacean saline. Hemolymph was collected from the crabs 2 h after injection and analysed for sugar levels.

In experiment 2, intact adult male crabs were injected with 25 μg retinoic acid/g live mass. This dose was selected based on the results of experiment 1 (see below). Hemolymph was withdrawn from injected crabs at different time points (30, 60, 120, and 360 min) and used for glucose quantification as described in experiment 1.

In experiment 3, both eyestalks were removed (in order to deprive the eyestalk hormones) from the crabs by cutting off the stalks at their bases at the arthrodial membrane without prior ligation, but with cautery of the wound after operation. We routinely achieve more than 95% survival of the crabs following the operation. Twenty four hours after eyestalk ablation, eyestalkless crabs (ESX) were then injected with 25 μg retinoic acid/g live mass. Hemolymph was collected from ESX crabs, 2 h after 13-CRA injection and used for glucose quantification. The dose of 13-CRA and time-point of hemolymph collection were selected based on the results of experiment 1 and 2 (see below).

In experiment 4, forty eight intact adult male crabs were divided into six equal groups. Crabs in group I served as control and the animals in group II served as saline injected controls. Animals in groups III and IV were injected with either saline (25 μl) or 25 μg retinoic acid/g live mass respectively. Two hours after injection, both eyestalks were collected from crabs in groups III and IV and eyestalk extracts were prepared by homogenizing the eyestalk neural tissue in crab physiological saline and then centrifuging at 10,000 × g for 10 min at 4°C. The supernatant was used for injections. Crabs in groups V and VI were injected with 25 μl of extract (two eyestalk equivalents per crab) prepared using eyestalks collected from crabs in groups III and IV respectively. Hemolymph was collected from control, saline injected control and eyestalk extract injected crabs 2 h after injection and glucose levels were determined.

In experiment 5, crabs were injected with either saline or 25 μg retinoic acid/g live mass. Two hours after injection, eyestalks were collected from saline injected and 13-CRA injected crabs and sinus glands were isolated. Following HPLC of sinus gland sample [17], collected fractions were assayed for CHH. CHH levels were measured using ELISA. The details of ELISA were described elsewhere [15,18]. The polyclonal antibodies for O. senex CHH raised in our laboratory were for the ELISA.

Hemolymph glucose determination

A 10 μl hemolymph sample was collected from the arthrodial membrane of coxa of third pair of walking leg using a micro-syringe and mixed with 250 μl of distilled water and then stored at -80°C.

Glucose concentration was measured enzymatically using the glucose oxidase assay kit (Sigma Chemical Co. St. Louis, MO, USA). The assay protocol essentially followed the one provided by the manufacturer.

Statistical analysis

Data were analyzed by one-way ANOVA followed by Tukey's test using SPSS (Student version 7.5, SPSS Inc, Chertsey, UK). Differences were considered to be significant when $p<0.05$. All data are reported as mean ± S.D.

Results

Effect of 13-*cis*-retinoic acid on hemolymph glucose levels of intact crabs

Injection of 13CRA into intact crabs resulted in significant hyperglycemia in a dose-dependent manner, whereas injection of physiological saline did not cause any significant effect on hemolymph glucose levels (Figure 1). At doses between 10 μg/g live mass and 25 μg/g live mass, the effect was statistically significant and dose-dependent. Doses lower than 10 μg/g live mass, however, retinoic acid did not elicit any hyperglycemic response and doses greater than 25 μg/g live mass exhibited a saturated response in inducing hyperglycemia. In the subsequent experiments, 25 μg/g live mass was selected as injection dose.

A time course action of 13CRA-induced hyperglycemia is presented in Figure 2. Hemolymph glucose levels increased significantly ($p<0.01$) within 30 min after 13-CRA injection and reached a highest peak in 2 h; thereafter, a decline in the hemolymph glucose levels were observed. The hemolymph glucose level was almost normal at 6 h post-injection.

Effect of eyestalk ablation and injection of 13-CRA on hemolymph glucose level in eyestalk-ablated crabs

Bilateral eyestalk ablation resulted in a significant decrease (-49.63%) in hemolymph glucose level in the crab O. senex (Figure 3). Injection of 13-CRA (25 μg/g live mass) into ESX crabs resulted in no significant change in the levels of hemolymph glucose when compared with ESX crabs (Figure 3).

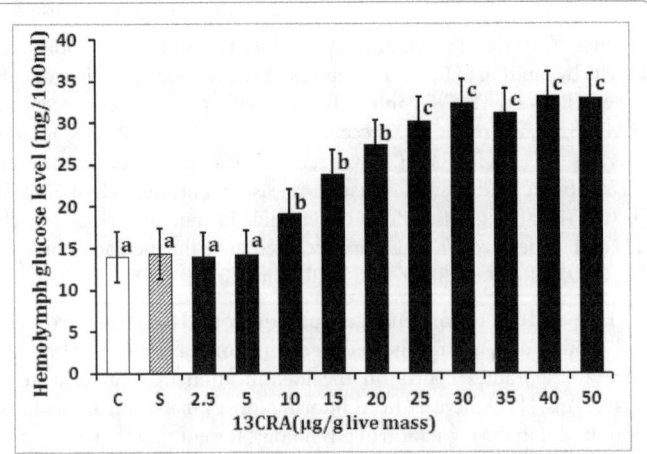

Figure 1: Dose dependent effects of 13CRA on the hemolymph glucose levels in intact crabs. Hemolymph was collected from animals for glucose quantification before (open bar; C: Control) and 2 h after injection with saline (striped bar; S) or 13CRA. Each bar represents a mean ± SD of 8 individuals. Bars with different superscript differ significantly from each other at $p<0.05$

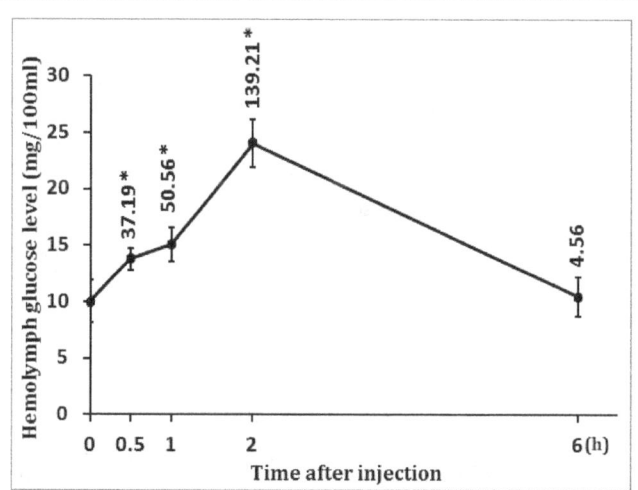

Figure 2: Time course action of 13-CRA-induced hyperglycemia in intact crabs. Hemolymph was collected from intact crabs after injection of retinoic acid (25 μg/g live mass), at the time points indicated for glucose quantification. Each point represents a mean ± S.D. of 8 individuals. Values in parenthesis represent percent change from control (0 h). *$p<0.05$.

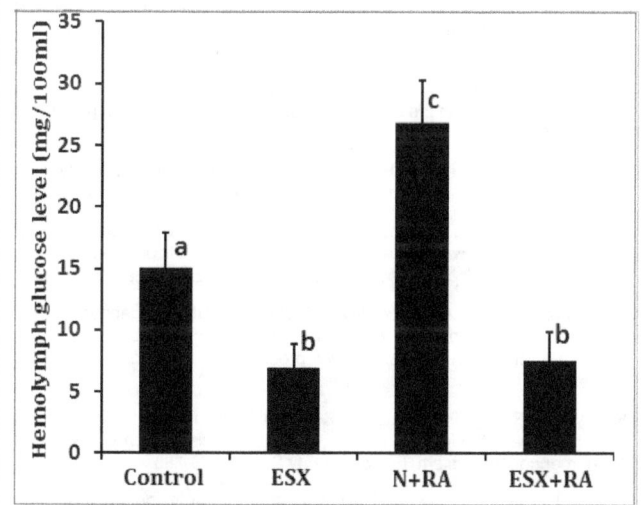

Figure 3: Effect of eyestalk ablation (ESX) and injection of 13-CRA in to normal and ablated crabs on hemolymph glucose levels in *O. senex*. Each bar represents a mean ± S.D. of 8 individuals. Bars with different superscripts differ significantly from each other at $p<0.05$.

Effect of 13-CRA on hyperglycemic activity and CHH content in the eyestalks

Injection of eyestalk extract of intact crabs produced significant hyperglycemia (83.05% above control) (Table 1). Injection of either saline (25 μl) or extracts of eyestalks collected from 13-CRA injected crabs did not cause any significant change in hemolymph glucose concentration when compared with uninjected crabs (Table 1).

Eyestalks of control crabs contained 76.43 ± 8.93 fmol CHH-immunoreactive peptide per sinus gland. Eyestalks of crabs injected with 13-CRA (25 μg/g live mass) contained significantly less (-62.98%) CHH-immunoreactive peptide than in the eyestalks of control crabs (Table 2).

Treatment	Hemolymph glucose level (mg/100 ml)
Control	15.75 ± 1.86
Saline injected	16.23 ± 1.69 (3.05)
Eyestalk extract from control injected	28.86* ± 2.35 (83.05)
Eyestalk extract from 13CRA injected	18.45 ± 2.46 (17.14)

Values are mean ± S.D of 8 crabs.

Values in parentheses are % change from control. *$p<0.05$.

Table 1: Effect of 13-CRA on hemolymph glucose concentrations in crabs injected with saline or various eyestalk extracts.

Treatment	CHH peptide level (fmol/sinus gland)
Control	76.43 ± 8.93
13-CRA injected	28.29* ± 7.86 (-62.98)

Values are mean ± SD n=5 for each group.

Values in parentheses are % change from control. *$p<0.05$.

Table 2: Effect of 13-CRA on levels of CHH peptides in eyestalks.

Discussion

In the crab, *O. senex*, bilateral eyestalk ablation resulted in significant decrease in the hemolymph glucose levels as compared to intact crabs. Injection of 13-CRA significantly elevated hemolymph glucose levels in intact crabs in a dose-dependent manner. An elevation in glucose concentration in the hemolymph of *O. senex* following 13-CRA injection is apparently mediated by the eyestalk CHH hormone, since an increase in the hemolymph glucose concentration following 13-CRA injection was absent after eyestalks were ablated. Zou and Bonvilliain [19] and Reddy and Sainath [15] also reported hyperglycemia in the crabs *U. pugilator* and *O. senex* respectively after 9-CRA administration. Both 13-CRA and 9-CRA most likely rendered their hyperglycemic action through stimulating the release of CHH from the eyestalk neuroendocrine cells to the circulation, since hyperglycemia was observed only in intact crabs but not in eyestalk-ablated crabs. From the results it can be hypothesized that CRA triggers the release of CHH from the sinus gland of the eyestalks. This hypothesis was further supported by the fact that hyperglycemic activity of the eyestalk extracts of the 13CRA injected crabs was less than that of the control crabs. Direct supporting evidence for this hypothesis is provided by the present studies, in which it was shown that the levels of CHH-immuno-reactive peptides were significantly reduced in the eyestalks of the crabs received 13CRA when compared to the CHH levels in the eyestalks of control crabs. These results strongly suggest that hyperglycemia caused by 13-CRA in intact crabs was due to the triggering release of CHH in the crab.

It is well known that alterations in the circulatory levels of CHH occurred during different stress conditions, including cold shock [20], parasite infection [21], exposure to pesticides [22], and heavy metals [23]. It is also known that elevated circulating titres of CHH were reported to occur following exposure to several environmental stressors [24,25] in intact but not in eyestalk ablated crabs, suggesting stress-induced hyperglycemia is CHH mediated response [26-29].

The role of RA in the regulation of glucose metabolism in vertebrates is well established [14]. Retinoids have been implicated in both stimulation of insulin secretion and expression of the glucose transporter 2 gene [10]. Further, retinoids are believed to exert their effects through the retinoic acid receptor/PPARγ heterodimer [11,

12] or retinoic acid receptor (RAR) homodimer [9]. However, such mechanisms in crustaceans are yet to be investigated.

In crustaceans, endogenous RA [13] and a nuclear RAR homologue have been detected [30, 31]. The RAR isolated from the crab has been found to bear a close similarity to vertebrate RARs in the ligand-binding domain. To date, RXR was identified from the ovaries of *C. pugilator*, *G. lateralis*, *M. japonicas*, *F. chinensis*, *C. maenas*, *Scylla serrata* and *Daphnia magna* [32-38]. Recently, RXR has been also identified in the eyestalks of *Callinectes sapidus* [39] *O. senex* [40] and *M. nipponence* [41]. Reduction in mRNA levels for RXR and for vitellogenin in the crab, *C. maenas* after treatment with RXR dsRNA [36] and fluctuations in expression of RXR mRNA in ovaries during reproduction [30] suggests RA acts as reproductive hormone. RAs in crustaceans are also identified as morphogens involved in the limb bud regeneration and morphogenesis [13,14]. RXR has also been identified for which terpenoids can serve as ligands [42]. It was also established that RXR binds with ecdysteroid receptor (EcR) and form a heterodimer [36,42]. This RXR-EcR complex may bind with RA or ecdysteroid (EcD) or methyl farnesoate (MF) and form a heterotrimeric complex (RXR-EcR-RA/EcD/MF) thereby regulate a range of physiological facets in crustaceans. The potential of retinoids in regulating various physiological aspects in crustaceans should now be open for analysis.

Conclusion

On the basis of the results obtained in this study it can be concluded that injection of 13-CRA resulted in significant hyperglycemia in intact crabs, but not in eyestalk ablated crabs suggesting that 13-CRA induced hyperglycemia is mediated through eyestalk hyperglycemic hormone. The levels of hyperglycemic hormone and the hyperglycemic effect was significantly lower in the eyestalks collected from 13-CRA injected crabs when compared with eyestalks from control crabs, strongly suggesting that retinoic acid act, at least in part, by triggering the secretion of hyperglycemic hormone from the eyestalk.

Acknowledgment

The authors are grateful to the Head, Department of Biotechnology, S.V. University, Tirupati, for providing facilities. We thank Professor K.V.S. Sarma, Department of Statistics, S.V. University, Tirupati, for the statistical analysis of data and Mr. S. Umasankar for maintaining the crabs in the laboratory. This research was supported by a research grant from DST to PSR.

References

1. Abramowitz AA, Hisaw FL, Papand DN (1944) The occurrence of a diabetogenic factor in the eyestalks of crustaceans. Biol Bull 86: 1-5.

2. Fanjal-Moles ML (2006) Biochemical and functional aspects of crustacean hyperglycemic hormone in decapod crustaceans. Comp Biochem Physiol 142: 390-400.

3. Van Herp F (1998) Molecular, cytological and physiological aspects of the crustacean hyperglycemic hormone family.

4. Chang ES, Chang SA, Mulder EP (2001) Hormones in the lives of crustaceans- An overview. Am Zool 41: 1090-1097.

5. Webster SG, Keller R, Dircksen H (2012) The CHH-super family of multifunctional peptide hormones controlling crustacean metabolism, osmoregulation, moulting, and reproduction. Gen Comp Endocrinol 175: 217-233.

6. Santos EA, Keller R (1993) Crustacean hyperglycemic hormone (CHH) and the regulation of carbohydrate metabolism- current perspectives. Comp Biochem Physiol A 106: 405-411.

7. Davidson VL, Sittman DM (1999) Biochemistry. Lippincott, William&Wilkins, Philadelphia, pp, 333-340.

8. Chambon P (1996) A decade of molecular biology of retinoic acid receptors. FASEB J 10: 940-954.

9. Chertow BS, Driscoll HK, Goking NQ, Primerano D, Cordle MB, et al. (1997) Retinoid-X receptor and the effects of 9-*cis*-retinoic acid on insulin secretion from RINm5F cells. Metabolism 46: 656-660.

10. Blumentrath J, Neye H, Verspohl EJ (2001) Effects of retinoids and thiazolidiones on proliferation, insulin release, insulin mRNA, GLUT 2 transporter protein and mRNA of INS-1 cells. Cell Biochem Funct 19: 159-169.

11. Cha BS, Ciaraldi TP, Carter L, Nikoulina SE, Mudaliar S, et al. (2001) Peroxisome proliferator activated receptor (PPAR) γ and retinoid X receptor (RXR) agonists have complementary effects on glucose and lipid metabolism in human skeletal muscle. Diabetologia 44: 444-452.

12. Singh AH, Liu S, Crombie DL, Boehm M, Leibowitz MD, et al. (2001) Differential effects of retinoids and thiazolidiones on metabolic gene expression in diabetic rodents. Mol Pharmacol 59: 765-773.

13. Hopkins P (2001) Limb regeneration in the fiddler crab, *Uca pugilator*- hormonal and growth factor control. Am Zool 41: 389-398.

14. Theodosiou M, Laudet V, Schubert M (2010) From carrot to clinic: an overview of the retinoic acid signalling pathway. Cell Mol Life Sci 67: 1423-1445.

15. Reddy PS, Sainath SB (2008) Effect of retinoic acid on hemolymph glucose regulation in the fresh water edible crab, *Oziotelphusa senex senex*. Gen Comp Endocrinol 155: 496-502.

16. Van Harreveld A (1936) A physiological solution for freshwater crustaceans. Proc Soc Exp Biol Med 34: 428-432.

17. Reddy PR, Reddy PS (2006) Isolation and characterization of three cDNAs encoding CHH-family peptides from the crab, *Oziotelphusa senex senex*. Biotechnology 5: 436-442.

18. Chang ES, Keller R, Chang SA (1998) Quantification of crustacean hyperglycemic hormone by ELISA in hemolymph of the lobster, *Homarus americanus*, following various stresses. Gen Comp Endocrinol 111: 359-366.

19. Zou E, Bonvilliain R (2003) Effects of 9-cis and all-trans-retinoic acids on blood glucose homeostasis in the fiddler crab, *Uca pugilator*. Comp Biochem Physiol C 136: 199-204.

20. Kuo CM, Yang YH (1999) Hyperglycemic responses to cold shock in the freshwater giant prawn, *Macrobrachium rosenbergii*. J Comp Physiol 169: 49-54.

21. Stentiford GD, Chang ES, Chang SA, Neil NM (2001) Carbohydrate dynamics and the crustacean hyperglycemic hormone (CHH)- effects of parasitic infection in Norway lobster *Nephrops norvegicus*. Gen Comp Endocrinol 121: 13-22.

22. Fingerman M, MHanumante MM, Deshpande UD, Nagabhushanam R (1981) Increase in the total reducing substances in the hemolymph of the fresh water crab *Barytelphusa guerini*, produced by a pesticide DDT and an indolealkylamine (serotonin). Experientia 37: 178-179.

23. Reddy, PS, Reddy, PR, Sainath, SB (2011) Cadmium and mercury-induced hyperglycemia in the fresh water crab, *Oziotelphusa senex senex*: Involvement of neuroendocrine system. Ecotoxicol Env Saf 74: 279-283.

24. Durand F, Devillers N, Lallier FH, Regnault M (2000) Nitrogen excretion and change in blood components during emersion of the subtidal spider crab *Maja squinado* (L). Comp Biochem Physiol A 127: 259-271.

25. Lorenzon S, Pasqual P, Ferrero EA (2002) Different bacterial lipopolysaccharides as toxicants and stressors in the shrimp, *Palaemon elegans*. Fish Shellfish Immunol 13: 27-45.

26. Reddy PS, Bhagyalakshmi A (1994) A change in oxidative metabolism in selected tissues of the crab *Scylla serrata* in response to cadmium toxicity. Ecotoxicol Environ Saf 29: 255-264.

27. Lorenzon S, Francese M, Ferrero EA (2000) Heavy metal toxicity and differential effects on the hyperglycemic stress responses in the shrimp *Palaemon elegans*. Arch Environ Contam Toxicol 39: 167-176.

28. Lorenzon S, Edoma P, Giulianini PG, Mettulio R, Ferrero EA (2004) Variations of crustacean hyperglycemic hormone (CHH) level in the eyestalk and hemolymph of the shrimp *Palaemon elegans* following stress. J Exp Biol 207: 4205-4213.

29. Lorenzon S, Edomi P, Giulianini PG, Mettulio R, Ferrero EA (2005) Role of biogenic amines and CHH in the crustacean hyperglycemic stress response. J Exp Biol 208: 3341-3347.

30. Durica DS, Wu X, Anilkumar G, Hopkins PM, Chung ACK (2002) Characterization of crab EcR and RXR homologs and expression during limb regeneration and oocyte maturation. Mol Cell Endocrinol 189: 59-76.

31. Chung AC, Durica DS, Clifton SW, Roe BA, Hopkins PM (1998a) Cloning of crustacean ecdysteroid receptor and retinoid-X receptor gene homologs and elevation of retinoid-X receptor mRNA by retinoic acid. Mol Cell Endocrinol 139: 209-227.

32. Chung AC, Durica DS, Hopkins PM (1998b) Tissue-specific patterns and steady-state concentrations of ecdysteroid receptor and retinoid-X receptor mRNA during the molt cycle of the fiddler crab, *Uca (Celuca) pugilator*. Gen Comp Endocrinol 109: 375-389.

33. Kim HW, Donald SGL, Mykles L (2005) Ecdysteroid-responsive genes, RXR and E75, in the tropical land crab, *Gecarcinus lateralis* - differential tissue expression of multiple RXR isoforms generated at three alternative splicing sites in the hinge and ligand-binding domains. Mol Cell Endocrinol 242: 80-95.

34. Asazuma H, Nagata S, Kono M, Nagasawa H (2007) Molecular cloning and expression analysis of ecdysone receptor and retinoid X receptor from the kuruma prawn, *Marsupenaeus japonicus*. Comp Biochem Physiol 148: 139-150.

35. Priya TAJ, Lia F, Zhanga J, Wanga B, Zhaoa C, Xiang J (2009) Molecular characterization and effect of RNA interference of retinoid X receptor (RXR) on E75 and chitinase gene expression in Chinese shrimp *Fenneropenaeus chinensis*. Comp Biochem Physiol B 153: 121-129.

36. Nagaraju GPC, Rajitha B, Borst DW (2011) Molecular cloning and sequence of retinoid X receptor in the green crab *Carcinus maenas* - a possible role in female reproduction. J Endocrinol 210: 379-390.

37. Girish BP, Swetha CH, Reddy PS (2015) Induction of ecdysteroidogenesis and methyl farnesoate synthesis and expression of ecdysteroid receptor and retinoid X receptor in the hepatopancreas and ovary of the giant mud crab *Scylla serrata* by melatonin. Gen Comp Endocrinol.

38. Wang YH, Wang G, LeBlanc GA (2007) Cloning and characterization of the retinoid X receptor from a primitive crustacean *Daphnia magna*. Gen Comp Endocrinol 150: 309-318.

39. Techa S, Chung JS (2013) Ecdysone and retinoid-X receptors of the blue crab, *Callinectes sapidus* - cloning and their expression patterns in eyestalks and Y-organs during the molt cycle. Gene 527: 139-153.

40. Girish BP, Swetha CH, Reddy PS (2015) Expression of RXR, EcR, E75 and VtG mRNA levels in the hepatopancreas and ovary of the freshwater edible crab, *Oziothelphusa senex senex* (Fabricius,1798) during differen t vitellogenic stages. Sci Nat 102:1272.

41. Li Z, Wang WQ, Zhang EF, Qiu GF (2014) Identification of spliced mRNA isoforms of retinoic acid receptor (RXR) in the oriental fresh water prawn *Macrobrachium nipponence*. Genet Mol Res 13: 3914-3926.

42. Hopkins PM, Durica D, Tracy W (2008) RXR isoforms and endogenous retinoids in the fiddler crab, *Uca pugilator*. Comp Biochem Physiol A 151: 602-614.

2

Development of Probiotics Based Culture System of *Macrobrachium rosenbergii* Using Different Stocking Densities

Istiaq Ahmad Chowdhury*, Jewel Das and Nani Gopal Das

Institute of Marine Sciences and Fisheries, University of Chittagong, Bangladesh

Abstract

Owing to the problem of antibiotic resistance and subsequent reluctance of using antibiotics, probiotics use in aquaculture is becoming popular day by day. One experimental design with 150 days culture period of *Macrobrachium rosenbergii* was conducted with 3 treatments maintaining stocking density of $02/m^2$, $03/m^2$ and $04/m^2$ in T1, T2 and T3 respectively. Each of the treatments was with 3 replicates where each replicate was segmented into two parts to separate probiotics and non-probiotics based culture system. The higher body weight of 63.7 g was recorded in lower SD of T1 in comparison to lower body weight of 55.7 g and 43.0 g in higher SD of T2 and T3 respectively for probiotics application segments. The average body weight of 55.7 g, 46.7 g and 37 g respectively were found for the same treatments in non-probiotics segments. The average survival rate of 69.3%, 62.7% and 58.3% were recorded in probiotics and 68.3%, 63% and 57.7% respectively in non-probiotics treatments. Average daily growth rate and gross production were found better in probiotics than that of non-probiotics segments in all the treatments. Average daily growth rate of T1 was found 0.41 g and 0.36 g respectively for probiotics and non-probiotics segments. Similarly, for T2 and T3 average daily growth rate were found 0.35 g and 0.27 g for probiotics and 0.30 g, 0.23 g for non-probiotics segments respectively. Gross average production showed better result of 103 g/m^2/crop in T2 probiotics treated segment than that of other two results of 87.23 g/m^2 and 98.10 g/m^2 in T1 and T3 treatments respectively where as 74.62 g, 87.23 g and 84.26 g/m^2/crop was recorded in T1, T2 and T3 respectively in non-probiotics treatments. Abiotic parameters in all segments of 3 treatments were within the optimum ranges for *M. rosenbergii* culture during the study period.

Keywords: Probiotics; *M. rosenbergii*; Stocking density; Treatment; Production

Introduction

The most widely cultured fast growing two giant prawns in the globe including Bangladesh are *Penaeus monodon* and *Macrobrachium rosenbergii*. Their culture techniques are traditional, extensive, improved extensive, and mainly based on the stocking density of post larvae and management techniques of the culture ponds. Both mono-culture as well as polyculture with fin fish (mullets, carps and tilapia) is the common practice in the South and Southeast Asian countries. Extensive and improved extensive polyculture are widely practiced in the freshwater environment throughout the country [1]. In Bangladesh, prawn farming can play an important role in poverty reduction by increasing job opportunity especially for the rural women that help to enhance their contribution to household income [2]. With increasing demand of environment friendly aquaculture, the use of probiotics in aquaculture is now widely accepted and recently, the use of probiotics to improve and maintain healthy environment for prawn culture has become popular in Bangladesh. Probiotics has been used to supply beneficial bacterial strains in rearing water that will help to increase microbial species composition in the environment and ultimately to improve water quality conditions [3]. This study attempted to make familiar with the fish farmer as well as farm owner's application of probiotics in the culture system of *Macrobrachium rosenbergii* and ultimately to minimize or totally stop the antibiotic application in farming system and finally draw the attraction of export market of the aquaculture products.

Materials and Methods

M. rosenbergii can be cultured in earthen ponds, concrete tanks, small reservoirs, cages, pens and paddy fields. In the present experiment, it was conducted in 9 ponds of two rural villages of Chittagong division, South-Eastern part of Bangladesh.

Probiotics selection, mode of action, doses and usage instruction

Methods of selection for probiotics use in *Macrobrachium* culture were considered the following steps:

1) Collection of background information

2) Acquisition of potential probiotics

3) Estimate the ability of potential probiotics to out-compete pathogenic strains

4) Assessment of pathogenicity of potential probiotics

5) Evaluation of effect of potential probiotics in culture and

6) An economic cost-benefit analysis and to check their availability in the market.

Selection of commercial probiotics

• Sanolife® Pro-1 was used by mixing with feed. Sanolife® Pro-1 contains specific mixture of strains of *Bacillus* spp. Minimum of 3×10^{10} cfu/g. Standard Sanolife® Pro-1 application rates instructed 5 g/kg feed.

• Biomin® Pondlife was another type of probiotics used in pond water. Biomin® Pondlife consists of *Bacillus sp., Enterococcus sp.,*

---

*Corresponding author: Istiaq Ahmad Chowdhury, Institute of Marine Sciences and Fisheries, University of Chittagong, Bangladesh
E-mail: istiaq.ahmad11@gmail.com

Thiobacillus sp. and *Paracoccus sp*. Organic carrier product contains a minimum 2×10^{10} cfu/g. Instructed initial stage dose was 0.175 g/m^2 and final stage dose was 0.3 g/m^2 of pond area.

• Sanolife® Pro-W was the 3rd type of probiotics used in the study. The product is a mixture strain of *Bacillus* spp. of minimum 5×10^{10} cfu/g. Standard Sanolife® Pro-W application rate was instructed 0.02 g/m^2.

Site selection and pond construction for freshwater prawn farming

The experiment was conducted in two different rural areas of Chittagong division, South-Eastern part of Bangladesh. Total nine ponds were selected for the experiment from two different villages and size varies from 1200 m^2 to 3600 m^2. Among the nine selected ponds seven were old and two were newly excavated. Aquatic vegetation and weeds were eradicated from the old ponds. Weed fishes as well as predatory fishes were eradicated by drying the old pond bottom.

Liming and fertilization

Required amount of lime was applied into the ponds to make the desired pH level. Depending upon soil pH, agricultural limestone was applied at 25 g/m^2 to 50 g/m^2. To increase the fertility and productivity of the ponds both organic and inorganic fertilizers were used. Raw cow dung was applied at 75 g/m^2 to 125 g/m^2 along with 6.25 g/m^2 triple super phosphate and 3.75 g/m^2 Urea. This application was continued for every 7 days interval up to the whole culture period.

Increasing surface area

Like many other crustaceans, *M. rosenbergii* shows strong territorial defense and hiding behavior. During molting, there is always a considerable risk of predation by larger dominant prawns. Hence, to increase habitat surface area and hiding, substrates are required to reduce predation in grow out ponds. Bushes, palm and coconut-leaves, bamboo, tiers, tree branches, etc. were used indigenously to provide substrates for prawns to cling and to hide in.

Juvenile collection and stocking

Good quality prawn juveniles were collected from a commercial private hatchery of Chittagong where the production of *M. rosenbergii* was done through probiotics based rearing system. The prawn juveniles were stocked at a size of 2 g to 3 g in the different ponds. In this experiment juveniles were stocked with three different stocking densities. In treatment 1 (T-1) juveniles were stocked at a rate of 2/m^2, in treatment 2 (T-2) at 3/m^2 and in treatment 3 (T-3) at 4/m^2.

Feed and feeding

In this experiment stocked juveniles were fed on a commercial pellet feed. Feeds were broadcasted evenly along the marginal areas of the pond. Pellet feeds were given at 3% to 15% of the body weight of prawn daily twice, where 60% to 75% of total feed should be fed at late evening of the day. Initially feeding was started at 15% of total biomass which was gradually reduced to 3% to 4% at the final stage of culture period.

Probiotics application with feed

In the present study, some replicates were practiced with probiotics mixed feed. At first the amount of feed was calculated required for those replicates for the day and then calculated the probiotics (Sanolife® Pro-1) amount at a rate of 5 g/kg feed. Probiotics was mixed with required amount of clean water to evenly spread over all the feed particles. Finally, probiotics mixed feed was kept for air drying and then applied to the pond.

Probiotics application in pond water

In this experiment threereplicates of each treatment was practiced with probiotics in pond water. Two commercial probiotics such as Sanolife® Pro-W and Biomin® Pondlife were used. Sanolife Pro-W powder was mixed with clean pond water and then evenly distributed directly to the pond water. For Biomoin® Pond life the calculated amount of powder was kept for germination in a 20-liter clean water and 50 g of sugar mixed solution for 3 hours as per given instruction and then evenly distributed directly to the pond water.

Statistical analysis

t- test was done to find out the value of 'p' by using software-R for the final average body weight, mean growth rate and gross production, where p=0.04018, p=0.03503 and p=0.00036 respectively. For all the three factors p value is smaller than 0.05, which indicates the results are highly significant.

Pearson's product-moment correlation was done for correlation analysis of survival with stocking densities and found, r=-0.98 and p=5.277e-06. r=-0.98 indicates that, there is a high degree of negative association between stocking density and survival rate. That is as the stocking density increases the survival rate decreases. P=5.277e-06 implies that the result is highly significant.

Observations and Results

Growth, survival and production performance

The experiment was carried out for 150 days. In the experiment T-1 (2/m^2), the final average weight gain was 63.7 g and 55.7 g in probiotics based and non-probiotics based culture systems respectively, in experiment T-2 (3/m^2), average weight gain were 55.7 g and 46.7 g and in experiment T-3 (4/m^2), average weight gain were 43 g and 37 g in probiotics and non-probiotics based culture system respectively. Mean growth rates were found higher (Table 1) in probiotics then in non-probiotics culture systems in all replicates.

In three treatments, the mean growth rates were recorded as 0.41 g, 0.36 g and 0.27 g per day in T-1, T-2 and T-3 treatments respectively for Sanolife® Pro-1 culture system. On the other hand, the average growth rates were recorded as 0.35 g, 0.30 g and 0.23 g per day in T-1, T-2 and T-3 treatments respectively with same stocking densities of non-probiotics culture system.

Survival rate of cultured *M. rosenbergii* found different with the same stocking densities among the probiotics and non-probiotics culture system. The highest average survival rates recorded were 69.3% and 68.3% (Table 2) in T-1 (2/m^2) for probiotics ponds and non-probiotics ponds respectively. The lowest average survival rate of 58.3% and 57.7% (Table 2) was found in T-3 (4/m^2) for probiotics and non-probiotics ponds respectively.

Gross production from different culture systems also differs in probiotics and non-probiotics culture system. The estimated average gross production of *M. rosenbergii* in T-1 at a stocking density of 2/m^2 was calculated 87.23 g/m^2 and 74.62 g/m^2 (Table 2) per crop for probiotics and non-probiotics based culture system respectively. On the other hand, the estimated gross productionin T-2 at a stocking density of 3/m^2 was found to be 103 g/m^2/crop and 87.23 g/m^2/crop

Experiment	Stocking density (ind/m²)	Initial average body wt. (g)	Final average body wt. (g)				Mean growth rate (g/day)			
			Sanolife® Pro-1	Average	Non-probiotics	Average	Sanolife® Pro-1	Average	Non-probiotics	Average
T-1	2	2.3	65		57		0.42		0.36	
	2	2.3	62	63.7	54	55.7	0.4	0.41	0.35	0.36
	2	2.3	64		56		0.41		0.36	
T-2	3	2	55		46		0.34		0.29	
	3	2	56	55.7	44	46.7	0.36	0.35	0.28	0.3
	3	2	56		50		0.36		0.32	
T-3	4	2	44		37		0.28		0.23	
	4	2	42	43	35	37	0.27	0.27	0.22	0.23
	4	2	43		39		0.27		0.25	

Table 1: Final body weight and mean growth rate of *M. rosenbergii* in probiotics and non-probiotics based culture system.

Experiment	Stocking density (ind/m²)	Survival (%)				Gross Production (g/m²/crop)			
		Sanolife® Pro-1	Average	Non-probiotics	Average	Sanolife® Pro-1	Average	Non-probiotics	Average
T-1	2	68		66		88.00		75.00	
	2	70	69.3	69	68.3	86.20	87.23	75.25	74.62
	2	70		70		87.50		73.60	
T-2	3	63		64		104.00		88.20	
	3	61	62.7	60	63.0	102.50	103.0	79.20	87.23
	3	64		65		102.50		94.30	
T-3	4	57		56		100.00		82.80	
	4	58	58.3	58	57.7	92.50	98.1	81.20	84.26
	4	60		59		101.75		88.78	

Table 2: Survival and gross production of cultured *M. rosenbergii* in different culture systems.

(Table 2) for probiotics and non-probiotics culture system respectively. In treatment three (T-3) at stocking density of 4/m² the final gross production was estimated to be 98.1 g/m²/crop for probiotics based culture system and 84.26 g/m²/crop (Table 2) for non-probiotics based culture system.

Influence on water quality

In the present study, the water parameters were recorded weekly basis in all ponds and found the recorded parameters within the range of Macrobrachium culture. Average water temperature range was from 29°C to 31°C in three replicates of all treatments pond which were found optimum for *Macrobrachium* culture. Average pH ranged from 8 to 8.2 in all ponds. Dissolved oxygen ranged from 4.8 mg/l to 5.5 mg/l, ammonia ranged from 0.1 mg/l to 0.19 mg/l. Nitrite concentration varied from 0.1 mg/l to 0.22 mg/l in all the replicates. Transparency also ranged from 33.5 cm to 35.25 cm. Alkalinity ranged from 77 mg/l to 88 mg/l. The variation of dissolved oxygen, pH and N-ammonia was recorded throughout culture period. Among all the three treatments, only some selected T1R3, T2R3 and T3R3 treatments were practiced with probiotics application in pond water. In those cases, ammonia and nitrite were variable results (Table 3) by the probiotics application. Due to the probiotics action waste feed and other waste matters were degraded more than other treatments those were practiced without probiotics. As a result, the concentration of toxic gases in those ponds was reduced in some extent. In all those three treatments ammonia and nitrite concentration found less (Table 3, highlighted value) compare to other treatments among the three different stocking density groups.

Discussion

Probiotics used in feed and pond water has a significant role in the growth of prawn culture. In the present experiment, it was found that in every replicate of the three different treatments higher growth and higher net gross production were found from the probiotics treated pond. The present may be compared with the following authors. Ziaei-Nejad et al. [4] and Wang [5] reported the benefits of the probiotics supplements improved feed value, enzymatic contribution to digestion, inhibition of pathogenic microorganisms, anti-mutagenic and anti-carcinogenic activity, growth promoting factors, and increase immune response. Dalmin et at. [6] reported that use of *Bacillus* sp. improved water quality, survival, growth rates and health status of the juveniles of *P. monodon* and reduced the pathogenic *vibrios*. Keysami et al. [7] mentioned that prawns gained higher weight and better FCR by using *B. subtilis* in feed.

In the present experiment, different survival rates were recorded in different treatments. Higher survival rate 69.3% was found where the stocking density was maintained at 2/m² and in this treatment average growth was also recorded higher in probiotics based treatment. In other two treatments, stocking density (3/m²) and (4/m²) showed lower survival as well as lower growth than the treatment with 2/m² stocking density. On the other hand, the average final body weights in three treatments were 55.7 g, 46.7 g and 37 g respectively for T-1, T-2 and T-3. Among three treatments the best result was found at treatments 2 where stocking density was maintained 3/m². Balcazar et al. [8]; Gullian et al. [9] reported that prawn's survival was also found significantly greater in *B. subtilis* treated groups than in the control group. It is possible that this phenomenon operates by the substitution of opportunist pathogens that reduced growth.

It was found that the final body weight gained 63.7 g and 55.7 g in Sanolife® Pro-1 and non-probiotics based culture systems respectively in T-1 treatment (2/m²). Similarly, the body weight gained 55.7 g and 46.7 g in T-2 (3/m²) and 43 g and 37 g in T-3 (4/m²) in probiotics and non-probiotics treatment respectively. It was found from the above results that body weight was higher in lower stocking densities and lower in higher stocking densities. Siddiqui et al. [10] and Karplus and Sagi [11] reported in their research that, the prawn response to higher

Experiments	Temp. (°C)	pH	DO (mg/l)	NH₃ (mg/l)	NO₂ (mg/l)	Transparency (cm)	Alkalinity (mg/l)
T1R1	29.05	8	5.49	0.125	0.1	35.25	77
T1R2	29.95	8.01	5.12	0.19	0.18	35.5	80
T1R3	30.4	8.1	4.96	0.15	0.1	36	83
T2R1	30.1	8.12	5.35	0.155	0.18	39.75	86
T2R2	30	8.16	5.09	0.13	0.22	34.5	88
T2R3	30.7	8.1	4.9	0.1	0.15	35.5	81
T3R1	29.7	8.11	5.41	0.165	0.18	36	82
T3R2	30.05	8.06	5.03	0.15	0.2	34.25	82
T3R3	30.85	8.2	4.8	0.12	0.15	35	81

Table 3: Average water parameters of different experimental ponds.

density uneven increase in number of small males. According to FAO [12], lower to higher yield of species depends on three different culture systems related to stocking density.

Improved water quality has an especial affinity with the pobiotics. It was reported by Dalmin et al. [6] that use of *Bacillus* sp. improves the water quality, survival, growth rates and the health status of juvenile *P. monodon* and reduced the pathogenic *vibrios*. Average water temperature ranges from 29°C to 31°C and pH from 8-8.2 in all three treatments of the experimental ponds are suitable for *Macrobrachium* culture. Timmons et al. [13] recommended pH level ranging from 6.5 to 8.5 for aquaculture, while Boyd and Zimmerman [14] recommended pH level from 7.0 to 8.5 as ideal for *M. rosenbergii* production. Dissolved oxygen is one of the important factors which limit the intensification of freshwater prawn culture. In the present study DO was recorded 4.8 mg/L to 5.49 mg/L which was in optimum level for prawn culture. Timmons et al. [13] recommended DO level above 5 mg/L for *M. rosenbergii* culture, while Boyd and Zimmerman [14] reported that *M. rosenbergii* can tolerate 2 mg/L with no stress, and level below 1 mg/L start to cause mortality. Ammonia ranges from 0.1 mg/L to 0.19 mg/L in different ponds in the present study. At the same time nitrite concentration varied from 0.1 mg/L to 0.22 mg/L in different water samples. Timmons et al. [13] recommended nitrite level below 1,000 mg/L for aquaculture systems, while New [15] stated that concentration under 2,000 mg/L is adequate for *M. rosenbergii*. But in this experiment, observed levels were lower than the results mentioned above. Low nitrite concentration indicates that released ammonia is rapidly assimilated by phytoplankton and nitrification is adequately occurred. Therefore, results suggest that both nitrification and photosynthesis were not negatively affected by intensification. According to Wang et al. [16] showed that the use of commercial probiotics in *P. vannamei* ponds can reduce the concentrations of nitrogen and phosphorus that increase the shrimp yields. In this study alkalinity ranges from 77 mg/L to 88 mg/L. Boyd and Tucker [17] stated that alkalinity has an indirect effect on organisms, making pH stable, increasing water fertility and decreasing toxic potential of metals. Timmons et al. [13] recommend alkalinity range from 50 mg/L to 300 mg/L in aquaculture ponds. On the other hand, Boyd and Zimmerman [14] and New [15] recommended the values between 20 mg/L and 60 mg/L for *M. rosenbergii* culture. Transparency is related to the quantity of suspended materials. Boyd and Zimmerman [14] and New [15] recommended transparency values from 25 cm to 60 cm for *M. rosenbergii* culture. In the present study, the transparency was recorded from 35 cm to 39.75 cm which is good for prawn culture.

Conclusion

The technical potentials of prawn (*M. rosenbergii*) polyculture using probiotics based system were investigated in this study. This study demonstrated that pond prawn production was significantly increased with the introduction of probiotics. Prawn stocking in monoculture has the potentials of increasing total yields as well as farm income with probiotics application. Some ecological advantages of probiotics based prawn culture, such as improved water quality help to increase the sustainability of this system of aquaculture.

References

1. New MB, D'Abramo LR, Valenti WC, Singholka S (2000) Sustainability of freshwater prawn culture. freshwater prawn culture: The farming of *Macrobrachium rosenbergii*, Blackwell Science Ltd, Oxford, UK.

2. Ahmed N (2005) The role of women in freshwater prawn farming in Southwest Bangladesh. Fish Farmer 28: 14-16.

3. Maeda M (1999) Microbial processes in aquaculture, National Research Institute of Aquaculture, Nansei, Mie 516-0193, Japan.

4. Ziaei-Nejad S, Rezaei MH, Takami GA, Lovett DL, Mirvaghefi AR, et al. (2006) The effect of *Bacillus* spp. bacteria used as probiotics on digestive enzyme activity, survival and growth in the Indian white shrimp *Fenneropenaeusindicus*. Aquaculture 252: 516-524.

5. Wang YB (2007) Effect of probiotics on growth performance and digestive enzyme activity of the shrimp *Penaeusvannamei*. Aquaculture 269: 259-264.

6. Dalmin G, Kathiresan K, Purushothaman A (2001) Effect of probiotics on bacterial population and health status of shrimp in culture pond ecosystem. Indian J Exp Biol 39: 939-942.

7. Keysami MA, Mohammadpour M, Saad CR (2012) Probiotic activity of *Bacillus subtilis* in juvenile freshwater prawn, *Macrobrachium rosenbergii* (de Man) at different methods of administration to the feed. Aquaculture Int 20: 499-511.

8. Balcazar JL, De Blas I, Ruiz-Zarzuela I, Vendrell D, Uzquiz JL (2004) Probiotics: A tool for the future of fish and shellfish health management. J Aquac Trop 19: 239-242.

9. Gullian M, Thompson F, Rodriguez J (2004) Selection of probiotic bacteria and study of their immune-stimulatory effect in *Penaeusvannamei*. Aquaculture 233: 1-14.

10. Siddiqui AQ, Al-Hinty HM, Ali SA (1995) Effects of harvesting methods on populations structure, growth and yield of freshwater prawn *Macrobrachium rosenbergii* cultured at two densities. J Appl Aquaculture 5: 9-19.

11. Karplus I, Sagi A (2010) The biology and management of size variation, Wiley-Blackwell Science, Blackwell Publishing, London, UK.

12. http://www.fao.org/fi/glossary/aquaculture/default.asp?lang=en

13. Timmons MB, Ebeling JM, Weathon FW, Summerfelt ST, Vinci BJ (2002) Recirculating aquaculture system, (2nd edn). Cayuga Aqua Ventures, Ithaca, New York, USA.

14. Boyd CE, Zimmermann S (2000) Grow out systems: Water quality and soil management, freshwater prawn farming: The farming of *Macrobrachium rosenbergii*, Blackwell Science, Oxford, UK.

15. New MB (2002) Farming freshwater prawns- A manual for the culture of the giant river prawn (*Macrobranchium rosenbergii*). Aqua Res 34: 1109-1110.

16. Wang YB, Xu ZR, Xia MS (2005) The effectiveness of commercial probiotics in Northern white shrimp (*Penaeusvannamei L.*) ponds. Fish Sci 71: 1034-1039.

17. Boyd CE, Tucker CS (1998) Pond aquaculture water quality management, Kluwer Academic Publishers, Netherlands.

3

Biological Treatment of Broad Bean Hulls and its Evaluation through Tilapia Fingerlings (*Oreochromis Niloticus*) Feeding

Malik M Khalafalla[1]* and EL-Sayed B[2]

[1]*Department of Aquaculture, Faculty of Aquatic and Fisheries Sciences, Kafrelsheikh University, 33516 Kafr El-Sheikh, Egypt*
[2]*Agricultural Botany Department (Agric. Microbiology), Faculty of Agriculture, Kafrelsheikh University, 33516, Kafr El-Sheikh, Egypt*

Abstract

Biological treatment of broad bean hulls by *Pleurotus ostreatus* and evaluation it through tilapia fingerlings (*Oreochromis niloticus*) feeding. The biological treatment show increased CP, NFE and GE content of broad been hulls comparing with the untreated as well as EE and CF decreased 27 and 43%, respectively. DM and OM content decreased with 3.5-1%, respectively. This bio-converted biomass was used as non-conventional feedstuff in diets of Nile tilapia fingerlings. The growth parameters recorded the highest values with nutrition of fish fed diet (3) containing (50%) biodegraded broad bean hulls. The growth parameters also, recorded lowest values with nutrition of fish fed diet (5) containing (100%) bio-degraded broad bean hulls. The same trend found on FCR, PER and PPV% for diet (3) comparing with the other fed diets. It was observed that, the fish fed diet containing bio-degraded broad bean hulls at level (100%) gave the lowest value comparing with the other treatments. Results showed no change in body composition content of Nile tilapia at the beginning and the end of the experimental for all the experimental diets.

Keywords: Biological treatment; Broad bean hulls; *Oreochromis niloticus*

Introduction

Tilapia is an ideal candidate for warm-water aquaculture. They spawn easily in captivity, use a wide variety of natural foods as well as formulated feeds, tolerate poor water quality, and grow rapidly at warm temperatures. These attributes, along with relatively low input costs, have made tilapia the most widely cultured freshwater fish in tropical and subtropical countries [1,2]. Tilapia is the most familiar and popular fishes in Egypt, as well as, in the Middle East and warm climate countries [3]. Nile tilapia is an important food fish that has been introduced to many different parts of the world by man [4]. Bridging the gap between the population and food production is one of the important tasks of developing countries. Expensive staple foods and policy constraints on food imports are the major factors worsening the food situation in developing countries [5]. Legumes have long shelf life and provide more proteins, abundant carbohydrates, high fiber, low fat (except oilseeds) and possess high concentration of polyunsaturated fatty acids. Legumes are also known for certain bioactive compounds, whose beneficial effects need to be explored for efficient exploitation. However, formulating economic tilapia feeds using unconventional, locally available feed resources remains a major challenge facing tilapia farmers, in general, and fish nutritionists, in particular. Several studies have been conducted to evaluate the incorporation of different unconventional animal and plant proteins and energy sources for farmed tilapia with varying results [6].

Many authors have shown that some fungi, particularly some species of *Pleurotus* are able to colonize different types of lignocellulsic wastes, increasing their digestibility [7,8]. Previous studies have shown the feasibility of using these kinds of wastes to produce animal feed [9], and as substrate for mushroom production [10].

The work was carried out to study the effect of biodegraded broad bean hulls by *Pleurotus ostreatus*, replacing there with 0, 25, 50, 75 and 100% of corn and wheat bran in the control diet as well as their efficacy on growth performance, feed utilization and body composition of Nile tilapia (*Oreochromis niloticus*) fingerlings.

Materials and Methods

This work was carried out at the Wet Fish Lab., Department of Animal Production and Botany Agricultural Department (Agricultural Microbiology), Faculty of Agriculture, Kafrelsheikh University. Feeding experiment was conducted for 14 weeks to study the effect of biodegraded broad bean hulls treated with *Pleurotus ostreatus* at five levels (0, 25, 50, 75 and 100%) instead of corn and wheat bran in the control diet on growth performance, feed utilization and body composition of Nile tilapia (*Oreochromis niloticus*) fingerlings.

Experimental fish

Nile tilapia, *Oreochromis niloticus* fingerlings were brought from a fresh water commercial farm in Kafr El-Sheikh governorate, Egypt. Prior to the start of the experiment, fingerlings were placed in a fiberglass tank and randomly distributed into glass aquaria to be adapted to the experimental condition until starting the experiment. Fish was fed on the control diet for two weeks, during this period healthy fish at the same weight replaced diet ones. All the experimental treatments were conducted under an artificial photo period equal to natural light/darkness period (12 h light: 12 h darkness).

Experimental design of rearing fish

A total of 150 fish with an average initial body weight were 10.25 g, were randomly stocked into 15 aquaria (70 liter). Each treatment was represented in three aquaria. Fresh tap water was stored in fiberglass tanks for 24 h under aeration for dechlorination. One third of water aquaria were replaced daily and totally once every week after removing

*Corresponding author: Malik M Khalafalla, Department of Aquaculture, Faculty of Aquatic and Fisheries Sciences, Kafrelsheikh University, 33516 Kafr El-Sheikh, Egypt
E-mail: malikkhalafalla@yahoo.com

the wastes. Nine air stones were used for aerating the aquaria water. Water temperature ranged between 27-28°C. Photoperiod was 14 hrs per day using florescent light. Fish feces and feed residue were removed daily by siphoning.

Experimental diets and feeding regime

The broad bean hulls was treated by *Pleurotous ostreatus* using solid state fermentation technique performed by Belal [11] and Belal and Khlafalla [8] to improve its nutritive value. Stock culture of *Pleurotous ostreatus* was maintained on Potato Dextrose Agar at 4°C.

Air dried milled biomass (50 gm from broad bean hulls) was placed in 500 ml Erlenmeyer flasks, moistened with distilled water (65%) and autoclaved at 121°C for 20 min., cooled overnight and inoculated with 10 agar disks (5 mm diam.) from culture *Pleurotous ostreatus* (7 day old). After incubation for 20 day at 25 ± 1°C, the sterilized bioconverted substrates were used as inoculum for non-sterilized air dried milled biomass with rate 10% as follow:-

Each air dried milled biomass (broad bean hulls) was placed in glass box (width 40 cm × 30 cm height), moistened (till 65%) with water, the substrate moistened when needed. The biomass materials were treated with rate 10% from the described bioconverted substrate (106 cfu/gm), mixed well after that was covered by polyethylene and incubated for 42 days at 25 ± 1°C [7,8].

Prior to the start of the experiment, the fish were adapted to a basal commercial diet [control diet (T1)] containing 30.11% crude protein and were consisted of herring fish meal, soybean meal, yellow corn and wheat bran for two weeks. Five experimental commercial diets were formulated to contain treated broad bean hulls at four levels (25, 50, 75 and 100%) instead of yellow corn and wheat bran (Table 1). Each diet was fed to three randomly assigned aquaria.

A basal diet was formulated using the commercial ingredients. The dry ingredients were finely grounded. The ingredients were mixed by a dough mixer for 20 minutes for homogeneity. Oil was gradually added and mixed. After homogenous mixture, forty ml water per hundred gm diet was slowly added to the mixture according to Shimeino et al. [12]. The diets were cooked on the water evaporator for 20 minutes. The diets were pelleted (3 mm) through fodder machine and the manufacture pellets were dried on dried oven at 70°C for 48 hrs. The diets were collected and tagged and stored in refrigerator at 4°C.

Ingredients	Diets[2]				
	D1 (control)	D2	D3	D4	D5
Fish meal	10	10	10	10	10
Soybean meal	45	45	45	45	45
Yellow corn	26	20	14	8	2
Wheat bran	14	10.5	7	3.5	0
Broad bean hulls	0	9.5	19	28.5	38
Sunflower oil	3	3	3	3	3
Vitam. and min. mix[1].	2	2	2	2	2
Total	100	100	100	100	100

[1]Vitamins composition/100 g mixture, VA (960000 IU), VD3 (160000 IU), VE (0.8 g), VK (0.16 g), VB_1 (80 mg), VB_2 (0.32 g), VB_6 (0.12 g), Pantothenic acid (0.8 g), VB_{12} (0.8 mg), Niacin (1.6 g), Folic acid (80 mg), Biotin (4 mg) and Choline chloride (40 g). Composition of mineral mixture (g\100 g mixture). $MgSO_4.7H_2O$ (12.75), $CaHPO_4 2H_2O$ (72.85), $ZnSO_4 7 H_2O$ (0.55), $Ca_2O6 H_2O$ (0.25), KCl (0.02), $FeSO_4 7H_2O$ (5), $CuSO_4 5H_2O$ (2.5), $CuSO_4 7H_2O$ (0.08), $CrCl_3 6H_2O$ (0.05), NaCl (0.01) and Folic acid (6). (Local market).

[2]Diets: Diet 1 (Control diet), diets 2, 3, 4 containing 25, 50, 75 and 100% treated broad bean hulls, respectively.

Table 1: Ingredients composition (%) of the experimental diets.

Fish in all treatment were fed daily on the experimental diets at a level of 3% of the fish biomass then the fish were weighed every two weeks, the amount of feed were adjusted according to body weight. The fish were given the feed two times daily at (9.0 am and 3.0 p.m.).

Analytical procedures

Moisture, crude protein (%N × 6.25), crude lipid, crude fiber and ash contents of diet ingredients and a sample of fish at the beginning and end of the experiment were determined in triplicate according to A.O.A.C [13] methods as follows: moisture was determined by oven-drying at 105°C for 24 h; lipid by extracting the residue with 40-60°C petroleum ether for 16 h; fiber as loss on ignition of dried lipid-free residues after digestion with 1.25% H_2SO_4 and 1.25% NaOH; ash by ignition at 550-600°C to constant weight; total nitrogen by micro-Kjeldahl method. Gross energy (GE) contents of the experimental diets and fish samples were calculated by using factors of 5.65, 9.45 and 4.22 kcal/g of protein, lipid and carbohydrates, respectively [14] for 14 weeks.

Measurements of water parameters: Water samples were taken each two days for ammonia and pH analysis. Analytical methods were done according to the American Public Health Association [15]. The pH values were determined by (A digital pH-meter). Water temperature and oxygen level were measured daily at 8 o'clock by (Oxygen meter model 9070). In all treatments water quality parameters for water temperature ranged between 27 to 28°C, pH (7 to 7.68); dissolved oxygen (6.11 to 6.75 mg/L) and water ammonia (0.05 to 0.0.08 mg/L). All the water quality parameters were within the acceptable ranges for fish growth [16].

Measurements of growth and feed utilization parameters

Body weight of fish in each aquarium was measured at start and every two weeks during the experimental period (14 weeks). Diet performance was evaluated as follows:

- Average weight gain AWG (g/fish)=Wt–W0.

- Average daily weight gain ADG (g/fish/day)=Wt–W0/t.

- Specific growth rate% day SGR (%/day) = 100 × (In Wt–InW0)/t

- **Where Wt is weight of fish at time t, W0 is weight of fish at time 0, and t is the experimental period in days.

- Feed conversion ratio, FCR=dry feed fed/wet weight gain.

- Protein efficiency ratio, PER=wet weight gain/Protein fed.

- Protein productive value, PPV (%)=100 × (protein gain/protein fed).

- Survival rate, SR=100 (Total No. of fish at the end of the experimental/Total No. of fish at the start of the experiment].

Statistical analysis:

The obtained numerical data were statistically analyzed using SPSS [17] for one-way analysis of variance. When F-test was significant, least significant difference was calculated according to Duncan [18].

Results and Discussion

Broad bean by-products

The chemical composition of treated and untreated broad bean hulls used in diets was shown in Table 2. The results revealed that, the CP content increased with 53% comparing with the untreated materials followed by NFE that increased 11% while, ash and GE

Ingredients	DM%	(On DM basis, %)						GE* Kcal/g
		OM	CP	EE	CF	Ash	NFE	
Untreated								
Broad bean hulls	93.78	85.24	9.52	9.87	20.47	12.52	47.62	4.22
Treated								
Broad bean hulls	90.54	84.33	14.57	7.20	11.67	13.44	53.12	4.12

*GE (Gross energy) was calculated according to NRC (1993) by using factors of 5.65, 9.45 and 4.22 K cal per gram of protein, lipid and carbohydrate, respectively.

Table 2: Proximate Chemical analysis (%) of untreated or treated broad bean hulls used in diets (% on DM basis).

Item	Control (D1)	Levels of treated broad bean hulls (%)			
		25 (D2)	50 (D3)	75 (D4)	100 (D5)
DM (%)	91.20	92.14	92.54	92.67	92.95
OM (%)	93.16	93.26	93.42	93.13	92.90
CP (%)	30.11	30.87	31.42	31.93	31.98
EE (%)	8.87	8.54	8.44	8.79	9.13
CF (%)	5.68	5.98	6.27	6.88	7.12
Ash (%)	6.54	6.74	6.58	6.87	7.10
NFE (%)	48.80	47.87	47.29	45.53	44.67
GE (Kcal/g)*	4.59	4.57	4.57	4.56	4.50

*GE (Gross energy) was calculated according to NRC (1993) by using factors of 5.65, 9.45 and 4.22 K cal per gram of protein, lipid and carbohydrate, respectively.

Table 3: Proximate chemical analysis (%) of the experimental diets used in the experiment.

Diet No.	Body weight		Total weight gain (g/fish)[2]	ADG (g/fish/day)[3]	SGR (%/day)[4]	SR%[5]
	Initial (g/fish)	Final (g/fish)				
1	10.25	39.47 ab	29.22 ab	0.30 ab	1.38 bc	100
2	10.26	41.84 ab	31.58 ab	0.32 ab	1.43 b	100
3	10.26	43.90 a	33.64 a	0.34 a	1.48 a	100
4	10.22	42.36 a	32.14 a	0.33 a	1.45 b	96.67
5	10.24	37.78 b	27.54 b	0.28 b	1.33 c	93.33
Mean	10.25	41.07	30.82	0.31	1.42	98
SE[1]	0.24	2.11	1.78	0.02	0.16	3.25

[1]Standard error of the mean derived from the analysis of variance.

[2]TWG (g/fish)=Average final weight (g)–Average initial weight (g).

[3]ADG (g/fish/day)= [ATG (g)/experimental period (d)].

[4]SGR (%/day)=100 (Ln final weight–Ln initial weight)/experimental period (d).

[5]SR=100[Total No of fish at the end of the experimental/Total No of fish at the start of the experiment].

Table 4: Effect of using treated broad bean on growth performance parameter of (*O. niloticus*) fingerlings.

slightly increased. On the other hand, EE and CF decreased by 27 and 40%, respectively. DM and OM content decreased 3.5 and 1%. The same trend was also reported by Valizadeh et al. [19] indicated CP content increased significantly (p<0.05) after mushroom growing, from 46.6 g/kg on zero day to 50.90 g/kg after 84 days. DM content was low for wheat straw. OM content significantly decreased from 91.14 to 84.47% at the end day of mushroom. NDF content of the wheat straw were 719, 690, 669, 612 and 550 g/kg for the day 0, 21, 42, 63 and 84 after seeding, respectively. A Similar tendency was found for ADF 474, 456, 443, 416 and 387 g/kg for the respective samples. These results are in agreement with the finding of Rzedzicki et al. [20] and Rzedzicki and Sobota [21]. Moreover, Broudiscou et al. [22] mushrooms are able to degrade between 25 and 60% of the dry weight of plant tissues although their efficiency varies according to the species, yield strain and the plant type. Williams et al. [23] found that biological treatment for straw produced more free sugars, more protein and gave less cellulose as well as lignin with an increase content of ash comparing with the untreated materials.

Chemical composition of diets

Chemical composition and calculated gross energy of different experimental diets are presented in Table 3. The chemical composition of the experimental diets showed limited variations among these diets, it contained nearly similar DM, OM, CP, EE, CF, Ash, NFE and GE content. The DM content ranged from 91.20 to 92.95; OM from 93.13 to 92.90; CP from 30.11 to 31.98; EE from 8.44 to 9.13; CF from 5.68 to 7.12; ash from 6.58 to 7.10 and NFE from 44.67 to 48.50%. The corresponding value of GE ranged from 4.50 to 4.57 kcal/g. The data revealed that both of CF and ash content increased by increasing levels of treated broad bean hulls, while the NFE content decreased. The relatively high fiber content is limiting in its use in tilapia feeds because these fishes are lack in their ability to secrete cellulose which is the main cellulose digesting enzyme [24,25]. The biological treatment reduces its fiber contents and improves its nutritive value for fish feeding. Similar results reported by Hassanen et al. [26]. These values were within the range suggested for tilapia by Jauncey and Ross [27] and NRC [14].

Growth performance and surviving rate:

The growth performance parameters of Nile tilapia (*Oreochromis niloticus*) fingerlings which fed diets contained with treated broad bean hulls (0, 25, 50, 75 and 100%) are shown in Table 4 as (initial and final) weights, average weight gain, average daily gain, specific growth rate and survival rate. Average of initial body weight of Nile tilapia fingerlings fed the experimental diets at the start did not differ, indicating that groups were homogenous. At the end of the experimental period (98 days), the final body weight of the fish groups fed on diets 3 and 4 had significantly (P<0.05) higher final body weight than the other groups. However, the lower growth performance found for the fish fed diet (5) contain 100% treated broad bean hulls; this may be due to the higher crude fiber content (7.12%). On the other hand, the fish fed on diets 3 and 4 had significantly (P<0.05) higher TWG and ADG than the different groups.

Abdolsamad et al. [28] reported that replacement of native starch by gelatinized starch improved faces removal rate, growth and digestibility (P<0.01), but reduced fermentation (P<0.05) at the end of the intestine. Addition of gelatinized starch did not change viscosity and dry matter of the digest at the end of the intestine. A high level of starch in the fish diet also increased digestibility, growth and faces removal percentage (P<0.05). Fermentation and dry matter content at the end of the intestine were not influenced by a high starch diet, but viscosity was higher at the high level of starch inclusion. Volatile fatty acid levels in the stomach of Nile tilapia were high in the treatments with gelatinized starch. Gouveia and Davies [29] evaluated the use of a pea seed derived meal in experimental diets for European sea bass fingerlings of initial weight 10 g. It was demonstrated that up to 40% pea seed meal inclusion was feasible in diets allowing for a 12% reduction in fish meal content and a 25% substitution of carbohydrate

Diet No.	Feed intake (g/fish)	Feed conversion ratio[2] (FCR)	Protein utilization	
			Protein efficiency ratio[3] (PER)	Protein productive value[4] (PPV, %)
1	47.91	1.64 ab	2.02 c	26.86 bc
2	47.63	1.51 b	2.15 ab	28.61 b
3	46.99	1.40 c	2.28 a	31.64 a
4	46.45	1.45 bc	2.17 ab	29.92 ab
5	47.55	1.73 a	1.80 cd	25.96 c
Mean	47.32	1.55	2.08	28.60
SE[1]	2.24	0.12	0.14	1.84

[1]Standard error of the mean derived from the analysis of variance.

[2]FCR=DM Feed Intake (g)/Live weight gain (g).

[3]PER=Live weight gain (g)/ Protein intake (g).

[4]PPV (%)=100 [Final fish body protein (g)–Initial fish body protein (g)]/crude protein intake (g).

Table 5: Effect of using treated broad bean hulls on feed and nutrients utilization parameter of (O. niloticus) fingerlings.

Diet No[1]	Dry Matter (%)	% On dry matter basis			Energy Content (Kcal/100 g)
		Crude Protein	Ether Extract	Ash	
Initial fish	19.52	60.12	15.17	20.15	502.28
1	23.73	54.19[b]	18.45	22.50	501.04
2	23.23	55.62[ab]	19.27	23.55	502.94
3	24.00	55.76[ab]	19.32	23.56	503.36
4	23.96	55.54[ab]	18.98	22.81	504.43
5	24.12	56.78[a]	19.11	23.41	504.35
Mean [2]	23.81	55.58	19.03	23.17	503.22

[1]Diet 1 (Control diet), diets 2, 3, 4 containing 25, 50, 75 and 100% treated broad bean hulls, respectively.

[2]The mean in the same column bearing different superscript are significantly different at (P<0.05).

Table 6: Body composition (%) of Nile tilapia affected by feeding different diets(On DM basis).

content without appreciable loss in growth performance of juvenile sea bass or diet utilization.

Feed intake and nutrient utilization

Nutrient utilization in terms of feed intake (FI), feed conversion ratio (FCR), protein efficiency ratio (PER) and protein productive value (PPV%) are illustrated in Table 5. Data showed that, there were no significant differences (P>0.05) between control group and all dietary experiments for FI, where it ranged between 46.45 and 47.91 g. Azaza et al. [30] indicated that there was no significant difference in feed intake among fish fed with the control feed and feeds containing 10 to 20% DSBM (dried soybean meal). The same trend was also reported by Gouveia and Davies [29] who indicated that the substitution of pea meal in the test diets between 20 and 40% and could be considered as an intermediate protein-energy supplement in the ration, there were no palatability problems and feed intake compared to favorably the reference fish meal diet.

The obtained results illustrated also that, the FCR, PER and PPV% were high significantly (P<0.05) for fish fed diet (3) containing 50% treated broad bean hulls comparing with the other fed diets. While fish fed diet (5) containing 100% treated broad bean hulls had the lowest value. Average of FCR of the different diets was 1.64, 1.51, 1.40, 1.45

and 1.73 kg feed for each kg gain for 0, 25, 50, 75 and 100% treated broad bean hulls, respectively. The average of PER and PPV% were 2.02, 2.15, 2.28, 2.17 and 1.80 for PER and PPV was 26.86, 28.61, 31.64, 29.92 and 25.96% for 0, 25, 50, 75 and 100% of treated broad bean hulls, respectively. The fish fed diet containing 100% treated broad bean hulls was the lowest values. While, the groups of fish fed diets containing 50% (D3) treated broad bean hulls higher values than other treatments.

Abdolsamad et al. [28] reported that increasing the starch content of the diets resulted in an increased growth (P<0.001) and a higher FCR (P<0.01). Moreover, Leary and Lovell [31] indicated that the excessive fiber in aquaculture diets may also lead to a decrease in feed utilization by obstructing the action of digestive enzymes and diluting nutrient density. Shalaby [32] studied that the effects of different levels of fenugreek seeds meal (FSM) 0, 2, 4, 6 and 8% of FSM on growth performance, feed and nutrients utilization as well as body composition. Results showed that fish fed diets containing 2% FSM had significantly higher (P<0.05) body weight, weight gain, SGR, FCR and PER than those of fed the control diet and the other supplemented fenugreek seed levels. However, FSM levels of 6 and 8% gave significantly (P<0.05) lower growth performance parameters, FCR, PER and PPV% than the control diet.

Body composition and energy content of Nile tilapia

Body chemical composition; DM, CP, EE, Ash and energy content of Nile tilapia at the beginning and the end of the experimental are shown in Table 6. There was no change in whole body composition among all experimental diets except CP content which significantly (P<0.05) increased by incorporation of treated broad bean hulls. Shalaby [32] studied the effects of different levels of fenugreek seeds meal (FSM) 0, 2, 4, 6 and 8% of FSM on body composition. Results showed that no significant differences (P>0.05) were observed in moisture, crude protein, ether extract, ash and energy content of Nile tilapia fed diets containing various levels of FSM

Azaza et al. [30] studied the possible use of broad bean meal (FBM) in juvenile Nile tilapia-practical diets by progressively increasing its inclusion level (12, 24 and 36%) at the expense of dehulled soybean meal (DSBM) in isonitrogenous (27.5%) and isoenergetic formulated diets. Carcass composition was not clearly affected by diet composition. There were no differences in carcass protein content among the treatments. Carcass water content was significantly higher in fish fed the diet contains 36% FBM than those fed the other diets, and the lowest values were recorded in fish fed the control diet. A tendency was noted for body lipid level to decrease as FBM content increased in the diets. The fish fed the diet containing 36% FBM had significantly lower body lipid than those fed the other diets (p>0.05). The body ash content did not vary significantly among treatments.

Conclusion

From the obtained results it could be found that, the biological treatment increased CP content of broad bean hulls by about more than 53% comparing with the untreated materials followed by NFE that increased about 11% while, ash and GE slightly increased. On the other hand, EE and CF decreased by proximally 27 and 40%, respectively. DM and OM content slightly decreased (3.5-1%). Treated broad bean hulls could be successfully replacement as a feed for feeding *Oreochromis niloticus* fingerlings especially at levels of 50% instead of corn and wheat bran without any adverse effects on their productive performance.

References

1. Tsadik GG, Bar AN (2007) Effects of feeding, stocking density and water-flow rate on fecundity, spawning frequency and egg quality of Nile tilapia, (*Oreochromis niloticus*). Aquaculture 272: 380-388.

2. Tahou AMA (2007) Studies on some factors affecting the production and reproduction of Nile tilapia.

3. El-Sherif MS, El-feky AMI (2009) Performance of Nile Tilapia (Oreochromis niloticus) Fingerlings .I. Effect of pH. Int J Agric Biol 11: 297-300.

4. Belal EB, Khalafalla MME, El-Hais AMA (2012) Use of Spirulina *(Arthrospira fusiformis)* for promoting growth of Nile Tilapia fingerlings. Afr J Microbiol Res 6: 6423-6431.

5. Weaver DS (1994) The chemical composition and nutritive value of Australian grain legumes. Grain Research and Development 18: 43-46.

6. Sayed AFM (1999) Alternative dietary protein sources for farmed tilapia, Oreochromis spp. Aquaculture 179: 149- 168.

7. Mukherjee R, Nandi B (2004) Improvement of in vivo digestibility through biological treatment of water hyacinth biomass by *Pleurotus* species. International Biodeterioration and Biodegradation 53: 7-12.

8. Belal EB, Khalafalla MME (2011) Biodegradation of *Panicum repens* residues by *Pleurotus ostreatus* for its use as a non-conventional feedstuff in diets of *Oreochromis niloticus*. Afr J Microbiol Res 5: 3038-3050.

9. Adamovic M, Grubic G, Milenkovic I, Jovanovic R, Protic R, et al. (1998) The biodegradation of wheat straw by *Pleurotus ostreatus* mushrooms and its use in cattle feeding. Anim. Feed Sci and Technol 71: 357-362.

10. Yildiz S, Yildiz UC, Gezer ED, Temiz A (2002) Some lignocellulosic wastes used as raw material in cultivation of the *Pleurotus ostreatus* culture mushroom. Process Biochemistry 38: 301-306.

11. Belal EB (2008) Biodegradation of wastepaper by Trichoderma viride and using bioprocessed materials in biocontrol of damping- off of pea caused by Pythium debarymanum. J Agric Res Kafrelsheikh Univ 34: 567-587.

12. Shimeino S, Masumoto T, Hujita T (1993) Alternative protein sources for fish meal diets of young yellowtail. Nippon Suisan Gakkaishi 59: 137-143.

13. AOAC (2000) Association of Official Analytical Chemists. Official Methods of Analysis. AOAC, Arlington, Virginia, USA.

14. NRC (1993) Nutrient requirements of fish. National Academy Press, Washington DC.

15. APHA American Public Health Association (1985) Standard methods for the examination of water and waste.

16. Boyd CE (1984) Water Quality in Warm water Fishponds. Auburn University Agriculture Experimental Station, Auburn, AL, USA (*Oncorhynchus tshawytscha*) and rainbow trout (*Oncorhynchus mykiss*). Aquaculture 161: 27-43.

17. SPSS (1997) Statistical package for the social sciences..

18. Duncan MB (1955) Multiple ranges and multiple F-tests. Biometrics 11: 1-42.

19. Valizadeh R, Sobhanirad S, Mojtahedi M (2008) Chemical Composition, Ruminal Degradability and in vitro Gas Production of Wheat Straw Inoculated by *Pleurotus ostreatus* Mushrooms. J Anim Vet Advances 7: 1506-1510.

20. Rzedzicki Z, Sobota A, Zarzycki P (2004) Influence of pea hulls on the twin screw extrusion-cooking process of cereal mixtures and the physical properties of the extrudate. Int Agrophysics 18: 73-81.

21. Rzedzicki Z, Sobota A (2006) Study on the process of single-screw extrusion-cooking of mixtures with a content of pea hulls. Int. Agrophysics 20: 327-336

22. Broudiscou LP, Agbagla-Dobnani A, Papon Y, Cornu A, Grenet E, et al. (2003) Rice straw degradation and biomass synthesis by rumen micro-organisms in continuous culture in response to ammonia treatment and legume extract supple mentation. Anim Feed Sci Technol 105: 95-108.

23. Williams BC, McMullan JT, McCahey S (2001) An initial assessment of spent mushroom compost as a potential energy feedstock. Bioresour Technol 79: 227-230.

24. Stickney RR, Shumway SE (1974) Occurrence of cellulose activity in the stomach of fishes. J Fish Biol 6: 779-790.

25. Buddington RK (1980) Hydrolysis-resistant organic matter as a reference for measurement of fish digestion efficiency. Trans Am Fish Soc 109: 653-656.

26. Hassanen GDI, Sherif MA, Hashem NA, Hanafy MA (1995) Utilization of some fermented waste food as a protein source in pelleted feeds for Nile tilapia (*Oreochromis niloticus*) fingerlings.

27. Jauncey K, Ross B (1982) A guide to tilapia feeds and feeding Ins.

28. Abdolsamad K, Amirkolaie A, Johan J, Johan W (2006) Effect of gelatinization degree and inclusion level of dietary starch on the characteristics of digesta and faeces in Nile tilapia (*Oreochromis niloticus* (L). Aquaculture 260: 194-205.

29. Gouveia A, Davies SJ (1998) Preliminary evaluation of pea seed meal (*Pisum satvum*) for juvenile European Sea bass (*Dicentrarchus labrax*). Aquaculture 166: 311-320.

30. Azaza MS, Wassim K, Mensi F, Abdelmouleh A, Brini B, Kraïem MM (2009) Evaluation of broad beans (Vicia broad L. var. minuta) as a replacement for soybean meal in practical diets of juvenile Nile tilapia *Oreochromis niloticus*. Aquaculture 287: 174-179.

31. Leary DF, Lovell RT (1975) Value of fiber in production type diets for channel catfish. Trans Anim Fish Soc 104: 328-332.

32. Shalaby Shymaa MM (2004) Response of Nile tilapia *Oreochromis niloticus*, fingerlings to diets supplemented with different levels of fenugreek seeds (Hulba). J Agric Sci Mansoura Univ 29: 2231- 2242.

cDNAs Encoding Chitin Synthase from Shrimp (*Pandalopsis Japonica*): Molecular Characterization and Expression Analysis

Md. Hasan Uddowla[1], Ah Ran Kim[2], Won-gyu Park[1] and Hyun-Woo Kim[1,2]*

[1]*Department of Marine Biology, Pukyong National University, Busan 608-737, South Korea*
[2]*Interdisciplinary program of Biomedical Engineering, Pukyong National University, Busan, 608-737, South Korea*

Abstract

Crustacean growth occurs via molting, the periodic shedding of the exoskeleton. Understanding the genes involved in chitin metabolism associated with the periodic molt cycle is important for various applications to decapod crustacean aquaculture. Chitin synthase is an important enzyme in the chitin biosynthetic pathway that plays a major role in synthesis of new cuticle after molting. In this study, we isolated a full-length cDNA encoding chitin synthase (PajCHS) from Pandalopsis japonica through a combination of PCR (Polymerase chain reaction)- based cloning and bioinformatics analysis. The identified PajCHS encodes a transmembrane protein with 1525 amino acid residues (175 kDa). Comparison with other CHSs from insects revealed that PajCHS contains three domains: N-terminal domain A, catalytic domain B, and C-terminal domain C. Three conserved motifs (EDR, QRRRW, and SWGTR) were also well conserved within and near the catalytic domain B, suggesting that Paj-CHS is functionally active. Variation in the transmembrane helix within the N-terminal and C-terminal domains suggested that the orientation of each CHS may be different. Phylogenetic analysis suggested that PajCHS is an ortholog of CHS1 group members from insect species. However, tissue expression profiles indicated that epidermis, hepatopancreas, intestine, and gill were the major production sites for PajCHS transcript, which is considerably different from insect CHS1. qPCR results showed that eye stalk ablation and 20 hydroxyecdysone (20E) injection increased the expression level of PajCHS mRNA, suggesting that the expression of PajCHS1 may be controlled by endogenous 20E.

Keywords: Decapod; *Pandalopsis japonica*; Chitin; Molting; Epidermis

Introduction

Chitin, a linear homopolymer of β-(1,4)-N-acetyl-D-glucosamine (GlcNAc), is a major component of the exoskeleton and peritrophic membrane (PM) in arthropods [1,2]. For decapods, the exoskeletons of which are mainly composed of chitin, its metabolism is one the most crucial processes for growth and development during which the periodic shedding of the old exoskeleton and replacement with a new one is required [3]. Understanding chitin metabolism has various industrial applications, including growth enhancement and the production of soft-shell products.

Chitin formation is catalyzed by the membrane protein chitin synthase (CHS, UDP-N-acetyl-D-glucosamine: chitin 4-β-N-acetylglucosaminyltransferase; EC 2.4.1.16), which is a glycosyltransferase family 2 protein [4,5]. Since the first cDNA encoding an insect CHS was identified in the sheep blowfly, Lucia cuprina [6], several CHS genes have been reported by various genome projects on insects, including Anopheles gambiae, Aedes aegypti [7], Drosophila melanogaster [8], Manduca sexta [9,10], and Tribolium castaneum [11]. Insect genomes possess two copies of CHS genes (CHS1 and CHS2) as a result of gene duplication from the ancestral CHS gene [12].

The expression and functions of CHS1 and CHS2 appear to be distinct from one another. CHS1 genes are exclusively expressed in the epidermis underlying the cuticular exoskeleton and are related to ectodermal cells, such as tracheal cells, whereas CHS2 genes are expressed during the synthesis of chitin in the peritrophic membrane (PM) of the gut [6,9,11,13,14]. CHS1 genes also contain exons that lead to the production of two alternatively spliced variants, CHS1A and CHS1B [9,11,14]. The distribution of these in tissue, and their expression, differ during development, suggesting that they may have different biological functions [9,11,14].

It is essential to understand the enzymes involved in chitin metabolism in crustaceans. At least four groups of chitinase genes have been identified and characterized in decapods [15], and the expression of some of these is influenced by the molting hormone ecdysone. However, there is little information on CHS genes and their functions in decapod crustaceans. In the present study, we identified and characterized the full-length cDNA (PajCHS) encoding a CHS from the decapod crustacean Pandalopsis japonica, which is an important fishery resource in East Asian countries including Korea and Japan and is a good model system in which to investigate the physiology of crustacean molting [16,17]. To determine the transcriptional effects of CHS gene expression during the molting cycle, both quantitative and qualitative expressional analyses were performed after eyestalk ablation (ESA) and 20-hydroxy ecdysone (20E) injection.

Materials and Methods

Experimental animals

Adult *P. japonica* were purchased from a local seafood market and acclimatized to culture tanks (50 L each). The physicochemical characteristics of the water, including salinity (33PPT), dissolved O$_2$ (~6 ppm), and water temperature (6°C), were maintained as described

*Corresponding author: Hyun-Woo Kim, Department of Marine Biology, Pukyong National University, Busan 608-737, South Korea
E-mail: kimhw@pknu.ac.kr

previously (NFRDI, 2009). Fifteen shrimp were kept in each tank and dead shrimp and molted exoskeletons were removed as quickly as possible to avoid water pollution. As a food source, peeled frozen shrimp (*L. vannamei*) and scallops were supplied once a day at a rate of 5% body weight. Every 2 days, waste and uneaten feed were removed by siphoning. Half of the water was exchanged by adding fresh seawater every 5 days (it was properly aerated and a similar temperature was maintained). Molting stages were identified based on setal development in the uropod [18] and intermolt shrimp were used for the experiments.

ESA and 20E injection

The animals were divided into three groups: control, ESA group, and 20E-injected group. ESA was performed as described previously [15]. Untreated animals were used as controls. Tissues obtained from individuals in each group were frozen directly in liquid nitrogen and stored at –70°C until used for RNA extraction. To determine the levels of CHS mRNA expression in different tissues, premolt [18]. *P. japonica* were injected with 20E (Santa Cruz Biotechnology, Inc., Santa Cruz, CA). Premolt animals were divided into two groups: one was administered 20E, and the other (control) received the same volume of PBS. A single injection of 20E at a concentration of 10 ng/g body weight was administered (hemolymph volume was assumed to be 30% of the wet body weight and ecdysteroid level was 30 pg/μL during the intermolt stage [19,20]. Injections (20 ng/μL stock solution of PBS containing 10% ethanol) were made into the sinus at the base of the fifth walking leg. After injection, the shrimp were kept in aerated plastic tanks and sacrificed at different time intervals: 12 h, 36 h, 72 h, day 5, and day 7. Tissues were collected and stored at –70°C until further use.

Cloning of the full-length PajCHS cDNA

A BLAST search of a shrimp EST database [21] identified a partial sequence (2237 base pairs [bp]) as being homologous to insect CHS. To obtain the full-length cDNA, a conventional cloning strategy and rapid amplification of cDNA ends (RACE) were applied. All primers used in the experiment were designed using the IDT SciTools program (http://www.idtdna.com/SciTools/SciTools.aspx) and were synthesized by Bioneer Corp. (Daejeon, Korea) (Table 1).

Total RNA was purified from different tissues using Trizol reagent according to the manufacturer's instructions (Invitrogen, Carlsbad, CA), quantified by spectrophotometry (Nanodrop Technologies, Wilmington, DE), and stored at –70°C. cDNA was synthesized in a reaction (10 μL) containing 2 μg total RNA, 0.5 μL DNaseI, and 1 μL 10×DNase buffer. The total reaction was kept at 37°C for 30 min followed by 70°C for 10 min. After adding 1 μL 3Race dT (TTTTTTTTTTTTTTTT) (20 μM) and 4 μL dNTPs (2.5 mM each), the reaction was terminated by heating at 70°C for 5 min and chilling on ice for 2 min. Then, 4 μL first-strand buffer (5×), 2 μL DTT (0.1 M), and 1 μL RNaseOUT (Invitrogen) were added, and the reaction mixture was incubated at 37°C for 2 min. Finally, 1 μL MMLV reverse transcriptase (Invitrogen) was added and the reaction mixture, which was incubated at 37°C for 50 min. The enzyme was inactivated at 70°C for 15 min. cDNA was quantified and stored at –20°C until use.

5' RACE was performed to obtain the 5' upstream region using a DNA Walking SpeedUp™ kit (Seegene, Seoul, Korea) until the full ORF was identified. PCR conditions were as described previously [16]. To obtain the 3' UTR sequence including the polyadenylation signals, 3' RACE was performed according to a previously described procedure [15]. Amplified PCR products were separated by 1.5% agarose gel electrophoresis and stained with ethidium bromide. Amplicons of the expected size were cut out from the gel with a razor blade and purified using a gel extraction kit (GeneAll Biotechnology, Seoul, Korea), ligated into the vector using a pGEM-T Easy Cloning Kit (Promega, Madison, WI), and transformed into One Shot Top 10 *E. coli* (Invitrogen). Fragmented sequences were joined together using the Align Sequences

Primer name	Sequences (5'-3')	Description
PajCHS PCF1	AACAATCTTGGAGCTGCTTGTGGAC	Forward primer for EST sequence confirmation
PajCHS PCF2	GGACTTATGGTATGGTATCAGATG	Forward primer for EST sequence confirmation
PajCHS PCR1	CCTAGACTGACGTCTTTCTTGGAC	Reverse primer for EST sequence confirmation
PajCHS PCR2	CGTCTTTCTTGGACCACAGTCTCG	Reverse primer for EST sequence confirmation
PajCHS5R R1	TGAACAGTCCTCTTGTAGTCTTGC	Specific reverse primers for 5' region
PajCHS5R R2	GTACATAATGACGAGCTTGGCTA	Specific reverse primers for 5' region
PajCHS5R R3	TAAGGAGTTGTACATTGGAGAGACAAA	Specific reverse primers for 5' region
PajCHS5R R4	TGCGATCCATGTTTGGGAAAGTAACCAAAA	Specific reverse primers for 5' region
PajCHS DEGF1	GARACNAARGGNTGG	Degenerate forward primers for 5' region
PajCHS DEGF1-1	GGNTGGGAYGTNTTY	Degenerate forward primers for 5' region
PajCHS DEGF2	TTYWSNTAYGCNTTYCCN	Degenerate forward primers for 5' region
PajCHS DEGF3	CARGGNTTYWSNTAYGCN	Degenerate forward primers for 5' region
PajCHS3R F2	CATTTATCCAGAGGAATGCTGTTG	Specific reverse primers for 3RACE
PajCHS3R F3	CAAGCAATCGGAACGTTGGACGTGGC	Specific reverse primers for 3RACE
PajCHS DW5R R5	CCCGAAACTAGCACAACACCGAAGGT	5'RACE reverse specific primer
PajCHS DW5R R6	CATCGAGGAACTTTTGGTTGTCCTC	5'RACE reverse specific primer
PajCHS DW5R R7	GTTGTCCTCCGAGAGAGAACCCGTGGAC	5'RACE reverse specific primer
PajCHS RTFD	TGTACACTGTTGCTTCAGCGAGGT	PajCHS Forward primer for qPCR
PajCHS RTRD	AGCTACAAAGGCACCCACCAACAT	PjCHS Forward primer for qPCR
PajCHS full con F1	GAAGTTACTGAGGAATTCTTAAAGGATC	Forward primer for ORF confirmation
PajCHS full con F2	AGGATCATTGTGCTCGGATGC	Forward primer for ORF confirmation
PajCHS full con R1	ATTGTACATATAAATAATGAAAAGCCTG	Reverse primer for ORF confirmation
PajCHS full con R2	ATAATGAAAAGCCTGTCTAGGAGAG	Reverse primer for ORF confirmation
M13F (-40)	CAGGAAACAGCTATGAC	Vector FWD primer for DNA sequencing
M13R (-20)	GTAAAACGACGGCCAG	Vector RVS primer for DNA sequencing

Table 1: List of primers used for the cloning of PajCHS and qPCR.

Nucleotide BLAST program (www.ncbi.nlm.nih.gov). A single PajCHS transcript was reconfirmed by RT-PCR using the primers for the 3' and 5' ends (Table 1), and nucleotide sequences were determined with an automated DNA sequencer (Applied Biosystems, Foster City, CA).

Bioinformatics analysis of PajCHS

The theoretical isoelectric point (pI) and molecular weight of PajCHS were calculated using the web-based program Compute PI/Mw (http://au.expasy.org/tools/pi_tool.html). Signal peptide and putative cleavage sites were predicted by SignalP 3.0 Server (http://www.cbs.dtu.dk/services/SignalP/). Multiple amino acid sequence alignment was performed using ClustalW, which is offered by the European Bioinformatics Institute (http://www.ebi.ac.uk/clustalw/), and the results were presented using the GeneDoc program (http://www.nrbsc.org/gfx/genedoc/index.html). Phylogenetic tree was constructed using the minimum-evolution method of Molecular Evolutionary Genetics Analysis (MEGA4) with 1000 bootstrapping replicates [22]. Coiled-coil domains were identified using the program COILS (http://www.ch.embnet.org/software/COILS_form.html). Protein sequences were analyzed for transmembrane helices (TMH) using the TMHMM v.2.0 software (available at http://www.cbs.dtu.dk/services/TMHMM/). NetNGlyc 1.0 Server was used to identify glycosylation sites of this enzyme (http://www.cbs.dtu.dk/services/NetNGlyc/).

Expression analysis of PajCHS

Expression analysis of PajCHS was performed by qualitative and quantitative PCR strategies. The tissue distribution of PajCHS transcripts was analyzed by end-point RT-PCR. Total RNA was isolated from gill, epidermis, gonad, hepatopancreas, deep abdominal flexor and extensor muscles, heart, thoracic and abdominal ganglia, brain, X-organ/sinus gland complex (XO/SG), and intestine. cDNA was synthesized under conditions similar to those used for cloning PajCHS except with random hexamers instead of the oligo-dT primer. Synthesized cDNA was treated with DNaseI (Promega), quantified, and aliquoted until it was used in reactions. PCR mixtures (20 µL) contained 1 µL cDNA (100 ng), 1 µL 4 µM sequence-specific primers (Table 1), 0.2 µL Ex Taq polymerase (Takara Bio Inc., Kyoto, Japan), 2 µL dNTPs (2.5 mM each), and 2 µL buffer (10×). PCR conditions were 3 min at 94°C, followed by 30 cycles of 94°C for 30s, 60°C for 30s, and 72°C for 30s. 18S rRNA primers were used as a positive control.

Quantitative PCR (qPCR) was carried out using the DNA Engine Chromo 4 Real-Time Detector (Bio-Rad, Hercules, CA) to compare PajCHS transcript levels between control and ESA groups. Six individual samples were analyzed in each group. SYBR Green premix Ex Taq™ (Takara Bio Inc.) was used with 50 ng cDNA as a template. qPCR was carried out under the same conditions used for end-point RT-PCR as described above, except 40 cycles were performed. Standard curves

were constructed to quantify copy numbers as described previously [23] and to confirm the efficiency of each primer set. Calculated copy number was normalized relative to the copy number of 18S rRNA. The relative copy number was calculated as follows: actual copy number of PajCHS/actual copy number of 18S rRNA. Statistical analysis of PajCHS was performed by a Student's t test using SigmaPlot (Systat Software, Inc., Richmond, CA). The effects of 20E on PajCHS gene expression in different tissues, such as the epidermis and gill, were quantified by qPCR as described above.

Results

Cloning and characterization of PajCHS

Based on the results of bioinformatics analysis and the traditional cloning strategy, the full-length chitin synthase cDNA was identified (PajCHS). The obtained PajCHS cDNA (4578 bp) encoded a protein of 1525 residues and had 81 bp and 402 bp untranslated regions (UTRs) at the 5' and 3' ends, respectively. PajCHS exhibited the highest sequence similarity (76%) to T. castaneum (TcCHS, NM_001039402) followed by 75% sequence similarity to A. gambiae (AgCHS, XP 321336) and L. migratoria (LmCHS, ACY 38589). BLAST analysis of PajCHS also showed 76% similarity to and 62% identity with an unannotated sequence from the cladoceran Daphnia pulex (EFX76951). To determine the evolutionary relationship between PajCHS and cladoceran CHS, we considered DaphCHS in our study.

PajCHS is a large protein with an estimated molecular mass of approximately 175 kDa and a slightly acidic pI of 6.12, similar to insect CHS (Table 2). To compare the structural characteristics between PajCHS and other known CHSs from insect species, multiple alignments were performed (Figure 1). Similar to other CHSs, PajCHS exhibited the conserved primary structure of chitin synthase composed of three domains, A, B, and C [6]. In domain A, a signal-anchoring sequence was identified between Gly14 and Ser15 residues, indicating that PajCHS is a membrane-bound protein. Insect signal-anchoring sequences range in length from 15 to 22 residues (data not shown). Secondary structure analysis predicted that PajCHS may contain 16 TMHs, each of which is composed of 19–23 amino acid residues. The number of TMHs in insect species varies from 14 and 17 (Table 2). Nine TMHs were predicted in domain A of PajCHS, whereas 7–9 TMHs were identified in the CHS from insect species (Figure 1). The B domain of PajCHS is similar in size to insect CHSs (~400 residues) and contains the catalytic center of the protein (Figure 1). Two unique conserved motifs (EDR and QRRRW) within the catalytic domain are present in all CHS and are therefore considered signature sequences. The catalytic activity of D. melanogaster CHS (DmCHS2) was confirmed by its expression in Schneider 2 cells with radiolabeled N-acetyl-D-glucosamine from uridine diphospho-N-acetyl-D-glucosamine. Two

	PajCHS	TcCHS1	TcCHS2	MsCHS1	MsCHS2	DmCHS1	DmCHS2	Daph CHS1	Daph CHS2
Organisms	P. japonica	T. castaneum	T. castaneum	M. sexta	M. sexta	D. melanogaster	D. melanogaster	D. pulex	D. pulex
GenBank Accession number	KC131026	AAQ55059	AAQ55061	AY062175	AY821560	NM_079509	NM_079485	EFX76951	EFX80669
Class	1	1	2	1	2	1	2	1	2
Molecular Mass (kDa)	175	177.76	167.67	178.56	174.42	182.8	161.39	182.46	
PI	6.12	6.66	5.93	6.56	6.02	6.37	6.66	6.27	6.19
TMH	16	16	14	16	16	15	17	16	16
Position of Coiled-coils	1080-1116 1151-1181	1061-1106 1133-1164	1098-1133	1063-1106 1134-1164	1058-1092	1079-1122 1153-1181	1347-1375	1069-1112 1138-1174	-

Table 2: Characteristics of some properties of insects and crustacean CHS. PI, isoelectric point, TMH, Transmembrane helices; (see, materials and methods, Bioinformatics analysis).

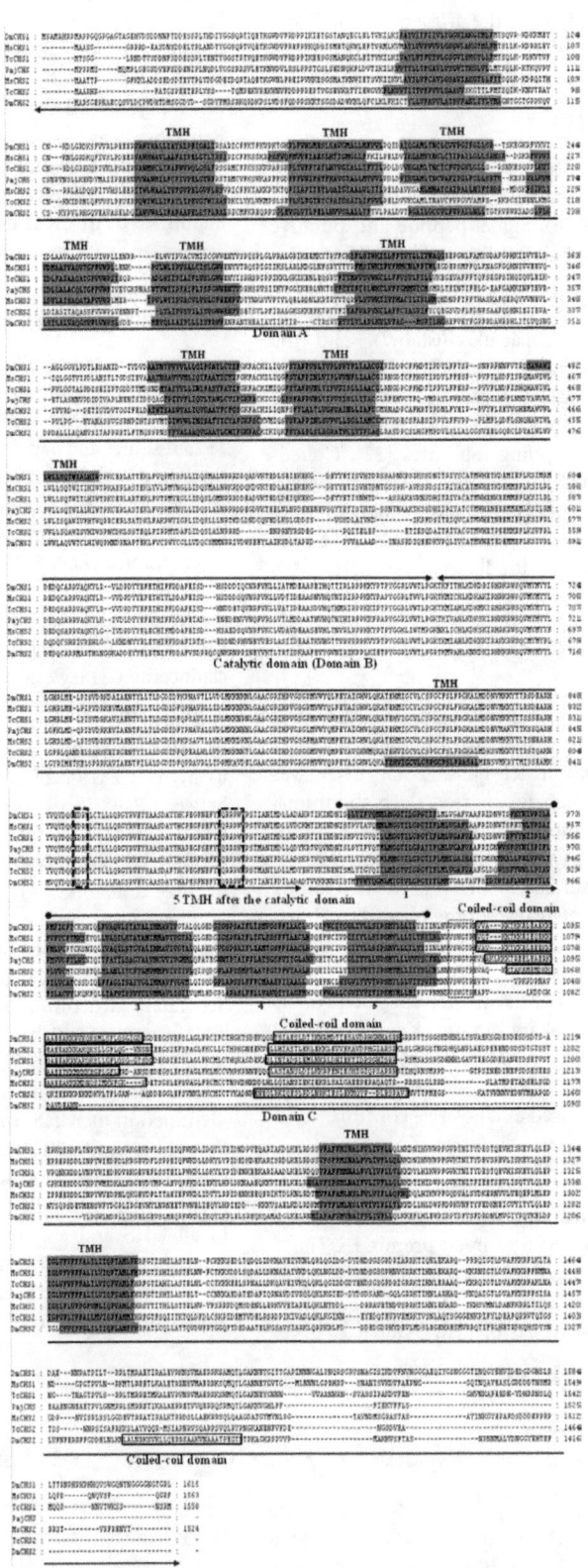

Figure 1: Multiple alignments of deduced amino acid sequences of PajCHS from *P. japonica* and representative insect species. Two types of CHSs from each of the diptera, *D. melanogaster*, DmCHS1 (NM_079509) and DmCHS2 (NM_079485); lepidopteran, *M. sexta*, MsCHS1 (AY062175) and MsCHS2 (AY821560); coleopteran, *T. castaneum*, TcCHS1 (AAQ55059) and TcCHS2 (AAQ55061) species. Putative transmembrane α helices (TMHs) are predicted by TMHMM server v. 2.0 (shaded). Three domains (A, B, C) are highlighted by pointed arrows and three conserved motifs (EDR, QRRRW, and SWGTR) are boxed with the broken line. Five TMHs near the catalytic domain are indicated by arabic numbers. Coiled-coil domains are boxed with solid line.

motifs in the catalytic domain, EDR and QRRRW, were predicted to be oriented to the cytoplasmic side, and a third motif in the C-terminal domain, SWGTR, was predicted to face the extracellular side of the membrane (Figure 1).

Five TMHs were predicted to be located just after the central catalytic domain (Figure 1). This is similar to that of cellular synthase, where it constitutes a pore in the membrane through which newly synthesized carbohydrate polymers may be extruded [24]. Seven potential N-glycosylation sites were identified throughout the sequence (data not shown) using PROSCAN [25], suggesting that this protein can be glycosylated. In addition, two coiled-coil sites were predicted from amino acid positions 1080–1116 and 1151–1181 (Figure 1), which are potential sites for protein–protein interactions and/or signals for vesicular trafficking [26]. TMHs in the N-terminus showed different patterns among the insect species, while the C-terminal TMHs were remarkably conserved both with respect to their locations and spacing between adjacent TMHs (Figure 1). Among the seven C-terminal TMHs, five were found in clusters immediately following the catalytic domain and two more sequences were located closer to the C-terminus (Figure 1). A phylogenetic tree was generated to determine the evolutionary relationships among PajCHSs (Figure 2). Three yeast

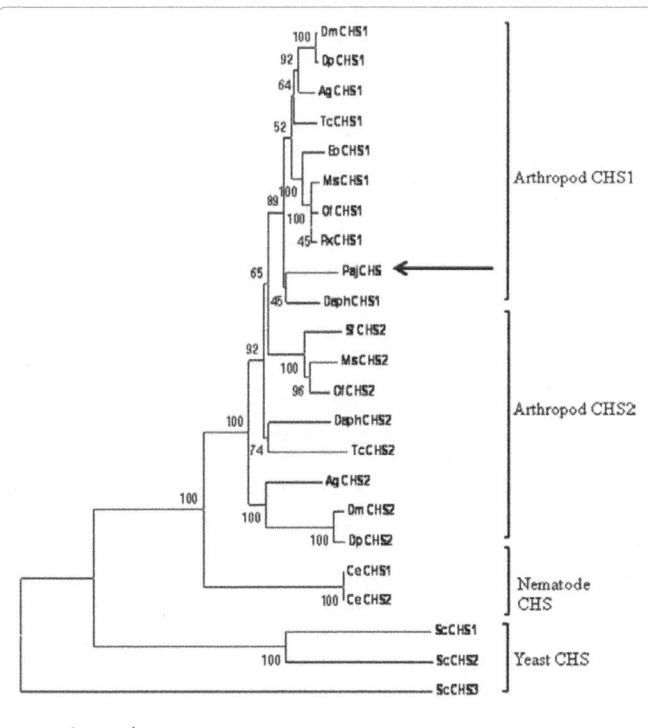

Figure 2: Phylogenetic tree of CHSs from various species. The Phylogenetic tree was constructed by minimum evolution algorithm using Molecular Evolutionary Genetics Analysis (MEGA4) software (see Materials and methods). Bootstrapping test was performed with 1000 replicates. The scale bar represents 0.2 amino acid substitution per site. Species and GenBank accession numbers: *P. japonica*, PajCHS (KC131026), *D. melanogaster*, DmCHS1 (NM_079509), DmCHS2 (NM_079485); *D. pseudoobscura*, DpCHS1 (XP_001359390), DpCHS2 (XP_001352881); *A. gambiae*, AgCHS1 (XM_321337), AgCHS2 (AY056833); *T. castaneum*, TcCHS1AAQ55059), TcCHS2 (AAQ55061); *E. obliqua*, EoCHS1 (ACA50098), *P. xylostella*, PxCHS1 (BAF47974), *M. sexta*, MsCHS1 (AY062175), MsCHS2 (AY821560); *S. frugiperda*, SfCHS2 (AAS12599); *Ostriniafurnacalis* OfCHS1 (ACB13821), OfCHS2 (ABB97082), *Caenorhabditiselegans*, CeCHS1 (Np_492113), CeCHS2 (Np_493682);*Saccharomyces cerevisiae,ScCHS1 (PO8004); ScCHS2 (AAA34493) ScCHS3 (P29465); D. pulex, DaphCHS1(EFX76951), DaphCHS2 (EFX80669).

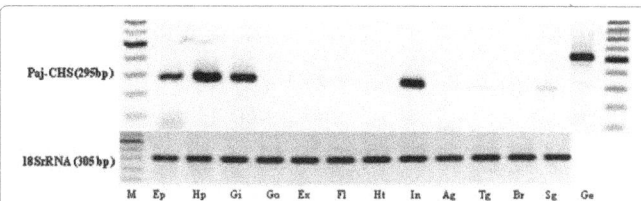

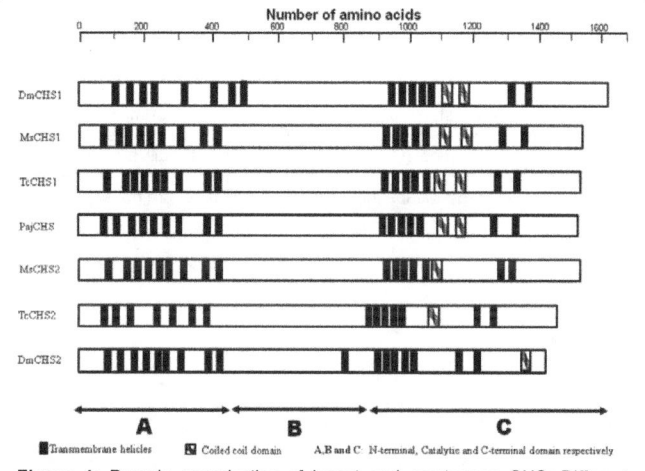

Figure 3: Tissue expression profile of PajCHS. End-point RT-PCR products of PajCHS were separated on 1.5% agarose gel and stained with ethidium bromide (inverted image). Amplification of an 18SrRNA product was used as a positive control. The larger size of genomic DNA PCR products (Ge) confirmed the PCR reaction was not contaminated with genomic DNA Abbreviations: M, molecular size marker; Ep, epidermis; Hp, hepatopancreas; Gi, gill; Go, gonad; Ex, extensor muscle; Fl, flexor muscle; Ht, heart; In, intestine; Ag, abdominal ganglion; Tg, thoracic ganglion; Br, brain; Sg, sinus gland.

Figure 4: Domain organization of insect and crustacean CHSs.Different numbers and location of transmembrane helices and coiled-coil domains in each CHS were presented. Length of each CHS protein was shown by scale bar above.

CHS clustered as outgroup members and nematode CHSs were also grouped together. PajCHS was clustered together with the insect CHS1 group, suggesting that an arthropod CHS gene duplication event may have occurred before the divergence of insects and crustaceans. Insect CHSs were further subdivided into CHS1A and CHS1B.

Transcriptional analysis of PajCHS

The distribution of the PajCHS transcript in tissue was determined by end-point RT-PCR and subsequent agarose gel electrophoresis of PCR products (Figure 3). Homogeneous 18S rRNA PCR products indicated that cDNA synthesis was successful, and the synthesized cDNA was free from genomic contamination. The transcript was mainly detected in epidermis, hepatopancreas, intestine, and gill (Figure 4). PCR product was also identified in the sinus gland/X-organ complex (SG); however, PajCHS transcript was not detected in the gonad, heart, brain, thoracic and abdominal ganglia, or deep abdominal flexor and extensor muscles.

qPCR was used to determine the effects of molt induction by ESA on the expression of PajCHS (Figure 5). The levels of PajCHS mRNA expression were significantly increased on day 7 after ESA in the epidermis and intestine compared to the control group. In the epidermis, mRNA expression level increased up to 2.8-fold on day 7 but reached 13.7-fold in the intestine compared to the control group. In the gill, PajCHS transcript level increased significantly on day 3 (1047-fold) and decreased to the basal level on day 7. However, no significant

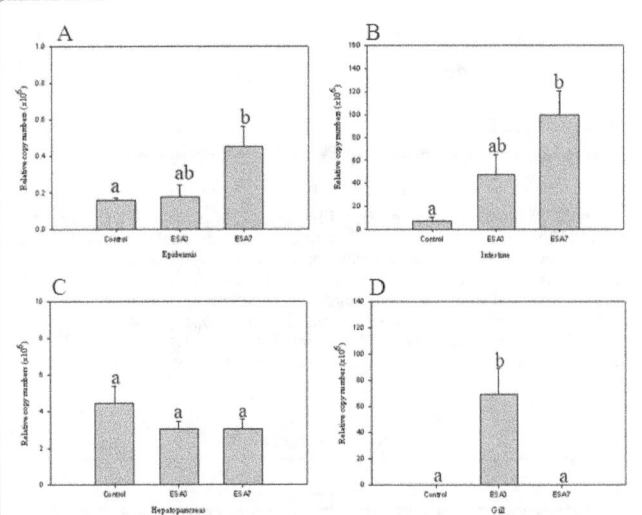

Figure 5: Effect of eyestalk ablation (ESA) on expression of mRNA levels of PajCHS. mRNA level was measured by qPCR from various tissues including (A) epidermis (B) intestine (C) hepatopancreas and (D) gill after ESA and data were normalized by the copy number of 18S rRNA. The data represents meanvalues ± SE (n=6). Means that share the same letter are not significantly different (P>0.05)

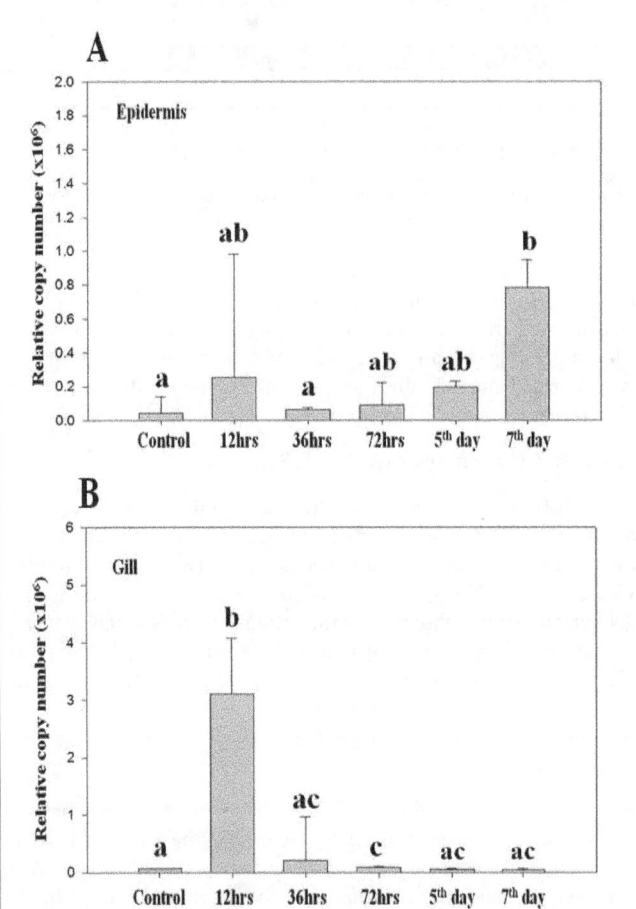

Figure 6: Effects of 20E on the PajCHS gene expression in (A) epidermis and (B) Gill. mRNA levels of PajCHS were quantified in epidermis and gill by qPCR and data were normalized by the copy number of 18S rRNA. Means that share the same letter are not significantly different (P>0.05).

effects of ESA were observed in the hepatopancreas. Similarly, PajCHS expression was also increased in the epidermis and gill after injection with 20E (Figure 6). In the epidermis, the expression was increased by 8.7-fold compared to the control group on day 7 after 20E injection and that in the gill was significantly increased to 53-fold at 12 h (Figure 5).

Discussion

Chitin degradation and synthesis are common phenomena in arthropods and are directly related to proper development and growth. Understanding genes involved in chitin metabolism is the first step for its applications in decapod aquaculture. Although CHS is a pivotal enzyme in the chitin biosynthesis pathway, no full-length CHS gene in a decapod crustacean has previously been reported, mainly due to their large size (~5 kb). To the best of our knowledge, PajCHS is the first full-length cDNA encoding CHS reported from a decapod crustacean. Although two partial sequences have been reported from *L. vannamei* [27] and *H. americanus* (NCBI database, GQ169704), they are too small to compare to PajCHS.

The primary structure and overall domain organization of PajCHS are similar to those from insect species (Figure 1). In addition, signature motifs (EDR, QRRRW, and SWGTR) are also well conserved. Site-directed mutagenesis of EDR and QRRRW motifs in fungal CHS results in loss of activity, suggesting that these two motifs are essential for CHS activity [28]. The (S/T)WGTKG motifs face the extracellular environment and play an important role in chitin translocation rather than synthesis [29]. These results indicate that PajCHS encodes the catalytically active chitin synthase. Despite the overall structural similarity, the number of TMHs varies among CHSs from different species (Figure 1); the N-terminal end contains 9–10 TMHs, whereas the C-terminal end contains 7–8 TMHs (Figure 1). Different numbers of TMHs have been proposed in some CHSs from insect species, including D. melanogaster and T. castaneum. Domains A and C of PajCHS were predicted to have 9 and 7 TMHs, respectively. The number of TMHs affects the orientation of CHSs, and further studies are required to determine whether differences in the number of TMHs may be responsible for the differences in protein orientation.

Due to the large sizes of CHSs (~160–180 kDa) and their membrane-bound functional forms, it has been difficult to study their structure and activity in both insect and crustacean species. Recent knockdown experiments using RNA-silencing techniques have revealed that each CHS gene plays an important role in growth and developmental stages [30,31]. In a previous study, injection of dsRNA corresponding to MsCHS1 cDNA into fifth-instar larvae (M. sexta) resulted in severe head deformities and death of pupae, whereas controls showed normal development [12]. In another study, the injection of dsRNA into Locusta migratoria corresponding to LmCHS1 or either of its two variants, LmCHS1A/LmCHS1B, showed a nymph mortality rate ranging from 50% to 95%; however, phenotypic deformities were similar to those observed when LmCHS1 or LmCHS1A were used, whereas the use of LmCHS1B led to a crippled-cuticle phenotype [31]. These results indicate that both of the variants perform essential functions for insect growth and development. Moreover, in vivo treatment of fourth-instar larvae of Spodoptera exigua with siRNA-silencing CHS1 causes abnormalities and disordered cuticle formation [30]. This indicates that CHS plays an important role in the proper growth and development of arthropods, and RNA-silencing techniques should be applied to estimate their functions in crustaceans.

Multiple sequence alignment (Figure 1), domain organization (Figure 4), and phylogenetic analyses (Figure 2) suggested that PajCHS is a class 1 CHS. PajCHS was more closely related to insect CHS1 than to CHS2. Both CHS1 and CHS2 exist in insect genomes as a result of gene duplication from the ancestral CHS gene [12]. These two copies are also found in the cladoceran *D. pulex*, suggesting that the presence of two copies of these genes may be common in arthropods [8,11,13]. A homologous sequence of PajCHS from Litopenaeus vannamei (LvCHS ~300 bp, FJ229468) reported in a previous study [27] was too small to compare to our results. However, PajCHS showed a substantially different tissue expression profile than those of CHS1s in insect species. In insects, CHS1s are expressed in epidermis-related ectodermal tissues, whereas CHS2 is expressed predominantly in the gut PM [6,11,14,32]. In case of M. sexta CHS genes, MsCHS1 and MsCHS2 show differential expression patterns in tissues and among developmental stages [14]. One of the major differences compared to insect CHS1 is the hepatopancreatic expression of PajCHS (Figure 3). Although the insect fat body is thought to correspond to the crustacean hepatopancreas, CHS transcripts have not been detected in the fat body in insects [9]. This suggests that CHSs in decapod crustaceans may have additional functions compared to those in insect species. In addition, expression of PajCHS in the intestine suggests that a single CHS transcript may perform the functions of both CHS1 and CHS2 in insects. A partial LvCHS sequence from *L. vannamei* showed a similar expression pattern to PajCHS, but it was too small (~300 bp) to allow a comparison, and it is still unclear whether LvCHS is an ortholog of PajCHS. There is no evidence to support the presence of two copies of CHS genes in decapod crustaceans.

The PajCHS transcript was upregulated by both ESA and 20E injection. ESA is one of the best established ways to induce the production of 20E in hemolymph to accelerate ecdysis in crustaceans [33]. In insect species, ecdysteroids (molting hormones) play pivotal roles in the expression of CHS genes [34-36]. However, CHS1 and CHS2 respond differently to increased ecdysteroid levels. The expression of MsCHS1 in tracheal cells is highest at the beginning of the molt when the ecdysteroid level remains high. In contrast, the MsCHS2 gene in columnar cells is down regulated despite high ecdysteroid levels [34-36]. We found that PajCHS mRNA transcription levels were significantly upregulated in the epidermis and intestine after ESA, indicating a relationship between increased ecdysteroid level and CHS expression. Upregulation of the transcripts suggested that ecdysone activates CHS genes by activating nuclear receptor heterodimer consisting of the EcR and the Drosophila retinoid X receptor homolog of ultra spiracle protein (USP) [37]. Temporal differences in PajCHS transcript levels associated with ESA and direct 20E injection may be caused by the slow increase in ecdysteroids by ESA rather than 20E injection (Figure 6). As a consequence, PajCHS appears to be an ecdysteroid-response gene.

In conclusion, we isolated the full-length cDNA encoding PajCHS in a decapod crustacean, and described its molecular characteristics. Its tissue distribution and transcriptional response to 20E induction suggest that PajCHS is an ecdysteroid-response gene and that its manipulation may be useful in the decapod aquaculture industry.

References

1. Kramer KJ, Hopkins TL, Schaefer J (1995) Applications of solids NMR to the analysis of insect sclerotized structures. Insect Biochem Mol Biol 25: 1067-1080.

2. Tellam R (1996) The peritrophic matrix. Biology of the insect midgut. Springer. 86-114.

3. Merzendorfer H, Zimoch L (2003) Chitin metabolism in insects: structure, function and regulation of chitin synthases and chitinases. J Exp Biol 206: 4393-4412.

4. Coutinho PM, Deleury E, Davies GJ, Henrissat B (2003) An evolving hierarchical family classification for glycosyltransferases. J Mol Biol 328: 307 317.

5. GLASER L, BROWN DH (1957) The synthesis of chitin in cell-free extracts of Neurospora crassa. J Biol Chem 228: 729-742.

6. Tellam RL, Vuocolo T, Johnson SE, Jarmey J, Pearson RD (2000) Insect chitin synthase cDNA sequence, gene organization and expression. Eur J Biochem 267: 6025-6043.

7. Ibrahim GH, Smartt CT, Kiley LM, Christensen BM (2000) Cloning and characterization of a chitin synthase cDNA from the mosquito Aedes aegypti. Insect Biochem Mol Biol 30: 1213-1222.

8. Gagou ME, Kapsetaki M, Turberg A, Kafetzopoulos D (2002) Stage-specific expression of the chitin synthase DmeChSA and DmeChSB genes during the onset of Drosophila metamorphosis. Insect Biochem Mol Biol 32: 141-146.

9. Hogenkamp DG, Arakane Y, Zimoch L, Merzendorfer H, Kramer KJ, et al. (2005) Chitin synthase genes in Manduca sexta: characterization of a gut-specific transcript and differential tissue expression of alternately spliced mRNAs during development. Insect Biochem Mol Biol 35: 529-540.

10. Zhu YC, Specht CA, Dittmer NT, Muthukrishnan S, Kanost MR, et al. (2002) Sequence of a cDNA and expression of the gene encoding a putative epidermal chitin synthase of Manduca sexta. Insect Biochem Mol Biol 32: 1497-1506.

11. Arakane Y, Hogenkamp DG, Zhu YC, Kramer KJ, Specht CA, et al. (2004) Characterization of two chitin synthase genes of the red flour beetle, Tribolium castaneum, and alternate exon usage in one of the genes during development. Insect Biochem Mol Biol 34: 291-304.

12. Merzendorfer H (2006) Insect chitin synthases: a review. J Comp Physiol B 176: 1-15.

13. Arakane Y, Muthukrishnan S, Kramer KJ, Specht CA, Tomoyasu Y, et al. (2005) The Tribolium chitin synthase genes TcCHS1 and TcCHS2 are specialized for synthesis of epidermal cuticle and midgut peritrophic matrix. Insect Mol Biol 14: 453-463.

14. Zimoch L, Hogenkamp DG, Kramer KJ, Muthukrishnan S, Merzendorfer H (2005) Regulation of chitin synthesis in the larval midgut of Manduca sexta. Insect Biochem Mol Biol 35: 515-527.

15. Salma U, Uddowla MH, Kim M, Kim JM, Kim BK, et al. (2012) Five hepatopancreatic and one epidermal chitinases from a pandalid shrimp (Pandalopsis japonica): cloning and effects of eyestalk ablation on gene expression. Comp Biochem Physiol B Biochem Mol Biol 161: 197-207.

16. Jeon JM, Lee SO, Kim KS, Baek HJ, Kim S, et al. (2010) Characterization of two vitellogenin cDNAs from a Pandalus shrimp (Pandalopsis japonica): expression in hepatopancreas is down-regulated by endosulfan exposure. Comp Biochem Physiol B Biochem Mol Biol 157: 102-112.

17. Lee SO, Jeon JM, Oh CW, Kim YM, Kang CK, et al. (2011) Two juvenile hormone esterase-like carboxylesterase cDNAs from a Pandalus shrimp (Pandalopsis japonica): cloning, tissue expression, and effects of eyestalk ablation. Comp Biochem Physiol B Biochem Mol Biol 159: 148-156.

18. de Oliveira Cesar JR, Zhao B, Malecha S, Ako H, Yang J (2006) Morphological and biochemical changes in the muscle of the marine shrimp (Litopenaeus vannamei) during the molt cycle. Aquaculture 261: 688-694.

19. Chan SM, Rankin SM, Keeley LL (1988) Characterization of the molt stages in Penaeus vannamei: setogenesis and hemolymph levels of total protein, ecdysteroids, and glucose. Biol Bull 175: 185-192.

20. Shechter A, Tom M, Yudkovski Y, Weil S, Chang SA, et al. (2007) Search for hepatopancreatic ecdysteroid-responsive genes during the crayfish molt cycle: from a single gene to multigenicity. J Exp Biol 210: 3525-3537.

21. Jeon JM, Kim BK, Lee JH, Kim HJ, Kang CK, et al. (2012) Two type I crustacean hyperglycemic hormone (CHH) genes in Morotoge shrimp (Pandalopsis japonica): Cloning and expression of eyestalk and pericardial organ isoforms produced by alternative splicing and a novel type I CHH with predicted structure shared with type II CHH peptides. Comp Biochem Physiol B Biochem Mol Biol 162 : 88-89

22. Tamura K, Dudley J, Nei M, Kumar S (2007) MEGA4: Molecular Evolutionary Genetics Analysis (MEGA) software version 4.0. Mol Biol Evol 24: 1596-1599.

23. Kim KS, Kim BK, Kim HJ, Yoo MS, Mykles DL, et al. (2008) Pancreatic lipase-related protein (PY-PLRP) highly expressed in the vitellogenic ovary of the

scallop, Patinopecten yessoensis. Comp Biochem Physiol B Biochem Mol Biol 151: 52-58.

24. Richmond T (2000) Higher plant cellulose synthases. Genome Biol 1: REVIEWS3001.

25. Hansen JE, Lund O, Rapacki K, Brunak S (1997) O-GLYCBASE version 2.0: a revised database of O-glycosylated proteins. Nucleic Acids Res 25: 278-282.

26. Melia TJ, Weber T, McNew JA, Fisher LE, Johnston RJ, et al. (2002) Regulation of membrane fusion by the membrane-proximal coil of the t-SNARE during zippering of SNAREpins. J Cell Biol 158: 929-940.

27. Rocha J, Garcia-Carreño FL, Muhlia-Almazán A, Peregrino-Uriarte AB, Yépiz-Plascencia G, et al. (2011) Cuticular chitin synthase and chitinase mRNA of whiteleg shrimp (Litopenaeus vannamei) during the molting cycle. Aquaculture 330: 111-115.

28. Nagahashi S, Sudoh M, Ono N, Sawada R, Yamaguchi E, et al. (1995) Characterization of chitin synthase 2 of Saccharomyces cerevisiae. Implication of two highly conserved domains as possible catalytic sites. J Biol Chem 270: 13961-13967.

29. Cohen E (2001) Chitin synthesis and inhibition: a revisit. Pest Manag Sci 57: 946-950.

30. Chen X, Tian H, Zou L, Tang B, Hu J, et al. (2008) Disruption of Spodoptera exigua larval development by silencing chitin synthase gene A with RNA interference. Bull Entomol Res 98: 613-619.

31. Zhang J, Liu X, Zhang J, Li D, Sun Y, et al. (2010) Silencing of two alternative splicing-derived mRNA variants of chitin synthase 1 gene by RNAi is lethal to the oriental migratory locust, Locusta migratoria manilensis (Meyen). Insect Biochem Mol Biol 40: 824-833.

32. Zimoch L, Merzendorfer H (2002) Immunolocalization of chitin synthase in the tobacco hornworm. Cell Tissue Res 308: 287-297.

33. Tamone SL, Adams MM, Dutton JM (2005) Effect of Eyestalk-Ablation on Circulating Ecdysteroids in Hemolymph of Snow Crabs, Chionoecetes opilio: Physiological Evidence for a Terminal Molt. Integr Comp Biol 45: 166-171.

34. Baker FC, Tsai LW, Reuter CC, Schooley DA (1987) In vivo fluctuation of JH, JH acid, and ecdysteroid titer, and JH esterase activity, during development of fifth stadium Manduca sexta. Insect Biochem 17: 989-996.

35. Bollenbacher WE, Smith SL, Goodman W, Gilbert LI (1981) Ecdysteroid titer during larval--pupal--adult development of the tobacco hornworm, Manduca sexta. Gen Comp Endocrinol 44: 302-306.

36. Riddiford LM (1994) Cellular and molecular actions of juvenile hormone. I. General considerations and premetamorphic actions. Adv. Insect Physiol 24: 3-74.

37. Yao TP, Forman BM, Jiang Z, Cherbas L, Chen JD, et al. (1993) Functional ecdysone receptor is the product of EcR and Ultraspiracle genes. Nature 366: 476-479.

5

Comparative Studies on the Nutrients, Sensory and Storage Qualities of Moon-Fish (*Citharinus citharus* Geoffery Saint-Hilaire 1809) Pre-Treated with Extracts from Two Spices

Agbabiaka LA[1]*, Kuforiji OA[2] and Egobuike CC[3]

[1]Department of Fisheries Technology, Federal Polytechnic Nekede Owerri, Imo State, Nigeria
[2]Department of Agricultural Technology, Federal Polytechnic Ekowe, Bayelsa State, Nigeria
[3]Department of Fisheries Technology, Imo State Polytechnic, Umuagwo-Ohaji, Imo State, Nigeria

Abstract

Comparative effects of soaking *Citharinus citharus* with extracts from two spices (Black pepper=*Piper guineense* and Ginger=*Zingiber officinale*) prior to oven drying using Nutrients and Organoleptic assessments as parameters were studied. Seventy two freshly caught Moonfish weighing 875 ± 25 g were killed, weighed, gutted and washed thoroughly with portable tap water and were divided into two batches of 36 fish each. Each batch was sub-divided into 3 groups of 12 fish per treatment made up of a triplicate, i.e., 4 fish per replicate in a completely randomized design. The fish in first batch was divided into 3 treatments of 12 fish and were immersed for 30 minutes in composite solution containing 3% brine and extract of *Piper guineense* at 0%, 2% and 5% concentration coded as BP_0, BP_2 and BP_5 respectively; the control treatment contained no Black pepper (BP_0). Similar operations were carried out on the second batch of fish soaked in composite solution containing same concentration of brine as hitherto but with varying concentration of ginger extract (*Zingiber officinale*) at 0%, 2% and 5% for 30minutes coded as GE_0, GE_2 and GE_5 respectively; the control treatment was soaked in 3% brine only without ginger extract (GE_0). All fish pre-treated were allowed to drain for five minutes, oven-dried for 5 hours at the temperature range of 80°C-90°C and allowed to cool at room temperature. Results from proximate analysis showed that the use of composite mixture of brine and spices extract increased the crude protein (CP) content of fish linearly (p<0.05), there was no significant differences (p>0.05) in sensory qualities between the two experiments pre-treated with black pepper and ginger. These data above indicated that fish pre-treated with composite solutions of brine and extract of two spices (*Piper guineense* and *Zingiber officinale*) prior to oven drying have potentials to improve nutrients assay and storage qualities on Moonfish.

Keywords: Moonfish; Comparative; Nutrients; Organoleptic; Spices and storage

Introduction

Fish contribute more than 60% of the world supply of protein, especially in the developing countries including Nigeria [1]. It has been reported that Nigeria as a maritime nation with a vast population estimated at over 160 million people and a coastline measuring approximately 853 kilometers [2], with huge fish production capacity to contribute significantly to the agricultural sector spends over $200 million on fish importation annually. The recent importation rate is over 750,000 metric tonnes [3,4] but the annual fish demand in the country was estimated at about 2.66 million metric tonnes, while a paltry domestic production is estimated to be 780,000 tonnes; the demand-supply gap stands at a staggering 1.8 million metric tonnes. Despite the popularity of farming in Nigeria, the fish farming industry can best be described as being at the infant stage when compared to the large market potential for its production and marketing [5].

The right step currently geared towards reducing the demand-supply deficit for fish (Aquaculture development) may not be realistic until appropriate measures towards curtailing post-harvest losses and quality depreciations are put in place through improved value-addition on processed products [6,7]. The high ambient temperature in Nigeria typical of the tropics that often encourages food/fish spoilage has made it imperative to search for alternative economical method(s) other than cold storage hindered by high energy cost occasioned by incessant power outage in developing nations to improve post-harvest/ shelf-life of fish. Many processing methods have been adopted with successes but use of spices on smoke-dried fish is yet to be commercialized in Nigeria.

Moonfish (*Citharinus citharus*) is a genus of lute fish from tropic Africa with six currently described species. Moonfish belongs to the family Citharinidae and are found in most habitats but they are particularly abundant in swamp, creeks and lakes in Nigeria [8]; where they spawn during the flood season. Their deep and flattened body earned them the popular name Moonfish [9]. The species is relatively cheap comparatively but a highly nutritious fish when cooked, smoked or dried. Proximate composition of the moonfish revealed that it contained 77.1% moisture, 0.98% Lipid, 20.4% crude protein and 1.5% ash [8].

Spices such as Black pepper (*Piper guineense*) and Ginger (*Zingiber officinale*) have been natural additives for more than 2000 years. Studies have shown that, the long term dietary intake of ginger has hypoglycemic and hypo-lipidaemic effect [10], while the seeds of black pepper are put into variety of uses, for instance in some parts of Nigeria, the seeds are consumed by women after child birth to enhance uterine contraction for expulsion of placenta and remains from the womb [11]. In order to guarantee abundant and sustainable fish supply, there is need to reduce

*Corresponding author: Agbabiaka LA, Department of Fisheries Technology, Federal Polytechnic Nekede Owerri, Imo State, Nigeria
E-mail: adegokson2@yahoo.com

spoilage and increase storage qualities/shelf life. Therefore, this work is aimed at determining the suitability of using spices (*Piper guineense* and *Zingiber officinale*) which have been reported to contain anti-microbial, anti-oxidant and anti-septic qualities/properties in improving the shelf life of moonfish and consumers' acceptability [12,13].

Materials and Method

Site of experiment

This experiment was carried out at the Department of Food Science and Technology, Imo State Polytechnic Umuagwo-Ohaji Nigeria. Umuago-Ohaji lies between latitude 5° 17[1] and 5° 19[1] N, longitude 7° 54[1] and 6° 56[1] E. It is twenty six kilometres from the State capital (Owerri) on the Port Harcourt road.

Collection of samples

Seventy two (72) freshly caught Moonfish (*Citharinus citharus*) weighing between 850-900 g were purchased from Swale Market in Yenegoa, Bayelsa State, Nigeria. Some quantity of dried black pepper seeds and ginger rhizomes were bought at Ekeonunwa Market in Owerri, Imo State.

Preparation of samples

Dried Black pepper seeds and Rhizome of Ginger were ground into powder using kitchen grinding machine respectively. Seventy two freshly caught Moonfish weighing 875 ± 25 g were killed, weighed, gutted and washed thoroughly with portable water and were divided into two batches of 36 fish each. Each batch was sub-divided into 3 groups of 12 fish per treatment in a triplicate, i.e., 4 fish per triplicate in a completely randomized design. First treatment was immersed in 3% brine without spice extract (control), second treatment was soaked in solution of 3% brine and 2% black pepper extract, while the third treatment was also immersed in the solution of 3% brine and 5% black pepper extract respectively for 30 minutes coded BP_0, BP_2 and BP_5. Similar procedure were carried out on the second batch of fish soaked in composite solution containing same concentration of brine and varying concentration of ginger extract (*Zingiber officinale*) at 0%, 2% and 5% for 30minutes coded as GE_0, GE_2 and GE_5 respectively; the control treatment was soaked in 3% brine only without ginger extract (GE_0). The spice extracts were obtained by soaking appropriate quantity of ground spices (black pepper and ginger) respectively in water overnight and sieved accordingly. Thereafter, fish samples soaked in the composite solutions of brine/spice extracts respectively for 30 minutes were removed and put into separate baskets covered with muslin cloth to drain for 5 minutes. Subsequently, the fish samples were arranged into labelled trays inside gas oven and allowed to dry at temperature of 80°C-90°C for 5 hours.

Processing techniques

Drying was conducted by using gas oven. The pre-treated fish samples were arranged on the oven trays and allowed to dry for 5 hours, during which turning over of the fish were done at interval to achieve uniformly dried product. Thereafter, the dried product were removed from the oven and arranged on separately labelled trays and were allowed to cool at room temperature before weighing in order to determine the moisture loss. Labelled samples were accordingly and carefully stored for 7 days in quality control room during which time, daily monitoring (8:00-9:00 hours) was conducted to check the effect of the spices on the shelf life and organoleptic qualities of the fish.

Proximate and storage analysis

Samples of oven-dried Moonfish were collected from the treatments on triplicate basis within 24 hours for the nutrients assay such as crude protein (Kjeldhal procedure), lipid (Soxhlet method), Ash (flame photometric) as described by AOAC [14]. The samples were stored 7days at ambient temperature and were monitored daily between the hours of 8:00-9:00 (CAT) in order to determine the effect of the spices on the shelf life.

Organoleptic evaluation

The samples of the oven-dried fish were served to 10 (ten) trained panelists made up of Staff and Students of Department of Nutrition and Food and Technology, Imo State Polytechnic Umuagwo; who were familiar with assessing the sensory attributes such as aroma, taste, appearance/colour, consistency and general acceptability. Nine points hedonic scale (9=excellent; 8=very good; 6=good; 4=fair; 2=poor; and 0=bad) was adopted according to Afolabi et al. [15] to evaluate changes in colour, aroma, taste and texture of fish.

The samples in batch A, pre-treated with black pepper at 0%, 2% and 5% concentration were coded as BP_0, BP_2 and BP_5 respectively while Ginger pre-treated samples were coded as GE_0, GE_2 and GE_5 respectively and presented in identical aluminum foil in trays simultaneously to avoid possibility of panelist re-evaluating a sample. Necessary precaution were taken to prevent bias of panelists during evaluation by ensuring that the panelists rinse their mouth after tasting each sample and were trained in each stage of organoleptic evaluation, it was also ensured that the panelists were ignorant of the actual samples represented by a code.

Statistical analysis

The data obtained from proximate and sensory evaluation were subjected to Statistical analysis using Analysis of variance (ANOVA) and Mean Separation [16] with SPSS Window 17.0 Version of Inc., USA.

Results and Discussion

Proximate analysis

Results showed that the use of the spices pre-treatment increased nutrients notably the percentage crude protein (CP) content of fish linearly ($p<0.05$) from 64.86 ± 0.01% in BP_0 (control) to 68.80 ± 0.40% in BP_5 for black pepper experiment while ginger treatments coded GE_0, GE_2 and GE_5 recorded Crude Protein values of 64.58 ± 0.03%, 68.37 ± 0.05% and 68.63 ± 0.01% respectively. Similar observation was recorded when smoked African Lungfish was treated with ginger [7] (Tables 1 and 2). The improved crude protein concentration linearly with spices fortification might be due to additional nutrients inherent in black pepper and ginger which contained crude protein (8.60%), Ether extract (15.21%) and minerals (Ash=5.21%) as reported by Nwinuka et al. [17] and Meadow [18]. Similarly, there were significant differences in the concentrations of crude fat, Ash and soluble carbohydrate between the control fish and those treated with extracts from the two spices respectively ($p<0.05$) but not significantly different in moisture content generally ($p>0.05$). Also, there was no significant difference ($p>0.05$) in percentage moisture loss in the fish products from the two spices which are within the range of 65.00% recommended by Cardinal et al. [19] as shown in Tables 3 and 4 below.

The Ash content of the oven-dried moonfish ranged from 4.06 ± 0.13% and 4.70 ± 0.01%. These values are higher than 0.40-1.35%

Samples	Live weight of fish (g)	Dressed Weight (g)	Weight after Total smoking (g)	Weight loss(g)	Weight loss (%)
BP_0	845	716	295	550.66	65.11
BP_2	860	725.6	265.31	594.69	69.15
BP_5	815	691.67	264.88	550.13	67.5

Average weight loss (%) =67.26
BP_0: Fish sample treated with 3% brine solution only (control)
BP_2: Fish sample treated with mixture of 3% brine and 2% black pepper
BP_5: Fish sample treated with mi8xture of 3% brine and 5% black pepper

Table 1: Mean weight Characteristics of Dried Moon-fish (*Citharinus citharus*) pretreated with black pepper.

Samples	Live weight of fish (g)	Dressed Weight (g)	Weight after smoking (g)	Total weight loss (g)	(%) weight loss
GE_0	890	751	242.26	647.74	72.78
GE_2	880	745.6	307.74	572.26	65.03
GE_5	865	737.67	302.49	562.51	65.03

Average weight loss (%) =67.13
GE_0: Fish sample treated with 3% brine solution only (control)
GE_2: Fish sample treated with mixture of 3% brine and 2% ginger extract
GE_5: Fish sample treated with mixture of 3% brine and 5% ginger extract

Table 2: Weight characteristics of *Citharinus citharus* pre-treated with ginger and oven-dried.

	Concentrations of Black pepper extract		
Parameters	BP_0	BP_2	BP_5
Moisture %	12.77 ± 0.03^a	12.64 ± 0.01^b	$12.29\ 0.01^c$
Protein %	64.86 ± 0.01^c	67.86 ± 0.02^b	$68.81\ 0.4^a$
Fat %	5.52 ± 0.02^a	4.62 ± 0.01^c	4.79 ± 0.2^b
Ash %	4.58 ± 0.02^a	4.06 ± 0.13^b	4.70 ± 0.01^a
NFE %	12.27 ± 6.02^a	10.82 ± 0.09^b	$9.41\ 0.40^b$

Mean and STD scores with different superscript at the same row are significantly different ($P<0.05$).
BP_0: Fish sample treated with 3% brine solution only (control)
BP_2: Fish sample treated with mixture of 3% brine and 2% black pepper
BP_5: Fish sample treated with mixture of 3% brine and 5% black pepper
STD: Standard Deviation
NFE: Nitrogen Free Extract

Table 3: Proximate composition of Moon Fish pre-treated with black pepper extract.

	Concentrations of Ginger extract		
Parameters	GE_0	GE_2	GE_5
Moisture (%)	13.22 ± 0.02^b	$11.70^c \pm 0.11^c$	13.81 ± 0.11^c
Protein (%)	65.58 ± 0.03^b	68.37 ± 0.05^a	$68.63\ 0.01^a$
Fat (%)	5.48 ± 0.02^a	5.41 ± 0.01^a	5.38 ± 0.01^a
Ash (%)	4.67 ± 0.01^a	3.98 ± 0.19^c	4.34 ± 0.01^b
NFE (%)	10.69 ± 0.04^a	10.53 ± 0.13^a	7.77 ± 0.01^c

Means plus STD within same row having different superscript differs significantly ($P>0.05$)
GE_0: Fish sample treated with 3% brine solution only (control)
GE_2: Fish sample treated with mixture of 3% brine and 2% ginger extract
GE_5: Fish sample treated with mixture of 3% brine and 5% ginger extract
STD: Standard Deviation
NFE: Nitrogen Free Extract

Table 4: Table of proximate composition of moonfish pre-treated with ginger extract.

reported by Abdullahi [20]. This suggests that the extracts of black pepper and ginger did not inhibit mineral content of the fish but improved it. The minerals especially calcium content in a fish product is an indication of the quality of the product [21].

The sensory evaluation revealed that products treated with Black pepper and Ginger extracts retained consumer acceptability in all attributes measured ($p>0.05$) up to the sixth day (Tables 5 and 6); though, samples treated with *Piper guineense* retained fairly good and better attributes on the six day comparatively ($p>0.05$). Nevertheless, the decline in qualities of the two samples from the fifth day may be probably due to the onset of microbes and degradation of protein/lipid in the muscle that leads to oxidative rancidity resulting to production of hypoxanthine and trimethylamine [22]. The improved shelf live up to 5th day may also be due to action of heat that reduced water activity and impaired the action of spoilage microbes [19,23]. The general acceptability in all the sensory parameters indicated that the spices might have inherent chemical compounds responsible for the pleasant colour, taste and flavor/aroma in oven-dried products [12,13].

Dressed weight=Carcass weight−weight of offals

Total weight Loss=Live Weight/Carcass weight−weight after smoking

$$\% \text{ weight loss} = \frac{\text{Total weight loss}}{\text{Live wieght of fish}} \times \frac{100}{1}$$

Parameters	Samples	Storage Period (Days)							
		1	2	3	4	5	6	7	LSD
Appearance	GE_0	7.13ab	7.88a	7.63ab	6.59bc	6.25c	3.63d	1.13e	0.57651
	GE_2	7.63a	7.88a	6.5bc	6.25b	5.88b	3.63c	1.13d	0.63735
	GE_5	8.13a	8.00a	7.50a	6.26b	5.88b	3.00c	1.13d	0.60288
Aroma	GE_0	6.50b	7.88a	6.76b	6.76b	4.88c	3.26d	3.38d	0.62504
	GE_2	7.26a	6.76ab	6.75a	6.00ab	4.88b	3.63c	3.13c	0.80571
	GE_5	7.50a	7.00ab	6.88ab	6.00bc	5.38c	3.75d	3.25d	0.61721
Taste	GE_0	7.38a	7.63a	7.00a	6.63a	5.13b	3.50c	1.75d	0.65835
	GE_2	7.50a	7.00ab	6.63ab	6.13b	5.50bc	3.88c	1.60d	0.69846
	GE_5	7.88a	7.38a	7.38a	6.13b	5.63b	4.13c	2.00d	0.61721
Texture	GE_0	6.88ab	7.50a	\7.38a	6.26ab	5.75b	2.88c	1.76d	0.57819
	GE_2	7.00a	7.50a	6.63ab	5.50bc	4.88c	3.25d	2.37d	0.60929
	GE_5	7.50a	6.38ab	6.63ab	5.75b	5.25b	3.25c	2.75c	0.67534

Means within rows with different superscripts differ significantly (P<0.05)
GE_0: Fish sample treated with 3% brine solution only (control)
GE_2: Fish sample treated with mixture of 3% brine and 2% ginger extract
GE_5: Fish sample treated with mixture of 3% brine and 5% ginger extract

Table 5: Organoleptic attributes of oven-dried moonfish with ginger extract pre-treatment.

Parameters	Samples	Storage Period (Days)							
		1	2	3	4	5	6	7	LSD
Appearance	BP_0	7.00ab	8.29a	7.57a	6.29b	6. 29b	4.00c	1.14d	0.52651
	BP_2	7.57a	7.71a	7.14a	6.86ab	7.00ab	5.00bc	1.14d	0.58703
	BP_5	7.57a	7.00ab	7.57a	7.00ab	7.29a	6.43b	1.14d	0.50218
Aroma	BP_0	5.86c	7.57a	6.86ab	6.43b	6.29b	4.00c	1.14d	0.65335
	BP_2	7.00ab	6.43b	7.29a	6.57ab	6.57ab	5.00c	1.86d	0.68346
	BP_5	7.43a	7.00ab	6.43b	6.43b	6.43b	5.42c	1.71d	0.67814
Texture	BP_0	6.14b	7.28a	7.00ab	6.29b	4.43c	4.00c	2.43d	0.62004
	BP_2	6.71ab	6.86ab	7.43a	6.14b	5.36bc	4.71c	1.71d	0.80117
	BP_5	6.71ab	7.00ab	7.00ab	6.14b	6.71ab	5.57bc	1.57d	0.79721
Taste	BP_0	6.14b	7.33a	7.29a	6.14b	5.14bc	4.71c	1.71d	0.52819
	BP_2	7.00ab	6.86ab	7.14a	6.43b	6.71ab	4.71c	1.71d	0.60428
	BP_5	6.86ab	7.85a	6.71ab	5.71c	6.43b	6.14b	1.43d	0.67036

Mean scores with different superscript along the columns are significantly different (P<0.05)
BP_0: Fish sample treated with 3% brine solution only (control)
BP_2: Fish sample treated with mixture of 3% brine and 2% black pepper
BP_5: Fish sample treated with mi8xture of 3% brine and 5% black pepper

Table 6: Organoleptic assessment of black pepper pre-treated oven dried moonfish.

Conclusion

However, the linear increment in nutrients assay with spice concentrations have further shown the positive value-addition attributes of these two spices. Hence, results from this study clearly recommend the use of these spices (Ginger and Black pepper) as value addition in fish processing to enhance storage and nutritional qualities.

References

1. Food and Agricultural Organization (2007) The state of World Fisheries and Aquaculture. FAO Fisheries Department, Rome, Italy.

2. http://www.thisdaylive.com/articles/aquaculture-as-path-to-thriving-agriculture/124614/.

3. United State Agency for International Development (2015) USAID funded project increases fish yield in Nigeria. Nigerian Vanguard Newspaper: Retrieved on Tuesday March 10th, 2015.

4. Oota L (2012) Is Nigeria committed to fish production. Accessed online 20th October.

5. Nwiro E (2012) Fish farming a lucrative business. Accessed online 20th October 2012.

6. Agbabiaka LA, Amadi AS, Madubuko CU, Nwankwo FC, Ojukannaiye AS (2012) Assessment of nutrients and sensory qualities of brine pre-treated catfish smoked with two different woods. African Journal of Food Science 6: 245-248.

7. Agbabiaka LA (2015) Assessment of nutrients and microbial evaluation of ginger pre-treated smoked African lungs-fish (Protopterus annecten OWEN, 1883) Tropical Agriculture (In press).

8. Omojowo FS, Ihuahi JA (2010) Preliminary report on the effects of different sauces on pouched Tilapia products. New York Science Journal 3: 83-86.

9. Gupta SK, Gupta PC (2006) General and Applied Ichthyology (Fish and Fisheries). S. Chand and Co. Ltd publishers, Ram Nagar, New Delhi, India.

10. Ahmed RS, Sharma SB (1997) Biochemical studies on combined effect of garlic (Allium satium Linn) and ginger (Zingiber officinale) in albino rats. Ind J Exp Biol 35: 841-843.

11. Negbenebor CA, I Nkama, P Sopade (1999) Raw material supply for fish and animal processing industries. In: Proceedings of a workshop organized by the North-East chapter of NIFST, Maiduguri, Nigeria. 12-24.

12. Bhandary CS (1993) Effect of spice treatment on lipid oxidation in smoked mackerel (Scomber scomber) FAO fisheries.

13. Amadi EC (2001) Antioxidant Effect of Oleoresin of some local Spices. Black pepper (Piper guineense), garlic (Allium satium), Nchuanwu (Ocimum gritininum) and Ehuru (Monodora myristica). Proceedings of 25th Annual Conference of NIFST.

14. AOAC (2000) Association of Official Analytical Chemist. Official methods of Analysis, 17th edition. Washington DC.

15. Afolabi OA, Arawomo OA, Oke OL (1984) Quality changes of Nigeria

Traditionally processed freshwater fish species. I. Nutritive and organoleptic changes. J Food Technol 19: 333-340.

16. Duncan DB (1955) Multiple range and multiple F tests. Biometrics 11:1-42.

17. Nwinuka NM, Ibeh GO, Ekeke GI (2005) Proximate Composition and levels of some toxicants in four commonly consumed spices. Journal of Applied Sciences and Environmental Management 9:150-155.

18. Meadows AB (1988) Ginger processing for food and Industry. Federal Institute of Indus. Res, Oshodi, Lagos, Nigeria.

19. Cardinal M, Knockaert C, Torrissen O, Sigurgisladottir S, Morkore T, et al. (2001) Relation of smoking parameters to the yield colour and sensory quality of smoked Atlantic salmon (Salmo salar). Food Res Int 34: 537-550.

20. Abdullahi SA (2001) Investigation of Nutritional status of Chrysichthys nigrodigitatus, Bagrus filamentous and Auchenoglanis occidentals: Family Barigdae. Journal of Arid zone Fisheries 1: 39-50.

21. Clucas IJ, Ward AR (1996) Postharvest fisheries development: a guide to Handling preservation, processing and Quality. Chatham Maritime, Kent ME44TB, United Kingdom.

22. Johnson WA, Nicholson FJ, A Roger (1994) Freezing and refrigerated storage in fisheries. Stroud Series. FAO Fisheries Technical Paper.

23. Abolagba OJ, Osifo SJ (2004) The effect of smoking on the chemical composition and keeping qualities of catfish (Heterobranchus bidorsalis) using two energy sources. Journal of Agric. Forestry and Fisheries 5: 27-30.

6

Advanced Techniques for Morphometric Analysis in Fish

Mojekwu TO[1]* and Anumudu CI[2]

[1]Department of Biotechnology Unit, Nigerian Institute for Oceanography and Marine Research, Nigeria
[2]Department of Zoology, University of Ibadan, Nigeria

Abstract

Information on the biology and population structure of any species is a prerequisite for developing management and conservation strategies. Morphometric characters of fish are the measurable characters common to all fishes. Some arbitrarily selected points on a fish body known as landmarks help the individual fish shape to be analyzed. A landmark is a point of correspondence on an object that matches between and within populations. Advanced techniques for morphometric analysis offers more efficient and powerful tools in identify differences between fish populations, detecting differences among groups and to differentiate between species of similar shape. Morphometric methods such as univariate comparisons, bivariate analyses of relative growth pattern and a series of multivariate methods have been developed and applied to discriminate stocks. The use of multivariate techniques such as principal components and discriminant analyses to quantify morphometric variables are also receiving increased attention in stock identification. Some of the advanced techniques developed for morphometric analysis in fish population are Truss network measurement, Image analysis- Univarite, Bivariate, and Multivariate, Principal Component Analysis (PCA).

Keywords: Biology; Fishery science; Morphometrics method; Fish recognition and monitoring

Introduction

The field of fishery science has employed many tools such as genetics and morphometric to differentiate fish population [1]. Morphometrics may be defined as a more or less interwoven set of largely statistical procedures for analyzing variability in size and shape of organs and organisms. Morphometric differences among stocks of a species are recognized as important for evaluating the population structure and as a basis for identifying stocks [2,3]. Morphometric measurements are widely used to identify differences between fish populations [4-6]. Truss network systems constructed with the help of landmark points are also a powerful tool for stock identification. Some arbitrarily selected points on a fish body known as landmarks help the individual fish shape to be analyzed. A landmark is a point of correspondence on an object that matches between and within populations [7].

Materials and Methods

Morphometric analysis offers more efficient and powerful techniques, such as image analysis, for detecting differences among groups and to differentiate between species of similar shape [8].

Image analysis systems played a major role in the development of morphometric techniques, boosting the utility of morphometric research. The characteristics may be more applicable for studying short term, environmentally induced disparities, stock identification [9,10], species differences, practical morphology and improved fisheries management [11].

Some morphometric methods have also been developed and applied to discriminate stocks such as univariate comparisons, bivariate analyses of relative growth pattern and a series of multivariate methods [12]. On the other hand, the use of multivariate techniques such as principal components and discriminant analyses to quantify morphometric variables is receiving increased attention in stock identification [13-16].

Advanced techniques for morphometric analysis in fish population includes:

Truss network measurements

Truss network systems constructed with the help of landmark points are powerful tools for stock identification of fish species [17]. A sufficient degree of isolation may result in notable morphological, meristic, and shape differentiation among stocks of a species which may be recognizable as a basis for identifying the stocks. The characteristics may be more applicable for studying short-term, environmentally induced disparities, and the findings can be effectively used for improved fisheries management [18-21].

The truss network system can effectively be used to distinguish between the hatchery and wild stocks. It has been also successfully utilized to differentiate and identify stock of the horse mackerel Trachurus trachurus and the Japanese threadfin bream Nemipterus japanicus [22]. It has been used to show more-significant differences between two completely different habitats i.e., in an open and closed water habitat [23].

Truss measurements are a powerful tool for the analysis of shape, and generally are designed to cover all, or most, of the animal's body [24]. According to Dwivedi and Dubey [25] the truss network is more useful and an effective strategy for the descriptions of shape; it has better data collection and diversified analytical tools in comparison to traditional morphometrics method. Thus it is able to discriminate phenotypic stock because the configuration of the constructed landmarks covers the entire fish body with no loss of information, and it is more sensitive to change [22].

Trust network measurement in fish involves anaesthetizing with

---

*Corresponding author: Mojekwu TO, Department of Biotechnology Unit, Nigerian Institute for Oceanography and Marine Research, Nigeria, E-mail: tonyystone@yahoo.com

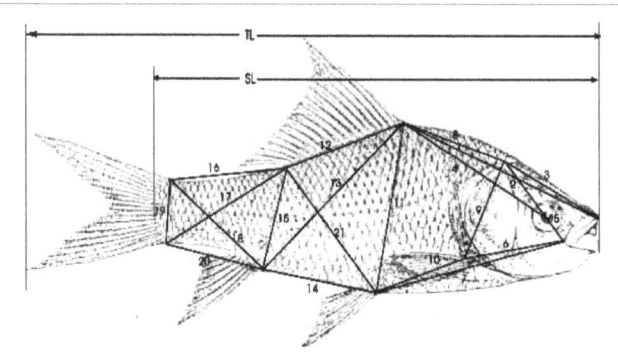

Figure 1: Typical Truss Morphometric Network (TMN) of Indian Major Carp.

benzocaine (ethyl-p-amino-benzoate) [26] before being weighed and measured Figure 1. The first step is to take and record the standard length (LS), post-orbital length (LPO) and maximum body width of the fish. Standard length will be taken from the tip of the upper jaw to the base of the caudal peduncle. Then a truss network is constructed between landmark points, homologous throughout the population, chosen because they describe the major features of the fish (Figure 1). The landmarks were linked closely to the skeletal structure of fish and were easily observed by eye.

Truss lengths measured between these landmark points should either lie on curved surfaces or be on straight lines lying on a flat plane [27,28]. The distances were measured with the help of vernier callipers (RS and Cam lab) or brass divider accurately to 0.1 mm [28]. Measuring four fishes repeatedly will help assess the accuracy of measurements and comparing results. Measurement errors were calculated by taking the mean of the standard deviation of the measurements for each fish. The relative error is the percentage of the mean truss length that the tabulated measurement error represents.

Legends: 1. Mouth tip to premaxilla (MTPM), 2. Mouth tip to dorsal fin (MTDF), 3. Mouth tip to operculum top (MTOT), 4. Pre maxilla to dorsal fin (PMDF), 5. Pre maxilla to operculum tip (PMOT), 6. Pre maxilla to pectoral fin (PMPC), 7. Pre maxilla to pelvic fin (PMPV), 8. Dorsal fin to operculum tip (DFOT), 9. Pectoral fin to operculum tip (PCOT), 10. Pectoral fin to pelvic fin (PCPV), 11. Dorsal fin to pelvic fin (DFPV), 12. Dorsal fin front to dorsal fin back (DFDB), 13. Dorsal fin to anal fin (DFAF), 14. Pelvic fin to anal fin (PVAF), 15. Dorsal back to anal fin (DBAF), 16. Dorsal fin back to caudal top (DBCT), 17. Dorsal back to caudal bottom (DBCB), 18. Anal fin to caudal top (AFCT), 19. Caudal top to caudal bottom (CTCB), 20. Anal fin to caudal bottom (AFCB), 21. Dorsal fin back to pelvic fin (DBPV).

Image analysis morphormetric methods

The development of image analysis systems has facilitated progress and diversification of morphometric methods and expands the potential for using morphometry as a tool for stock identification. Traditional multivariate morphometrics, accounting for variation in size and shape, have successfully discriminated many fish stocks. However, traditional methods have been enhanced by image processing techniques, through better data collection, more effective descriptions of shape, and new analytical tools. (Properly calibrated coordinates of morphometric locations, or landmarks', are generally more efficient and precise than manual distance measurements. Truss networks' of distances between landmarks coordinates provide more comprehensive coverage of form for greater discriminating power [9,10]. Image analysis systems

also allow more advanced geometric morphometrics, which include outline methods and landmark methods [29]. Unfortunately, few stock identification studies have applied outline methods, and landmark methods have not been used to discriminate fishery stocks.

Image analysis is the extraction of meaningful information from images; mainly from digital images by means of digital image processing techniques [30]. Image analysis tasks can be as simple as reading bar coded tags or as sophisticated as identifying a person from their face. Computers are indispensable for the analysis of large amounts of data, for tasks that require complex computation, or for the extraction of quantitative information. There are many different techniques used in automatically analysing images. Each technique may be useful for a small range of tasks, however there still aren't any known methods of image analysis that are generic enough for wide ranges of tasks, compared to the abilities of a human's image analysing capabilities. Examples of image analysis techniques in different fields include: 2D and 3D object recognition, image segmentation, motion detection e.g. tracking, video, flow, analysis, Estimation, automatic [31].

From the previous works, there are a lot of papers discussing methods to measure the various sizes of fish [32-34] from a digital image. Naiberg and Little [32] developed a size assessment system underwater using model-based recognition and stereoscopic vision. Model-based recognition is used to locate the object and stereo vision system to determine distance and sizes given stereo video input. However, the stereo vision system is very expensive and the matching procedures also still have error and poor image quality that can affect the accuracy of measurement.

Another method is automated Fish Recognition and Monitoring (FIRM) proposed by Lee et al. [33]. He focuses on comparison of technique to shape matching but did not focus to obtain the size of fish. This method has five main processing steps: image acquisition, object detection algorithm detects the presence of an object (DMA), identifying the fish with object contour extraction used Canny Edge operator and identification, Tracking of the fish object determines the location of the fish , triggering the recognition process when an image of the whole fish can be acquired. The measurement size and species of fish used shape-based recognition. This method is suitable for fish in an aquarium. Besides that, White et al., [34] measured species and fish length by implementing computer vision for sorting fish in industrial area. Such method must have laboratory equipped with a conveyor belt and other hardware such as pc, lamp and sensor as in Figure 2.

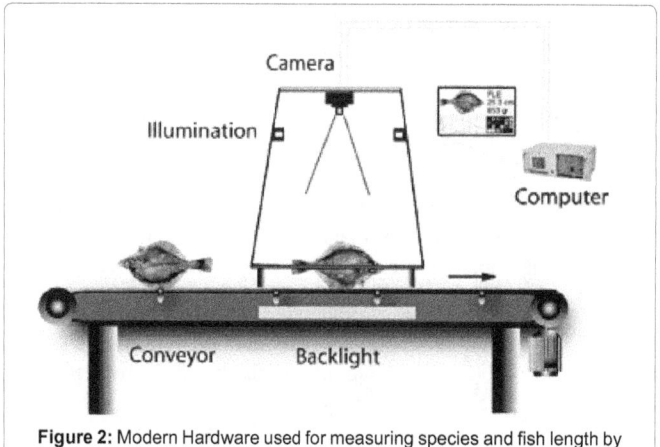

Figure 2: Modern Hardware used for measuring species and fish length by implementing computer vision for sorting fish in industrial areas.

Figure 3: Method of analysing digital object using Object Reference by Pickle.

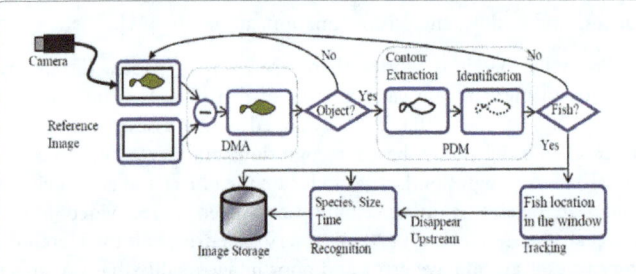

Figure 4: Fish Recognition and monitoring processing algorithm flowchart.

This method may not be suitable for studies in which one needs to measure the fish length rapidly without going to laboratory because the fisherman sells the fish as soon as possible. It is also an inexpensive, accurate and short time method of measuring length of fish.

Hsiu [35] solve the problem of Pickle's method which measure size of object from digital without object reference (Figure 3). Hsiu methods obtain size of object directly from digital image by using equation magnification and software image viewing program [35]. The equation magnification is used to obtain ratio for actual object value and software image viewing program to detect edge automatically.

The object is a very simple feature (Figure 4). Detecting fish image is usually difficult because head and tail of fish have curve and corner which must be detected accurately in calculating fish size. Fish can be detected using maximizing bending ratio and curvature to detect corner as done by Serkan et al. [36].

From the Hsiu's and Serkan's method, a new method was developed for measuring fish length from digital image by combining both of the method. Hsiu's method has been used to obtain the actual length of fish and Serkan's method has been used for image processing. The new method is used to measure length of fish from digital image. Whole fish features including fish contour and significant landmark points can then be extracted for analysis. Shape-based recognition can be performed to obtain size and species information.

After the image is processed to obtain corner detection at head and tail, the image will be processed to obtain number of pixel from corner head to corner tail [37]. The image of fish length will be obtained when the number of pixels is multiplied with pixel size which is gotten from image. The actual fish length is calculated with equation of "object distance/object height=image distance/image height". Wherein actual of fish length=object height by image height=image of fish length and image distance=focus length.

This new method is an automated method of measuring the length of a fish using equation magnification and image processing which has high potential to be commercialized given its high reliability, durability and accuracy factors; as well as minimizing cost and time needed for such task. This means that it has to be able to measure length of the fish without having a person holding the fish. The idea is to capture the image of the fish using a camera that uploads the picture to the software to evaluate the length of the fish. The impact of the contribution of this new method will ensures stability and security of a country's main source of protein.

Univariate, bivariate, and multivariate morphormetric analysis

The earliest analyses of morphometric variables for stock identification were univariate comparisons, but these were soon followed by bivariate analyses of relative growth to detect ontogenetic changes and geographic variation among fish stocks. As the field of multivariate morphometrics grew, a set of multivariate methods was applied to quantify variation in growth and form among stocks.

Univariate analysis is the simplest form of quantitative (statistical) analysis [38]. The analysis is carried out with the description of a single variable in terms of the applicable unit of analysis. Univariate analysis contrasts with bivariate analysis – the analysis of two variables simultaneously – or multivariable analysis – the analysis of multiple variables simultaneously [38]. Univariate analysis is commonly used in the first, descriptive stages of research, before being supplemented by more advanced, inferential bivariate or multivariate analysis [39,40]

In univariate analysis, the measure of central tendency is an average of a set of measurements, the word "average" being variously construed as (arithmetic) mean, median, mode or another measure of location, depending on the context. For a variable measured on an interval scale, such as temperature on the Celsius scale, or on a ratio scale, such as temperature on the Kelvin scale, the median or mean can also be used [38].

Bivariate analysis is one of the simplest forms of quantitative (statistical) analysis [41]. It involves the analysis of two variables often denoted as (X, Y), for the purpose of determining the empirical relationship between them. In order to see if the variables are related to one another, it is common to measure how those two variables simultaneously change together. Bivariate analysis can be helpful in testing simple hypothesis of association and causality- checking to what extent it becomes easier to know and predict a value for the dependent variable if we know a case's value of the independent variable. The major differentiating point between univariate and bivariate analysis, in addition to the latter's looking at more than one variable, is that the purpose of a bivariate analysis goes beyond simply descriptive: it is the analysis of the relationship between the two variables. Bivariate analysis is a simple (two variable) special case of Multivariate analysis (where multiple relations between multiple variables are examined simultaneously) [41]. Common forms of bivariate analysis involve creating a percentage table or a scatter plot graph and computing a simple correlation coefficient. The type of analyses that are suited to particular pairs of variables vary in accordance with the level of measurement of the variables of interest (e.g. nominal/categorical, ordinal, interval/ratio). If the dependent variable-the one whose value is determined to some extent by the other, independent variable- is a categorical variable, such as the preferred species of fish, then probit or logit regression (or multinomial probit or multinomial logit) can be used. If both variables are ordinal, meaning they are ranked in a sequence as first, second, etc., then a rank correlation coefficient can be computed. If just the dependent variable is ordinal, ordered probit or particular type or causality known as Granger causality can be tested for, and vector auto regression can be performed to examine the intertemporal linkages between variables.

Multivariate statistics encompasses the simultaneous observation and analysis of more than one outcome variable [42]. The application of multivariate statistics is multivariate analysis. Multivariate statistics concerns understanding the different aims and background of each of the different forms of multivariate analysis, and how they relate to each other [42]. The practical implementation of multivariate statistics to a particular problem may involve several types of univariate and multivariate analyses in order to understand the relationships between variables and their relevance to the actual problem being studied.

In addition, multivariate statistics is concerned with multivariate probability distributions, in terms of both how these can be used to represent the distributions of observed data; how they can be used as part of statistical inference, particularly where several different quantities are of interest to the same analysis [43]. Certain types of problem involving multivariate data, for example simple linear regression and multiple regression, are not usually considered as special cases of multivariate statistics because the analysis is dealt with by considering the (univariate) conditional distribution of a single outcome variable given the other variables [43].

Principal component analysis (PCA)

PCA is a multivariate statistical technique that uses orthogonal transformation to convert a set of correlated variables into a set of orthogonal uncorrelated axes called principal components [44-46]. PCA enables condensation of data on a multivariate phenomenon into its main, representative features by projection of the data into a two- dimensional presentation. The two created resource axes are independent, and although they reduce the number of dimensions-i.e. the original data complexity –they maintain much of the original relationship between the variables: i.e., information or explained variance [47]. This is helpful in focusing attention on the main characteristics of the phenomenon under study. It is convenient that ,if the first few principal components (PCs) explain a high percentage of variance,environmental variables that are not correlated with the first few PCs can be disregarded in the analysis [48]. In addition , applying PCA has become relatively user- friendly because of the numerous programs that assist in carrying out the computational procedure with ease [49-52].

PCA has been widely used in various fields of investigation and for different tasks. Many authors have used PCA for its main purpose i.e., to reduce strongly correlated data groups or layers. These studies concern either environmental variation [53-57]; the investigated species or community characteristics Youlatos [58]; Kitahara and Fujii [59]; Xiaoyun et al. [60]; Zhao et al. [61]; Noura et al. [62] or both, sometimes in combination with Detrended Correspondence Analysis (DCA), Canonical Correspondence Analysis (CCA), and other ordination methods [63-65].

The application of PCA has helped in aquatic habitat studies; it has been applied to evaluate aquatic habitat suitability, its regionalization, analysis of fish abundance, their seasonal and spatial variation, lake ecosystem organization change etc. [66-68]. However, it has been often applied in analyzing farming system changes [69]. In many cases, PCA has been used as a source or supporting analysis in the performance of more complex analysis, such as the study of adaptive fish radiation, strongly influenced by trophic niches and water depth [70], predicting the potential spatial extent of species invasion [71] and multi-trait analysis of intra- and interspecific variability of plant traits [72].

Conclusion

Morphometrics adds a quantitative element to descriptions, allowing more rigorous comparisons. It enables one to describe complex shapes in a rigorous fashion and permits numerical comparison between different forms. Advancements in morphometrics uses powerful tools for testing and displaying differences in shape, isolate shape from size variation and identifying stocks of specie with unique morphological characteristics. This has enabled a better management of species subunits and ensures better management of the fishery resources. These morphormetric advancements techniques analysis in fish population includes: Truss network measurement, Image analysis-Univarite, Bivariate, and Multivariate, Principal Component Analysis (PCA).'Trust Network System' has emerged as a new tool with more effective strategies for descriptions of shape, better data collection and diversified analytical tools. Recent developments made in the discipline of morphormetric differentiation in body shape among fish populations showed that the truss based techniques is more effective than manual distance measurement for the management of fishery resources throughout the world. When combined with multivariate statistical methods (e.g. Principal Component Analysis, Cluster Analysis etc) they offer powerful tool for testing and displaying differences in shape.

References

1. Mir JI, Sarkar UK, Dwivedi AK, Gusain OP, Jena JK (2013) Stock structure analysis of Labeorohita (Hamilton, 1822) across the Ganga basin (India) using a truss network system. Journal of Applied Ichthyology 29: 1097 -1103.

2. Turan (2004a) Stock identification of Mediterranean horse mackerel (Trachurusmediterraneus) using morphometric and meristic characters. Journal of Mar. Science 61:774-781.

3. Cadrin SX, Friedland KD (1999) The utility of image processing techniques for morphometric analysis and stock identification. Fisheries Research 43: 129-139.

4. Cheng Q, Lu D, Ma L (2005) Morphological differences between close populations discernible by multivariate analysis: A case study of genus Coilia (Teleostei: Clupeiforms). Aquatic Living Resources 18: 187-92.

5. Buj I, Podnar M, Mrakovcic M, Caleta M, Mustafic P, et al. (2008) Morphological and genetic diversity of Sabanejewiabalcanicain Croatia. Folia Zooology 57:100-110.

6. Torres RGA, Gonzalez PS, Pena SE (2010) Anatomical, histological and ultraestructural description of the gills and liver of the Tilapia (Oreochromisniloticus). International Journal of Morphology 28: 703-12.

7. Hossain MAR, Nahiduzzaman M, Saha D, Khanam MUH, Alam MS (2010) Landmark- based morphometric and meristic variations of the endangered carp,Kalibauslabeocalbasu,from stocks of two isolated rivers,thejamuna and Halda, and a hatchery. Zoological studies 49: 556 -563.

8. Costa C, Loy A, Cataudella S, Davis D, Scardi M(2006) Extracting Fish Size Using Dual Underwater Cameras. Aquacultural Engineering 35: 218-227.

9. Bronte CR, Moore SA (2007) Morphological Variation of Siscowet Lake Trout in Lake Superior. Transactions of the American Fisheries Society 136: 509-517.

10. Shao Y, Wang J, Qiao Y, He Y, Cao W (2007) Morphological variability between wild populations and inbred stocks of a Chinese minnow, Gobiocyprisrarus. Zoological Science 24: 1094-1102.

11. Turan C, Erguden D, Gurlek M, BasustaN,Turan F (2004) Morphometric structuring of the anchovy (Engraulisencrasicolus L.) in the Black, Aegean and

northeastern Mediterranean Seas. Turkish Journal of Vetinary Animal Science 28:865-871.

12. Cadrin SX (2000) Advances in morphometric identification of fishery stocks. Reviews in Fish Biology and Fisheries 10: 91-112.

13. Kusznierz J ,Kotusz J, Kazak M , Popiolek M, Witkowski A (2008) Remarks on the morphological variability of the Arctic charr, Salvelinusalpinus(L.) from Spitsbergen. Polish Polar Research 29: 227-236.

14. Specziar A, Bercsenyi M, Muller T (2009) Morphological characteristics of hybrid pikeperch (Sander luciopercax Sander volgensis) (Osteichthyes, Percidae). Acta Zoo. Hung 55: 39-54.

15. Yakubu A, Okunsebor SA (2011) Morphometric differentiation of two Nigerian fish species (Oreochromisniloticusand Latesniloticus) using principal components and discriminant analysis. International Journal of Morphology29:1429-1434.

16. Cronin Fine L, Stockwell JD, Whitener ZT, Labbe EM, Willis TV, et al. (2013) Application of Morphometric Analysis to Identify Alewife stock structure in the Gulf of Maine. Marine and Coastal Fisheries. Dynamics, Management, and Ecosystem Science 5: 11-20.

17. Turan C, Erguden D and Gurlek M (2004b) Genetic and morphologic structure of Liza abu(Heckel, 1843) populations from the Rivers Orontes, Euphrates and Tigris. Turkish Journal of Vetinary Animal Science 28: 729-734.

18. Sajina AM, Chakraborty SK, Jaiswar AK, Pazhayamadom DG, Sudheesan D (2011) Stock structure analysis of Indian Mackerel, Rastrelligerkanagurta (Cuvier,1816) along the Indian coast. Asian Fisheries Science 24: 331-342.

19. Merz JE, Garrison TM, Bergman PS, Blankenship S, Garza JC (2014) Morphological Discrimination of Genetically Distinct Chinook Salmon Populations: an Example from California's Central Valley. North American Journal of Fisheries Management 34: 1259-1269.

20. Pazhayamadom DG, Chakraborty SK, Jaiswar AK, Sudheesan D, Sajina AM (2015) Stock structure analysis of Bombay duck (Harpadonnehereus Hamilton,1822) along the Indian coast using truss network morphometrics. Journal of Applied Ichthyology 31: 37- 44.

21. Wiadnya DGR, Widodo D, Soemarno S (2015) Intra- species variations of photopectoralisbindus (Family:Leiognathidae) collected from two geographical areas in East Java,Indonesia. Journal of Biodiversity and Environmental Sciences 6:160-168.

22. Lim TY (2008) Biology of the Bettapucnaxgroup from Johore. Universiti Sains Malaysia, Penang.

23. Hossain MAR, Nahiduzzaman M, Saha D, Khanam MUH, Alam MS (2010) Landmark- based morphometric and meristic variations of the endangered carp,Kalibauslabeocalbasu, from stocks of two isolated rivers, the jamuna and Halda, and a hatchery. Zoological studies 49: 556 -563.

24. Strauss RE, Bookstein FL (1982) The truss: body form reconstruction in morphometrics. Systematic Zoology 31:113-135.

25. Dwivedi AK, Dubey VK (2013) Advancements in morphormetric differentiation: a review on stock identification among fish populations. Review in Fish Biology and Fisheries 23:23-39.

26. Ross LG, Ross B (1999) Anaesthetics and Sedative Techniques for Aquaculture Animals.

27. Hockaday S, Beddow S, Stone M, Hancock P, Ross LG (2000) Using truss networks to estimate the biomass of Oreochromisniloticus and to investigate shape Characteristics. Journal of Fish Biology 57: 981-1000.

28. Ujjania NC, Kohli MPS (2011) Landmark-based morphormetric analysis for selected species of Indian Major Carp (Catlacatla, Ham. 1822). International Journal of Food, Agriculture and Veterinary Sciences 1:64-74.

29. Toh YH, Ng TM, Liew BK (2009) Computational Intelligence and Software Engineering. IEEE International Conferenceat Wuhan 3: 1-5.

30. Solomon CJ, Breckon TP (2010) Fundamentals of Digital Image Processing: A Practical Approach with Examples in Matlab.

31. Hay GJ, Castilla G (2008) Geographic Object-Based Image Analysis (GEOBIA): A new name for a new discipline.

32. Naiberg A, Little JJ (1994) A Unifed Recognition and Stereo Vision Systemfor size assessment of fish.

33. Lee DJ, Schoenberger R, Shiozawa D, Xu X, Zhan P (2008) Contour Matching for a Fish Recognition and Migration Monitoring System. SPIE Optics East, Two and Three-Dimensional Vision System for Inspection, Control, and Metrology.

34. White DJ, Svellingen C, Strachan NJC (2006) Automated Measurement of Species and Length of fish by Computer Vision. Fisheries Research 80: 203-210.

35. Hsu HO (2008) Method for Calculating Distance and Actual Size of Shot Object.

36. Serkan K, Hantao L, Miguel F, Moncef G (2007) An Efficient Approach for Boundary Based Corner Detection by Maximizing Bending Ratio and Curvature, 9th International Symposium onSignal Processing and Its Applications.

37. Chanda B (2008) Morphological Algorithms for Image Processing. IETE Technical Review 25: 9-18.

38. Babbie E R(2009) The Practice of Social Research. Wadsworth Publishing,USA.

39. Bernard HR (2006) Research methods in anthropology: qualitative and quantitative approaches.

40. Cooper A, Weekes TJ (1983) Data, models, and statistical analysis, Barnes& Noble Books, USA.

41. Earl RB (2009) The Practice of Social Research. Wadsworth Publishing,USA.

42. Hidalgo B, Goodman M (2013) Multivariate or multivariable regression?. American journal of public health 103:39- 40.

43. Johnson RA, Wichern DW (2007) Applied Multivariate Statistical Analysis. Prentice Hall6:77-153.

44. Robertson MP, Caithness N,Villet MH (2001) A PCA-based modeling technique for predicting environmental suitability for organisms from presence records. Diversityand Distributions 7: 15-27.

45. Legendre P, Legendre L (1998) Numerical Ecology. Elsevier 17: 853.

46. Gotelli NJ, Ellison AM (2004) A Primer of Ecological Statistics.

47. Litvak MK, Hansell RIC (1990) A community perspective on the multidimensional niche. J Animal Ecology 59: 931-940.

48. Toepfer CS, Williams LR, Martinez AD, Fisher WL (1998) Fish and habitat heterogeneity in four streams in the central Oklahoma/Texas plains ecoregion. Proceedings of the Oklahoma Academy of Science 78: 41-48.

49. Dolédec S, Chessel D, Gimaret-Carpentier C (2000) Niche Separation in Community Analysis: A New Method. Ecology 81: 2914-2927.

50. Guisan A, Zimmerman NE (2000) Predictive habitat distribution models in ecology. Ecological Modelling 135:147-186.

51. Rissler LJ, Apodaca JJ (2007) Adding more ecology into species delimitation: Ecological niche models and phylogeography help define cryptic species in the black salamander (Aneidesflavipunctatus). Systematic Biology 56: 924-942.

52. Marmion M, Parviainen M, Luoto M, Heikkinen RK, Thuiller W (2009) Evaluation of consensus methods in predictive species distribution modelling. Diversity andDistributions 15: 59-69.

53. Glor RE, Warren D (2010) Testing ecological explanations for biogeographic boundaries. Evolution 65: 673-683.

54. Novak T, Thirion C, Janžekovič F (2010) Hypogeanecophase of three hymenopteran species in Central European caves. Italian J Zoology 77: 469-475.

55. Faucon MP, Parmentier I, Colinet G, Mahy G, Luhembwe M N (2011) May rare metallophytes benefit from disturbed soils following mining activity? The Case of the Crepidorhopalontenuisin Katanga (D. R. Congo). Restoration Ecology 19: 333-343.

56. Adamu M (2012) Identification of Anthropogenic influences on water quality of Jakara River, northwest Nigeria. J Applied Sciences in Environmental Sanitation 7: 11 -20.

57. Andem AB, Okafor KA, Ekanem SB (2015) Factor Analysis and Physico-Chemical characteristics of Calabar River, Southern Nigeria. International Journal of scientific and Technology Research 4:25- 28.

58. Youlatos D (2004) Multivariate analysis of organismal and habitat parameters in two neotropical primate communities. American J Physical Anthropology 123: 181-194.

59. Kitahara M, Fujii K (2005) Analysis and understanding of butterfly community

composition based on multivariate approaches and the concept of generalist/specialist strategies. Entomological Sci 8: 137 140.

60. Xiaoyun FB, Cui H, Zhao Z, Zhiming Z, Honggang Z (2010) Assessment of river water quality in Pearl River Delta using multivariate statistical techniques. Procedia Environ Sci 2: 1220-1234.

61. Zhao J, Fu G, Lei K, Li Y (2011) Multivariate analysis of surface water quality in the three Georges area of China and implications for water management. J Environ Sci 23: 1460-1471.

62. Noura S, Mustapha B, Moncef B (2015) Taxonomic diversity and benthic community structures of watersheds of Medjerda and Joumine (North east of Tunisia).Inter Res J Earth Sci 3: 24 -30.

63. Warner BG, Asada T, Quinn NP (2007) Seasonal influences on the ecology of testateamoebae (Protozoa) in a small Sphagnum peatland in Southern Ontario, Canada. Microbial Ecology 54: 91-100.

64. Gonzalez-Cabello A, Bellwood DR (2009) Local ecological impacts of regional biodiversity on reef fish assemblages. J Biogeography 36: 1129-1137.

65. Mezger D, Pfeiffer M (2011) Partitioning the impact of abiotic factors and spatial patterns on species richness and community structure of ground ant assemblages in four Bornean rainforests. Ecography 34: 39-48.

66. Ahmadi-Nedushan B, St-Hilaire A, Bérubé M, Robichaud E, Thiémonge N, et al. (2006) A review of statistical methods for the evaluation of aquatic habitat suitability for instream flow assessment. River Research and Applications 22: 503-523.

67. Catalan J, Barbieri MG, Bartumeus F, Bitušk P, Botev I, et al. (2009) Ecological thresholds in European alpine lakes. Freshwater Biology 54: 2494-2517.

68. Eze EB, Efiong J (2010) Morphometric parameters of calabar River basin: Implication for Hydrologic Processess. J Geography and Geology 2: 1916-1920.

69. Amanor KS, Pabi O (2007) Space, time, rhetoric and agricultural change in the transition zone of Ghana. Human Ecology 35:51-67.

70. Clabaut C, Bunje PM, Salzburger W, Meyer A (2007) Geometric morphometric analyses provide evidence for the adaptive character of the Tanganyikan cichlid fish radiations. Evolution 61: 560-578.

71. Broennimann O, Treier U A, Müller-Schärer H, Thiuller W, Peterson AT, et al. (2007) Evidence of climatic niche shift during biological invasion. Ecology Lett. 10: 701-709.

72. Albert CH, Thuiller W, Yoccoz NG, Douzet R, Aubert S, et al. (2010) Amultitrait approach reveals the structure and the relative importance of intra- vs -interspecific variability in plant traits. Functional Ecology 24: 1192-1201.

7

Digestibility of Soybean Cake, Niger Seed Cake and Linseed Cake in Juvenile Nile Tilapia, *Oreochromis niloticus* L.

Akewake Geremew[1]*, Abebe Getahun[2] and Krishen Rana[3]

[1]*Department of Biology, Dilla University, P. O. Box 419, Dilla, Ethiopia*
[2]*Department of Zoological Sciences, Addis Ababa University, P. O. Box 1176, Addis Ababa, Ethiopia*
[3]*Department of Sustainable Aquaculture, Stellenbosch University, Stellenbosch, South Africa*

Abstract

The apparent digestibility coefficients (ADCs) of dry matter, protein, lipid and energy for soybean cake (SBC), Niger seed cake (NSC) and linseed cake (LSC) were determined in juvenile Nile tilapia. The ADCs were determined using faeces collected with a settling chamber attached to the fish rearing tank. Test diets contained 70% reference diet and 30% test ingredients, with Cr_2O_3 as an inert marker. All treatments were triplicated. There was significant difference in Apparent Dry Matter Digestibility (ADMD), Apparent Protein Digestibility (APD) and Apparent Energy Digestibility (AED) between the test ingredients. However, there was no significant difference (P>0.05) in Apparent Lipid Digestibility (ALD) between the test ingredients. Of the three ingredients tested, SBC produced the highest nutrient digestibility coefficients (P<0.05) while LSC showed the lowest nutrient digestibilities (P<0.05). The NSC, which was the cheapest plant protein source, was a good feed ingredient for Nile tilapia diets in terms of overall nutrient composition and acceptable digestibility coefficients enabling more accurate and economical feed formulation.

Keywords: Juvenile nutrition; Plant proteins; African aquaculture; Fish meal; Ethiopian oilseeds

Introduction

With the increase in intensive aquaculture, demand for more efficient aquafeed is rising. Feed comprises the principal operating cost in fish production and the main protein source has traditionally been fish meal [1]. Fishmeal, the conventional protein source in aquaculture feeds, supports good fish growth because of its protein quality and palatability [2]. However, fish meal is often scarce and expensive, due to limited availability and high demand, which often leads to high fish production costs [3,4]. According to Ng and Romano [2], cost-effective, practical aquaculture feeds can be produced without the use of fish meal with no resulting or apparent loss in fish growth in some species. Hence, replacing fish meal with cheaper ingredients of either animal origin or protein-rich plant sources is a necessary priority for nutrition research [1,2]. In view of this, oilseed meals have been found to have considerable economic potential [2,5]. While grain legumes have not been widely used within aquaculture feeds, oilseeds and their by-products frequently constitute a major source of dietary protein within aquaculture feeds for warm water fish species such as those commonly used in African aquaculture, including tilapias (*Oreochromis* spp.) and African catfish [6].

A feed ingredient may appear from its chemical composition to be an excellent source of nutrients but will be of little actual value unless it can be ingested, digested and absorbed in the target species. Only a proportion of ingested food is digested and its nutrients absorbed, the rest is voided as faeces. By definition, digestibility is a relative measure of the extent to which ingested food and its nutrient components have been digested and absorbed by the animal. Knowledge of nutrient digestibility is, therefore, important to establish the potential of an ingredient for use in diets of aquaculture species [2,7]. Determining the digestibility of nutrients in feedstuffs is important not only to enable formulation of diets that maximize the growth of cultured species, by providing appropriate amounts of available nutrients, but also to limit the wastes produced by the fish and reduce costs [1,2,7,8].

For tilapia feeds typical protein sources examined have included cereal grain products [9], defatted soybean meal, full-fat toasted soybean, lupin seed meal and faba bean meal [10], cottonseed meal, sunflower meal [11], fish and poultry meals, corn gluten, rapeseed meal, sorghum, barley [12], anchovy meal, corn gluten meal, soybean meal, gammaridmeal and crayfish exoskeleton meal [13]. Among the plant protein sources, soybean meal has been used most widely because it has a good amino acid profile, which, as the main source of protein, supports fish growth [14]. Soybeans, however, are not grown widely in Ethiopia; hence there is a need to evaluate soybeans together with other more locally available plant proteins. According to Lovell [15] feed ingredients containing 20% or more crude protein are considered protein sources. In the present study soybean cake (SBC), linseed cake (LSC) and Nigerseed cake (NSC) were selected as dietary protein sources on the basis of their high protein content, availability and use in animal feeds for Ethiopia. Studies conducted on SBC, LSC and NSC showed they have good protein contents (30-40%), depending on processing methods [16,17].

Niger seed is the most important oil crop of Ethiopia, providing 50–60% of the country's indigenous edible oil. It is also minor oil crop in India, Kenya, Uganda, Sudan, Malawi and other African and Indian sub-continent countries. Its seeds are inexpensive to process, and the cake remaining after oil extraction is used as a protein supplement in animal diets [16]. Niger seed cake contains few or no known antinutritional factors [18].

Ethiopia ranks among the top five world producers of linseed and it is the second most important oil crop in the country next to Niger seed [19]. The usefulness of linseed as an ingredient in the diets of fish has been studied by different authors [20-23]. Nutrient and energy

---

*Corresponding author: Akewake Geremew, Dilla University, Department of Biology, P.O.Box 419, Dilla, Ethiopia, E-mail: khaliger@yahoo.com

Digestibility of Soybean Cake, Niger Seed Cake and Linseed Cake in Juvenile Nile Tilapia, Oreochromis niloticus L.

35

digestibility studies have been conducted more extensively on soybean for many fish species than on LSC. However, digestibility of NSC in fish diets has not been researched into at all, probably because it is restricted to Eastern Africa, mainly Ethiopia.

This study was conducted to evaluate the apparent digestibility coefficients (ADCs) of dry matter (DM), crude protein (CP) and gross energy (GE) for SBC, LSC and NSC for Nile tilapia, O. niloticus.

Materials and Methods

Fingerlings of Nile tilapia of Lake Hora origin with an average weight of 8.9 ± 1.6 g were stocked at 10 fish per tank (60 l tanks) in a water recirculation system established at Ziway Fisheries Resources Research Center by Addis Ababa University Department of Zoological Sciences. All treatments were conducted in triplicate. Fish were fed by hand, twice a day (10:00, 16:00 hours) at a rate of 6% body weight per day. The experiment took 2-3 weeks. The recirculation system was supplied with aerated water from a sump tank thermoregulated at 28 ± 1°C and a constant photoperiod of 12 hours light/12 hours darkness was maintained. Water quality parameters measured during the experiment averaged (± SD): temperature, 28.9 ± 0.4°C; pH, 7.3 ± 0.1; ammonia, 0.17 ± 0.08 mg l[-1]; nitrite, 0.20 ± 0.1 mg l[-1]; Nitrate, 50 ± 23.2 mg l[-1] and dissolved oxygen, 5.4 ± 0.4 mg l[-1] and they were within acceptable ranges for tilapia.

Diet Formulation

A reference diet (Table 1) was formulated to satisfy the nutrient requirements of Nile tilapia [24]. It contained 320 g kg[-1] crude protein, 100 g kg[-1] lipid and 18 kJg[-1]. The test ingredients for apparent digestibility were soybean cake (SBC), linseed cake (LSC) and Nigerseed cake (NSC). All test feed ingredients were obtained from commercial sources in Ziway, Ethiopia with the exception of soybean cake which was acquired from Addis Ababa oil processing factory outlet.

Three test diets were formulated using 70% reference diet and 30% of each of the test ingredients as described by Cho et al. [25]. This method assumes that there are no interactions among the components of the diet during digestion [26]. Chromic oxide was used as an inert marker at a concentration of 0.5% in the diets. Other supplements used in the diet are indicated in Table 1.

Diet preparation

Fishmeal was processed from waste obtained from a local fish processing plant known as "ZiwayFish Processing Plant". The filleting residues were purchased from the processing plant at a price of 0.15 birr kg[-1]. The freshly collected filleting residues of tilapia were minced using an electrical meat mincer and then dried in an oven for 48 hours at 75°C. The dried residue was ground into a fine powder using an electrical smashing machine, sieved (0.5 mm mesh size sieve) and then stored in a plastic bag at -18°C in a deep freeze.

The diets were formulated on as anfed basis. Fish meal as the main dietary protein source and wheat and corn grains (milled) as main carbohydrate sources were used in the experiment. A poultry grade vitamin/mineral premix (Table 1) at 50 g kg[-1] and a binder (carboxymethyl cellulose, high viscosity) at 20 g kg[-1] were added. The vitamin/mineral premix was purchased from the local market in Addis Ababa. This premix is prepared for egg laying hens by an Israeli company called Koffolk Animal Health and Nutrition. Soybean oil was used as the source of lipid in the diets. Chromic oxide was added as an indigestible marker for digestibility study [27].

Diets were prepared by wet extrusion using a meat mincer (Model TJ 22). All ingredients were finely ground and sieved through a 500 μm sieve to obtain a homogenous mixture. The dry ingredients were then weighed out according to the formulation, placed in an aluminum bowl and mixed until uniformly blended using a modified mixer. The resulting homogenate was moistened after addition of water (20%-30%) slowly with continuous stirring until a dough was formed before passing through an electrical meat mincer with a 2.5 mm die. The expelled strands produced from meat mincer were dried in an oven with a convector fan at 35-40°C for 24 hours. They were then crushed in to crumbles and sieved with 1 mm mesh size sieve. The resulting pellets were packaged in polythene bags and stored in a deep freeze at -18°C. Prepared diet samples were analyzed for proximate composition, energy and chromic oxide.

Faecal collection system

In this study a settling column system was employed for faeces collection, but it was adapted to the 60 l cylindrical tanks used. This collection system employed pipes fitted to the bottom of the rearing tanks with a vertical column and transparent hoses connected to a valve system at the bottom ends, where the faeces were deposited after settling. At the top end of the vertical column an overflow was provided to get rid of excess water flowing through the system. Deposited faeces were collected by opening the valve at the tip end and carefully draining the faeces into centrifuge bottles. The collectors were fixed to the rearing tanks the night before and faeces collected early the next morning. Faeces were immediately centrifuged at 4,300×g for 10 min and the supernatant discarded. Wet settled solids of faeces were frozen at -20°C to retard bacterial decomposition. Faecal samples were later defrosted and oven dried at 60°C, ground and analyzed for crude protein (CP), crude lipid (CL), gross energy (GE) and chromic oxide contents.

Analytical techniques

Ingredients, diets and faeces were analyzed in triplicates for proximate composition according to standard methods [28], and chromic oxide of diets and faeces analyzed by acid digestion with molybdate reagent followed by DPC (Diphenylcarbazide) colorimetry following the procedure in Divakaran et al. [27]. Energy was determined using an Adiabatic Autobomb Calorimeter with benzoic acid as a standard.

The apparent digestibility coefficients (ADC) for the nutrients of the diets were calculated as follows [29]:

Ingredients	Reference diet	Test diets
Test ingredient	-	298.5
Fish waste meal	407.6	285.3
Soybean meal	100	70
Wheat grain	20	14
Corn grain	392.4	274.7
Soybean oil	5.0	3.5
Vitamin mineral premix[1]	50	35
Carboxymethyl cellulose	20	14
Chromic oxide	5.0	5.0

[1]Vitamin mineral premix (providing per kg): vitamin A (retinol), 14000 mg; vitamin D₃ (chole-calciferol), 4000 mg; vitamin E (tocopheryl acetate),10000 I.U; vitamin K₃, 2000 mg; thiamine, 1000 mg; riboflavin, 4000 mg; niacin, 10000 mg; pantothenic acid, 5000 mg; pyridoxine, 750 mg; folic acid, 250 mg; vitamin B12, 8 mg; vitamin H as Biotin, 30 mg; betain, 100000 mg; Antioxidant, 125000 mg. Minerals: Manganese, 80000mg; Zinc, 50000 mg; Iron, 20000 mg; Copper, 5000 mg; Iodine, 1200 mg; Cobalt, 200 mg; Selenium, 200 mg.

Table 1: Composition of reference and test diets (g kg[-1]) for the digestibility study

$$ADC = 100 \times \left[1 - \left(\frac{F}{D} \right) \times \left(\frac{D_i}{F_i} \right) \right]$$

Where D=% nutrient of diet; F=% nutrient of faeces; Di=% Cr_2O_3 of diet; F_i=% Cr_2O_3 of faeces and ADC of ingredients as,

$$ADC_{test\,ingredient} = ADC_{test\,diet} + \left[(ADC_{test\,diet} - ADC_{ref.diet}) \left(\frac{0.7 \times D_{ref}}{0.3 \times D_{ingr}} \right) \right]$$

Where D_{ref} = % nutrient (or kJg^{-1} gross energy) of reference diet (as fed); D_{ingr} = % nutrient (or kJg^{-1} gross energy) of test ingredient (as fed).

Digestible protein and energy were calculated as follows:

Digestible protein (DP, gkg^{-1}) = dietary crude protein (gkg^{-1}, dry weight basis) × $ADC_{protein}$

Digestible energy (DE, kJg^{-1}) = gross energy (kJg^{-1}, dry weight basis) × ADC_{energy}

Statistical analyses in this study were conducted using Minitab Statistical Package (Version 15.0). Differences among dietary treatment means were tested by analysis of variance (ANOVA), and means compared using Tukey's Multiple Comparison Test to test for significance of variation between the means and differences were considered significant at $p<0.05$. All percentages were arcsine transformed before analysis [30].

Results

Chemical composition and prices of ingredients

Proximate composition and energy contents of the ingredients used in the study are given in Table 2. Crude protein for oilseed cakes ranged from 310 to 393.8 g kg^{-1} with SBC the highest and LSC the lowest. In contrast, crude lipid was highest for LSC (108.2 g kg^{-1}) and lowest for SBC (74.4 g kg^{-1}). NSC had the highest crude fibre (201.1 g kg^{-1}) level, about three times higher than SBC which had the lowest fibre content (64.8 g kg^{-1}). Gross energy values for ingredients ranged from 17.8 to21.8 kJ g^{-1}.

The prices of ingredients used in the study are shown in Table 2. Fish meal was the least expensive (0.5 birr kg^{-1}) ingredient as the cost for it is directly converted from the cost of fresh offal (0.15 birr kg^{-1}) and 3.33 kg of offal dried in an oven can make approximately 1 kg of dried fishmeal. SBC and wheat grain were the most expensive ingredients, about double the price (4 birr kg^{-1}) of NSC which was the least expensive among the oilseed cakes.

Chemical composition of test diets

Proximate and energy compositions of the reference and test diets used in the digestibility study are presented in Table 3. Analyzed crude protein, crude lipid, NFE, dry matter, ash and energy contents of test diets showed little variation. However, crude fibre contents of diets varied considerably. Crude fibre of test diets followed similar trend as the test ingredients. Energy contents of the diets ranged between 18.6 and 18.9 kJ g^{-1}.

Nutrient and energy digestibility

Apparent digestibility coefficients (ADCs) of protein, lipid, dry matter and energy in selected test ingredients for Nile tilapia are shown in Table 4. The results indicate that ADCs of the nutrients and energy studied were significantly different between the test ingredients

except for crude lipid digestibility. Generally SBC had the highest ADC coefficients followed by NSC with LSC having the least ADC for energy and nutrients.

Discussion

The suitability of three oilseed by-products (SBC, NSC and LSC) available in Ethiopia were evaluated for their proximate composition and ADC values with the aim of providing information that aids improved formulation of balanced diets for Nile tilapia. The ADC values for dry matter, protein and energy were significantly different between the three test ingredients. However, lipid digestibility was not significantly different between the test ingredients.

In this study the test ingredients used had a high crude protein content and their values were close to previously reported values by Assaminew et al. [17], except for relatively higher crude protein (32.4% Vs 28.1%) and lower crude fibre (25.3% Vs 20.1%) values reported for Niger seed cake in this study. This variation between nutrient compositions of NSC could be due to differences in the origin, state and processing methods used to produce the cakes. The high crude fibre content of Nigerseed cake could limit the inclusion of this ingredient at higher levels in the diets of fish. It has been reported that dietary fibre is not utilized by fish [2].

The digestibility of ingredients provides insight concerning nutrient utilization and should enable better ingredient substitutions in diets designed for target species. The nutrient digestibility will vary depending on the composition of ingredients used [2,8]. The results of this study showed that ADC for dry matter, crude protein and energy in test ingredients were affected by test ingredients. These differences can be explained by the differences in chemical composition, origin and processing of these feed ingredients. The results of the present study indicated that Nile tilapia fingerlings have the capacity to digest protein and lipid satisfactorily in the oilseed by-product ingredients tested.

The overall dry matter digestibility of the test ingredients in the present study ranging from 59% to 78% is in the range reported for

Ingredients	DM	CP	CL	CF	Ash	NFE	GE	Price
Linseed cake	908.8	310	108.2	136.3	82.7	233.4	18.6	7.5
Niger seed cake	928	324.2	92	201.1	90.7	220	18.1	4.0
Soybean cake	938	393.8	74.7	64.8	54	350.7	19.3	9.0
Fish waste meal	950	610.9	187.1	0	220.4	0	21.8	0.5
Wheat grain	875	96	16.54	57.9	13.6	690.9	17.9	8.0
Corn grain	882.9	78.1	42.6	27.1	13.0	722.2	19.0	5.0

*DM (dry matter), CP (crude protein), CL (crude lipid), NFE (nitrogen free extract) and GE (gross energy).

Table 2: Proximate composition (g kg^{-1} as fed), energy (kJ g^{-1}) and prices (birr kg^{-1}) of individual feed ingredients used in this study.

Components	Reference diet	Test diets*		
		SBC	NSC	LSC
Dry matter	922.8	927.8	924.6	918.9
Crude protein	321.0	341.0	320.3	316.1
Crude lipid	105.8	96.0	101.1	106.0
Crude fibre	18.6	32.2	72.8	53.5
Ash	126.1	98.8	109.8	107.4
NFE	351.5	359.5	320.6	336.0
Chromic oxide	5.2	4.8	4.7	4.9
Gross energy (kJ g^{-1})	18.9	18.9	18.6	18.7

*SBC= soybean cake, NSC= Niger seed cake, LSC= linseed cake

Table 3: Proximate composition (g kg^{-1}) and energy of reference and test diets

Digestibility of Soybean Cake, Niger Seed Cake and Linseed Cake in Juvenile Nile Tilapia, Oreochromis niloticus L.

37

Components	Soybean	Niger seed	Linseed
Dry matter	78.0 ± 2 6[a]	70.7 ± 3.0[b]	59.0 ± 0.5[a]
Crude protein	87.9 ± 3.2[a]	72.6 ± 2.0[b]	62.4 ± 4.2[c]
Crude lipid	81.4 ± 2.2[a]	78.9 ± 2.8[a]	79.4 ± 3.0[a]
Gross energy (kJ g[-1])	86.0 ± 2.5[a]	72.9 ± 1.8[b]	53.7 ± 3.5[c]
Digestible protein	36.9	25.3	21.3
Digestible energy	17.7	14.2	11.0

Table 4: Apparent digestibility coefficients (%) of protein, lipid, dry matter, energy and digestible protein and energy (g kg[-1] and kJ g[-1] respectively, dry weight basis) in the test ingredients for Nile tilapia. Coefficients in each row with a different letter are significantly different (P<0.05).

plant protein-rich products (46-86.2%) in the diets of Nile tilapia [8-10]. Dry matter digestibilities in this study were generally lower than those reported for Nile tilapia elsewhere [13]. For example, the lower dry matter ADC of soybean in the present study could be explained by the higher crude fibre content of the product evaluated in this study of 69 g/kg, compared with 39 g/kg in the study by Köprücü and Özdemir [13]. Other studies on fish have also indicated the negative correlation between crude fibre content and dry matter ADC [31-33]. In general, results of dry matter ADC can be used to estimate the amount of solid waste released to the environment and to help determine the environmental impacts of aquaculture production [7,32].

Generally, the protein quality of dietary ingredients is one of the leading factors (apart from palatability) affecting fish performance and protein digestibility (digestible protein) is the first measure of its availability to fish. Protein quality of dietary protein sources depends on the amino acid composition and their digestibility. In the present study, the values obtained for protein digestibility for SBC corroborate previous findings (87.4%-96.2%) for soybean meal in tilapia diets [8,10,12,13,24,34]. Protein digestibility of test ingredients LSC (62.4%) and NSC (72.6%) for Nile tilapia in this study was lower than the reported digestibility coefficients of various other oilseed meals for this species. For example, reported APD in tilapia were 78.5% for cottonseed meal [34], 85% for rapeseed meal [12], 77.6% for peanut meal, 77.8% for canola meal, and 84% for de-gossypoled cottonseed meal [8].

Lower values of LSC protein digestibility in this study could be explained by other dietary factors present in plant protein products such as: i) suboptimal amino acid balance [24]; ii) presence of antinutritional factors [35]; and iii) inadequate levels of energy in linseed meals [36]. Plant products, especially oilseed cakes, usually have poor amino acid profiles and a certain amount of antinutritional factors (ANFs) which could affect nutrient utilization and, consequently, animal growth performance in different degrees depending on the type and amount of the compound [5,20-23]. Although linseed meal has been reported to have one of the best amino acid profiles after soybean meal and the composition fulfills the requirements of amino acid for Nile tilapia, the biological availability of amino acids in linseed to tilapia is less [37]. Linseed contains mucilage (5-8%) which has a large capacity to bind to water and increases intestinal viscosity, thus reducing nutrient digestibility [38]. Major antinutritional factors known to be present in linseed include: cyanogens, phytic acid, tannins, estrogenic factors, antithiamine factor and antipyridoxine factor [5]. For example, the ANF phytic acid has the ability to non-selectively bind to proteins, carbohydrates and minerals (divalent cationssuch as Ca^{2+}, Fe^{2+} Mg^{2+}) and inhibit activities of a number of digestive enzymes such as pepsin, trypsin and alpha-amylase [35].

The lower protein digestibility coefficients obtained for the two test ingredients (LSC and NSC) in the present study could not only be attributed to the ANFs but also to the higher levels of crude fibre that interfere In protein digestion of the diets that contain LSC and NSC. Previous studies [12,31] indicated that feeds with high crude fibre contents have poor nutrient digestibility due to reduced enzymatic access to potential substrates or due to the direct interaction between crude fibre components and the digestive process. Fibre levels as high as 8-12% are tolerated by most fish, but such levels often result in growth depression [39,40]. Fish fed diets high in indigestible fibre increase their feed intake and gastric evacuation time, but the extent to which fish can compensate in this manner is limited [41].

ADC values of fats in fish range from 85% to 95% when administered routinely either alone or in a mixed diet [42]. Reported fat digestibility in other species ranged from 70% to 90% [12,43] and similar values were found for tilapia in this study (78.8-81.3%). The ADCs of energy (53.6%-85.9%) in test ingredients for Nile tilapia in this study are generally in agreement with that reported (39-89%) and (54.8-92.1%) by Sklan et al. [12] and Köprücü andÖzdemir [13], respectively. Variation in apparent GE digestibility coefficients of ingredients in this study followed the same trend as that of protein and DM digestibility.

In the present study all three oilseed cakes tested proved valuable as protein sources in the diets of Nile tilapia as indicated by their ADCs. However, best values were observed for SBC. NSC, which was less than half the cost of SBC, appeared to be a good protein feed ingredient for Nile tilapia diets on balance in terms of overall nutrient composition and acceptable digestibility coefficients despite the highest crude fibre content. The LSC generally performed poorly, although it contained 310 g/kg CP. The nutrient and energy digestibilities were very low except for lipid digestibility. It seems that the ANF present in linseed cake may be responsible for the low ADC values. However, further research is required to establish the effect of dietary inclusions of LSC and NSC on productivity and on the various potential methods of increasing their utilization in fish diets before considering these ingredients in production feeds. The results of this digestibility study should contribute towards a better understanding of the nutrition of this species, especially in the grow-out stages.

Acknowledgments

This work was financed by the Department for International Development (DFID) from Development Partnership for Higher Education (DelPHE) project.

References

1. Glencross BD, Booth M, Allan GL (2007) A feed is only as good as its ingredients - a review of ingredient evaluation strategies for aquaculture feeds. Aquaculture Nutrition13: 17-34

2. Ng WK, Romano N (2013) A review of the nutrition and feeding management of farmed tilapia throughout the culture cycle. Reviews in Aquaculture 5: 220-254.

3. El-Sayed AFM (2004) Protein Nutrition of Farmed Tilapia: Searching for Unconventional Sources.

4. Hardy RW (2010) Utilization of plant proteins in fish diets: effects of global demand and supplies of fishmeal. Aquaculture Research 41: 770-776.

5. Tacon AGJ (1997) Fishmeal replacers: review of antinutrients within oilseeds and pulses—a limiting factor for the aquafeed Green Revolution?.

6. Hecht T (2007) Review of feeds and fertilizers for sustainable aquaculture development in sub-Saharan Africa. Study and analysis of feeds and fertilizers for sustainable aquaculture development, Rome.

7. Allan GL, Parkinson S, Booth MA, Stone DAJ, Rowland SJ, et al. (2000) Replacement of fish meal in diets for Australian silver perch, *Bidyanusbidyanus*: I. Digestibility of alternative ingredients. Aquaculture 186: 293-310.

8. Zhou QiC, Yue YiR (2012) Apparent digestibility coefficients of selected feed ingredientsforjuvenilehybridtilapia, *Oreochromis niloticus× Oreochromisaureus*. AquacultureResearch 43: 806-814.

9. Guimaraes IG, Pezzato LE, Barros MM, Tachibana L (2008) Nutrient digestibility of cereal grain products and by-products in extruded diets for Nile tilapia. Journal of the World Aquaculture Society 29: 781-789.

10. Fontainhas-Fernandes A, Gomes E, Reis-Henriques MA, Coimbra J (1999) Replacement of fish meal by plant proteins in the diet of Nile tilapia: digestibility and growth performance. Aquaculture International 7: 57-67.

11. El-Saidy DMSD, Gaber MMA (2003) Replacement of fish meal with a mixture of different plant protein sources in juvenile Nile tilapia, Oreochromis niloticus (L.) diets. Aquaculture Research 34: 1119-1127.

12. Sklan D, Prag T, Lupatsch I (2004) Apparent digestibility coefficients of feed ingredients and their prediction in diets for tilapia, Oreochromisniloticus x Oreochromisaureus (Teleostei, Cichlidae). Aquaculture Research 35: 358-364.

13. Köprücü K, Özdemir Y (2005) Apparent digestibility of selected feed ingredients for Nile tilapia (Oreochromis niloticus). Aquaculture 250: 308-316.

14. El-Sayed AFM (1999) Alternative dietary protein sources for farmed tilapia, Oreochromis spp. Aquaculture179: 149-168.

15. Lovell T (1998) Nutrition and feeding of fish. Kluwer Academic Publishers, Massachusetts.

16. Tadelle D, Nigusie D, Alemu Y, Peters KJ (2002) The feed resource base and its potentials for increased poultry production in Ethiopia. World's Poultry Science Journal 58: 77-87.

17. Assaminew K, Waidbacher H, Zollitsch W (2012) Proximate composition of selected potential feedstuffs for small-scale aquaculture in Ethiopia. Livestock Research for Rural Development 24.

18. Getinet A, Sharma SM (1996) Niger. Guizotiaabyssinica (L. f.) Cass. Promotingthe conservation and use of underutilized and neglected crops. 5. Institute of Plant Genetics and Crop Plant Research, Gatersleben/International Plant Genetic Resources Institute, Rome.

19. Wijnands J, Biersteker J, Hiel R (2007) Oilseeds business opportunities in Ethiopia. Ministry of Agriculture, Nature and Food Quality, Netherlands.

20. Hasan MR, Macintosh DJ, Jauncey K, (1997) Evaluation of some plant ingredients as dietary protein sources for common carp (CyprinuscarpioL.) fry. Aquaculture151: 55-70.

21. Mukhopadhyay N, Ray AK (2001). Effects of amino acid supplementation on the nutritive quality of fermented Linseed meal protein in the diets for rohu, Labeorohita, fingerlings. J ApplIchthyol 17: 220-226.

22. Mukhopadhyay N, Ray AK (2005) Effect of fermentation on apparent total and nutrient digestibility of Linseed, Linumusitatissimum, meal in rohu, Labeorohita, fingerlings. Actaichthyologicaetpiscatoria 35: 73-78.

23. Latif KA, Alam MT, Sayeed MA, Hussain MA, Sultana S, et al. (2008) Comparative study on the effects of low cost oil seed cakes and fish meal as dietary protein sources for Labeorohita (Hamilton) fingerling. Univ j ZoolRajshahiUniv 27: 25-30.

24. National Research Council (NRC) (1993) Nutrientrequirementsoffish. National Academic Press, Washington DC, USA.

25. Cho CY, Cowey CB, Watanabe T (1985) Finfish Nutrition in Asia. Methodological approaches to research and development. International Development Research Centre, Ottawa.

26. Cho CY, Slinger SJ, Bayley HS (1982) Bioenergetics of Salmonid fishes: energy intake, expenditure and productivity. CompBiochemPhysiol 73: 25-41.

27. Divakaran S, Leonard GO, Ian PF (2002) Note on the methods for determination of chromic oxide in shrimp feeds. J Agric Food Chem 50: 464-467.

28. AOAC (Association of Official Analytical Chemists) (1995) Official methods of analysis AOAC, Arlington, Virginia, USA.

29. Bureau DP, Harris AM, Cho CY (1999) Apparent digestibility of rendered animal protein ingredients for rainbow trout (Oncorhynchusmykiss). Aquaculture 180: 345-358.

30. Zar JH (2010) Biostatistical analysis. Prentice-Hall, Inc, New Jersey, USA.

31. Maina JG, Beames RM, Higgs D, Mbugua PN, Iwama G, et al. (2002) Digestibility and feeding value of some feed ingredients to tilapia Oreochromisniloticus (L.). Aquaculture Research 33: 853-862.

32. Guimaraes IG, Pezzato LE, Barros MM, Fernandes RDN (2012) Apparent nutrient digestibility and mineral availability of protein-rich ingredients in extruded diets for Nile tilapia. R Bras Zootec 41: 1801-1808.

33. Asad F, Rehman T, Qureshi NA, Tahir N (2013) Estimation of apparent digestibility coefficient of plant feed ingredients (soybean and sunflower meal) for LabeoRohita. American Journal of Biomedical and Life Sciences 1: 8-11.

34. Guimaraes IG, Pezzato LE, Barros MM (2008) Amino acid availability and protein digestibility of several protein sources for Nile tilapia, Oreochromis niloticus.Aquaculture Nutrition.14: 396-404.

35. Liener IE (1994) Implications of antinutritional components in soybean foods. Critical Reviews Food Science and Nutrition 34: 31-67.

36. El-Saidy DMS, Gaber MMA (2001) Linseed meal- its successful use as a partial and complete replacement for fish meal in practical diets for Nile tilapia Oreochromis niloticus.

37. Hanafy MA (2006) Effect of replacement of soybean meal by linseed meal on growth performance, and body composition of Nile tilapia Oreochromisniloticus (L) cultured in concrete ponds. Egypt J AquatBiol& Fish 10: 185-200.

38. Fedeniuk RW, Biliaderis CG (1994) Composition and physicochemical properties of linseed (Linum usitatissiumum) mucilage. J Agric Food Chem 42: 240-247.

39. Edwards DJ, Austreng E, Risa S, Gjedrem T (1977)Carbohydrate in rainbow trout diets. I. Growth of fish of different families fed diets containing different proportions of carbohydrate. Aquaculture 11: 31-38.

40. Leary DF, Lovell RT (1975) Value of fibre in production diets for channel catfish. Trans Am Fish Soc 104: 328-332.

41. Leenhouwers JI, Ortega RC, Verreth JAJ, Schrama JW (2007) Digesta characteristics in relation to nutrient digestibility and mineral absorption in Nile tilapia (Oreochromisniloticus L.) fed cereal grains of increasing viscosity. Aquaculture 273: 556-565.

42. Aksnes A, Opstvedt J (1998) Content of digestible energy in fish feed ingredients determined by the ingredient-substitution method. Aquaculture 161: 45-53.

43. Lupatsch I, Kissil GW, Sklan D, Pfeffer E (1997) Apparent digestibility coefficients of feed ingredients and their predictability in compound diets for gilthead seabream, Sparusaurata L. Aquaculture Nutrition 3: 81-89.

8

Comparative Structural Organization of Skin in Red-Tail Shark (*Epalzeorhynchos Bicolor*) and Guppy (*Poecilia Reticulata*)

Doaa M Mokhtar*

Department of Anatomy and Histology, Faculty of Veterinary Medicine, Assuit University, Egypt

Abstract

The surface architecture and histological organization of the skin of two ornamental fish; red-tail shark (*Epalzeorhynchos bicolor*) and guppy (*Poecilia reticulata*) were the main focus of this study. The skin of the two species was composed of epidermis, dermis and hypodermis, although the epidermis showed great variations in their components in the two species. The epidermis of red-tail shark was consisted of epidermal cells, mucous goblet cells, serous goblet cells, club cells, rodlet cells and melanocytes. While, the epidermis of guppy was composed of epidermal cells, mucous goblet cells, eosinophilic granular cells, lymphocytes and melanocytes. The skin of red-tail shark included a variety of sense organs as tuberous receptor organs in the head, superficial neuromasts on the lower lips and the head, canal neuromast in the operculum and the head, and taste buds on the lips, operculum, dorsum of the head and lateral regions of the trunk. However, the skin of guppy was characterized by presence of ampullary organ on the dorsal side of the head, superficial neuromasts on the lips and the head, canal neuromast in the operculum and the head, and taste buds on the operculum, dorsum of the head and trunk regions. These structural peculiarities with histochemical features indicate additional physiological role of the skin of the two species, as the mucous goblet cells in the two species contained a considerable amount of glycoconjugates, whereas the other unicellular gland types, the serous goblet cells and club cells in red tail shark were proteinous in nature. The dermis and hypodermis was consisted of connective tissue, mainly collagenous fibers. Scanning electron microscopy indicated presence of fingerprint like- patterns of microridges of the epidermal cells, pores for lateral canal system, openings of mucous cells and taste buds with specific sensory organs in each fish species.

Keywords: Epidermis; Guppy; Red tail shark; Scanning electron microscopy; Histochemistery; Lateral line system

Introduction

The skin of a fish is a multifunctional organ and may serve important roles in protection, communication, sensory perception, locomotion, respiration, excretion, osmoregulation and thermal regulation [1]. The skin is self active secretory organ that their cellular components provide many useful products. Goblet cells secrete mucus that keep the body surfaces moist and protect it from stressors [2], club cells produce the alarm substances that initiate the alarm reaction [3] and melanocytes produce pigments to provide the fish with specific colorations [4]. The skin is also a vehicle for coetaneous sense organs that allowed the fish for detection of predators and foods. Among them are taste buds and lateral line system that include neuromasts and electroceptive organs [5].

The skin of fish shows various inter-species difference, as some species have scales and others have special cells. The skin is composed of dermis and epidermis. The comparative anatomy of fish skin, especially the epidermis, has been investigated by light and electron microscopy [6,7]. However, less consideration have been given to the skin of both red-tail shark and guppy fish. Red tail shark (*Epalzeorhynchos bicolor*) is one of freshwater fish, belongs to family Cyprinidae that originates from the streams and waterways of Thailand. It is characterized by black body and orange tail and its skin is covered with transparent scales [8]. The guppy (*Poecilia reticulata*), also known as million fish and rainbow fish is one of the most popular freshwater aquarium fish species. It is a member of the Poeciliidae, whose natural range is in South America, were introduced to many habitats and are now found all over the world. The body of guppy is transparent and is covered with colorless scales and has ornamental dorsal and caudal fins. Guppies are used as a model organism in the field of ecology, evolution, and behavioural studies [9].

The aim of this study was to assess the morphology of skin at

different parts of the body focusing on patterns of distribution of the sensory organs of red-tail shark and guppy fish with special reference to histochemistery of glycoconjugates and protein secreting cells (Figure 1).

Materials and Methods

Samples collection

The materials employed in this study were consisted of randomly obtained 16 adult specimens of males and females red- tail shark (*Epalzeorhynchos bicolor*) and guppy (*Poecilia reticulata*) that purchased from shops of ornamental fish. The mean standard length was 4.80 ± 1.01 cm for red-tail shark and 3.05 ± 0.8 cm for guppy. The fish were

Figure 1: Showing red tail shark (A) and guppy (B). (Scale bar = 1cm)

***Corresponding author:** Doaa M Mokhtar, Department of Anatomy and Histology, Faculty of Veterinary Medicine, Assuit University, Egypt
E-mail: doaamokhtar33@yahoo.com

deeply anaesthetized with benzocaine (4 mg/L) and decapitated. Pieces of skin from dorsum sides of head, snout, operculum, lips, lateral parts of the trunk regions and tail region were excised and rinsed in physiological saline.

Histological and histochemical examination

Samples for histological technique were dissected at 1×1×.05 cm and were immediately fixed in Bouin's fluid for 22 hours. The fixed materials were dehydrated in an ascending series of ethanol, cleared in methyl benzoate and then embedded in paraffin wax. Serial sections of the entire fish at 5-8 µm in thickness were cut and stained with Harris haematoxylin and Eosin for general structure, Crossmon's trichrome for collagenous fibers. For carbohydrates histochemistry, sections were stained by combined Periodic Acid–Schiff (PAS) technique and Alcian blue (pH 2.5). Representative sections were stained with bromophenol blue, performic acid-methylene blue and Mallory triple stain for detection of proteins [10].

Morphometery

Mean thickness of the epidermis (µm) and density of mucous cells/100 µm of the epidermis in all selected regions in the two fish species were recorded ± SE.

SEM preparation

The skin of all head regions of both species were immediately washed by 0.1 M Na-cacodylate buffer. Then they were fixed in a mixture of 2.5% paraformaldehyde and 2.5% glutaraldehyde in 0.1 M Na-cacodylate buffer, pH 7.3 for 4 hours at 4°C. Thereafter, they were washed in the same buffer used and post-fixed in 1% osmic acid in 0.1 M Na-cacodylate buffer for further 2 hours at room temperature. The samples were then dehydrated by aceton followed by isoamyl acetate and then subjected to critical point drying method with a polaron apparatus. Finally, they were coated with gold and observed with JEOL scanning electron microscope (JSM-5400 LV) at KV 10.

Results

The skin was composed of epidermis of non-keratinized stratified squamous epithelium with goblet cells, followed by dermis of dense regular connective tissue.

Red-tail shark

At the lower lip region: The epidermis was 51 µm in thickness (Table 1) and consisted of stratified squamous epithelium non-cornified that formed of a simple columnar epithelium lie on wavy thick basement membrane, on the top of these layers were several layer of eosinophilic club cells followed by squamous cell layers. Pear-shaped taste buds were observed at the top of the epidermis (Figure 2A). Large number of PAS-AB-positive mucous goblet cells was distributed at the superficial layers. The density of mucous cells was 12/100 µm (Table 1). Club cells were large cells varied from spherical to saccular in shape and were negative to this stain (Figure 2B). The dermal layer was formed of dense compact parallel bundles of collagenous fibers. Aggregated melanocytes were located at the dermal margin as a continuous layer. The hypodermis was clearly organized layer of loose connective tissue (Figure 2A and 2B).

At the upper lip: The epidermis was thicker; 65 µm (Table 1). Taste buds were distributed in this region as it occurred more frequently in the anterior parts of the body than posteriorly (Figure 2C and D). Mucous goblet cells occupied the superficial layers of epidermis that opened directly to the surface and stained positive with combined PAS-

AB (Figure 2D). The density of mucous cells was 10/100 µm (Table 1).

At the snout: The epidermis was of less thickness; 44 µm (Table 1) and characterized by large number of club cells in mid epidermal level and numerous PAS-positive mucous goblet cells. The density of mucous cells was 8/100 µm (Table 1). Tuberous receptor organ was distributed at this region and appeared as spherical structure of sensory cells surrounded by a cellular capsule. Melanocytes were randomly distributed in the dermis (Figure 2E).

At the operculum: The epidermis was 42 µm in thickness (Table 1). The mucous goblet cells were restricted on the superficial layer. The cells were positive with PAS-AB stain. The density of mucous cells was 8/100 µm (Table 1). Taste buds were observed as pale staining, pear-shaped structures consisting of columnar sustentacular cells with dark nuclei alternated with fusiform sensory cells with light nuclei and associated with small pyramidal basal cells and surrounded by marginal cells. Melanocytes were observed as a continuous layer at the dermal

Body regions	Red tail shark		guppy	
	Thickness of the epidermis (µm)	Density of mucous cells/100 µm	Thickness of the epidermis (µm)	Density of mucous cells/100 µm
Lower lip	51.3 ± 4.1	12.2 ± 2.4	36.1 ± 2.3	14.8 ± 2.3
Upper lip	65.2 ± 5.0	10.5 ± 2.7	58. 2 ± 3.6	12.0 ± 2.3
Snout	44. 5 ± 3.9	8.1 ± 2.8	38. 6 ± 3.0	8.6 ± 1.9
Operculum	42.2 ± 2.8	8.9 ± 2.6	30. 8 ± 1.9	8.5 ± 2.8
Dorsum of head	38.6 ± 3.6	8.3 ± 1.4	36. 5 ± 2.8	10 ± 1.8
Trunk	18.0 ± 2.2	4.5 ± 1.0	11. 6 ± 2.4	6.3 ± 1.3
Tail	40. 5 ± 3.7	2.2 ± 1.5	39. 4 ± 4.4	14.7 ± 3. 7

Table 1: The mean thickness of epidermis (µm) at different body region and the density of mucous cells/100 µm of the epidermis in both fish species.

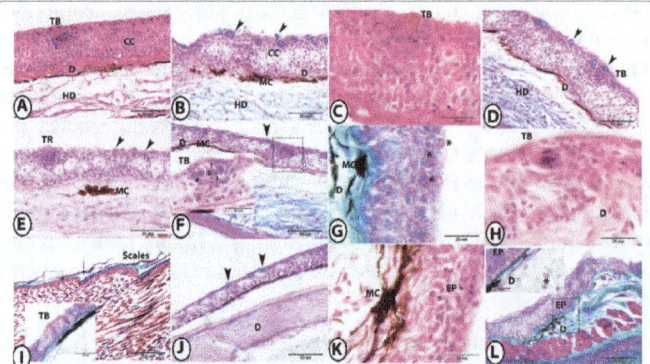

Figure 2: Organization of skin of red tail shark at different body regions. **A and B:** The skin of the lower lip stained by HE and PAS-AB-HX respectively showing taste bud (TB), club cells (CC), mucous cells (arrowheads) and melanocytes (MC) at the dermis (D) followed by hypodermis (HD). **C and D:** The skin of the upper lip stained by HE and PAS-AB-HX respectively showing taste bud (TB), mucous cells (arrowheads) and thin dermis (D) followed by hypodermis (HD). **E:** The skin of the snout stained by PAS-AB-HX showing PAS positive mucous cells (arrowheads), tuberous organ (TR) and melanocytes (MC). **F:** The skin of the operculum stained by PAS-AB-HX showing mucus cells (arrowhead), taste bud (square, TB, inserted figure) that consisted of sensory cells (1), supporting cells (2), basal cells (3) and mentle cells (4). Note, melanocytes (MC) on the dermis(D). **G and H:** The skin of the dorsum of the head stained by Crossmon's Trichrome and HE respectively showing rodlet cells (asterisk), taste bud (TB), melanocytes (MC) on the dermis (D) of collagenous fibers. **I:** The skin of the trunk stained by Crossmon's Trichrome showing scale pockets (arrow), taste bud (square, TB, inserted figure). **J:** Tigher magnification of square of figure. I showing the skin of trunk stained by PAS-AB-HX indicates positive stained mucous cells (arrowheads). Note parallel bundles of collagenous fibers in the dermis (D). **K and L:** The skin of the tail stained by HE and Crossmon's Trichrome respectively showing melanocytes (MC) in the deeper layers of epidermis (EP). The dermis (D, inserted figure in L) contained many rodlet cells (asterisk).

margin. The dermis was composed of undulated compactly arranged collagenous bundles (Figure 2F).

At the dorsal aspect of the head: The mean thickness of the epidermis was 38 µm and the density of mucous cells was 8/100 µm (Table 1). Many rodlet cells were observed at the superficial layers as ovoid cells with basal nucleus and numerous cytoplasmic inclusions (rodlets). Large numbers of melanocytes were condensed at the superficial region of the dermis (Figure 2G). Taste buds were observed at the superficial epidermal layer. The dermal region showed compactly arranged collagenous bundles (Figure 2H).

At the trunk region: This region was covered with scales that originated from dermal scale-pockets and protrude posteriorly where they were covered by the epidermis. The epidermis was thinner (18 µm) (Table 1) above the free portion of the scales. Bulb-shaped taste bud was demonstrated at the superficial region (Figure 2I). Few small spherical PAS-AB positive mucous goblet cells were observed. The density of mucous cells was 4/100 µm (Table 1). Melanocytes were observed as a continuous sheath. The dermis was formed of dense white regular collagenous connective tissue (Figure 2J).

At the tail region: The epidermis was thick; 40 µm and the density of mucous cells was 2/100 µm (Table 1). Highly branched melanocytes were distributed randomly throughout the deeper layers of the epidermis and in the dermis. These cells were filled with dark brown to black coarse granules (Figure 2K). The dermis was composed of collagenous bundles with many rodlet cells (Figure 2L).

The guppy

At the lower lip: The epidermis was 36 µm in thickness and the density of mucous cells was 14/100 µm (Table 1), it was consisted of stratified squamous epithelium non-cornified with highly mitotically active epidermal cells. The dermal region was formed of compact parallel bundles of collagenous fibers (Figure 3A).

At the upper lip: The epidermis was a thick (58 µm, Table 1), composed of stratified squamous epithelium non-cornfied. Mucous goblet cells were observed at the superficial layer of the epidermis. The density of mucous cells was 12/100 µm (Table 1). Melanocytes were observed at the dermal margin (Figure 3B).

At the snout: The epidermis was 38 µm in thickness and the density of mucous cells was 8/100 µm (Table 1). Eosinophilic granular cells were spherical cells that occurred in the superficial layers of the epidermis. Their nuclei were eccentrically placed and their cytoplasmic granules stained bright red by HE. Lymphocytes were frequently distributed in the intercellular spaces of the deeper epidermal layers of this region and stained deep blue in HE (Figure 3C).

At the operculum: The epidermis was in 30 µm in thickness (Table 1). The mucous cells were restricted on the superficial layer and were positive with PAS-AB stain. The density of mucous cells was 8/100 µm (Table 1). Taste bud was observed in the superficial layer. The opercular core was formed of loose connective tissue and supported by strands of osteoid tissues. It was provided with well developed wide lateral canal (canal neuromast) (Figure 3D).

At the dorsal side of the head: The epidermis was 36 µm in thickness (Table 1). The mucous cells were PAS-AB-positive and were restricted to the superficial layer of the epidermis (Figure 3E). The density of mucous cells was 10/100 µm (Table 1). Taste bud was also observed at this region. The dermis was formed of collagenous bundles (Figure 3F).

At the trunk: The trunk region was covered with scales that originated from scales pockets in the dermis. The epidermis above the scales was thin; 11 µm (Table 1). It contained large number of PAS-AB- positive mucous goblet cells with taste bud interspersed between the epidermal layer (Figure 3G). The density of mucous cells was 6/100 µm (Table 1).

At the tail region: The mean thickness of the epidermis was 39 µm (Table 1). Numerous PAS-AB- positive mucous goblet cells and large number melanocytes were observed (Figure 3H and I). The density of mucous cells was 14/100 µm (Table 1).

The lateral line system

Red tail shark: It was divided into 2 subsystems; mechanoreceptive neuromasts and electroreceptive tuberous receptor organs. The tuberous organ was more frequently occurred in the head region. The upper region of tuberous organ was covered by simple squamous epithelial cells. The tuberous organs contained 4-5 elongate nonciliated receptor cells within a cellular capsule. A single layer of supporting cells was present between the base of the receptor cells and the base of the capsule. A single thin nerve fiber innervates each group of organs (Figure 4A and 4B).

2 types of neuromast present; canal (deep) and superficial neuromasts. Canal neuromasts were embedded in the dermis in form of tunnel-like canals. Large canal neuromasts were observed in the superficial layer of epidermis of the head and the operculum and

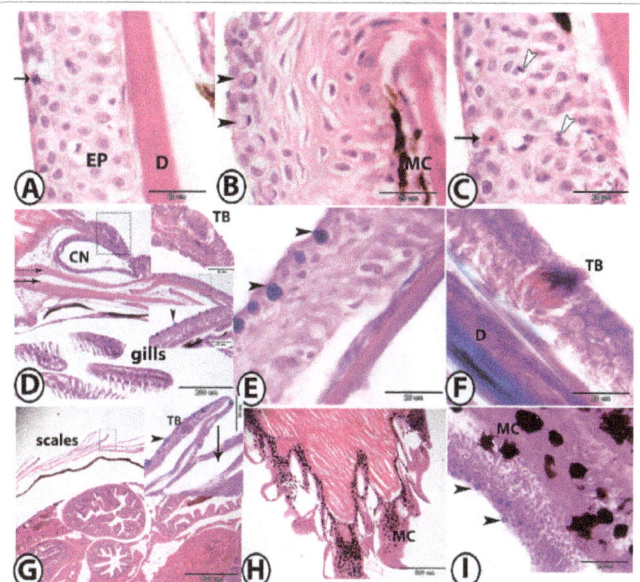

Figure 3: Organization of skin of guppy at different body region. **A:** The skin of the lower lip stained by HE showing mitotic cells (arrow) in stratified squamous epidermis (EP), followed by the dermis (D). **B:** The skin of the upper lip stained by HE showing mucous cells (arrowheads) and melanocytes (MC). **C:** The skin of the snout stained by HE showing eosinophilic granular cells (arrow) and lymphocytes (arrowheads). **D:** The skin of the operculum stained by HE showing taste bud (large square, TB). The small square indicates mucous cells (arrowhead) in the epidermis of the inserted figure stained by PAS-AB-HX. Note presence of canal neuromast (CN) surrounded by osteoid tissues (arrows). **E and F:** The skin of the dorsum of the head stained by PAS-AB-HX and Crossmon's Trichrome respectively showing mucous cells (arrowheads) and taste bud (TB). Note the dermis (D) is formed of collagenous fibers. **G:** The skin of the trunk stained by HE showing scale pockets (arrow). The square indicated the inserted figure stained with PAS-AB-HX showing taste bud (TB) and mucous cells (arrowhead). **H and I:** The skin of the tail stained by HE and PAS-AB-HX respectively showing numerous black melanocytes (MC) and many mucous cells (arrowheads).

invested by dense fibrous wall without any cartilaginous support. It lined by stratified cuboidal epithelium and opened directly to the surface epithelium by a pore (Figure 4C and 4D). Large number of superficial neuromasts located at the lower lips and the head. Four main cell types encountered in the superficial neuromast; sensory hair cells, supporting cells, basal cells and surrounded by crescent shaped mantle cells. The superficial neuromast in many cases were covered by the cupula (dome of gelatinous materials) extending into the surface (Figure 4E and 4F).

Guppy: The lateral line system was divided into 2 subsystems; ampullary organ and neuromasts. The ampullary organs localized in the head region behind the eye, it ranged in size from 200-250 μm in length and 100-125 μm in diameter. It formed of specialized receptors. These receptors were generally ampoule- shaped, and their lumen opened to the surface by means of a short channel, the upper part of sensory cells was covered by a mucus-like substance called cupula. The sensory epithelium was found in a chamber situated in the basal portion of the epidermis or somewhat depressed into the dermis. The sensory epithelium of the organ was formed by 8 sensory cells. The sensory cells were pyriform in shape with dark cytoplasm with a large rounded centrally located nucleus and had microvilli on their apical surface with axon nerve endings in their basal part that arise from the underlying conjunctive tissue. Supporting cells were tall and narrow cells that completely enveloping the sensory cells with basally located

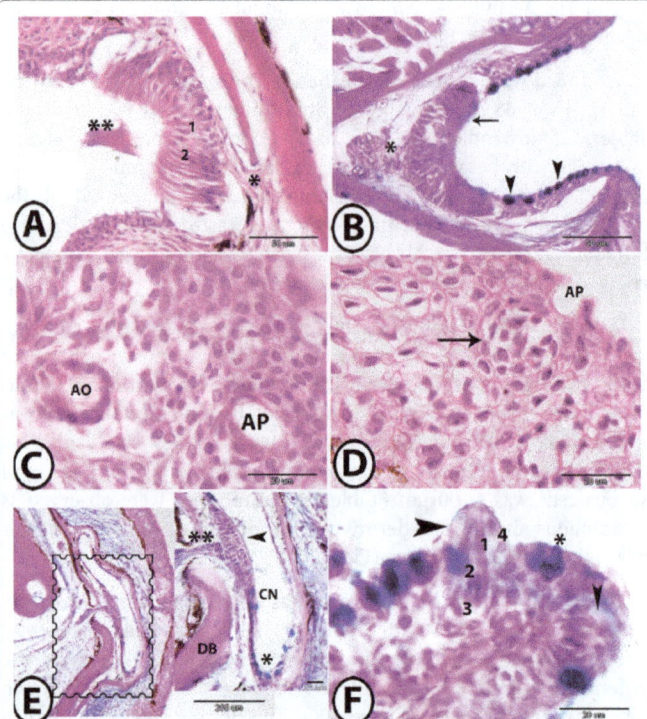

Figure 5: The lateral line system of guppy. **A:** the ampullary organ is formed of sensory cells (1), supporting cells (2) and covered by cupula (2 asteriks). Note presence of nerve fibers (asterisk) basally. **B:** ampullary organ stained with PAS-AB-HX showing AB-positive apical border of supporting cells (arrow), basal bundle of nerve fibers (asterik). Note that the ampullary channel contained PAS-AB positive mucous cells (arrowheads). **C:** a horizontal skin section of the epidermis stained by HE showing an ampullary pore (AP) in close proximity to an ampullary organ (AO). **D:** a vertical skin section stained with HE showing cross section of the ampullary organ (arrow) under an ampullary pore (AP). **E:** canal neuromast (square, inserted figure) stained with PAS-AB-HX showing the dermal bone (DB) that surrounded the canal neuromast (CN), which lined by sensory cells (arrowhead), AB-positive mucous cells (asterisk). Note presence of nerve fibers (two asterisks) below the neuromast. **F:** Superficial neuromasts (arrowheads) stained by PAS-AB- HX is formed of sensory cells (1), supporting cells (2), basal cells (3) and surrounded by mentle cells (4). Note presence of mucous cells (asterisk) near neuromasts.

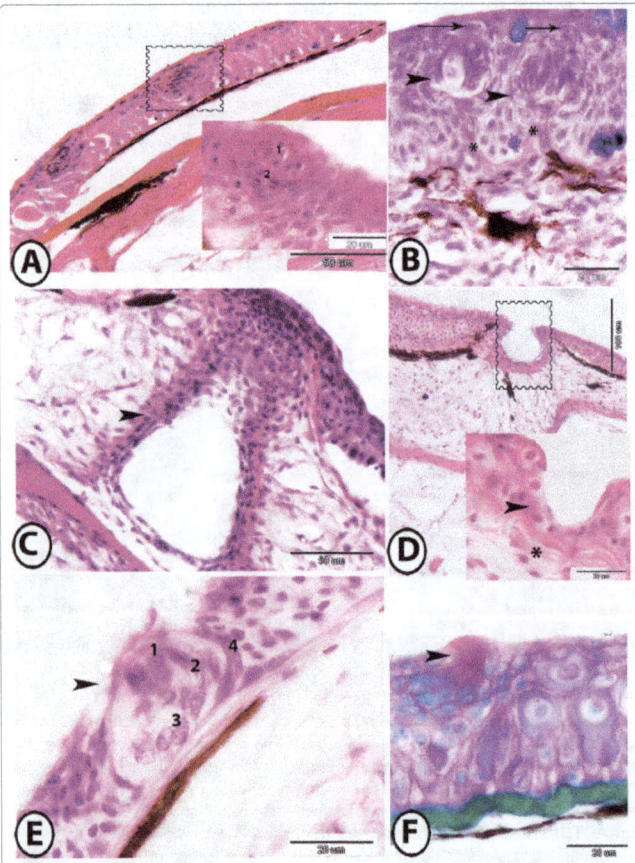

Figure 4: The lateral line system of red tail shark. **A:** Tuberous organ (square, inserted figure) stained by HE that composed of sensory cells (1), supporting cells (2). **B:** Tuberous organs (arrow heads) stained by PAS-AB-HX are covered by flattened cells (arrows) and innervated by nerve fibers (asterisk). **C and D:** Canal neutomast (arrowheads) stained by HE surrounded by fibrous tissue (asterik). **E:** superficial neuromast (arrowhead) stained by HE formed of sensory cells (1), supporting cells (2), basal cells (3) and mantle cells (4). **F:** Superficial neuromast stained with Crossmon's trichrome showing attached cupula (arrowhead).

nucleus (Figure 5A and 5B). The canal of the ampullary organ contained mucous-cells that stained slight violet with PAS-AB stain, also theapical border of support cells were AB-positive (Figure 5B). The ampullary pores were distributed throughout the head and lined with a single layer of squamous cells (Figure 5C). Cross section of ampullary organ that formed of a sac of fusiform receptor cells was observed below the ampullary pore (Figure 5D).

The superficial and deep neuromasts differed from those present in red tail shark. The canals neuromast was widely distributed in the head and operculum and open at intervals to the exterior and surrounded by dermal bone. The canal neuromast was lined by two layers of sensory cells with apical microvilli and the opposite site of the canal was lined with a simple squamous epithelium containing AB-positive mucus cells. Bundle of nerve fibers lie below the canal and passed between the dermal bones (Figure 5E). The superficial neuromast were concentrated in lips and the head. They rose above the surface epithelium and lined with sensory, supporting, basal cells and surrounded by mantle cells. Some mucus cells were present in the epithelium adjacent to the neuromast (Figure 5F).

Histochemical analysis to protein cells in red tail shark: Sacciform cells (serous goblet cells or protein sectreting cells): were encountered in red- tail shark and they resembled the ordinary goblet cells and widely distributed in head and snout. They were pear-in shape and their cytoplasm stained deep red with Mallory triple stain, while the mucous goblet cells appeared blue with the same stain (Figure 6A and 6B). Also, sacciform cells appeared blue with bromophenol blue (Figure 6C) and deep blue with performic acid-methylene blue and the cell membrane appeared having an intensive reaction (Figure 6D). Histochemical analysis detected prevalence of protein within the cytoplasm of club cells rather than carbohydrates, as they were positive to bromophenol blue (Figure 6C).

SEM

Red- tail shark: Lateral line system and their associated pores based on their position around the head were categorized into infraorpital, supraorbital and otic lines (Figure 7A). Each canal neuromast opened as a pore in the skin of the head region (Figure 7B). The tuberous receptor organ formed of group of sensory cells that found in pits but was not covered by epidermis (Figure 7C). Microvilli projecting from the surfaces of the receptor cells were observed (Figure 7D). Superficial neuromast were noticed in the upper lips and dorsum of head localized mainly within the anteorbital area and may present as isolated elements or found in rows (Figure 8A). Each neuromast was sitting on an epidermal protrusion (Figure 8B). The epidermal protrusion was encircled with epithelial ring that showed shallow depression or invaginations to form a characteristic ring-like pattern. The top of the protrusion possessed hair bundles (Figure 8C). Neuromast with attached cupula was also observed (Figure 8D). Group of elevations were demonstrated in the dorsal surface of the head and appeared conical in shape (Figure 8E). The apex of each elevation was characterized by the presence of taste buds that protruded above the epithelial surface. At the summit of this elevation, numerous closely packed microvilli were originated and short cilia were observed in the center (Figure 8F). The surface of the epidermal cells appeared polyhedral in shape bearing finger-print like microridges and interspersed by openings of mucous cells and melanocytes (Figure 8G). The broken surface of the epidermis indicated presence of large club cells and melanocytes with cytoplasmic processes (Figure 8H).

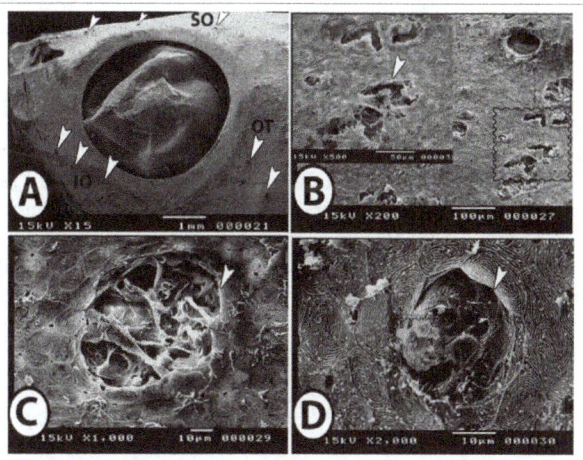

Figure 7: SEM of the lateral line system of red tail shark. **A:** The lateral line (arrowheads) branch in the head into supraorbital (SO), infraorbital (IO) and otic (OT). **B:** Pores of lateral lines (large square, inserted figure, arrowhead). **C:** Higher magnification of small square in Figure B showing tuberous receptor organ (arrowhead). **D:** Tuberous receptor organ (arrowhead) showing microvilli projecting from the receptor cells (square).

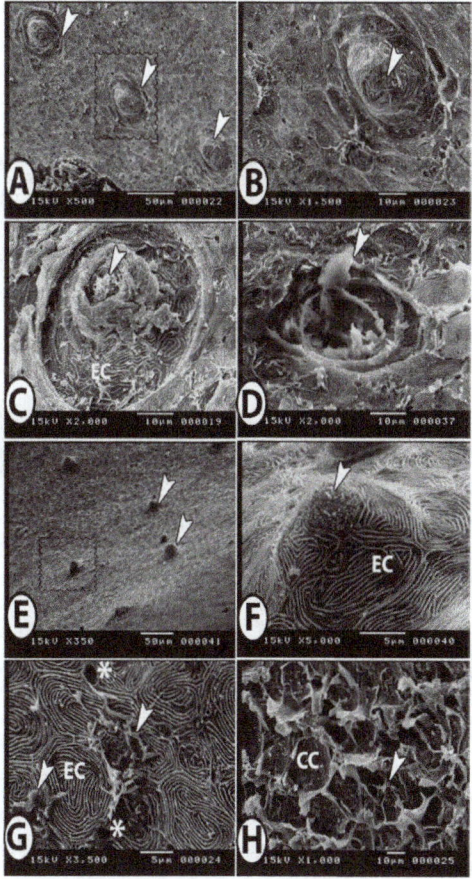

Figure 8: Neuromasts and taste buds of red tail shark. **A:** Superficial neuromasts arranged in rows (arrowheads). **B:** Higher magnification of the square in Figure A showing superficial neuromast (arrowhead). **C:** Ring like pattern of neuromast formed of epidermal cells (EC) with hairs from sensory cells at their apex (arrowhead). **D:** Superficial neuromast with attached cupula (arrowhead). **E:** Epidermal elevations (arrowheads, square). **F:** Conical non-keratinized protuberance of epidermal cells (EC) with a taste bud at a summit (arrowhead). **G:** surface of the epidermis showing epidermal cells (EC) with microridges, interspersed by opening of mucous cells (stars) and melanocytes (arrowheads). **H:** Broken surface of the epidermis showing club cells (CC) and melanocyte (arrowhead).

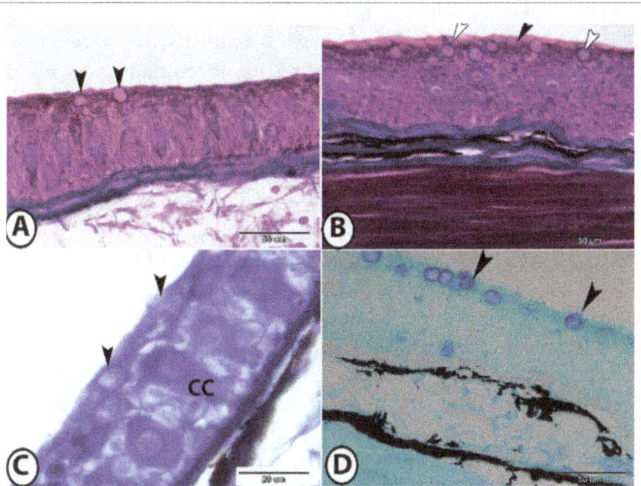

Figure 6: Histochemistery of protein secreting cells in red tail shark. **A and B:** The skin stained with Mallory triple stain showing red serous goblet cells (black arrowheads) and blue mucous goblet cells (white arrowheads). **C:** The skin stained with bromophenol blue showing serous goblet cells (arrowheads) and positive stained club cells (CC). **D:** The skin stained with Performic acid-methylene blue showing positive stained serous goblet cells (arrowheads).

The guupy: Lateral line system and their associated pores of guppies based on their position around the head were categorized into supraorbital, otic and preopercular lines (Figure 9A). The ampullary organ was localized within grooves or minute pores in the head region behind the eye beside the lateral line channels (Figure 9B). From the surface view, the ampullary organ appeared spherical with attached cupula (Figure 9C). In the lateral view, it appeared spherical with alveoli-like appearance (Figure 9D). Ampullary pore distribution patterns were relatively unique, with the majority of pores occurring on the dorsal region of the head, which were analogous in location to the ampullary organs (Figure 9E). The diameter of ampullary pores ranged from 5-10 μm. Superficial neuromasts with attached cupula were observed along the line of ampullary organ (Figure 9F). The surface of the epidermis was composed of polyhedral epithelial cells that were covered by fingerprint- like pattern microridges. The boundaries between cells were clearly defined (Figure 9E and 9F).

Discussion

The epidermis of both red- tail shark and guppy was consisted mainly of stratified squamous epithelium interspersed with mucous cells. The primary function of the epidermis is protection against environmental hazards. In fish, mucogenic cells generally provide this function by secreting their contents on the surface. The distribution of the mucus cells in the epidermis varied greatly in this study. In red tail shark, they were condensed at anterior body regions, reduced in trunk region and decreased in tail. Similar findings were found in the eels [11]. While in

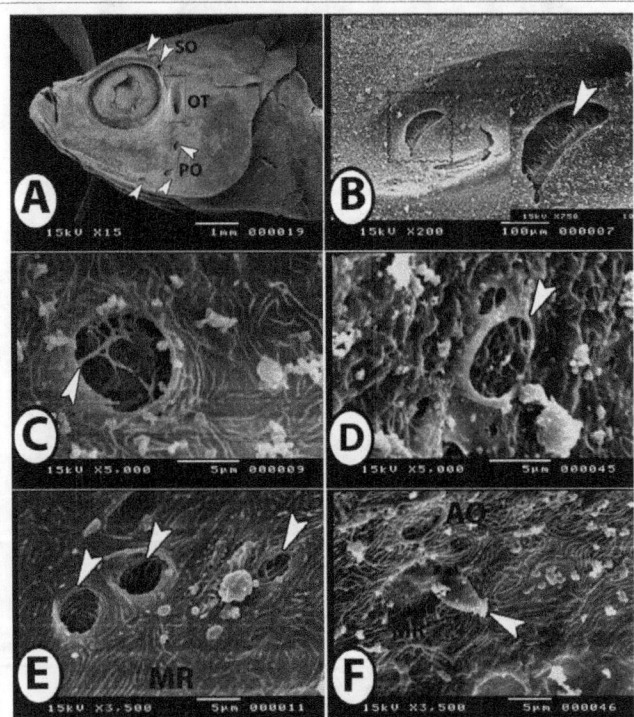

Figure 9: SEM of lateral line system of guppy. **A:** The lateral line branch in the head into supraorbital (SO), otic (square) and preopercular (PO). **B:** Higher magnification of the square in Figure A showing ampullary organ (square, inserted figure) in a pit in the skin of the head that formed of receptor cells (arrowhead). **C:** Surface view of ampullary organ with attached cupula (arrowhead). **D:** Lateral view of ampullary organ (arrowhead) with alveolar-like appearance of the organ. **E:** Different size and shapes of ampullary pores (arrowheads). Note presence of fingerprint-like microridges (MR) of the surrounding epidermal cells. **F:** Superficial neuromast with attached cupula (arrowhead) that occur near ampullary organ (AO) among microridges (MR) of epidermal cells.

the guppy, the mucous cells condensed in anterior region and tail and decreased in trunk. The condensation of mucoid cells in the anterior body regions was very important in lubrication and protection of fish against abrasive injuries during searching of the foods from bottom. Furthermore, the mucus cells of both species were PAS-AB positive as the sulphate groups provide an acidification of glycoproteins, which was effective to prevent bacterial and viral invasion [12].

In the guppy, the mucous cells were numerous and are well developed, club cells are either rare or absent, whereas in red- tail shark the mucous cells were smaller in number and the club cells are numerous and well developed. Singh and Mittal [7] suggested that the low density of mucous cells is compensated by the high density of club cells as an effective defense mechanism. The club cells of red- tail shark secret protinous substances as they stained positive with bromophenol blue. Other authors believed that the club cells are related to production, storage and release of the alarm substance, leading to alarm reaction in phylogenetically close species [13], or suggested a phagocytic function [14]. Chondroitin and keratin were also found in cytoplasm of club cells of some fish, suggesting a healing function, thus helping on repair of damaged tissue [15]. Sacciform or serous goblet cells were encountered in the epidermis of red tail shark and were positive to protein stains. Sacciform and club cells have been interoperated recently as a storehouse of biologically active substances and have also been associated with specific functions such as defense [16].

Eosinophilic granular cells were demonstrated in the snout of guppy. The marked histological and biochemical similarity between fish eosinophils and mammalian mast cells has been noted by Powell, et al. [17] and suggested that they release toxic proteins and oxygen radicals onto the body surface of multicellular parasites in areas of inflammation.

Melanocytes increased in tail, lips, dorsum of the head in both species. Toledo and Jared [18] working on amphibian skin, suggested that the dermal chromatephore units provide patterns of coloration and may also function to absorb or reflect radiations (thereby contributing to regulation of body temperature). Their roles in coloration has been attributed to the presence of large numbers of a variety of pigment-cells types (*chromatophores*) that are present at different levels within the dermis, which include *melanophores, xanthophores, erythrophores* and *iridophores*. *Melanophores* are the most common class of fish pigment cells. The pigmented material of *melanophores*, called melanin, is deep brown in color. The *melanophores* are often star-shaped, of *neuroectodermal* origin. The coloration in fishes performs adaptive functions and is useful to the animal in a variety of ways such as camouflage, aggressive purpose and courting patterns [19].

Rodlet cells were encountered in the epidermis of the head region and dermis of red tail shark. Rodlet cells are commonly observed in tissues and organs of fish include kidney, gills, heart and gut [20]. The accurate function of these cells is unknown. Vickers [21] thought that they may be modified goblet cells. While Mattey, et al. [22], supposed a regulatory role of these cells in ion transportation and osmoregulation. These may also considered as non-specific immune cells, involved in immunity as their number is increased in parasitic infection [23]. Doaa and Hanan [24] supposed a secretory function to these cells with a holocrine mode of secretion.

The thickness of the epidermis is depending mainly on the number and size of the layers of the cells that constitute the epidermis at different body regions. In the present study, the lips and tail of both species have the highest epidermal thickness. SEM revealed presence of well organized fingerprint-like microridges of the surface of epidermal cells in both species that may provide structural integrity to the epithelium and increase the surface area as well as providing mechanical flexibility to the wall during distortion or stretching, retaining the mucus, facilitate the movement of materials over the surfaces and prevent physical abrasions. These observations confirm the finding of [25]. Taste buds are chemoreceptive organs and are characteristic features of fish skin and assist them in the recognition and location of food. Fishes are characterized by presence of widely distributed taste buds over various parts of the body include snout, lips, barbels, skin, gill rakers, oropharyngeal cavity and oesophagus of many fish species [26,27]. The present study revealed that in red tail shark, they were frequently observed in lips, operculum, head and trunk region and in guppy, they were found in the head, operculum and trunk region.

The lateral line is sensory system that allows fish to sense objects and motion in their local environment [28]. The lateral line system divided into 2 subsystems; mechanoreceptive neuromasts and electroreceptive ampullary and tuberous organs. 2 types of neuromast present in both species; superficial and canal neuromasts. Superficial one is sit on the skin surface and in direct contact with water, may present as isolated elements or found in rows that follow the path of lateral line system, while canal or deep neuromast occur under the skin within canals that contact the surrounding water via a series of small pores in the skin. Superficial neuromasts contain two cell types, hair cells or sensory cells and supporting cells [29]. Hair cells are basic transducers for sound, vibration, and in determining position in vertebrates [30]. The support cells encircle the hair cells and secrete a gelatinous-like material or "cupula" which covers the whole neuromast. The cupula enables the organ to communicate with the exterior [31] and it is involved in perception of the local water movements [32]. The results revealed that the canal neuromast of guppies were surrounded by ossois coat, while that of red tail shark were supported by fibrous coat and this may be adaptive mean against squeezing the contents of the canal by high water pressures in the bottom. Canal neuromasts of guppies contained AB positive mucous cells and this agreed with findings of Wark and Peiche [28] as these canals are filled with fluid (water and mucous secretion), which allow transmission of vibrations to the neuromasts through the skin pores. Several types of lateral line organs may thus exist in the same fish.

Tuberous organ was observed in red tail shark. It also encountered in some fish species as gymnotid [33] that play a significant role in sensation of weak electrical stimuli. They were consisted of sensory cells and covered by flattened cells to prevent the current from passing to the organ. Ampullary organs (or called electroreceptors) (formerly considered as neuromast) are found in the head region of guppy and also found in some fish and represent a morphologically and physiologically heterogeneous group of lateral line tonic receptors specialized in detecting the electromagnetic fields as well as temperature gradients [34]. The presence of the neutral mucosubstances that fill the ampulla along with the acid mucosubstances secreted by abundant mucous cells of the epidermis suggest that it may modulate the intensity of signals from receptor cells and/or give some protection and these materials have excellent electric conductivity. Several different biological functions of the ampullary electrosense have been proposed, including prey detection [35], detection of predators [36], social communication [36], detection of mates [37]. Moreover, the majority of morphological studies have assessed ampullary morphology via cross-section, while the use of scanning electron microscopy has been used very little. Ampullary organs have been classified into two different types based

on the size and the length of the canals [38]. Wobbegong sharks possess 'macro-ampullae', which are characterised by large, visible pores and long canals of up to several centimeters in length that are common in marine environment. Whereas the 'mini-ampullae' of freshwater rays, with their microscopic size and short canals, are thought to be an adaptation to low conductivity within the freshwater environment [39]. SEM revealed alveolar like appearance of the ampullary organ of guppies. Within the different ampullary groups, variations in alveolar morphology exist and have been classified into five types based on alveolar arrangement: single-alveolate, multi-alveolate, branched alveolate, centrum cap and club-shaped. Ampullary organ differed in structures from species to species and this difference resulted from different environmental conductivity. The morphological analysis and a comparison of the different distribution patterns are used to infer how electroreception may participate in natural feeding behaviour, as well as possible adaptations of electroreception to the respective habitat types of each species.

References

1. Hawkes JW (1974) The structure of fish skin I. general organization. Cell Tissue Res 149: 147-158.

2. Gona O (1979) Mucous glycoprotein of teleostean fish. A comparative histochemical study. Histochemistery 11: 709-718.

3. Damasceno EM, Monteiro JC, Duboc LF, Dolder H, Mancini K (2012) Morphology of the Epidermis of the Neotropical Catfish Pimelodella lateristriga (Lichtenstein, 1823) with Emphasis in Club Cells. Plos One 7: 1-7.

4. Takeuchi IK (1967) Electron microscopy of two types of reflecting chromatophores (iridophores and leucophores) in the guppy, Lebistes reticulatus Peters. Cell and tissue research 173: 17-27.

5. Jakubowski M (1974) Structure of lateral system and related bones in Bery-coid fish (hoplostetus mideteraneous L.). Acta Anatomica 87: 261- 274.

6. Imaki H, Chavin W (1984) Ultrastructure of mucous cells in the sarcopterygian integument. Scanning Electron Microscopy 1: 409-422.

7. Singh SK, Mittal AK (1990) A comparative study of the epidermis of the common carp and the three Indian major carp. J Fish Biology 36: 9-19.

8. Yang JX, Winterbottom R (1998) Phylogeny and zoogeography of the cyprinid genus Epalzeorhynchos Bleeker (Cyprinidae: Ostariophysi). J Copeia 1: 48-63.

9. Anne ME, Dawn PAT (2001) Evolutionary implications of large-scale patterns in the ecology of Trinidadian guppies, Poecilia reticulata. Biol J Linn Society 73: 1-9.

10. Bancroft JD, Gamble M (2002) Theory and Practice of Histological and Histochemical Techniques.

11. Gattas SM, Yanai T (2010) Light microscopical study on the skin of European eel (Anguilla Anguilla). World J Fish Marine Sci 2: 152-161.

12. Mittal AK, Ueda T, Fujimori O, Yamada K (1994) Histochemical analysis of glycoproteins in the unicellular glands in the epidermis of an Indian fresh water fish Mastacembelus panculus (Hamilton). Histochemistry J 26: 666-677.

13. Smith RJF (1992) Alarm signals in fishes. Rev Fish Biology and Fish 2: 33-63.

14. Lufty RG (1964) Studies on the epidermis of the catfish Synodontis schall. Sch Ain Shams Sci Bull Cairo 10: 153-163.

15. Iger Y, Abraham M (1990) The process of skin healing in experimentally wounded carp. J Fish Biology 36: 421-437.

16. Yashpal M, Mittal AK (2014) Serous goblet cells: The protein secreting cells in the oral cavity of a catfish, Rita rita (Hamilton, 1822) (Bagridae, Siluriformes). Tissue and cell 46: 9-14

17. Powell MD, Wright GM, Burka JF (1993) Morphological and distributional changes in the eosinophilic granular cells (EGC) population of the rainbow trout (Oncorhynchus mykiss Walbaum) intestine following systemic administration of capsacin and substance P. J Exp Zool 266: 19-30.

18. Toledo RC, Jared C (1993) Cutaneous adaptations to water balance in

amphibians. Comp. Biochem. Physiology 105: 593-608.

19. Kottler VA, Koch I, Flo¨tenmeyer M, Hashimoto H, Weigel D (2014) Multiple Pigment Cell Types Contribute to the Black, Blue, and Orange Ornaments of Male Guppies (Poecilia reticulata). PLoS One 9: 85647.

20. Manera M, Dezfuli BS (2004) Rodlet cells in teleosts: a new insight into their nature and functions. Fish Biology 65: 597-619.

21. Vickers T (1962) A study of the intestinal epithelium of the goldfish Carassius auratus: its normal structure, the dynamics of cell replacement and the changes induced by salts of cobalt and manganese. Q J Microsc Sci 103: 93-110.

22. Mattey DL, Morgan M, Wright DE (1979) Distribution and development of rodlet cells in the gills and pseudobranch of the bass, Dicentrarchus labrax (L). J Fish Biology 15: 363-370.

23. Reite OB (2005) The rodlet cells of teleostean fish: their potential role in host defense in relation to the mast cells/eosinophilic granule cells. Fish Shellfish Immunolgy 19: 253-267.

24. Mokhtar DM, Abd-Elhafeez HH (2014) Light- and electron-microscopic studies of olfactory organ of Red-tail shark, Epalzeorhynchos bicolor (Teleostei: Cyprinidae). J Micr Ultras 2: 182-195

25. Gamal AM, Elsheikh EH, Nasr ES (2012) Morphological adaptation of the buccal cavity in relation to feeding habits of the omnivorous fish Clarias gariepinus: A scanning electron microscopic study. Basic Appl zool 65: 191-1980

26. Harvey R, Batty RS (1998) Cutaneous taste buds in cod. Fish Biology 53: 138-149.

27. Enas A Abd El Hafez, Mokhtar DM, Abou-Elhamd AS, Hassan AHS (2013) Comparative histomorphological studies on oesophagus of catfish and grass carp.

28. Wark AR, Peiche CL (2010) Lateral line diversity among ecologically divergent three spine stickleback Populations. J Expl Biol 213: 108-117

29. Ghysen A, Dambly-Chaudie`re C (2004) Development of the zebrafish lateral line. Curr Opin Neurobiology 14: 67-73.

30. Kornblum HI, Corwin JT, Trevarrow B (1990) Selective labeling of sensory hair cells and neurons in auditory, vestibular, and lateral line systems by a monoclonal antibody. J Comparative Neurology 301: 162-170.

31. Cernuda-Cernuda R, Garcia-Fernandez JM (1996) Structural diversity of the ordinary and specialized lateral line organs. Micr Res Tech 34: 302-312.

32. Mukai Y, Yoshikawa H, Koboyashi H (1994) The relationship between the length of the cupulae of free neuromasts and feeding ability in larvae of the willow shiner Gnathopogon elongatus caerulescens (Teleostei. Cyprinidae). J Experimental Biology 197: 399-403.

33. Wachtel AW, Bruce- Szamier R (1966) Special cutaneous receptor organs of fish: The tuberous organs of Eigenmannia. J Morphology 119: 51-80.

34. Whitehead DL, Tibbetts IR, Daddow LYM (2003) Microampullary Organs of a Freshwater Eel-Tailed Catfish, Plotosus (tandanus) tandanus. J Morphology 255: 253-260.

35. Kalmijn AJ (1971) The electric senses of sharks and rays. J Experimental Biology 55: 371-383.

36. Sisneros JA, Tricas TC, Luer CA (1998) Response properties and biological function of the skate electrosensory system during ontogeny. J Comparative Physiology A 183: 87-99

37. Tricas TC, Michael SW, Sisneros JA (1995) Electrosensory optimization to conspecific phasic signals for mating. Neuroscience Lett 202: 129-132

38. Andres KH, Van-During M (1988) Comparative anatomy of vertebrate electroreceptors. Prog Brain Res 74: 113-131.

39. Szabo T, Kalmijn AJ, Enger PS, Bullock TH (1972) Microampullary organs and a submandibular sense organ in the freshwater ray, Potamotrygon. J Comparative Physiology 79: 15-27

9

Does Tobacco (*Nicotiana tabacum*) Leaf Dust Save the Life of Rohu (*Labeo rohita*) Fingerlings During Transport?

Dinesh R[1]*, Chandra Prakash[1], Chadha NK[1], Nalini Poojary[2] and Sherry Abraham[1]

[1]Division of Aquaculture, Central Institute of Fisheries Education, Mumbai, India
[2]Division of Aquatic Environment and Health Management, Central Institute of Fisheries Education, Mumbai, India

Abstract

Fishes undergo a multi-phase of stress due to multiple stressors involved during transportation which is an inexorable and indispensable operation in aquaculture. The aim of this study is to promote a novel, inexpensive, active and eco-friendly sedative to replace expensive and toxic sedatives used in aquaculture. With this intention, we investigated the efficacy of tobacco leaf dust as a sedative for the transport of rohu (*Labeo rohita*) fingerlings. The experiment of sedative efficacy and simulated transportation was conducted for 12 h in glass tanks (30 L capacity) and plastic bags (75 cm length × 45 cm wide), respectively with different concentrations of tobacco leaf dust such as 0 ppm, 25 ppm, 50 ppm, 75 ppm, 100 ppm and 125 ppm among which 0 ppm was used as control. The fingerlings (6.45 ± 0.68 cm and 3.29 ± 0.52 g) were stocked at a stocking density of 10 fishes/ tank and 30 fishes/ plastic bag in triplicates. The induction and recovery times observed in the anesthetic bath significantly ($p < 0.05$) decreased and increased with increase in the concentrations of tobacco leaf dust. The lowest effective dose found to produce induction (≤ 15 min) and recovery (≤ 5 min) was 25 ppm and the same was effective in inducing light sedation in rohu during the behavioral response observation. Mortality rate (15% to 40%) of fingerlings during transportation was significantly higher in control (without sedative) than the sedative doses of tobacco. Also, poor water quality was noticed in control group with the serious changes in hemogram and leukogram of fingerlings. The experimental results revealed the efficacy of tobacco in minimizing the metabolic activity of the fishes and thereby reducing the water quality deterioration and stress during transportation. Therefore, the present study reveals that tobacco leaf dust (25 ppm) could be a futuristic sedative for safe and successful transportation of *L. rohita* fingerlings.

Keywords: Rohu; Fish transportation; Sedative; Tobacco; Water quality parameters; Hematological values

Introduction

Owing to its richness in fish biodiversity that has been distributed widely among different freshwater ecosystems, India is considered to be one of the affluent nations in the world. Aquaculture in India is highly promising and has grown over the last two decades with freshwater aquaculture contributing over 95% of the total aquaculture production. The country shared 4.39 million tons to world freshwater fish production in 2014 [1]. The aquaculture systems in India and its neighboring countries mainly constitute Indian major carps namely Catla (*Catla catla*), Rohu (*Labeo rohita*) and Mrigal (*Cirrhinus mrigala*). Among the three Indian major carps, rohu is the species of significance preferred in poly-culture systems of carps because of its high growth potential. It is a fast-growing freshwater fish species which belongs to the family Cyprinidae and genus, Labeo and is an important species of great demand by the consumers due to its texture and taste of the carcass.

To maximize and sustain the production, a healthy and quality seed is an essential input. Fish seeds are produced in hatcheries and are supplied to the farms for grow-out culture. But many times the hatcheries are located far away from the culture sites. Therefore, transportation of fish is one of the most significant operations in aquaculture which plays a crucial role in the supply of seeds from hatcheries to grow out farms. The duration of transport varies depending on the distance of farms to which the fishes are transported and is divided into short term ($\leq$ 8 h) or long term ($\geq$ 8 h) with the threshold of 8 h as the distinction between short and long transport [2]. During transport, the metabolic activities of fish are three times more than that of the normal [3]. Stress is an indispensable factor during the transportation of fishes. Rohu is a fish which is highly stressed as it is quite sensitive to the transport stress. Consequently, there is a high mortality rate (about 90%) among transported rohu fingerlings [4], since packing those at high densities get resulted in confinement [5]. Transport duration and the physico-chemical parameters of the transport media determine the severity of transport stress in fishes but it can be minimized with the help of light sedation, i.e., low concentration of an anesthetic. Therefore, in order to reduce the stress-induced mortality, natural or synthetic sedatives and anesthetics can be used to sedate and immobilize fish [6,7].

Sedation is a mild form of anesthesia which puts the fish to sleep by calming and immobilizing their activities. The application of sedatives silences the activity of fishes and is beneficial in lowering the stress-induced mortality during transportation [8]. Sedative and anesthetic substances used for fishes must have rapid induction and recuperation period, safe margin to fish and humans, no physiological and residual effect, local availability and low cost [9]. Several anesthetics have been evaluated and used so far in which some are toxic and inexpensive while some are expensive and less toxic. As of now, MS-222 is the only anesthetic that has a USFDA approval to be used as an anesthetic in food fishes irrespective of its demerits like low pH, low efficacy on plasma cortisol control and expensiveness [10]. To avoid adverse stress effects from anesthetics, knowledge on the optimum concentration of

---

*Corresponding author: Dinesh R, Division of Aquaculture, Central Institute of Fisheries Education, Mumbai-400061, India
E-mail: dinesh.albe@gmail.com

an anesthetic for different fish species is more significant [11,12]. For practical use in aquaculture, only a few studies have been conducted in commonly cultured tropical freshwater food fishes.

Tobacco, *Nicotiana tabacum*, is an herbaceous annual or perennial plant in the family Solanaceae (Night shade), grown for its leaves. It is a medicinal plant of remarkable benefits and history of use in traditional Indian medicine as a sedative, antispasmodic, and vermifuge with high alkaloid content mainly nicotine and it is under the utmost care of humans [13]. It has a potential to heal and protect when used effectively but has the ability to harm when abused. India stands 3rd in tobacco production globally [14]. The expensive anesthetics are scarce in developing and under developed countries whereas inexpensive anesthetics are toxic and harmful to fish and humans with more deleterious effects. In search of safe, effective and less expensive sedatives and anaesthetics, tobacco which is of natural origin, cheap, easily available and eco-friendly narcotic could be a novel and futuristic sedative and anaesthetic for fishes during transport and surgeries.

No literature on use of tobacco as sedative for fish transport is available and it appears that experimental studies on this subject are rare. The change in water quality parameters such as temperature, pH, dissolved oxygen (DO), dissolved free carbon dioxide (CO_2), ammonia (NH_3) and nitrite (NO_2) is one of the prime factors in causing mortality among the transported fishes [15]. Also, the commercial use of sedative could be enhanced by perceiving the water quality changes occurring when fish are under its influence. Hematological parameters such as total erythrocyte count (TEC), hemoglobin concentration (Hb), hematocrit value (Hct), erythrocyte indices, total leukocyte count (TLC) and differential leucocyte count provide detailed information on general metabolism and physiological status of fish which are usually affected by transportation [16]. Hoseini et al. [11] also demonstrated that blood parameters particularly stress indicators are affected by anesthetic concentrations and their exposure period. This study therefore, determined the efficacy of tobacco leaf dust as a feasible sedative in *L. rohita* and its effect on water quality and hematological parameters of the fish during transport.

Material and Methods

Animal housing and design

1000 nos. of *L. rohita* fingerlings of 6.46 ± 0.68 cm mean length and 3.29 ± 0.52 g mean weight were obtained from Maharashtra State Fish Seed Farm, Aarey, Goregaon, Mumbai and the same were used for the experiment. The fishes were disinfected with 5 ppm $KMnO_4$ and were acclimatized in 300 L capacity FRP tanks at the wet laboratory unit, Division of Aquaculture, ICAR- Central Institute of Fisheries Education, Andheri (West), Mumbai, India. The fishes were kept under natural photoperiod and fed twice a day (0900 h and 1700 h) with commercial floating type pellet diet. A completely randomized design was used to carry out the experiment. The experiment was conducted in triplicates for both the transit times of five treatments and for the control as well.

Tobacco leaf dust preparation

Good quality tobacco leaves were procured from a retailer in Kerala. The authentication of the specimen was accomplished in Department of Botany, St. Xavier's College, Mumbai, Maharashtra (India). The leaves were sun-dried for 7 days and were ground into fine dust (powder) with the help of mixer. The fine ground tobacco leaf dust (7% to 10% moisture content) was then stored in an air-tight container and used for the experiment.

Experimental setup

Sedation dose for transportation: Five concentrations of tobacco leaf dust (25 ppm, 50 ppm, 75 ppm, 100 ppm and 125 ppm) as predetermined earlier from lethal toxicity studies were tested for sedative potential on *L. rohita* fingerlings. The known quantities of tobacco leaf dust were weighed according to each treatment dose and were added to glass tanks with fresh water. The glass tanks with known amounts of tobacco leaf dust were agitated and mixed vigorously. To evaluate the time required for induction, 10 fishes, each of which were placed in individual aquaria, were used for each concentration tested, and each fish was used only once. The maximum observation time was 30 min. The induction and recovery times were studied based on the stages described by Keene et al. [17] and were calculated using a stopwatch. The behavioral response of fingerlings was monitored and noted down for every 2 h interval up to 12 h and simultaneously the survival rate was also checked. The fingerlings were further monitored in recovery tanks for another 24 h to check the survival rate.

Protocol for simulated transportation: Prior to the transportation experiment, the fishes were starved for 24 h in FRP tanks. Five concentrations of tobacco leaf dust were followed (25 ppm, 50 ppm, 75 ppm, 100 ppm and 125 ppm) with two transportation times (6 h, 12 h). Only freshwater was added in control without sedatives (0 ppm). The plastic bags (75 cm length × 45 cm wide) were filled with 2 L freshwater and were mixed with each sedation dose in triplicates. 30 fingerlings were stocked (approximately biomass weight of 100 g) per plastic bag. Medical-grade oxygen was filled inside the plastic bag after the squeezing of air from the plastic bag. Water quality was maintained under controlled conditions during the start of the experiment. 12 h was maintained as the total duration of stress. After 6 h and at the termination of the transit time (12 h), all the plastic bags were opened and mortality rate was checked. The fingerlings thus survived were taken from each bag and then were transferred to FRP tanks of 300 L capacity with ample aeration and was further monitored for 24 h to evaluate post-transport mortality. Stress was studied by analyzing the water quality and hematological parameters.

Water quality parameters

All the bags from each treatment group were examined to assess water quality at 6 h intervals until the end of the experiment. During each sampling, the bags were opened, dead fingerlings were counted and water samples were collected for analysis of different parameters. Temperature and pH was measured using digital thermometer (Fisher Scientific) and pH tester (Eutech Instruments) respectively, while DO (Winkler's method), CO_2, total alkalinity and total hardness were measured by titrimetric methods [18]. Total ammonia nitrogen (TAN) (Indophenol blue method), nitrite-nitrogen (NO_2-N) and nitrate-nitrogen (NO_3-N) were measured following standard methods [18].

Hematological parameters

Blood from the fishes were drawn with the help of a sterilized 1 ml hypodermal syringe and 28 gauge needles directly from the caudal vein containing 2.7% EDTA (Qualigens, India) as an anticoagulant. TEC and TLC were determined using Neuber's hemocytometer (Feinoptik, Germany) with appropriate diluting fluids (Himedia, India) for TEC and TLC. The Hb concentration was analyzed following the cyanmethemoglobin method using Drabkins fluid (Qualigens, India). Hematocrit capillary tubes were two-third filled with the whole blood and centrifuged in a hematocrit centrifuge at 12000 rpm for 5 min and the percentage of the packed cell volume was determined by the

Treatment (ppm)	Induction (min)			Recovery (min)
	Stage 1	Stage 2	Stage 3	Stage 6
25	14.36 ± 0.03[a]	-	-	0.57 ± 0.02[e]
50	13.70 ± 0.08[b]	28.51 ± 0.11[a]	-	0.88 ± 0.01[d]
75	11.76 ± 0.04[c]	20.48 ± 0.11[b]	29.82 ± 0.03[a]	1.23 ± 0.01[c]
100	8.18 ± 0.06[d]	12.50 ± 0.04[c]	19.60 ± 0.17[b]	1.75 ± 0.04[b]
125	6.17 ± 0.03[e]	9.50 ± 0.04[d]	15.30 ± 0.11[c]	2.26 ± 0.02[a]

Values in the same column with different superscripts are significantly (P<0.05) different for each stages of anaesthesia and tobacco leaf dust concentration. One-way ANOVA was used following Tukey's HSD post hoc test in SPSS 16.0.

Table 1: Time (min) required for induction and recovery (stage 6) (Mean ± SE) of the anaesthesia using tobacco leaf dust in rohu fingerlings (Maximum observation time - 30 min).

Treatment (ppm)	Behavioral responses						Survival Rate (%)
	2 h	4 h	6 h	8 h	10 h	12 h	
Control	N	N	N	N	N	N	100
25	LS	LS	LS	LS	LS	LS	100
50	DS	DS	DS + 5% PLE	DS + 5% PLE	DS + 5% PLE	DS + 5% PLE	100
75	DS + 20% PLE	DS + 20% PLE	DS + 30% PLE	DS + 30% PLE	DS + 30% PLE	DS + 40% PLE	100
100	DS + 30% PLE	DS + 50% PLE	DS + 50% PLE	70% PLE	80% PLE	80% PLE	100
125	PLE	PLE	PLE	PLE	PLE	PLE	100

N: Normal; LS: Light Sedation; DS: Deep Sedation; PLE: Partial Loss of Equilibrium (Hyperactive phase).

Table 2: Behavioral responses of *rohu* fingerlings to different tobacco leaf dust concentrations at 2 h interval during 12 h exposure period in anaesthetic bath.

Treatment (ppm)	Mortality rate (%)		
	6 h	12 h	24 h
Control	20.83 ± 1.14[a]	38.33 ± 1.52[a]	15.83 ± 0.59[a]
25	0.00 ± 0.00[b]	0.00 ± 0.00[b]	0.00 ± 0.00[b]
50	0.00 ± 0.00[b]	0.00 ± 0.00[b]	0.00 ± 0.00[b]
75	0.00 ± 0.00[b]	0.00 ± 0.00[b]	0.00 ± 0.00[b]
100	0.00 ± 0.00[b]	0.00 ± 0.00[b]	0.00 ± 0.00[b]
125	1.66 ± 0.56[b]	3.33 ± 0.95[b]	0.00 ± 0.00[b]

Values in the same column with different superscripts are significantly (P<0.05) different. One-way ANOVA was used following Tukey's HSD post hoc test in SPSS 16.0.

Table 3: Mortality rate (%) (Mean ± SE) of rohu fingerlings found in polyethylene bags at different durations during simulated transportation experiment at 50 g L^{-1} loading density.

hematocrit tube reader [19]. Erythrocyte indices including mean corpuscular volume (MCV), mean corpuscular hemoglobin (MCH) and mean corpuscular hemoglobin concentration (MCHC) were calculated according to Dacie and Lewis [20]. Differential leukocyte count (Lymphocytes and Neutrophils) was performed with blood smears stained with Giemsa solution. The smears were examined by light microscopy (Olympus, Tokyo, Japan) under oil immersion at 100 x magnification.

Statistical analysis

All the data were represented as Mean ± SEM (Standard error of mean). The treatment means for induction and recovery stages, mortality rate, water quality and blood parameters were compared using one-way ANOVA followed by Tukey's HSD post hoc for multiple comparisons. Data were analyzed using statistical software SPSS version 16.0 with a level of significance of P<0.05.

Results

Induction and recovery, behavioral response and mortality rate percentage

The induction and recovery times (Table 1) observed in anesthetic bath significantly decreased (p<0.05) with increase in concentrations among all treatments. Treatment group with 25 ppm tobacco showed

slight delay in induction but a rapid recovery whereas 125 ppm showed quick induction and prolonged recovery times. While observing the behavioral responses (Table 2) of rohu fingerlings in anesthetic bath, treatment with 25 ppm concentration showed light sedation whereas treatment with 125 ppm concentration expressed a phase of excitement and hyperactivity. Control (without sedative) showed normal swimming patterns while intermediate responses were found in other treatments. Mortality rate (%) (Table 3) observed during simulated transportation and post transportation was significantly higher (p<0.05) in control group. No mortality was noticed among the treatments with sedative doses except in treatment with 125 ppm concentration that showed low mortality rates.

Water quality parameters

The water quality parameters of transport water showed a varying trend throughout the simulated transportation period. Each parameter is represented in Table 4 comprising all the treatments and their values of the respective parameter at different time intervals. The water quality deterioration was highly pronounced in control and to a lesser extent in other treatment groups. Temperature of transport water was significantly higher (p<0.05) in control at both the durations (6 h and 12 h) followed by 125 ppm treatment group. pH levels observed at different time intervals showed a significant difference (p<0.05) with control showing the lowest pH and 25 ppm treatment showing the highest pH respectively. There was a significant reduction (p<0.05) in DO levels in control during the simulated transportation while high DO levels were found in all treatments. During the transportation experiment, the increase in CO_2 level was found in all treatments with a significant increase (p<0.05) in control group. Low carbon dioxide levels were found in 25 ppm compared to other treatments. Control showed higher alkalinity values and were significantly different (p<0.05) from other treatments. Water hardness was found high in 100 and 125 ppm treatments and were significantly different (p<0.05) from other treatments. Low levels of ammonia and nitrite were recorded in 25 ppm with significantly (p<0.05) higher amounts of control. High nitrate levels were exhibited by 125 ppm and were significantly different (p<0.05) while low nitrate values were observed in control group.

Concentration (ppm)	Water quality parameters								
	Temperature (°C)	pH	DO (ppm)	CO_2 (ppm)	Alkalinity (ppm)	Hardness (ppm)	Ammonia (ppm)	Nitrite (ppm)	Nitrate (ppm)
0 h transport									
Control	27.35 ± 0.20[a]	7.92 ± 0.02[a]	16.3 ± 0.04[a]	00 ± 0.0[a]	73 ± 1.0[a]	52 ± 0.20[a]	0.25 ± 0.03[a]	0.004 ± 0.004[a]	0.12 ± 0.03[a]
25	27.35 ± 0.10[a]	7.90 ± 0.04[a]	16.2 ± 0.05[a]	00 ± 0.0[a]	72 ± 1.0[a]	52 ± 0.10[a]	0.25 ± 0.02[a]	0.004 ± 0.002[a]	0.12 ± 0.01[a]
50	27.35 ± 0.30[a]	7.94 ± 0.02[a]	16.3 ± 0.02[a]	00 ± 0.0[a]	72 ± 2.0[a]	54 ± 0.10[a]	0.25 ± 0.02[a]	0.004 ± 0.001[a]	0.12 ± 0.00[a]
75	27.35 ± 0.05[a]	7.92 ± 0.05[a]	16.2 ± 0.15[a]	00 ± 0.0[a]	72 ± 1.0[a]	54 ± 0.50[a]	0.25 ± 0.01[a]	0.004 ± 0.002[a]	0.13 ± 0.02[a]
100	27.33 ± 0.20[a]	7.94 ± 0.04[a]	16.4 ± 0.01[a]	00 ± 0.0[a]	74 ± 1.0[a]	52 ± 0.20[a]	0.25 ± 0.01[a]	0.005 ± 0.001[a]	0.13 ± 0.01[a]
125	27.34 ± 0.10[a]	7.94 ± 0.01[a]	16.3 ± 0.05[a]	00 ± 0.0[a]	72 ± 2.0[a]	52 ± 0.30[a]	0.25 ± 0.02[a]	0.005 ± 0.002[a]	0.13 ± 0.02[a]
6 h transport									
Control	28.75 ± 0.05[a]	7.32 ± 0.01[c]	9.1 ± 0.30[b]	18 ± 0.0[a]	114 ± 4.0[a]	52 ± 0.10[a]	0.76 ± 0.01[a]	0.066 ± 0.003[a]	0.32 ± 0.02[d]
25	28.15 ± 0.05[b]	7.48 ± 0.02[a]	13.5 ± 0.10[a]	12 ± 0.0[c]	89 ± 1.0[b]	52 ± 0.50[b]	0.27 ± 0.00[d]	0.004 ± 0.002[c]	0.45 ± 0.02[c]
50	28.30 ± 0.10[b]	7.45 ± 0.01[ab]	13.2 ± 0.10[a]	14 ± 0.0[bc]	94 ± 2.0[b]	57 ± 0.65[ab]	0.37 ± 0.00[c]	0.009 ± 0.002[bc]	0.65 ± 0.00[b]
75	28.35 ± 0.05[b]	7.44 ± 0.01[ab]	13.1 ± 0.10[a]	15 ± 1.0[b]	97 ± 1.0[b]	62 ± 0.30[a]	0.43 ± 0.01[bc]	0.011 ± 0.002[bc]	0.68 ± 0.00[b]
100	28.40 ± 0.10[ab]	7.43 ± 0.01[ab]	13.0 ± 0.20[a]	16 ± 0.0[ab]	97 ± 3.0[b]	62 ± 0.60[a]	0.47 ± 0.01[b]	0.016 ± 0.001[bc]	0.73 ± 0.01[ab]
125	28.45 ± 0.05[ab]	7.40 ± 0.02[b]	12.8 ± 0.20[a]	16 ± 0.0[ab]	97 ± 3.0[b]	63 ± 0.20[a]	0.50 ± 0.01[b]	0.018 ± 0.001[b]	0.79 ± 0.03[a]
12 h transport									
Control	29.75 ± 0.05[a]	6.91 ± 0.01[d]	5.4 ± 0.20[c]	32 ± 2.0[a]	139 ± 1.0[a]	52 ± 0.20[b]	0.92 ± 0.02[a]	0.082 ± 0.002[a]	0.47 ± 0.00[d]
25	28.75 ± 0.05[d]	7.28 ± 0.04[a]	12.9 ± 0.10[a]	14 ± 0.0[c]	101 ± 3.0[d]	52 ± 0.20[b]	0.37 ± 0.01[c]	0.004 ± 0.001[d]	0.54 ± 0.01[d]
50	28.85 ± 0.05[cd]	7.23 ± 0.01[b]	12.3 ± 0.10[ab]	16 ± 0.0[bc]	107 ± 1.0[cd]	57 ± 0.25[b]	0.42 ± 0.00[bc]	0.009 ± 0.002[cd]	0.71 ± 0.01[c]
75	29.05 ± 0.05[bc]	7.17 ± 0.03[c]	11.8 ± 0.20[b]	19 ± 0.5[bc]	114 ± 4.0[bc]	66 ± 0.25[a]	0.51 ± 0.00[b]	0.013 ± 0.001[bc]	0.77 ± 0.01[bc]
100	29.20 ± 0.20[b]	7.15 ± 0.03[c]	11.3 ± 0.10[b]	21 ± 1.0[bc]	121 ± 1.0[b]	66 ± 0.30[a]	0.54 ± 0.01[b]	0.018 ± 0.002[b]	0.83 ± 0.01[ab]
125	29.25 ± 0.05[b]	7.14 ± 0.02[c]	11.3 ± 0.30[b]	22 ± 2.0[b]	124 ± 2.0[b]	68 ± 0.60[a]	0.55 ± 0.01[b]	0.020 ± 0.001[b]	0.90 ± 0.02[a]

Concentration (ppm)	Hematological parameters								
	TEC (10^6 mm^{-3})	Hb (g dL^{-1})	Hct (%)	MCV (fl)	MCH (pg)	MCHC (g dL^{-1})	TLC (10^3 mm^{-3})	Lymphocytes (%)	Neutrophils (%)
0 h transport									
Control	0.82 ± 0.01[a]	3.40 ± 0.15[a]	7.40 ± 0.10[a]	90.1 ± 0.4[a]	40.6 ± 0.1[a]	46.2 ± 0.6[a]	81.80 ± 0.60[a]	95.10 ± 0.35[a]	3.50 ± 0.05[a]
25	0.82 ± 0.01[a]	3.30 ± 0.10[a]	7.35 ± 0.10[a]	89.1 ± 0.5[a]	40.5 ± 0.0[a]	45.4 ± 0.2[a]	80.80 ± 0.40[a]	94.40 ± 0.05[a]	3.55 ± 0.15[a]
50	0.82 ± 0.01[a]	3.35 ± 0.15[a]	7.50 ± 0.05[a]	89.1 ± 0.4[a]	40.3 ± 0.2[a]	45.2 ± 0.2[a]	82.00 ± 1.00[a]	94.20 ± 0.50[a]	3.60 ± 0.15[a]
75	0.82 ± 0.02[a]	3.30 ± 0.10[a]	7.30 ± 0.20[a]	90.1 ± 0.4[a]	40.3 ± 0.2[a]	45.2 ± 0.4[a]	81.50 ± 0.60[a]	94.50 ± 0.10[a]	3.75 ± 0.05[a]
100	0.82 ± 0.00[a]	3.30 ± 0.20[a]	7.30 ± 0.10[a]	89.1 ± 0.6[a]	40.4 ± 0.2[a]	46.2 ± 0.2[a]	82.50 ± 0.40[a]	95.40 ± 0.10[a]	3.60 ± 0.20[a]
125	0.82 ± 0.02[a]	3.30 ± 0.15[a]	7.40 ± 0.10[a]	89.4 ± 0.5[a]	40.4 ± 0.2[a]	46.2 ± 0.3[a]	81.80 ± 0.60[a]	94.80 ± 0.15[a]	3.60 ± 0.15[a]
6 h transport									
Control	0.23 ± 0.01[b]	0.85 ± 0.05[d]	1.55 ± 0.15[c]	65.8 ± 1.2[d]	36.1 ± 0.1[d]	55.0 ± 1.1[a]	16.40 ± 1.10[c]	65.55 ± 2.95[c]	23.15 ± 0.65[a]
25	0.80 ± 0.01[a]	3.30 ± 0.10[c]	7.05 ± 0.15[b]	88.0 ± 0.7[c]	41.2 ± 0.2[c]	46.7 ± 0.4[b]	80.00 ± 1.00[b]	94.00 ± 0.40[a]	3.60 ± 0.10[d]
50	0.84 ± 0.02[a]	3.50 ± 0.10[bc]	7.65 ± 0.25[ab]	90.5 ± 0.3[c]	41.3 ± 0.4[c]	45.7 ± 0.2[b]	82.40 ± 0.60[b]	91.50 ± 0.60[ab]	5.05 ± 0.05[d]
75	0.85 ± 0.03[a]	3.60 ± 0.10[abc]	7.75 ± 0.35[ab]	91.1 ± 0.9[bc]	42.3 ± 0.3[bc]	46.4 ± 0.8[b]	85.05 ± 2.65[b]	89.35 ± 1.55[ab]	9.15 ± 0.25[c]
100	0.86 ± 0.00[a]	3.80 ± 0.15[ab]	8.40 ± 0.10[a]	97.1 ± 0.6[ab]	43.8 ± 0.4[ab]	45.2 ± 0.5[b]	95.25 ± 1.05[a]	86.55 ± 1.25[ab]	12.05 ± 0.85[b]
125	0.88 ± 0.01[a]	4.00 ± 0.10[a]	8.70 ± 0.10[a]	98.2 ± 0.5[a]	45.1 ± 0.6[a]	45.9 ± 0.6[b]	98.10 ± 0.20[a]	84.65 ± 0.85[b]	13.65 ± 0.15[b]
12 h transport									
Control	0.44 ± 0.03[c]	1.55 ± 0.05[e]	3.30 ± 0.20[d]	75.0 ± 0.6[c]	35.2 ± 0.8[d]	46.9 ± 0.7[a]	4.10 ± 2.50[d]	39.40 ± 0.90[d]	26.25 ± 0.55[a]
25	0.80 ±0.01[bc]	3.35 ± 0.15[d]	7.15 ± 0.25[c]	88.7 ± 1.4[b]	41.5 ± 0.4[c]	46.7 ± 0.4[a]	81.40 ± 1.40[c]	94.15 ± 0.75[a]	3.90 ± 0.20[d]
50	0.86 ±0.01[ab]	3.70 ± 0.20[cd]	7.95 ± 0.05[bc]	92.4 ± 0.5[b]	43.0 ± 0.5[bc]	46.5 ± 0.3[a]	86.20 ± 1.60[bc]	88.45 ± 0.65[ab]	7.00 ± 0.30[c]
75	0.90 ±0.02[ab]	3.95 ± 0.15[bc]	8.40 ± 0.10[b]	93.3 ± 1.0[b]	43.8 ± 0.7[abc]	46.9 ± 0.4[a]	91.50 ± 1.10[bc]	86.45 ± 1.35[abc]	9.80 ± 0.40[c]
100	0.93 ± 0.01[a]	4.35 ± 0.05[ab]	9.50 ± 0.20[a]	102 ± 1.0[a]	46.7 ± 0.0[ab]	45.7 ± 0.8[a]	104.90 ± 1.90[ab]	82.90 ± 2.70[bc]	13.40 ± 0.10[b]
125	0.94 ± 0.00[a]	4.55 ± 0.05[a]	9.80 ± 0.10[a]	103.6± 2.1[a]	48.1 ± 0.5[a]	46.4 ± 1.0[a]	118.40 ± 3.40[a]	80.15 ± 0.95[c]	14.75 ± 0.95[b]

Control without sedative (0 ppm), 25, 50, 75, 100 and 125 ppm = tobacco leaf dust concentrations in transport water during simulated transportation.
DO: Dissolved Oxygen, CO_2 = dissolved free carbon dioxide; TEC: Total Erythrocyte Count; Hb: Hemoglobin; Hct = Hematocrit; MCV: Mean Corpuscular Volume; MCH: Mean Corpuscular Hemoglobin; MCHC: Mean Corpuscular Hemoglobin Concentration (MCHC); TEC: Total Leucocyte Count
Values in the same column with different superscripts are significantly (P<0.05) different for each parameter. One-way ANOVA was used following Tukey's HSD post hoc test in SPSS 16.0.

Table 4: Water quality parameters and haematological parameters observed at different time intervals during simulated transportation of rohu fingerlings.

Hematological parameters

Blood parameters (Table 4) showed a drastic variation between control and treatments and among treatments as well. The poor hematologic profile was observed in control group fishes at different time intervals during simulated transportation. There was a high reduction in TEC of control group and were significantly different (p<0.05) from other treatments. TEC increased insignificantly (p>0.05) in all other treatments. High Hb and Hct levels were found in 125 ppm whereas low levels were found in control. MCV levels were high in 125 ppm followed by 100 ppm and low in control and were significantly different (p<0.05) from other treatments. A significant (p<0.05) increase and decrease in

MCH level was identified in 125 ppm and control group, respectively. At the end of 6 h, control group showed the highest MCHC and is significantly different ($p<0.05$) from other treatments whereas at the termination of the experiment, all the treatments were insignificantly different ($p>0.05$). TLC was found low in control group and high in 125 ppm. There was no significant difference ($p>0.05$) in other treatments. High lymphocyte count was noticed in 25 ppm while low count was observed in control during transportation and were found significantly different ($p<0.05$). Neutrophils were found more in control group and less in 25 ppm exhibiting a significant difference ($p<0.05$) from other treatments.

Discussion

In this present study, the induction and recovery times significantly decreased and increased, respectively with increase in concentrations of tobacco leaf dust. The results indicate that the increase in tobacco concentration reduced the induction time and prolonged the recovery time in fishes. There is a highly significant correlation between the concentrations of tobacco and time of induction and recovery with larger doses having shorter induction times and prolonged recovery [21]. The recovery time is usually faster at lower concentrations of anesthetic and it becomes more prolonged as the concentration increases [7]. Several workers have also obtained the above results using different sedatives and anesthetics in other fish species [5,12,22-24]. The findings from the present investigation reveal that application of tobacco leaf dust have notable effects on the fish behavior during induction and recovery. The behavioral responses observed in *L. rohita* fingerlings during 12 h exposure period in the anesthetic bath were in line with the results reported in Nile tilapia [21] and Thai magur [25] at higher anesthetic doses. Several studies have envisaged that sedation can decrease transport and post-transport mortality [26-28]. The highest mortality of rohu fingerlings occurred when no sedative was used. Fish mortality in control resulted from poor water quality and poor hematologic profile of the fingerlings in confined space. There was no immediate and delayed mortality in tobacco leaf dust treatment groups due to the effectiveness of sedatives that have helped in maintaining better environmental conditions which in turn assisted in less consumption of oxygen by reducing metabolic activity and hyperactivity and thus eliminating undue injuries. However, little mortality was observed in 125 ppm, due to the continuous hyperactive phase caused by the sedative.

The water quality parameters were kept constant and were within the acceptable levels as reported by Bhatnagar and Devi [29] indicating that the experimental condition was suitable for transport. After the experiment, variation in the reported result of monitored parameters was noticed which may be associated with the transportation in a closed container (polyethylene bag). A significant increase in water temperature was observed for control group after transport, which was attributed to the transport condition as well as the heat generated by the fishes as a result of metabolic activity. *L. rohita* fingerlings transported without sedative (control group) showed low pH as a result of CO_2 accumulation. The production of more CO_2 from respiration lead to the formation of more carbonic acid and subsequent dissociation yield more hydrogen ions turning water acidic [30]. Low pH for the control group compared with that for the sedative treated groups indicates the efficacy of sedative in reducing carbon dioxide excretion and therefore metabolic activity of the fish. Kutty [31] and Das et al. [15] stated that elevated respiration rate due to the hyperactivity of fishes during initial phase contribute to reduced DO tension in water. Low DO observed in control group could be due to high consumption of

oxygen resulted from increased temperature and metabolic rate of the fingerlings over the duration. The reduction in oxygen consumption could be due to either a decrease in the rate of oxygen uptake and/or a decrease in metabolism caused by the effect of tobacco leaf dust. The CO_2 content in transport water showed a gradual increase with transport duration. Hypercapnia in control group was attributed to the concomitant increase in metabolic rate of the fishes during transport. Next, to control, CO_2 was also found high in 125 ppm and 100 ppm which could be due to the hyperactivity phase of the fishes caused by the sedative. The CO_2 increase in other sedative treatments during the initial hours may be due to the stress caused by capture and handling before the start of the experiment but declined and subsided towards the end of the experiment indicating the efficacy of the sedative. The results corroborate with the previously simulated transport experiments conducted in Tiger barb [28], Southern Platyfish [32] and Winter Flounder [33]. Boyd [30] indicated that ammonia released by fish into water reacts with water molecules to form ammonium (NH_4^+) and hydroxyl (OH^-) ions and further, the hydroxyl ion reacts with CO_2 to produce HCO_3 which results in alkalinity. The results obtained from the present experiment showed a significant increase of total alkalinity in control groups which confirms the phenomena explained by Boyd [30] as there had been a continuous addition of ammonia and CO_2 to the ambient water from the excretion and respiration of the fingerlings, respectively. Hardness values in control group and in all treatment groups were within the optimum range of fishes as reported by Bhatnagar and Devi [29]. Control group transported without sedative showed a significant increase in ammonia which was attributed to the increase in metabolic rate of the fishes and release of excretory products. Sedative treatments decreased the metabolic activity of *Labeo rohita* fingerlings and hence ammonia was found low compared to control which conforms with the results obtained by Park et al. [33], Pramod et al. [28], Becker et al. [34] and Husen and Sharma [5] using other sedatives. The increase in metabolic activities coupled with the exciting phase of the fishes caused by the higher concentrations of tobacco leaf dust led to the increase of ammonia in the respective treatment groups. An increase in NO_2-N concentration in water is considered toxic to fish [15]. The increased NO_2-N concentration observed in control group was found lethal. The results indicate that the increase could be due to the release of excretory products and metabolites by the fishes caused by transport stress coupled with the nitrification of ammonia. Nitrite was found low in treatment groups due to the sedative efficacy in reducing the release of fecal matter by the fishes and thereby reducing the NO_2 concentrations in transport water. Where ammonia and nitrite are toxic to the fish, Nitrate is harmless and is produced by the autotrophic *Nitrobacter* bacteria combining oxygen and nitrite. Major chemical constituents identified in tobacco are nicotine, nor-nicotine, anabasine, myosmine, anatabine, nitrate, sorbitol [35]. The results from the current study indicate that the increase in NO_3-N concentrations in the sedative treatments could be attributed to the increased nitrate content found in tobacco leaf dust which increased the NO_3-N concentrations in water over the duration. Low NO_3 was found in control which may be due to poor bacterial degradation of toxic metabolites such as ammonia and nitrate.

Hematological profile changes of peripheral blood have often been used as a stress indicator, though results are equivocal. Serious changes in hemogram of fishes were found in control group due to stress caused by transportation. Erythrocyte value is a function of oxygen absorption and transportation within a cell and depletion in count may weaken the fish and lead to death. The significant decrease in circulating erythrocytes in control group may be due to the poor water quality found which agrees with the results reported in Gilthead Sea Bream [36] and in *Tilapia zilli* [37]. Similar results were also acquired by Ahmed et al. [38] and Ishikawa et al. [39] in *O. niloticus* when exposed to poor water quality. These findings underline that water quality influences hematological parameters. In stressed fish, an increase in RBC is often observed [40]. The significant increase in the no. of red blood cells in treatment groups with higher concentrations of tobacco leaf dust (125 ppm and 100 ppm) may be due to the higher metabolic demand stress caused by the hyperactivity of the fishes in the respective treatments. The release of the catecholamine is primary stress response causing erythrocytes to swell and spleen to release new erythrocyte to blood [41]. Hyperactivity initiated the release of catecholamine and cortisol to promote the increase in oxygen demand in the tissues, leading to a quick differentiation and proliferation of erythrocytes in fishes of higher treatment groups. Certain blood variables like hemoglobin and hematocrit are considered as auxiliary stress response indicators [42]. The results showed a significant reduction of hemoglobin content in control group which could be due to handling and transport stress as well as due to methemoglobinemia [43,44] caused by the increased nitrite, found in control group. During stress situations, elevated hemoglobin contents increase the oxygen carrying capacity of blood and thus, supply oxygen to the major organs, in response to higher metabolic demands [45]. Thus, the results reveal that hyperactivity of fishes enhanced the blood hemoglobin level in 125 ppm and 100 ppm in order to meet the higher metabolic demands. As stress is an outcome of transport procedures, it could have decreased the Hct values in the control group. These considerations suggest a hemodilution caused by osmoregulatory disturbance [42,46]. Hct values increased significantly during the transportation in treatment groups with higher concentrations of tobacco leaf dust (125 ppm and 100 ppm) which were similar to that reported in common carp [47]. Hrubec and Smith [48] stated that Hct values increase was a result of splenic contraction and RBC swelling. Hct can also vary in fish according to their swimming performance. Wilhelm Filho et al. [49] observed that active fish species presented higher Hct, Hb and RBC when compared to less active fish species. Therefore, the results reveal that increased no. of erythrocytes and hyperactivity of the fishes was the reason for the increased Hct values of fishes in 125 ppm and 100 ppm treatments. Erythrocyte indices namely, MCV, MCH and MCHC measure the volume, weight and the concentration of hemoglobin respectively [50]. Fluctuation in these indices corresponded with values of TEC, Hb and Hct observed. Stress is thought to be responsible for leucopenia in fish [51]. The significant decline in the circulating WBC in control group of the present study may be due to an increase of plasma cortisol concentration which is a glucocorticoid hormone that can act as an immunosuppressive. In the present investigation, the increase in WBC (leukocytosis) may have resulted from the excitation of the defense mechanism of the fish to counter the acute stress effect caused by the higher concentrations of the leaf dust. Pulsford et al. [52] also detected an increased number of leukocytes, particularly phagocytes and damaged cells, in peripheral blood of dab, *Limanda limanda*, when subjected to an acute stress. Lymphocytes are the most common leukocytes found in a healthy teleost, and they represent an important function in the cell immunity of fish [53]. Lymphocytes and neutrophils from the present study showed a significant decrease and increase in control group, respectively which resembles the results reported in slender seahorse [23], *Cyprinus carpio* [54], and Channel catfish [55]. The decline of lymphocytes in control group may be due to the stress of capture and transport which led to increasing in plasma cortisol inducing lymphocyte migration from blood to tissues and thereby decreasing the circulating lymphocytes in the blood. Wiik et al. [56] suggested that transport and handling stress

leads to the elevation of plasma cortisol that is reported to have a direct cytolytic effect on lymphocytes. Also, a significant decrease and increase of lymphocytes and neutrophils in the treatment group (125 ppm and 100 ppm) was observed confirming the results reported in Rainbow trout [57] using a high dose of clove oil. It could be due to increase in stress hormones caused by the hyperactivity of the fishes during transportation. No significant changes occurred in treatments up to 75 ppm concentration indicating that the sedative is beneficial to reduce stress. Neutrophilia in control group could be due to the secondary effects of stress in fish, as a consequence of the stress-related release of catecholamine during transportation. In general, acute stress induces both neutrophilia and lymphopenia in fish and that these changes are due to cortisol and norepinephrine, which induce leukocyte migration from blood to tissues and vice-versa [52]. Thus our findings are in agreement with Wendelaar Bonga [41], who has reported that stress causes a rapid increase in neutrophils and a reduction of lymphocytes in peripheral blood. It is noteworthy that throughout the experiment, there were no significant changes in water quality and blood parameters in tobacco leaf dust concentration of 25 ppm which implies the efficacy of the sedative at the particular concentration. Hence, tobacco use in the transport of rohu and other fishes is suggested because of its stress reducing capacity during transport even at low concentrations.

Conclusion

On the whole, the present study suggests that tobacco leaf dust of up to 75 ppm concentration can be used as a sedative for rohu fingerlings transport with a slight modification of fish behavior and its hemogram. But the tobacco leaf dust is effective in inducing light sedation in rohu at a concentration of 25 ppm. Therefore, it has been concluded that *L. rohita* fingerlings can be successfully transported up to 12 h using 25 ppm tobacco leaf dust as a tranquilizing agent without any change in the transport environment and health of fish. In general, tobacco is less expensive compared to other synthetic sedatives. Therefore, the study would be helpful to fish farmers to prevent the mortality of seed during transportation using *N. tabacum* leaf dust as an alternative, eco-friendly and novel natural product to synthetic, harmful and expensive sedatives, due to its cost effectiveness, easy availability, narcotic at low dose and biodegradability.

Acknowledgement

The authors are thankful to the Director, ICAR – Central Institute of Fisheries Education Mumbai for the necessary support and encouragement.

References

1. FAO (2016) The State of World Fisheries and Aquaculture: Contributing to food security and nutrition for all. Rome.

2. Stieglitz JD, Benetti DD, Serafy JE (2012) Optimizing transport of live juvenile cobia (*Rachycentron canadum*): effects of salinity and shipping biomass. Aquaculture 364-365: 293-297.

3. Froese R (1998) Insulating properties of styrofoam boxes used in transporting live fish. Aquaculture 159: 283-292.

4. Lewis DJ, Wood GD, Gregory R (1996) Trading the silver seed. University Press Limited. Dhaka-1000, Bangladesh.

5. Husen MA, Sharma S (2015) Immersion of rohu fingerlings in clove oil reduced handling and confinement stress and mortality. International Journal of Fisheries and Aquatic Studies 2: 299-305.

6. King WV, Hooper B, Hillsgrove S, Benton C, Berlinsky D (2005) The use of clove oil, metomidate, tricaine methanesulphonate and 2-phenoxyethanol for inducing anaesthesia and their effect on the cortisol stress response in black sea bass (*Centropristis striata* L.). Aquaculture Research 36: 1442-1449.

7. Ross LG, Ross B (2008) Anaesthetic and sedative techniques for aquatic animals. (3rdedn), Blackwell Science Ltd., Oxford, London, UK.

8. Strange RJ, Schreck CB (1978) Anaesthetic and handling stress on survival and cortisol concentration in yearling chinook salmon (Oncorhynchus tshawytscha). Journal of the Fisheries Research Board of Canada 35: 345-349.

9. Marking LL, Meyer FP (1985) Are better anaesthetics needed in fisheries? Fisheries 10: 2-5.

10. Coyle SD, Durborow RM, Tidwell JH (2004) Anaesthetics in aquaculture. SRAC Publications, USA.

11. Hoseini SM, Hosseini SA, Nodeh AJ (2011) Serum biochemical characteristics of Beluga, Huso huso (L.), in response to blood sampling after clove powder solution exposure. Fish Physiology and Biochemistry 37: 567-572.

12. Hoseini SM, Ghelichpour M (2012) Efficacy of clove solution on blood sampling and hematological study in Beluga, Huso huso (L.). Fish Physiology and Biochemistry 38: 493-498.

13. Binorkar SV, Jani DK (2012) Traditional Medicinal Usage of Tobacco-A Review. Spatula DD 2: 127-134.

14. http://www.tobaccoboard.com.

15. Das PC, Mishra B, Pati BK, Mishra SS (2015) Critical water quality parameters affecting survival of Labeo rohita (Hamilton) fry during closed system transportation. Indian Journal of Fisheries 62: 39-42.

16. Vázquez GR, Guerrero GA (2007) Characterization of blood cells and haematological parameters in Cichlasoma dimerus (Teleostei, Perciformes). Tissue and cell 39: 151-160.

17. Keene JL, Noakes DLG, Moccia RD, Soto CG (1998) The efficacy of clove oil as an anaesthetic for rainbow trout, Oncorhynchus mykiss (Walbaum). Aquaculture Research 29: 89-101.

18. APHA (2012) Standard methods for the examination of water and wastewater. (22nd edn), American Public Health Association, Washington DC, USA.

19. Smith CE (1967) Haematological changes in Coho Salmon fed folic acid deficient diet. Journal of the Fisheries Research Board of Canada 25: 151-156.

20. Dacie JV, Lewis SM (2001) Practical Haematology. (9th edn), Churchill Livingstone, London, UK.

21. Agokei OE, Adebisi AA (2010) Tobacco as an anaesthetic for fish handling procedures. Journal of Medicinal Plants Research 4: 1396-1399.

22. Hseu JR, Yeh SL, Chu YT, Ting YY (1998) Comparison of efficacy of five anaesthetics in goldlined sea bream, Sparus sarba. Acta zoologica Taiwanica 9: 35-41.

23. Cunha MAD, Silva BFD, Delunardo FAC, Benovit SC, Gomes LDC, et al. (2011) Anaesthetic induction and recovery of Hippocampus reidi exposed to the essential oil of Lippia alba. Neotropical Ichthyology 9: 683-688.

24. Balamurugan J, Kumar TTA, Prakash S, Meenakumari B, Balasundaram C, et al. (2016) Clove extract: A potential source for stress free transport of fish. Aquaculture 454: 171-175.

25. Jegede T (2014) Anaesthetic potential of tobacco (Nicotiana tobaccum) on Clarias gariepinus (Burchell 1822) Fingerlings. Journal of Agricultural Science 6: 86-90.

26. Inoue LAKA, Afonso LOB, Iwama GK, Gilberto MG (2005) Effects of clove oil on the stress response of matrinxã (Brycon cephalus) subjected to transport. Acta Amazonica 35: 289-295.

27. Iversen M, Eliassen RA, Finstad B (2009) Potential benefit of clove oil sedation on animal welfare during salmon smolt, Salmo salar L. transport and transfer to sea. Aquaculture Research 40: 233-241.

28. Pramod PK, Sajeevan, TP, Ramachandran A, Thampy S, Pai SS (2010) Effects of two anesthetics on water quality during simulated transport of a tropical ornamental fish, the Indian tiger barb, Puntius filamentosus. North American Journal of Aquaculture 72: 290-297.

29. Bhatnagar A, Devi P (2013) Water quality guidelines for the management of pond fish culture. International Journal of Environmental Sciences 3: 1980-2009.

30. Boyd CE (1990) Water quality in ponds for aquaculture. Alabama Agricultural Experiment Station, Auburn University, Alabama, USA.

31. Kutty MN (1987) Transport of fish seed and brood fish, Seed Production-Working paper for senior Aqua-culturist Course at African Regional Aquaculture Centre, Port Harcourt, Nigeria.

32. Guo FC, Teo LH, Chen TW (1995) Effects of anaesthetics on the water parameters in a simulated transport experiment of platyfish, Xiphophorus maculatus (Günther). Aquaculture Research 26: 265-271.

33. Park IS, Park MO, Hur JW, Kim DS, Chang YJ, et al. (2009) Anesthetic effects of lidocaine-hydrochloride on water parameters in simulated transport experiment of juvenile winter flounder, Pleuronectes americanus. Aquaculture 294: 76-79.

34. Becker AG, Parodi TV, Heldwein CG, Zeppenfeld CC, Heinzmann BM, et al. (2012) Transportation of silver catfish, Rhamdia quelen, in water with eugenol and the essential oil of Lippia alba. Fish Physiology and Biochemistry 38: 789-796.

35. Fowles J, Phillips D, Kaiserman M (2003) Chemical composition of tobacco and cigarette smoke in two brands of New Zealand cigarettes. Prepared as part of a New Zealand Ministry of Health contract. Porirua: ESR, final report revised 25: 3-4.

36. Fazio F, Marafioti S, Filiciotto F, Buscaino G, Panzera M, et al. (2013) Blood haemogram profiles of farmed onshore and offshore gilthead sea bream (Sparus aurata) from Sicily, Italy. Turkish Journal of Fisheries and Aquatic Sciences 13: 415-422.

37. Gbore FA, Oginni O, Adewole AM, Aladetan JO (2006) The effect of transportation and handling stress on haematology and plasma biochemistry in fingerlings of Clarias gariepinus and Tilapia zillii. World Journal of Agricultural Sciences 2: 208-212.

38. Ahmed NA, El-Serafy SS, El-Shafey AAM, Abdelhamide NAH (1992) Effect of ammonia on some haematological parameters of Oreochromis niloticus. Proceedings of the Zoological Society AR Egypt 23: 155-160.

39. Ishikawa NM, Ranzani-Paiva MJT, Lombardi JV, Ferreira CM (2007) Haematological parameters in Nile Tilápia, Oreochromis niloticus exposed to sub-lethal concentrations of mercury. Brazilian Archives of Biology and Technology 50: 619-626.

40. Svobodová Z, Vykusová B, Machová J (1994) The effect of pollutants on selected haematological and biochemical parameters in fish. Sub lethal and Chronic Effects of Pollutants on Freshwater Fish. FAO – Fishing News Books, Cambridge, UK.

41. Bonga SW (1997) The stress response in fish. Physiological Reviews 77: 591-625.

42. Morgan JD, Iwama GK (1997) Measurements of stressed states in the field. Fish stress and health in aquaculture. Society for experimental biology seminar series. Cambridge University Press, Cambridge, UK.

43. Knudsen PK, Jensen FB (1997) Recovery from nitrite induced methemoglobinemia and potassium balance disturbances in carp. Fish Physiology and Biochemistry 16: 1-10.

44. Jensen FB (2003) Nitrite disrupts multiple physiological functions in aquatic animals. Comparative Biochemistry and Physiology - Part A: Molecular & Integrative Physiology 135: 9-24.

45. Ruane NM, Bonga SW, Balm PHM (1999) Differences between rainbow trout and brown trout in the regulation of the pituitary inter-renal axis and physiological performance during confinement. General and Comparative Endocrinology 115: 210-219.

46. Houston AH, Dobric N, Kahurananga R (1996) The nature of haematological response in fish. Studies on rainbow trout, Oncorhynchus mykiss exposed to stimulated winter, spring and summer conditions. Fish Physiology and Biochemistry 15: 339-347.

47. Dobšíková R, Svobodová Z, Blahová J, Modrá H, Velíšek J (2006) Stress response to long distance transportation of common carp (Cyprinus carpio L.). Acta Veterinaria Brno 75: 437-448.

48. Hrubec TC, Smith SA (2010) Hematology of fishes. Schalm's veterinary hematology. (6th edn), Ames, Iowa, Wiley-Blackwell, USA.

49. Wilhelm FD, Eble GJ, Kassner G, Caprario FX, Dafré AL, et al. (1992) Comparative haematology in marine fish. Comparative Biochemistry and Physiology - Part A: Molecular & Integrative Physiology 102: 311-321.

50. Wedemeyer GA, Gould RW, Yasutake WT (1983) Some potentials and limits of the leucocrit test as a fish health assessment method. Journal of Fish Biology 23: 711-716.

51. Wedemeyer GA (1970) Stress of anesthesia with MS 222 and benzocaine in rainbow trout (Salmo gairdneri). Journal of the Fisheries Research Board of Canada 27: 909-914.

52. Pulsford AL, Lemaire-Gony S, Tomlinson M, Collingwood N, Glynn PJ (1994) Effects of acute stress on the immune system of the dab, Limanda limanda. Comparative Biochemistry and Physiology - Part C: Toxicology & Pharmacology 109: 129-139.

53. Thrall MA, Baker DC, Campbell TW, DeNicola D, Fettman MJ, et al. (2007) Hematologia e bioquímica clínica veterinária. São Paulo, Roca.

54. Sopinska A (1984) Effect of physiological factors, stress, and disease on haematological parameters of carp, with a particular reference to leukocyte pattern II. Hematological results of stress in carp. Acta Ichthyologica Et Piscatoria 14: 121-139.

55. Ellsaesser CF, Clem LW (1986) Haematological and immunological changes in channel catfish stressed by handling and transport. Journal of Fish Biology 28: 511-521.

56. Wiik R, Andersen K, Uglenes I, Egidius E (1989) Cortisol-induced increase in susceptibility of Atlantic salmon, Salmo salar, to Vibrio salmonicida, together with effects on the blood cell pattern. Aquaculture 83: 201-215.

57. Kanani HG, Mirzargar SS, Soltani M, Ahmadi M, Abrishamifar A, et al. (2011) Anaesthetic effect of tricaine methanesulfonate, clove oil and electro-anesthesia on lysozyme activity of Oncorhynchus mykiss. Iranian Journal of Fisheries Sciences 10: 393-402.

10

Biological Treatments of Fish Farm Effluent and its Reuse in the Culture of Nile Tilapia (*Oreochromis niloticus*)

Bamidele Oluwarotimi Omitoyin[1]*, Emmanuel Kolawole Ajani[1], Oluwabusayo Israel Okeleye[1], Benjamin Uzezi Akpoilih[1] and Adeniyi Adewale Ogunjobi[2]

[1]*Department of Aquaculture and Fisheries Management, University of Ibadan, Ibadan, Nigeria*
[2]*Department of Microbiology, University of Ibadan, Ibadan, Nigeria*

Abstract

Aquaculture wastewater collected from a catfish farm in Ibadan metropolis was treated with duckweed, *Lemna minor* (Td) for two weeks and thereafter used in the culture of Nile tilapia (*O niloticus*). The performance of *O niloticus* raised in *Lemna minor* treated waste water was compared with bacteria-treated waste water, *Bacillus sp.* (Tb) and well water (Tc) as control (untreated). The *Bacillus sp.* was isolated from the catfish wastewater, and was positive to Gram's staining, catalase and glucose fermentation test. Nile tilapia juveniles (n=54) of an average initial weight of 10.43 ± 0.04 g were stocked in triplicates per treatment and fed to satiation twice daily for 8 weeks. There was significant difference (P<0.05) in the quality of waste water in all the treatments. Compared to initial waste water, Td showed a significant reduction in biological oxygen demand, BOD (1.23 ± 0.03 mg/L vs. 36.80 ± 1.89 mg/L), chemical oxygen demand, COD (2.20 ± 0.06 mg/L vs. 58.81 ± 1.89 mg/L), sulphate (0.50 ± 0.06 mg/L vs. 5.53 ± 0.33 mg/L) and phosphate (5.40 ± 0.31 mg/L vs. 18.43 ± 0.78 mg/L) after 2 weeks of treatment. The level of phosphate, BOD, COD, nitrate, and TSS were lowest in Td compared to Tb and Tc (P<0.05). The lowest level of ammonia was obtained in Tc (0.15 ± 0.10 mg/L), compared to Td (0.15 ± 0.10 mg/L) and Tb (0.66 ± 0.28 mg/L). The highest percentage weight gain (WG) of 34.37 ± 0.60% and the lowest feed conversion ratio (FCR) of 1.59 ± 0.03were recorded in fish raised in Td (P<0.05). *Oreochromis niloticus* juveniles raised in Td also had the highest specific growth rate (SGR) of 0.23 ± 0.01% compared to 0.19 ± 0.00% recorded in fish raised in both Tb and Tc. Fish raised in Tc had the highest survival rate (100 ± 0.00%) compared to the fish cultured with Tb (77.80 ± 2.30%) and Td (72.20 ± 1.95%). The research findings suggest that *Lemna minor* could be used in fish culture with positive effect on water quality and growth performance.

Keywords: Duckweed; *Bacillus sp.*; Wastewater; Fish culture

Introduction

The aquaculture industry is one of the fastest growing agriculture sector globally. With a total production of 66.6 million metric tonnes in 2012, it provides almost half of all fish production for human consumption [1]. However, the long-term sustainability of aquatic environment has raised concerns over the environmental impact of this vital sector, due to its negative impact on aquatic ecology and systems [2,3]. This is because intensification of aquaculture involves the use of highly nutritious feeds and other chemical products, which generate wastes that, in most cases, are difficult to curtail and toxic to aquatic lives [4-6]. Effluent water containing wastes are discharged in all aquaculture systems [7]. The amount of wastes generated from aquaculture practices depends on the culture system characteristics, choice of species, feed quality and management practices [8].

The discharge of wastewater in the form of effluents into aquatic ecosystems could lead to the alterations of the receiving environments. High organic load in aquaculture wastewater can result in the eutrophication of receiving water bodies, which causes a lot of havoc on the biodiversity in aquatic ecosystems [9,10]. Nitrogenous wastes, which are the major component of aquaculture waste, are highly toxic to macro-fauna in the open water body. Stephen and Farris [11,12] reported that an increase in ammonia concentrations could elevate blood ammonia, which is highly toxic to fish. Suspended solids in aquaculture wastes in receiving water bodies cause interstitial clogging and substrate embeddedness [13]. The deposition of solids and sediments could enhance the growth of heterotrophic bacteria and increase the formation of colony-forming units, leading to additional interstitial clogging and deoxygenation [14].

The use of microorganisms to degrade and reduce harmful wastes in contaminated sites has been reported in several studies. Bio-remediation offers the possibility of rendering harmless various contaminants in wastewater. Different microorganisms, including bacteria, fungi, algae, and plants have been used to decontaminate polluted environments [15-17]. Under controlled conditions, organic wastes are degraded by microbes to levels that are harmless, or below concentration limits [18,19]. Bio-remediation techniques are cheaper than traditional methods such as incineration; and some pollutants can be treated on site, which reduces exposure risks of cleanup personnel as a result of transportation accidents [20]. It also provides an alternative for effective management of wastewater for the purpose of reuse, thereby reducing pressure on limited freshwater resources.

Microbes exist in diverse environmental conditions, which make them useful in waste management. Prescott et al. [21,22] reported that microorganisms, indigenous (native) or extraneous (introduced), are prime agents in any bio-remediation system. Indigenous bacteria are crucial to bio-remediation processes, due to the important role they play in the biogeochemical cycle of nutrients [23]. The potentials for

*Corresponding author: Bamidele Oluwarotimi Omitoyin, Department of Aquaculture and Fisheries Management, University of Ibadan, Ibadan, Nigeria
E-mail: bam_omitoyin@yahoo.co.uk

bio-remediation have been reported for different organisms, with microbes showing the highest efficiency in several studies [24]. Dead microbial cells are also useful in bio-remediation technologies [25].

Several studies have shown *Bacillus pumifus* to be a good candidate for bio-remediation [26]. Other *Bacillus sp*, including *B. cereusmycoides, B. megaterium, B. mucosis, B. agglomerates, B. cartilaginous* could be used for bio-remediation because they possess antagonism, proteolysis and catalytic activity characteristics [27]. In a study by Quieroz and Boyd [28-30], commercially prepared Bacillus species mixed with rearing water of channel catfish (*Ictalurus punctatus*) improved production and survival rate of fish. High capacity for bio-remediation of organic sediments has also been reported for *Bacillus sp* such as *Bacillus subtilis, B. licheniformes, B. cereus, B. coagulans* and *Phenibacillus polymyxa*, have been shown capacity for bio-remediation. However, they are present in large amount in sediments.

Phytoremediation has been found to be well suited for use for sites with low concentration of pollutants and which require expensive technology for bio-remediation [31]. The potential of plants to reduce high load of harmful wastes and tolerate harsh environmental conditions has been reported [32]. Duckweeds are small aquatic plants belonging to the family Lemnaceae [33]. They are reported to have a high potential to absorb and remove nutrients in wastewater, such as nitrate, phosphate, calcium, sodium, potassium, magnesium, carbon, and chloride. These nutrients are permanently removed from the system when the plants are harvested. More so, the use of duckweeds in waste treatment has been shown to reduce harmful substances such as total suspended solids (TSS), biochemical oxygen demand (BOD), and chemical oxygen demand (COD) in wastewater significantly, and have reported to tolerate ammonia level as high as 240 mg/L.

Smith and Moelyowati [34] stated that wastewater treatment systems are feasible for developing countries in hot climates to provide low-cost treatment of domestic sewage particularly in rural areas. The sustainability of aquaculture industry would, therefore, depend on the availability of cheap and affordable technology for waste treatment. The full potential of duckweed as a cheap and low cost method of waste treatment, as opposed to high cost technologies, has not been exploited, particularly in Nigeria, which is one of the largest producers of cultured fish species in Africa [1].

Therefore, the study was carried out to evaluate the effectiveness of duckweed and microorganism (*Bacillus sp*) in bio-remediating wastewater from a catfish farm and the effect of the bio-remediated water on the performance of Nile tilapia (*Oreochromis niloticus*).

Materials and Methods

Wastewater sampling and analysis

The bio-remediated aquaculture wastewater used in this study was obtained from a reputable fish farm (SDC Farm, Ibadan, Oyo State, Nigeria, located on coordinate of N7°35`38.69``, E3°85`42.79``) in active fish production in Ibadan metropolis, Nigeria. The sampled fish farm operated a semi-intensive production system. The wastewater were collected at point of discharge between 6.30-7.00 and transported in 25 litres plastic containers immediately to the Department of Aquaculture and Fisheries Management laboratory, University of Ibadan, Nigeria (N7°26`27.98`` E3°54`8.99``). Two litres of aquaculture wastewater were collected in sampling bottles at the point of discharge and were analyzed for the following physicochemical parameters: Dissolved Oxygen (DO), Chemical Oxygen Demand (COD), Biological Oxygen Demand (BOD), phosphate, sulphate, Total Ammonia Nitrogen

(TAN), nitrate, Total Suspended Solid (TSS), pH and temperature were monitored weekly before the commencement and after the bio-remediation process according to APHA [35] standard procedure.

Microbial remediation

Isolation of the microorganism: The bacteria strain (*Bacillus sp.*) used for the bio-remediation experiment was isolated from the wastewater collected using sterile sampling bottles. Isolation of *Bacillus sp.* was carried out on nutrient agar (NA, Oxoid CM3). The media was prepared by suspending 28.0g of NA in 1 litres of distilled water. The dissolved agar solution was autoclaved at temperature of 121°C for 15 minutes. The medium plates were inoculated by pour plate method with 1 mL aliquot of the diluents pipetted aseptically into labeled sterile Petri dish after serial dilution of the wastewater samples in 9 mL sterilized distilled water from 10^{-9} to 10^{-10} dilution. The inoculated plates were incubated in an incubator at 37°C for 24 hours. Discrete colonies from plate prepared by pour plate methods were sub-cultured into sterile NA agar plates (incubated for 24hrs) without contaminants aseptically by streaking, using a wire loop sterilized with spirit-lamp flame to obtain pure cultures of the isolates. The isolated pure bacteria colonies were characterized using standard morphological and biochemical tests, such as Gram staining, catalase, oxidase and sugar fermentation tests, after 24 hours of incubation as described in Berge's Manual of Bacteriology.

Inoculation of the wastewater with the prepared bacterial inoculums: The inoculum was prepared in nutrient broth and the concentration of the bacterial cells was adjusted to a 10^5 colony-forming unit using sterile physiological saline to correspond to 0.5 MacFarland standards. Fifteen (15) milliliter of the bacterial inoculums was introduced into 15 litres of wastewater in experimental tanks (0.39 m × 0.28 m × 0.26 m), using a sterile needle and syringe. Mosquito net was used to screen the wastewater treatments to prevent insect infestation. The bio-remediation experiment lasted for two weeks. The water quality parameters were recorded before and after bio-remediation.

Phytoremediation of the wastewater: Duckweed, *Lemna minor*, used for phytoremediating the wastewater were obtained from University of Ibadan Botanical Garden, Nigeria (N7°26`46.59``, E3°54`11.72``). Fifteen (15) litres of aquaculture wastewater sample collected was bio-remediated in plastic tanks (0.39 m × 0.28 m × 0.26 m) with about 49.53 ± 0.25 g (mean wet weight) of fresh duckweed (*Lemna minor*) plants, enough to cover the entire surface of the water with approximately a single layer of fronds to avoid direct contact with sunlight (Al-Nozaily, 2001). This was done to prevent the formation of green algae in the experimental setup. The experimental tanks were arranged outside the laboratory to have adequate access to sunlight. Harvesting and weighing of the plants followed after two weeks of bio-remediation to determine change in the plants biomass. AOAC [36] provided analytical method the proximate compositions of the plant before and after bio-remediation.

Culture of fish in bio-remediated wastewater: The University of Ibadan Fish Farm, Ibadan Oyo state, Nigeria (N7°26`27.2472``, E3°53`58.1532``), provided ninety (90) juveniles of Nile tilapia (*Oreochromis niloticus*) of an average weight of 10.44 ± 0.90 g used in the study. The fish were acclimatized in well water (control treatment) for two weeks before being used in the study. Triplicate plastic tanks (0.39 m × 0.28 m × 0.26 m), with six fish per tank were randomly allocated to the treatments consisting of 7 litres of bio-remediated wastewater with duckweed (Td), bacteria (Tb) and well water as the control treatment (Tc) and fed to satiation twice (morning, 8:00 and evening, 16:00) per day. Imported floating feed (ME-2, Skretting, France) was used for the

Treatment	Phosphate (mg/L)	Nitrate (mg/L)	Sulphate (mg/L)	BOD (mg/L)	COD (mg/L)	DO (mg/L)	pH (mg/L)	Temp (oC)	TAN (mg/L)	TSS (mg/L)
RAW	18.43 ± 0.78	9.93 ± 0.36	5.53 ± 0.33	36.80 ± 1.89	58.81±1.89	4.00 ± 0.14	6.83 ± 0.67	24.63 ± 0.22	1.17 ± 0.48	2136.75 ± 332.37
Tc	16.57 ± 0.23	6.48 ± 0.02	3.74 ± 0.17	31.90 ± 0.21	58.17±0.20	4.09 ± 0.06	7.46 ± 0.03	24.38 ± 0.03	0.15 ± 0.10	2015.00 ± 2.89
Tb	16.47 ± 0.03	6.77 ± 0.04	4.17 ± 0.09	27.27 ± 0.09	52.10±0.31	4.81 ± 0.02	7.86 ± 0.08	24.26 ± 0.07	0.66 ± 0.28	2034.00 ± 3.06
Td	5.40 ± 0.31	5.94 ± 0.47	0.50 ± 0.06	1.23 ± 0.03	2.20±0.06	4.45 ± 0.06	8.11 ± 0.21	27.96 ± 0.26	0.26 ± 0.11	1347.33 ± 1.45
Sig-values	0	0.01	0.01	0.01	0.01	0.01	0.01	0.01	0.01	0.01

Note: Mean values on the same column with Sig-values > 0.05 are not significantly different ($\alpha_{0.05}$).

Abbreviations: BOD: Biological Oxygen Demand, COD: Chemical Oxygen Demand, DO: Dissolved Oxygen, Temp: Temperature, TAN: Total Ammonia Nitrogen, TSS: Total Soluble Solids.

Table 1: Mean water quality parameters of raw aquaculture wastewater (RAW); bioremediated wastewater with bacteria (Bacillus sp.) (Tb) and duckweed (Td); and non-bioremediated wastewater (Tc) after two weeks.

feeding the fish throughout the experiment duration, which lasted for 8 weeks. Water quality was analysed biweekly for a period of eight weeks.

Growth performance and feed utilization of experimental fish: Growth performance measured every two weeks throughout the experiment enabled quantification of growth and nutrient utilization parameters. Standardized metre rule and sensitive scale provided measurement for length and fish weight, respectively. Mean weight gain (MWG) and specific growth rate (SGR) were determined from the mean initial and final weight of fish at the end of the experiment (8 weeks), while feed conversion ratio (FCR) was determined from mean data of feed consumed and weight gain. Gross feed conversion efficiency ratio(GFCE) was derived from the reciprocal of feed conversion ratio and expressed in percentage. Protein efficiency ratio (PER) was derived from mean values of weight gain and protein intake (PI), while survival rate, which was calculated from the initial number of fish and mortality after the experiment was terminated.

Data analysis: Data collected were subjected to statistical analysis using Statistical Package for Social Sciences (SPSS) version 20 software. Non-parametric statistics were used in analyzing the data generated. Descriptive statistics was used in estimating the mean and standard deviation while non-parametric analysis of variance (ANOVA) of Kruskal-Wallis and the Mann & Whitney's U tests were used to determine the level of significant difference (p<0.05) observed between groups in the ANOVA analysis. Spearman's Rank Order Correlation analysis was used to establish relationship between nutrient utilization, growth performance and water quality parameters.

Results and Discussion

Water quality parameters of experimental water samples

The result of analysis of water physicochemical parameters indicated that there was significant difference in the raw aquaculture wastewater, duckweed bio-remediated (Td), microbial (Bacillus sp.) bio-remediated (Tb) and untreated wastewater (Tc) after two weeks (Table 1). The mean values of phosphate, nitrate, sulphate, TAN, BOD, COD and TSS measured in the raw aquaculture wastewater (RAW) were significantly higher than the values observed in the treatment. These differences could be explained by the ability of bio-remediators (duckweed and microbes) in removing nutrients and other related pollutants in aquaculture wastewater. This is an indication that pollutants in aquaculture wastewater are biodegradable. Similar observation was also made by Martinez-Cordova [37] in a related study. Throughout the period of study, the least DO value was recorded in RAW within the range of 3.91-4.21 mg/L. This lower DO level in the wastewater probably shows high aerobic microbial activities (BOD) and chemical oxidation demand which are indicators of pollution. Lower DO ranged of 3.00 mg/L to 3.20 mg/L was reported by Ling [38] in wastewater drained from shrimp ponds.

According to Wang [8] the quality and quantity of waste from aquaculture depend on the culture system characteristics, culture species, feed quality and management practices.

Higher BOD value was recorded in RAW when compared to the untreated wastewater after two weeks and bio-remediated water samples. This is reflection of higher biodegradable organic substances from uneaten feed, fish fecal wastes and metabolites from microbial activities in the discharged wastewater. Similarly observation was also made by Lee [39]. The range of BOD (35.20-39.50 mg/L) recorded in the sampled wastewater was higher than the range of 5.90-18.70 mg/L recorded by Ling [38] in shrimp pond while Babatunde and Woke [40] reported higher BOD of 78.04 mg/L in effluent from fish pond.

Although the mean values of COD in RAW was lower than that of untreated wastewater after two weeks, however there was no significant difference in the COD values. This is an indication that temporal variation does not have significant influence on the amount of dissolved oxygen require to chemically oxidize organic materials in the wastewater as the rate this process occurs naturally is slow. This could be as a result of high organic contamination in aquaculture wastewater. In related studies, Amirkolaie [41,42] and Ogwo and Ogu [43] ascribed high COD in aquaculture wastewater due to the presence of high organic matter.

The pH of the RAW ranged between 6.25-7.75 with an average value indicating slight acidic condition. This is an indication of septic condition of the wastewater resulting from putrefaction of organic matters resulting to production of acidic substances such as humic acids which reduces the pH below 7. Soonnenholzner and Boyd [44] reported that oxidation of sulfide produced from wastewater during microbial decomposition process lead to production of sulfuric acid creating acidic condition which could harm the culture fish species. This result corroborates the observation of Babatunde and Woke [40] in wastewater from fish ponds.

Similar to the observed trend in other parameters, the phosphate, sulphate and nitrogenous pollutants (nitrate and TAN) in the RAW were higher than the mean values in untreated and bio-remediated wastewater. Elevated level of these pollutants in the wastewater may be as a result of leached nutrients from fish feed which are rich in proteinaceous feed components. The phosphate level in the RAW ranged between 17.55 mg/L to 19.45 mg/L which was higher than the phosphate values of 0.11 mg/L and 0.16 mg/L recorded in wastewater from extensive and intensive aquaculture farms by Bowley and Allan [45]. Ganczarczyk [46] and McCasland [47] reported that phosphate levels above 1.00 mg/L could prevent coagulation of wastewater in water treatment system. The mean nitrate level of 9.93 ± 0.36 mg/L recorded in RAW was lower than the value recorded by Babatunde and Woke [40] while TAN of 1.17 ± 0.48 mg/L observed in RAW was higher than 0.03 mg/L recorded Martinez-Cordoval [37] in effluent water from shrimp culture system. Ammonia, nitrate and nitrite are primary forms of nitrogen in inorganic form in wastewater [48]. These inorganic nitrogen forms are indicators of bacterial (such as nitrifiers) contamination [49,50] which could result to anoxic condition while ammonia is oxidized to nitrate [51]. Excess phosphate and nitrogenous pollutant content in wastewater could also result to algal bloom (eutrophication) in the receiving waterbody [52,53].

According to the Federal Ministry of Environment in Nigeria, the permissible pH, BOD, COD, TSS, sulphate, phosphate, nitrate and TAN in effluent wastewater discharge into surface water is 6.00-9.00, 30 mg/L, 80.00 mg/L, 30.00 mg/L, 500.00 mg/L, 5.00 mg/L, 20.00 mg/L and 0.10 mg/L respectively [54-56]. The results of this study showed that the aquaculture wastewater from the sampled fish farm had BOD, phosphate, TSS and TAN above the permissible limits. This could be due to accumulative effect of uneaten fish feed and metabolic wastes from culture fish on the culture water prior discharge from production unit. This result was in line with the observation of Babatunde and Woke [40,41] of wastewater from fish ponds in southwestern Nigeria.

Bio-remediation experiment

Identification of the bacteria candidate use in bio-remediation experiment: The microorganism used for the microbial bio-remediation experiment was tested positive to Gram staining test and appeared rod-like in shape when viewed under microscope. In colony morphology test, the microbe colonies were large with undulating circular margins. Bubbles were produced when the microbes were exposed to hydrogen peroxide indicating the production of oxygen and water due to production of the enzyme catalase. The bacterial was able to ferment fructose completely and maltose partially and other sugars (mannitol, lactose, sucrose, galactose and glucose) were not fermented. The probable micro-organism used in this study was *Bacillus sp* which is in agreement with the observation of Turnbull [57].

Percentage efficiency in reduction and improvement of water quality parameters: Presented in Figures 1 and 2 are percentage reduction of phosphate, nitrate, sulphate, BOD, COD and TSS; and percentage improvement of DO and pH in bio-remediated aquaculture wastewater after two week. The highest percentage reduction of phosphate, sulphate, BOD, COD, nitrate and TSS of 70.70%, 90.96%, 96.66%, 96.26%, 40.18% and 36.94% respectively and highest percentage of improvement of pH (18.74%) were recorded in wastewater bio-remediated with duckweed. Meanwhile, the *Bacillus sp*. bio-remediated wastewater tends to have the highest ammonia reduction percentage of 87.18% and DO improvement of 20.25%. Untreated wastewater had the lowest phosphate, BOD, COD and TSS reduction efficiency.

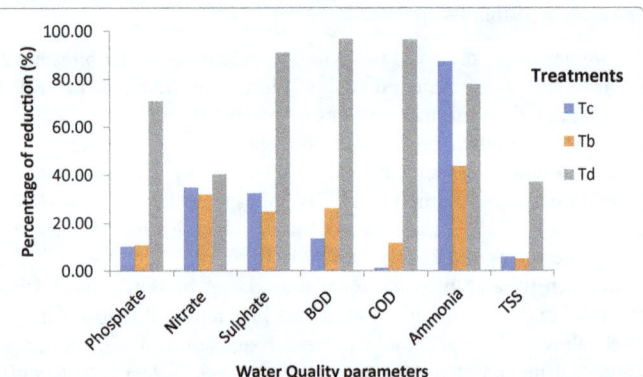

Figure 1: Percentage of reduction of pollutants in non-bioremediated wastewater and bioremediated wastewater with *Bacillus sp*. (Tb) and duckweed (Td) after two weeks of bioremediation.

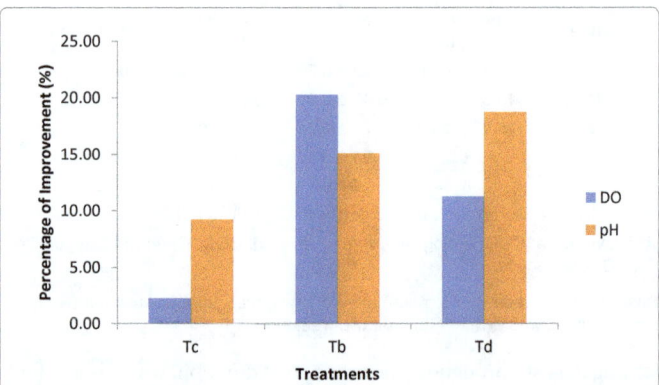

Figure 2: Percentage of improvement of DO and pH in non-bioremediated wastewater and bioremediated wastewater with *Bacillus sp*. (Tb) and duckweed (Td) after two weeks of bioremediation.

Indicative in the results of this study was the change in the water quality parameters of untreated water (Control treatments) after two weeks. Although the Phosphate, nitrate, sulphate, BOD, pH, TAN and TSS in untreated wastewater after two weeks was significantly lower (P<0.05) than the RAW collected from the sampled fish farm. This reduction could be as a result of the biochemical activities of microorganisms and chemical processes associated with the wastewater that have the potential of utilizing pollutants in the wastewater over time. Martinez-Cordova [37] suggested that sedimentation could have resulted to the same result recorded in their study.

The results of the bio-remediation experiment indicated that duckweed had the highest nutrient removal efficiency of phosphate, sulphate, nitrate and ammonia. The high affinity for nutrient uptake in aquaculture wastewater by duckweed was an indication that nutrients uptake improved biomass production of duckweed. Therefore, duckweed has been identified to be an important bio-remediation tool in reducing the nutrient content in aquaculture wastewater before discharge into open environment. Alaerts also demonstrated that the duckweed sewage stabilization pond system achieved 74% and 77% removal of nitrogen and phosphorus respectively. Similar observations were also reported by Körmer and Vermaat [58,59]. Duckweed plants typically contain more phosphorus in its tissue than other floating plants, which makes them suitable for phosphorus removal. High removal efficiency of COD and BOD of the wastewater by duckweed plant after the bio-remediation experiment was similar with the observation of Chaudhary and Sharma [33] and Ugya [60]. This could be due to the ability of the plant to remove organic compounds as well as degradation of organic materials by microbes [61,62]. The reduction efficiency of TSS by duckweed observed in this study was an indication that duckweed has the potential to reduce total suspended solids (TSS) which was similar to the observation of Zirschky and Reed and Ugya [60,61]. Meanwhile, the removal efficiency observed in this study was lower than the values reported by Zirschky and Reed. This could be as a result of high level of total suspended solids in the wastewater which was beyond the capacity of the plant to reduce within the period of study. However, the reduction in TSS shows the efficiency of duckweed in solid removal which was similar with the observation of Ugya [60].

Bacillus sp. exhibited potency in removal of nitrogenous wastes (nitrate and ammonia) and sulphate level in the wastewater. The result is in line with the study of Bhutto and Dahot [63] who reported that some *Bacillus sp*. utilized nitrogen from ammonium nitrate, ammonium sulphate, among other sources, in the production of an enzyme called amylase that is of industrial importance. There was also reduction in the sulphate and nitrate level of untreated wastewater after two weeks. This could be as a result of biochemical activities of the indigenous microbes within the wastewater which tend to use up the pollutants in the wastewater. This was in agreement with the observation of Sarmila [23] whose study revealed that biological treatments of aquaculture wastewater are carried out by mixed microbial cultures to decompose and remove toxic wastes.

The temperature range measured in the phytoremediated wastewater was within temperature tolerance limit for duckweed growth. Culley [64] reported that the upper temperature tolerance limit for duckweed growth was around 34°C with a slight decrease in growth below 10°C. It was also proved that duckweed survived in outdoor wastewater treatment tanks [65].

The pH level in all the treatments tend to increase after two weeks of bio-remediation but the highest increase in pH was recorded in the duckweed. The high pH value also enhances the process of ammonia volatilization and this means that the duckweed treatment in this case functioned similarly to an algal bio-remediated wastewater pond with major ammonia removal attributed to volatilization [66]. Ammonia volatilization is mainly linked to pH and temperature.

Duckweed plant was more effective in bio-remediation of aquaculture wastewater than *Bacillus sp.* This could be as a result of combined effects of plant uptake and bacteria (endophytic and rhizospheric bacteria) associated with duckweed in phytoremediation process compare with the microbial remediation where only bacteria are involved in the bio-remediation of the wastewater. Similar result was observed by El-Kheir [67-69] and Farrell [70] in bio-remediation of wastewater. Presence of such bacteria in plants leads to more efficient phytoremediation activity, and reduces the need for additional fertilization [71-74].

It should be noted from this that bio-remediation of aquaculture wastewater does not result to complete removal of pollutants in the wastewater. However, bio-remediation system has expressed a great potential in treatment of aquaculture wastewater by reducing the level of nutrient and solid pollutants in the wastewater. Martinez-Cordova [37] study on bio-remediation of effluent from shrimp culture system also supports this observation.

Proximate analysis and biomass yield of duckweed: The results of proximate composition of the duckweed before and after bio-remediation are presented in Table 2. Significant difference was recorded in the crude protein, crude fibre and carbohydrate composition of the duckweed samples before and after bio-remediation. The crude protein value of the duckweed increased from 17.65% to 18.47%; ether extract, from 4.41% to 4.46%; ash, from 13.24% to 13.28%; crude fibre, from 16.18% to 19.11%; while the level of carbohydrate of the plant declined from 48.53% to 44.59%. The average wet of duckweed increased from 49.53g to 98.92 after bio-remediation.

Proximate composition (%)	Before Bioremediation	After Bioremediation	Sig-value
Crude Protein	17.65 ± 0.19	18.47 ± 0.09	<0.05
Ether Extract	4.41 ± 0.92	4.46 ± 0.13	>0.05
Ash	13.24 ± 0.04	13.28 ± 0.07	>0.05
Crude Fibre	16.18 ± 0.12	19.11 ± 0.10	<0.05
Carbohydrates	48.53 ± 0.16	44.59 ± 0.17	<0.05

Note: Mean values on the same row with Sig-values >0.05 are not significantly different ($\alpha_{0.05}$).

Table 2: Proximate composition of duckweed (% Dry matter) before and after bioremediation of aquaculture waste water for two weeks.

Water Parameters	Treatments			Sig.-values
	Tc	Tb	Td	
Phosphate (mg/L)	0.38 ± 0.05	15.25 ± 0.67	7.23 ± 0.40	0.00
Nitrate (mg/L)	1.28 ± 0.04	7.07 ± 0.46	5.84 ± 0.42	0.00
Sulphate (mg/L)	0.33 ± 0.06	5.10 ± 0.34	0.53 ± 0.03	0.00
BOD (mg/L)	3.52 ± 0.39	27.48 ± 0.33	2.77 ± 0.45	0.00
COD (mg/L)	32.13 ± 1.75	71.47 ± 1.09	12.33 ± 0.33	0.00
DO (mg/L)	3.60 ± 0.30	4.03 ± 0.10	4.82 ± 0.28	0.00
pH	6.66 ± 0.18	8.18 ± 0.30	8.54 ± 0.27	0.00
Temperature (°C)	27.71 ± 0.15	26.14 ± 1.02	26.25 ± 0.71	0.03
TAN (mg/L)	0.06 ± 0.03	0.66 ± 0.24	0.26 ± 0.09	0.00
TSS (mg/L)	23.39 ± 0.88	2142.19 ± 61.88	1405.08 ± 13.21	0.00

Note: Mean values on the same row with Sig.-value >0.05 are not significantly different ($\alpha_{0.05}$).

Table 3: Water quality parameters of bioremediated aquaculture wastewater with *Bacillus sp.* (Tb) and duckweed (Td) and well water use in production from the sampled fish farm (Tc) used in culturing the experimental fish.

The crude protein of the duckweed sample showed an increase after two weeks of bio-remediation. This observation corroborates the report of Ansal [75]. Nelson [76] inferred that through transformation, absorbed ammonia is converted to plant protein, which may be utilized for growth; resulting in an increase in biomass yield of duckweed as observe in this study. This may partly explain the reduction in ammonia level of duckweed-treated water and a significant increase in crude protein in the plant. The crude protein value recorded in the duckweed samples prior bio-remediation was similar to the value reported by Solomon and Okomoda [77]. The increase in biomass yield is in line with the work of Edward [78] who observed that pond water with less than 3 mg/L TKN and 0.3 mg/L total phosphate (TP) did not support normal growth of *Lemna perpusilla* and *Spirodela polyrrhiza*. The limiting factor in waters for Lemnaceae growth is mainly phosphorus [79]. In this study, the high value of phosphate in wastewater prior to exposure may indicate the sufficiency in the water for uptake, resulting in a reduction post-exposure (Table 1). Several factors inhibit duckweed growth rates. Growth rate decreases due to overcrowding as biomass accumulates to the point that fronds start overlapping each other [29,53,64] and decline in nutrient level in the wastewater [80].

Environmental condition of experimental fish in bio-remediated wastewater: Significant difference was recorded in the water quality parameters of bio-remediated aquaculture wastewater with duckweed and *Bacillus sp.* and the control treatment (well water used in production from sampled fish farm) as shown in Table 3. The highest mean values of phosphate, nitrate, sulphate, BOD, COD, pH, TAN and TSS were recorded in *Bacillus sp.* bio-remediated wastewater used in culturing Nile tilapia while the highest mean pH value was recorded in duckweed bio-remediated wastewater used in culturing Nile tilapia. However, the lowest mean concentrations of phosphate, nitrate, sulphate, DO, pH, TAN and TSS were observed in the control treatment (well water used in production from the sampled fish farm). Higher physicochemical parameters measured in the bio-remediated (recycled) wastewater used in culturing Nile tilapia could be as a result of incomplete removal of pollutants in the water which is furthered exacerbated by feed used during the feeding trial experiment. This result contradicts the observation of Martinez-Cordova [37] in a related study.

Growth performance and nutrient utilization: There was a significant difference in mean weight gain of Nile tilapia juveniles raised in bio-remediated aquaculture wastewater with duckweed and

Parameters	Treatments			Sig-values
	Tc	Td	Tb	
Mean initial weight(g)	10.44 ± 0.01	10.42 ± 0.02	10.43 ± 0.01	0.47
Mean final weight(g)	13.29 ± 0.04	13.39 ± 0.01	14.01 ± 0.05	0.00
Mean weight gain (g)	2.85 ± 0.03	3.58 ± 0.06	2.97 ± 0.03	0.00
Mean daily weight gain (g/day).	0.92 ± 0.02	1.15 ± 0.03	0.95 ± 0.03	0.00
Percentage weight gain (%)	27.33 ± 0.31	34.37 ± 0.60	28.46 ± 0.04	0.00
Specific growth rate (%)	0.19 ± 0.00	0.23 ± 0.01	0.19 ± 0.05	0.00
Condition Factor	0.07 ± 0.01	1.36 ± 0.03	0.08 ± 0.01	0.00
Mean feed intake (g/fish)	6.92 ± 0.09	6.13 ± 0.01	5.71 ± 0.10	0.00
Food conversion ratio	2.42 ± 0.02	1.59 ± 0.03	2.06 ± 0.06	0.00
Gross Feed conversion efficiency ratio	41.27 ± 0.26	62.90 ± 1.19	48.57 ± 1.42	0.00
Protein intake (g)	41.51 ± 0.02	27.57 ± 0.03	36.73 ± 0.03	0.00
Protein efficiency ratio	0.07 ± 0.00	0.13 ± 0.00	0.08 ± 0.00	0.00
Survival rate (%)	100.00 ± 0.00	72.20 ± 1.95	77.80 ± 2.30	0.00

Note: Mean values on the same row with Sig.-value >0.05 are not significantly different ($\alpha_{0.05}$).

Table 4: Growth performance and nutrient utilization of Nile tilapia (*Oreochromis niloticus*) cultured in bio-remediated aquaculture wastewater with *Bacillus sp.* (Tb) and duckweed (Td) and well water use in production from the sampled fish farm (Tc).

Bacillus sp. and the control treatment (well water used in production from sampled fish farm) as presented in Table 4. The mean weight gain, mean daily gain, specific growth rate, percentage weight gain, Gross Feed conversion efficiency ratio and condition factor recorded in Nile tilapia cultured in duckweed bio-remediated aquaculture wastewater were significantly higher than the mean values recorded in Nile tilapia raised in wastewater bio-remediated with *Bacillus sp.* and well water. With respect to FCR, fish cultured in bio-remediated wastewater had FCR values (1.59 ± 0.03 and 2.06 ± 0.06 in duckweed and *Bacillus sp.* treatments respectively) which were significantly lower than FCR of 2.42 ± 0.02 recorded in fish cultured in the control treatment (well water). Base on the results of this study, it can be hypothesized that bio-remediated aquaculture wastewater tend to be more productive in terms of growth and nutrient utilization of culture fish than non-bio-remediated water. This may be attributed to the presence of beneficial microfloral and fauna colonizing the bio-remediated wastewater due to its richness in supporting nutrients as well as improved water quality. Therefore, higher efficiency in conversion of feed to biomass observed in fish cultured in bio-remediated aquaculture wastewater implies better economic returns from utilization of bio-remediated wastewater in aquaculture production system. In a similar context, Martinez-Cordova [37] observed better productive response of shrimp reared in bio-remediated effluents than untreated wastewater. Juarez [13] also

Parameters		WG	FI	FCR	GFCE	M	K	PO₄³⁻	NO₃⁻	SO₄²⁻	BOD	COD	DO	pH	Temp	TAN	TSS
WG	R	1															
	Sig.	.															
FI	R	0.279	1														
	Sig.	0.159	.														
FCR	R	-.641**	.498**	1													
	Sig.	0	0.008	.													
GFCE	R	.685**	-.457*	-.973**	1												
	Sig.	0	0.017	0	.												
M	R	-.476*	.529**	.923**	-.884**	1											
	Sig.	0.012	0.005	0	0	.											
K	R	.771**	-0.175	-.807**	.873**	-.686**	1										
	Sig.	0	0.382	0	0	0	.										
PO₄³⁻	R	-.704**	0.129	.808**	-.818**	.780**	-.697**	1									
	Sig.	0	0.52	0	0	0	0	.									
NO₃⁻	R	-.807**	-.434*	.489**	-.457*	.435*	-.579**	.510**	1								
	Sig.	0	0.024	0.01	0.017	0.023	0.002	0.007	.								
SO₄²⁻	R	-.670**	-.428*	.384*	-0.355	.494**	-.441*	.537**	.866**	1							
	Sig.	0	0.026	0.048	0.069	0.009	0.021	0.004	0	.							
BOD	R	-.498**	.455*	.878**	-.850**	.945**	-.694**	.735**	.468*	.507**	1						
	Sig.	0.008	0.017	0	0	0	0	0	0.014	0.007	.						
COD	R	-.500**	.452*	.877**	-.851**	.943**	-.696**	.736**	.469*	.504**	1.000**	1					
	Sig.	0.008	0.018	0	0	0	0	0	0.014	0.007	0	.					
DO	R	-0.095	-.904**	-.526**	.554**	-.436*	0.325	-0.223	.436*	.537**	-.403*	-.405*	1				
	Sig.	0.639	0	0.005	0.003	0.023	0.099	0.263	0.023	0.004	0.037	0.036	.				
pH	R	.559**	-.493**	-.858**	.884**	-.783**	.868**	-.651**	-0.324	-0.29	-.704**	-.703**	.537**	1			
	Sig.	0.002	0.009	0	0	0	0	0	0.099	0.143	0	0	0.004	.			
Temp	R	.846**	.444*	-.468*	.498**	-.403*	.661**	-.605**	-.864**	-.825**	-0.341	-0.34	-.390*	.500**	1		
	Sig.	0	0.02	0.014	0.008	0.037	0	0.001	0	0	0.082	0.082	0.044	0.008	.		
TAN	R	-0.342	-.786**	-0.259	0.297	-0.307	0.15	-0.151	.630**	.490**	-0.245	-0.244	.797**	.475*	-.406*	1	
	Sig.	0.081	0	0.191	0.132	0.12	0.456	0.453	0	0.009	0.218	0.22	0	0.012	0.035	.	
TSS	R	-.720**	-.442*	.431*	-.405*	.494**	-.447*	.587**	.916**	.950**	.553**	.554**	.488**	-0.207	-.776**	.581**	1
	Sig.	0	0.021	0.025	0.036	0.009	0.019	0.001	0	0	0.003	0.003	0.01	0.3	0	0.001	.

Note: **Correlation is significant at the 0.01 level (2-tailed).
*Correlation is significant at the 0.05 level (2-tailed);
Abbreviations: WG: Weight Gain; FI: Feed Intake; FCR: Feed Conversion Ratio; GFCE: Gross Feed Conversion Efficiency Ratio; PO₄³⁻: Phosphate; NO₃⁻: Nitrate; SO₄²⁻: Sulphate; BOD: Biological Oxygen Demand; COD: Chemical Oxygen Demand; DO: Dissolved Oxygen; Temp: Temperature; TAN: Total Ammonia Nitrogen; TSS: Total Suspended Solid.

Table 5: Spearman correlation matrix indicating the relationship between the culture water quality and growth parameters of Nile Tilapia.

considered lower FCR to be a profitable value for commercial purposes in fish production.

Effect of physicochemical parameters of bio-remediated wastewater on growth performance and nutrient utilization of Nile tilapia: Regarding the influence of physicochemical characteristics of culture water on growth and nutrient utilization performance of Nile Tilapia, the correlation matrix (Table 5) indicates that phosphate, nitrate, sulphate, BOD, COD and TSS had a significant strong negative influence on the weight gained of the experimental fish while nitrate, sulphate, TAN and TSS tend to exhibit a strong significant negative influence on the feed intake of the culture fish. This is an indication that excessive of the enlisted water quality variables above (indicators of pollution) as a result of indiscriminate discharge of aquaculture wastewater in aquatic ecosystem will hamper productivity of fish in the affected waterbody. This therefore necessitates the need for treatment (bio-remediation) of aquaculture wastewater before discharge. In line with Redner and Stickney, [30] and El-Sherif and El-Feky [68] observed nitrate (nitrogenous pollutant) has a negative influence on the survival rate of the experimental fish. It depresses feed intake and growth at concentrations as low as 0.1 mg/L [68]. The relatively higher level of ammonia recorded for bacteria-treated water (0.66 mg/L) may partly explain the lower growth performance compared to duckweed. However, the recorded value of ammonia in duckweed bio-remediated wastewater was higher than the control treatment which had the highest fish survival rate. The lowest survival rate recorded for duckweed compared to bacteria and control treatments may stem from elevated level of pH and temperature. High pH could increase the toxicity of ammonia lower dissolved oxygen level which reduces fish survival [74]. This may be due to species differences in optimum levels for fish growth. The optimum concentration of ammonia for Nile tilapia was estimated to be below 0.05 mg/L [68]. This may suggest that 0.26 mg/L ammonia did not affect growth performance; survival rate could be reduced following prolong exposure. The correlation analysis of water quality and growth performance established a strong negative linear relationship between nitrogenous compounds (TAN and nitrate) and growth as well as feed intake of the culture fish. Frias-Espericueta and Ray also observed that exposure of penaeid shrimps to high concentrations of nitrogenous compounds and suspended solids have negative impact on their growth and food intake.

Conclusion

It was evident from all the results obtained from the experiment that duckweed is highly effective in the removal of high toxic organic waste components of wastewater discharge from catfish farm. The technology involved in the use of duckweed is very simple and at the lower cost when compare to the use of *Bacillus sp.* in bio-remediation of waste water, which require high level of expertise in identification, isolation, mass production and application. The use of duckweed in bio-remediation was effective in reducing high phosphate, sulphate, ammonia, nitrate, biological oxygen demand and chemical oxygen demand in aquaculture wastewater. The reuse of Duckweed-treated wastewater is suitable for fish culture without affecting growth performance, and dissolved oxygen level as most polluting substances were reduced significantly. *Bacillus sp.* was only effective in removal of sulphate, ammonia and nitrate in the wastewater. Future research effort should investigate long term growth studies under different culture conditions and fish species to assess the use of duckweed as an effective approach to sustain aquaculture development.

References

1. Food and Agriculture Organization (FAO) (2014) State of the world fisheries. FAO, Rome.

2. Fernandes TF, Eleftheriou A, Ackerfors H, Eleftheriou M, Eruik A, et al. (2001) The scientific principles underlying the monitoring of the environmental impacts of aquaculture. Journal of Applied Icthyology 17: 181-193.

3. Hasan MR (2001) Nutrition and feeding for sustainable aquaculture development in the third millennium. Aquaculture in the Third Millennium. Technical Proceedings of the Conference on Aquaculture in the Third Millennium, Bangkok, Thailand, NACA, Bangkok and FAO, Rome.

4. Pandey A, Satoh S (2006) Effects of organic matter on growth and phosphorus utilization in rainbow trout *Onchorhynchus mykiss*. Fisheries Science 74: 867-874.

5. Livestock Research for Rural Development of Ponds (2001) An ecologically and economically viable integrated approach for rural development through aquaculture.

6. Population of Ferguson's Gulf, Lake Turkana, Kenya. J Fish Biol 33: 181-188.

7. Tacon AGJ, Forster IP (2003) Global trends and challenges to aquaculture and aqua feed development in the new millennium. Middlesex, UK.

8. Wang YB, Xu ZR, Guo BL (2005) The danger and renovation of the deteriorating pond sediment. Feed Industry 26: 47-49.

9. Hardy RW, Gatlin DM (2002) Manipulations of diets and feeding to reduce losses of nutrients in intensive aquaculture In: Aquaculture and the environment in the United States, World Aquaculture Society, Baton Rouge, Louisiana.

10. Lazzari R, Baldisserotto B (2008) Nitrogen and phosphorus waste in fish farming. B Inst Pesca São Paulo 34: 591-600.

11. Stephen WW, Farris JL (2004) Stream community assessment of aquaculture effluents. Aquaculture 231: 148-162.

12. Stewart KM (1988) Changes in condition and maturation of the *Oreochromis niloticus* L. Journal of Fish Biology 33: 181-188.

13. Magni P, Rajagapal S, Vandervelde G, Perel G, Kasserberg J, et al. (2008) Sediment features, macroziobathic assemblages and trophic relationship following a dystrophic event with anoxia and sulphide development in the Santa Giuta Lagoon. Marine Pollution Bulletin 57: 125-136.

14. Carr OJ, Goulder R (1990) Fish farm effluents in Rivers: Effects on bacterial populations and alkane phosphatase activity. Water Research 24: 631-638.

15. Vidali M (2001) Bioremediation: An overview. Pure and Applied Chemistry 73: 1163-1172.

16. Leung M (2004) Bioremediation: Techniques for cleaning up a mess. Journal of Biotechnology 2: 18-22.

17. Levent S, Mustafa A, Erhan A (2007) Weight-Length relationships for 39 fish species from the North-Eastern Mediterranean Coast of Turkey. Turk J Fish Aquat Sci. 7: 37-40.

18. Mueller JG, Cerniglia CE, Pritchard (1996) Bioremediation of Environments by contaminated Polycyclic Aromatic Hydrocarbons. In: Bioremediation: Principles and Applications, Cambridge University Press, Cambridge.

19. Nayyef MA, Amal AS (2012) Efficiency of *Lemna minor* L. in the phytoremediation of waste water pollutants from Basrah oil refinery. Journal of Applied Biotechnology in Environmental Sanitation 1: 163-172.

20. Sharma S (2012) Bioremediation: Features, Strategies and Applications. Asian Journal of Pharmacy and Life Science 2: 2231-4423.

21. Prescott LM, Harley JP, Klein DA (2002) Microbiology. (6th edn). McGraw Hill Publishers.

22. *Pseudomonas* sp. strain ADP: Gene sequence, enzyme purification, and protein characterization. J Bacteriol 178: 4894-4900.

23. Sarmila M, Vikineswary S, Geok-Yuan AT, Ving CC (2015) Identification of indigenous bacteria isolated from shrimp aquaculture wastewater with bioremediation application: Total ammonia nitrogen (TAN) and nitrite removal. Sains Malaysiana 44: 1103-1110.

24. Watanabe K, Kodoma Y, Stutsubo K, Harayama S (2001) Molecular characterization of bacterial populations in petroleum contaminated ground water undergoing water discharge from crude oil storage cavities. Applied and Environmental Microbiology 66: 4803-4809.

25. Sasikumar CS, Papinazath T (2003) Environmental management: Bioremediation of polluted environment. Proceedings of the third International conference on environment and health, Chennai, India.

26. De Souza ML, Sadowsky MJ, Wackett LP (1996) Atrazine chlorohydrolase from *Pseudomonas* sp. strain ADP: Gene sequence, enzyme purification, and protein characterization. J Bacteriol 178: 4894-900.

27. Chandrika V, Nair PVR (1992) Studies on bacterial flora on Trivandrum Coastal Waters. J Mar Biol Assoc India 34: 47-53.

28. Queiroz JF, Boyd CE (1998) Effects of bacterial inoculum in channel catfish ponds. J World Aquaculture Society 29: 67-73.

29. Reddy KR, Debusk WF (1985) Nutrient removal potential of selected aquatic macrophytes. Journal of Environmental Quality 14: 459-462.

30. Redner BD, Stickney RR (1979) Acclimation to ammonia by Tilapia aurea. Trans Am Fish Soc 108: 383-388.

31. Jerald LS (1997) Technology Evaluation Report: Phytoremediation. TE-98-01: 1-6.

32. Schnoor JL, Licht LA, McCutcheon SC, Wolfe NL, Carriera LH (1995) Phytoremediation: Uptake and metabolism of organic compounds: Green-Liver Model, in McCutcheon. Biorem Jou 4: 17.

33. Chaudhary E, Sharma P (2014) Use of Duckweed in wastewater treatment. International Journal of Innovative Research in Science, Engineering and Technology 3: 13622-13624.

34. Smith MD, Moelyowati I (2001) Duckweed based wastewater treatment (DWWT). Design guidelines for hot climates. Water Sci Technol 43: 291-299.

35. American Public Health Association (APHA) (2005) Standard methods for the examination of water and wastewater. (21stEdn), APHA, AWWA and WEF.

36. Association of Official Analytical Chemists (AOAC) (1990) Official Methods of Analysis. (15thEdn), Association of Official Analytical Chemists, Inc. Virginia, USA.

37. Martinez-Cordova LR, Lopez-Ellias JA, Leyva-Miranda G, Armenta-Ayoin L, Martinez-Porchas M (2011) Bioremediation and reuse of shrimp aquaculture effluents to farm whiteleg shrimp, *Litopenaeus vannamei:* A first approach 42: 1415-1423.

38. Ling TY, Buda D, Nyanti L, Norhadi I, Emang JJJ (2010a) Water quality and loading of pollutants from shrimp ponds during harvesting. Journal of Environmental Science and Engineering 4: 13-18.

39. Lee N, George B, Ling TY (2011) Shrimp Pond effluent quality during harvesting and pollutant loading estimation using Simpson's Rule. International Journal of Applied Science and Technology 1: 208-213.

40. Babatunde BB, Woke GN (2015) Analysis of the Physicochemical Burden of Oyo State Fish Pond, Ibadan, Southwest Nigeria. J Appl Sci Environ Manage 19: 259-264.

41. Bagenal TB, Tesch FW (1978) Methods for assessment of fish production in freshwaters. Oxford, Blackwell Scientific Publication.

42. Amirkolaie AK (2008) Environmental impacts of nutrient discharged by aquaculture waste water on Haraz River. Journal of Fisheries and Aquatic Science 3: 275-279.

43. Ogwo PA, Ogu OG (2014) Impact of industrial effluents discharge on the quality of Nwiyi river Enugu South Eastern Nigeria. IOSR Journal of Environmental Science, Toxicology and Food Technology 8: 22-27.

44. Soonnenholzner S, Boyle CE (2000) Chemical and physical properties of shrimp pond bottom soils in Ecuador. Journal of the World Aquaculture Society 31: 358-375.

45. Bowley DG, Allan GL (2012) Nutrients in pond based aquaculture discharge water used for irrigation.

46. Ganczarczyk JJ (1983) Activated sludge process. Marcel Dekker, Inc., New York, USA.

47. McCasland M, Trautmann N, Porter K, Wagenet R (2008) Nitrate: Health effects in drinking water.

48. Hurse JT, Connor AM (1999) Nitrogen removal from wastewater treatment lagoons. Water Sci Technol 39: 191-198.

49. CDC (2002) US Toxicity of Heavy Metals and Radionucleotides. Department of Health and Human Services, Centers for Disease Control and Prevention. Savannah river-site health effects subcommittee (SRSHES) meeting.

50. Chandrakant SK, Shwetha SR (2011) Role of microbial enzymes in the bioremediation of pollutants: A review. Enzyme Research 2011: 1-11.

51. Kurosu O (2001) Nitrogen removal from wastewaters in micro-algal bacterial-treatment ponds.

52. Akpor OB, Muchie M (2011) Review: Environmental and public health implications of wastewater quality. African Journal of Biotechnology 10: 2379-2387.

53. Al-Nozaily FA (2001) Performance and Process analysis of duckweed-covered sewage lagoons for high strength sewage. Doctoral dissertation. Delft University of technology. International Institute of Hydraulic and Environmental Engineering. Delft-Holland.

54. Federal Environmental Protection Agency (FEPA) (1991) "National Environmental Protection Regulations (Effluent Limitation)". Regulations S. 1. 8. Federal Republic of Nigeria Official Gazette. Lagos, Nigeria.

55. World Health Organization (2004) Guidelines for drinking water quality. (3rd edn) Recommendation. WHO: Geneva, Switzerland.

56. Zar JH (1996) Bio-statistical analysis. Prentice Hall, Upper Saddle River, New Jersey.

57. Turnbull PCB (1996) *Bacillus.* In: Barron's Medical Microbiology, (4th edn). University of Texas Medical Branch.

58. Koermer S, Vermaat IE (1998) The relative importance of *Lemna gibba* L., bacteria and algae for nitrogen and phosphorous removal in duckweed-covered domestic wastewater. Water Res 33: 3651-3661.

59. Kosh R (1883) Isolation of individual bacterial colonies on solid media.

60. Ugya YA (2015) The efficiency of *Lemna minor* L. in the phytoremediation of Romi Stream: A case study of Kaduna Refinery and petrochemical company polluted stream. J App Biol Biotech 3: 011-014.

61. Umran TU, Sadettin EO (2015) Removal of heavy metals (Cd, Cu, Ni) by electrocoagulation. International Journal of Environmental Science and Development 6: 425-429.

62. Zimmon OR, Van Der Steen NP, Gijzen HJ (2005) Effect of organic surface load on process performance of pilot scale algae and duckweed based waste stabilization ponds. J Environ Engr 131: 587-594.

63. Bhutto MA, Dahot MU (2010) Effect of alternative carbon and nitrogen sources on production of alpha-amylase by *Bacillus megaterium.* World Applied Sciences Journal 8: 85-90.

64. Culley DD, Rejmankova E, Kvet J, Frey JB (1981) Production, chemical quality and use of duckweeds (Lemnaceae) in aquaculture, waste management and animal feeds. J World Maric Soc 12: 27-49.

65. Classen JJ, Cheng J, Bergmann BA, Stomp AM (2000) *Lemna gibba* growth and nutrient uptake in response to different nutrient levels. In: Animal, Agriculture and Food Processing. Proceedings of the 8th International Symposium, Des Moines Iowa.

66. Blier R, Laliberte G, de La Noue J (1995) Tertiary treatment of cheese factory anaerobic effluent with Phormidium bohneri and Micractinium puspillum. Bioresource Technol 22: 151-155.

67. El-Kheir WA, Ismail G, El-Nour A, Tawfik T, Hammad D (2007) Assessment of the efficiency of duckweed (Lemna gibba) in wastewater treatment. International Journal of Agriculture and Biology 5: 681-689.

68. El-Sherif MS, El-Feky AMI (2009) Performance of Nile tilapia (*Oreochromis niloticus*) fingerlings. I. Effect of pH. Int J Agric Biol 11: 297-300.

69. Fagade SO (1979) Observation of the biology of two species of Tilapia from Lagos lagoon Nigeria. Bull Inst Fond Afr Norc (Scr A) 41: 627-658.

70. Farrell JB (2012) Duckweed uptake of phosphorus and five pharmaceuticals: Microcosm and Wastewater Lagoon Studies. All Graduate Theses and Dissertations 1212: 57-120.

71. Afzal M, Khan QM, Sessitsch A (2014) Endophytic bacteria: Prospects and applications for the phytoremediation of organic pollutants. Chemosphere 117: 232-242.

72. Akan JC, Abdulrahman FI, Dimari GA, Ogugbuaja VO (2008) Physicochemical determination of pollutants in wastewater and vegetable samples along the Jakara wastewater Channel in Kano Metropolis, Kano State, Nigeria. European Journal of Scientific Research 23: 122-133.

73. Akinrotimi OA, Abu OMG, Ansa EJ, Edun OM, George OS (2009) Haematological responses of Tilapia guineensis to acute stress. International Journal of Natural and Applied Sciences 5: 338-343.

74. Akpoilih BU, Ajani EK, Omitoyin BO (2015) Dietary phytase improves growth and water quality parameters for juvenile *Clarias gariepinus* fed soyabean meal-based diets. International Journal of Aquaculture 5: 1-20.

75. Ansal MD, Dhawan A, Kaur VI (2010) Duckweed based bio-remediation of village: An ecologically and economically viable integrated approach for rural development through aquaculture. Livestock Research for Rural Development 22.

76. Nelson SG, Smith BD, Best BR (1981) Kinetics of nitrate and ammonia uptake by the tropical fresh water macrophyte *Pista stratiotes* L. Aquaculture 24: 11-19.

77. Solomon SG, Okomoda VT (2012) Growth performance of *Oreochromis niloticus* fed duckweed *(Lemna minor)* based diets in outdoor hapas. International Journal of Research in Fisheries and Aquaculture 2: 61-65.

78. Edwards P, Hassan MS, Chao CH, Pacharaprakiti C (1992) Cultivation of duckweeds in septage loaded earthen ponds. Bioresource Technol 40: 109-117.

79. Landolt E (1996) Duckweeds (Lemnaceae): Morphological and ecological characteristics and their potential for recycling nutrients. In: Environmental research forum Vols. 5-6: Recycling the resource, ecological engineering for wastewater treatment, Transtec Publications, Switzerland.

80. Hassan MS, Edwards P (1992) Evaluation of duckweed *(Lemna perpusilla* and *Spirodela polyrrhiza)* as feed for Nile tilapia (*Oreochromis niloticus*). Aquaculture 104: 315-326.

11

Abundance and Distribution of Freshwater Eels in Pangi River, Maitum, Sarangani Province

Valdez ASM* and Castillo TR

Mindanao State University, General Santos City, Philippines

Abstract

A study on the abundance and distribution of Freshwater Eels was conducted in Pangi River, Maitum Sarangani Province. Three stations, 50 meters away from each other, were set starting at the mouth of the river with coordinates 06.02278° N, 124.52069°. In every station are three substations at the right side near the bank, at the center and at the left side near the bank. The study was done from a period of 5 months starting from October 2014 to February 2015. Sampling was done every Full moon and New moon of the month during the low tide and high tide fluctuation. Salinity and temperature were also monitored during sampling.

The freshwater eels were identified using morphological characteristics. A total of 4,262 individuals had been collected for five (5) months of sampling, Freshwater eels were found abundant during new moon with a total of 4,249 individuals and only 13 individuals were collected during full moon.

Results showed that two species of Freshwater eels has been observed on the area out of 18 species of Freshwater Eels, the Anguilla marmorata and the Aguilla bicolor pacifica. The study further showed that the Anguilla marmorata is most abundant with abundance index of 96.62%, frequency index of 100% and dominance index of 196.62%, while Anguilla bicolor pacifica had the abundance index of 3.26%, frequency index of 42.86% and dominance index of 46.12%.

There are two types of bottom substrates observed in the area Station 1 had sandy while Station 2 and 3 had rocky substrate. Anguilla marmorata is more often seen in the rocky bottom since they prefer to hide under the rocks and gravely bottom.

Catch per unit effort (CPUE) was computed to range from 0.075gm/hour to 500 gm/hour. Temperature was monitored to be within the range of 23°C-25°C for the entire sampling period.

Further study is recommended to monitor the freshwater eels that migrating upstream the river of Pangi, Maitum, Sarangani Province, beyond Station 3.

Keywords: Abundance; Distribution; Eels; Pangi river

Introduction

Eel is the common name for any fish of the 10 families constituting the Order Anguilliformes, it is characterized by a long snakelike body covered with minute scales embedded in the skin. Eels lack the hind parts of fins, adapting them for wriggling in the mud and through the crevices of reefs and rocky stones.

According to Schrank in year 1798, freshwater eels are catadromous, that spawns in tropical ocean waters, and has a peculiar leptocephalus larval stage that is unique to elopomorph fish. They constitute a single Genus, Anguilla.

The fishing for freshwater eels, along with environmental pollution and other human impacts, have all contributed to a significant decline in eel numbers over the last 25-30 years. Total volume of glass eels collected on an annual basis is around 150 tonnes which satisfies the current aquaculture needs of approximately 100 tonnes/year with the excess going to human consumption in Spain [1]. Many people are involved in the eel collection, transportation and distribution, from glass eel fishers to the eel farmer and processor [1].

In the Philippines, freshwater eels have different populations scattered in various regions in country in which they are considered important food fish especially to the indigenous people [2]. There has been documented on the evolutionary history and phylogenetic studies of the eels in the country and by traditional molecular methods, where this fishes are quite abundant and geographically near to their spawning ground [2]. Knowledge and background of the natural life history and specification are the basis for sound fishery management or aquaculture industry development of eels.

The study provides baseline information on the abundance and distribution of the freshwater eels that present in Pangi River Maitum, Sarangani Province. This serves as basis for action plan of the Local and National Government Unit for protection and conservation of Freshwater Eels in the area [3-10].

Materials and Methods

Sampling area and sampling stations

Pangi River can be found at Brgy. Pangi, Maitum, Sarangani Province (Figure 1). The sampling stations stretch from 50 to 100 meters which is further subdivided into tree substations, where two (2)

---

*Corresponding author: Valdez ASM, Mindanao State University, General Santos City, Philippines, E-mail: enairashane.zedlav@yahoo.com.ph

is designated at both sides and 1 substation at the center.

There are two (2) types of substrates observed in the area: sandy and rocky. The mouth of the river or the station 1 had its sandy bottom while the station 2 and 3 had its rocky bottom.

The Sampling Stations will be as follows:

Station 1-at the mouth of the river with the coordinates of 06.02278° N, 124.52069° E

Substation 1: at the right side near the bank

Substation 2: at the center

Substation 3: at the left side near the bank

Station 2-50 meters ahead from the station 1 with the coordinates of 06.02435° N, 124.52196° E

Substation 1: at the right side near the bank

Substation 2: at the center

Substation 3: at the left side near the bank

Station 3-100 meters ahead from the station 1 with the coordinates of 06.02479° N, 124.52287° E

Substation 1: at the right side of the bank

Substation 2: at the center

Substation 3: at the left side near the bank

The first substation is placed on the mouth of the river which expands from 31 meters at low tide to 100 meters up at high tide. The second station will be 50 meters away from the mouth of the river, the third station is 50 meters away from the second station.

Sampling methods and sampling frequency

Sampling is done by collecting the anguillid eels that is trapped using net traps as collecting gear on the 1st, 2nd station and 3rd station. This is made of 8 meters mosquito net, rope and bamboo sticks. A set of three net traps are set together in three stations to intercept as many as possible eels over a wide area. The net traps are set in the afternoon and timed by the incoming high tide and get the trapped glass eels using the (sigpaw) Scoop net.

Sampling is done every full moon and new moon of the month to observe the tidal fluctuation in collecting freshwater eels.

Data collection

The information that are noted and gathered during sampling is the following:

- Date and time of sampling
- Length of fishing period
- Water parameters like temperature, salinity
- Biomass of animals caught
- Morphometric measurements of elvers and adults that will be collected.

The eels that will be collected will be identified through its morphological characteristics.

Determining the species using morphological characteristics (Figure 2).

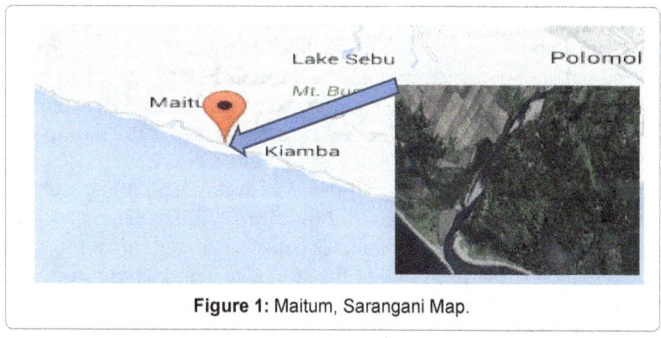

Figure 1: Maitum, Sarangani Map.

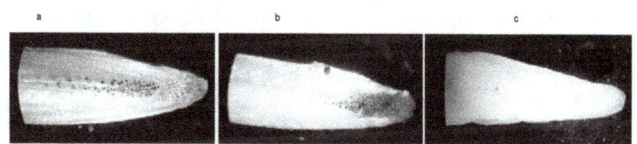

Figure 2: Morphological chracteristics of glass eel species showing the tail bud and caudal fin cutaneous pigmentation patterns;(a) *Anguilla marmorata* and/or *Anguilla luzonensis* (b) *Anguilla bicolorpacifica* (c) *Anguilla japonica*.

The following data are computed using the information gathered during sampling:

Catch per unit effort

The information on biomass of animals caught and length of fishing are used to compute for the Catch per unit effort using the formula:

$$CPUE \ (g/hour) = \frac{(Total \ weight \ of \ catch)}{(No. \ of \ hour \ in \ catching)}$$

$$Abundance \ Index = \frac{Total \ no. \ of \ particular \ species}{Total \ no. \ of \ all \ species} \times 100$$

$$Frequency \ Index = \frac{No. \ of \ sampling \ the \ particular \ species \ found}{Total \ no. \ of \ sampling} \times 100$$

Dominance Index = Abundance Index + Frequency Index

The Collected anguillid eels were fixed in 95% ethanol alcohol just after weighing and the specimens were counted.

Results and Discussion

Results showed that there are two types of Freshwater eels which can be found on Pangi River, Maitum Sarangani Province those are the *Anguilla marmorata* species and the *Anguilla bicolor pacifica*. A total of 4,262 individuals were collected and a total Catch per unit effort of 65.98 g/hr for the duration of five (5) months for the month of October-February (2014-2015). The *Anguilla marmorata* species are the highly observed in the area of sampling with the total number of 4,118 individuals out of the 4,262 individual collected for five (5) months. It had an Abundance Index of 96.62%, Frequency Index 100% and Dominance Index 196.62 and the *Anguilla bicolor pacifica* species with the total number of 139 individuals, had an Abundance Index of 3.26%, Frequency Index 42.86%, and Dominance Index of 46.12 (Figure 3).

The largest catches of the freshwater eels were in the month of November during the New Moon phase and with the zero catches during the month of December and January Full moon. It is also showed that on the station 2, substation 1 right side near the bank is more applicable area on setting the trap than station 1 and station 3, since the water depth and current is more severe on the area of Station

1 and Station 3. As well as visual observation indicated that glass eels were most abundant on the side of the river bank. The trap which is placed in station 1 which is found on the mouth of the river had been destroyed by the waves due to the strong winds and drifted trees during sampling, and for the trap which is set on the center of the station 1 is not comfortable for the collector and trap due to the high level of the water that occur on the center of the river during high tide it reaches 10 to 20 feet depth at the center of the river. While on station 3 it is hard to set the trap since station 3 possesses larger rocks and also not suitable for the traps due to the strong current coming from the upstream of the river, it gives damage to the net trap. Data also shows that freshwater eels is more abundant during the new moon phase than the full moon phase, due to the moon light which is present every full moon, freshwater eels is more seen during dark hours such as new moon, glass eels prefer to migrate during those times, suggesting that the lunar phase has a strong effect on inshore migration mechanism of glass eel at the mouth of the Pangi River, there were 4,249 individuals collected caught during the new moon samples obtained in five (5) months (from 2014-2015) used for comparison during the full moon phase which are 13 individuals collected during those times. Weather condition also had been noted and observed during the sampling period (Figures 4 and 5).

Temperature is also a factor that affects the migration of Freshwater Eels they are more likely to stay on colder part of the river which is more observed on the study, they prefer to migrate upstream. Other study also who has been conducted on other countries said that temperature affects the survival of the Freshwater eels (Figure 6).

Abundance, frequency, and dominance of freshwater eels in Pangi river

The following Figures 7-9 showed that Anguilla marmorata is the

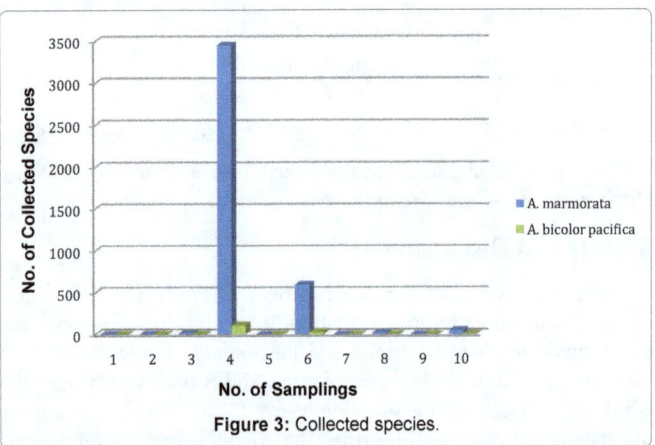

Figure 3: Collected species.

Figure 4: Station 1 destroyed by the waves and drifted trees.

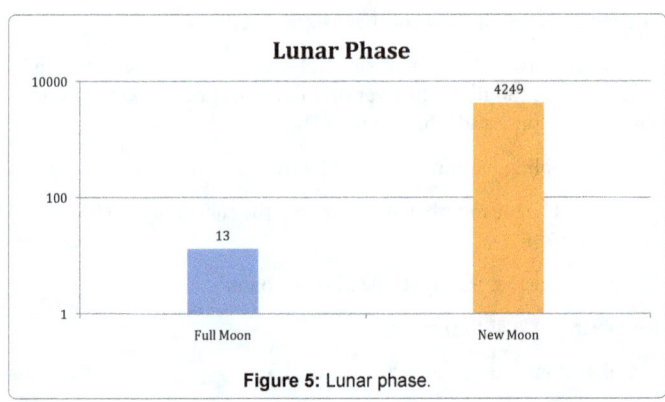

Figure 5: Lunar phase.

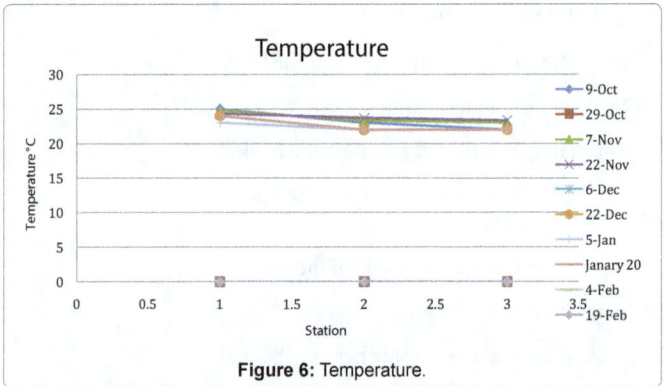

Figure 6: Temperature.

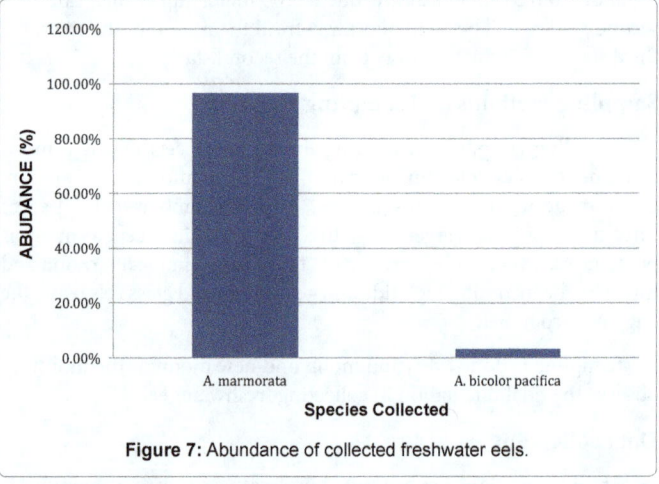

Figure 7: Abundance of collected freshwater eels.

most abundant, frequent, and dominant species of freshwater eels that can be found in the area.

In terms of abundance *A. marmorata* has an average of 96.62%, while the *A. bicolor pacifica* has an average of 3.26%. This could be attributed to their habitat and migration pathways as mentioned earlier.

In terms of frequency, *A. marmorata* has an average of 100%, while *A. bicolor pacifica* has an average of 42.86%.

In terms of Dominance, *A. marmorata* has an average of 196.62, *while A. bicolor pacifica* has an average of 46.12.

Table 1 represents the abundance, frequency and dominance of the two species that has been collected on the study as general.

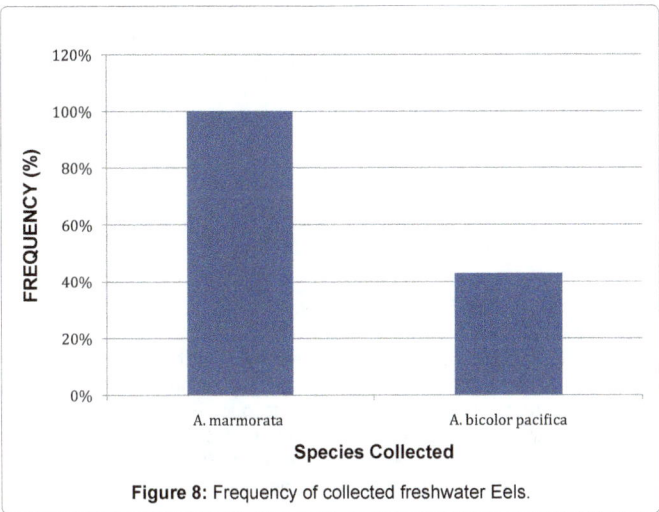

Figure 8: Frequency of collected freshwater Eels.

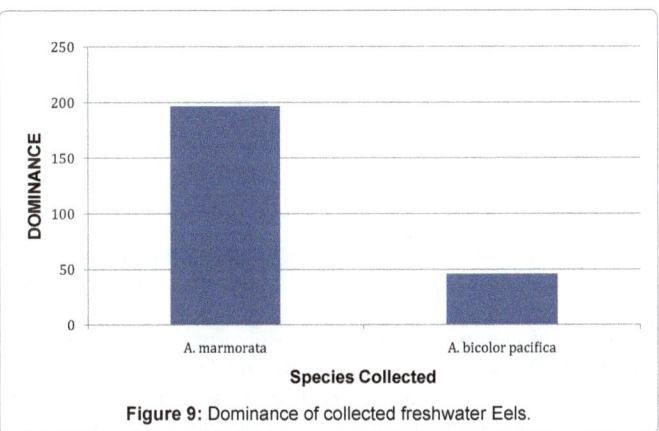

Figure 9: Dominance of collected freshwater Eels.

Species collected	Abundance (%)	Frequency (%)	Dominance	No. of individual
A. marmorata	96.62%	100%	196.62	4,118
A. bicolor pacifica	3.26%	42.86%	46.12	139

Table 1: Table presenting the abundance, frequency ad dominance of the collected freshwater eels.

Conclusion and Recommendation

The Abundance and Distribution of Freshwater Eels in the area of Pangi River, Maitum Sarangani Province revealed that the river had its source for wild eels especially the *Anguilla marmorata* species that suits for culturing process in the near future, since the river possesses a rocky substrate which is good for the habitat of the said species. In conclusion the findings of this study confirm that there are two species which are present on the Pangi River which is the *Anguilla marmorata* and *Anguilla bicolor pacifica*, the good quality of the water and maintaining its cleanliness is another key factor why those wild eels keep on migrating on the area, due also to the abundance of other food that can be found on the area.

Studies in the laboratory or with different types of sampling gear fished in the Pangi River itself will be helpful to further examine the abundance and distribution of freshwater eels in Pangi River, Maitum Sarangani Province.

Acknowledgement

The researcher would like to dedicate her work to those who shared their support and guidance.

To her adviser, Prof. Tersa R. Castillo for the time and supervision, for the significant ideas she had contributed for the success of the study.

To the members of the defense panel, Prof. Glennville A. Castrence and Prof. Ronald P. Sombero for their suggestions and contributions for the best outcome of the study. To Dr. Apolinario A. Yambot for the knowledge he had shared and for his time and guidance throughout the study.

The researcher would like to express her gratitude to Mr. Rey Bonrustro a.k.a uncle illing, Bon Bon, Jannet and Chikoy for the assistance, to her friends, Julius, Jacob, Katelyn, KH Family, and classmates, CFC-Youth for Christ and MBH family for giving her moral support you have made this experience exceptional.

All of these are impossible without her parents Mr Allan A. Valdez and Mrs Ailen M. Valdez, to his siblings Ezra and Dwin and her special someone Rein who had been her inspiration to finish this study.

And lastly, she is most grateful to the Almighty God, for He made all this work in order. Thank you.

References

1. Nielsen T, Prouzet P (2008) Capture-based aquaculture of the wild European eel (Anguilla anguilla). In: Lovatelli A, Holthus PF (eds). Capture-based aquaculture. Global overview. FAO Fisheries Technical Paper. No. 508. Rome, FAO.

2. Jamandre BWD, Shen KN, Yambot AV, Tzeng WN (2007) Molecular phylogeny of Philippine freshwater eels Anguilla spp. (Actinopterygi: Anguilliformes: Anguillidae) inferred from mitochondrial DNA. Raffles Bull Zool 14: 51-59.

3. Yambot AV (2007) Determining the species of glass eels. Molecular Biology and Biotechnology Laboratory College of Fisheries, Central Luzon State University.

4. Ege V (1939) A revision of the genus Anguilla Shaw, a systematics, phylogenetic and geographical study. Dana Rep 16: 1-256.

5. Miller MJ, Tsukamoto K (2004) The worldwide distribution of anguillid leptocephali. In: Aida K, Tsukamoto K, Yamauchi K (eds). Eel Biology. Springer Verlag, Tokyo.

6. Rellon LP (2013) Classification of Pangi River. Program for Public Hearing.

7. Tzeng WN (1985) Immigration timing and activity rhythms of the eel, Anguilla japonica elvers in the estuary of northern Taiwan, with emphasis on environmental influences. Bull Jap Soc Fish Oceanography 47: 11-28.

8. Tzeng WN (2004) Modern Research on the natural life history of the Japanese eel Anguilla japonica. Journal of the Fisheries Society of Taiwan 31: 8-39.

9. Tsukamoto K, Arai T (2001) Facultative catadromy of the eel Anguilla japonica between freshwater and seawater habitats. Mar Ecol Prog Ser 220: 265-276.

10. Watanabe S, Aoyama J, Tsukamoto K (2009) A new species of freshwater eel Anguilla luzonensis (Teleostei: Anguillidae) from Luzon Island of the Philippines. Fisheries Science 75: 387-392.

12

EA Preliminary Study on Domestication of Bluespotted Snakehead (*Channa Lucius,* Channidae) in Concrete Tank

Azrita[1]*, Yuneidi Basri[2] and Hafrijal Syandri[2]

[1]*Department of Biology Education, Faculty of Education of Bung Hatta University Padang, Indonesia*
[2]*Department of Aquaculture, Faculty of Fisheries and Marine Sciences of Bung Hatta University Padang, Indonesia*

Abstract

This study aim to gain insight into the domestication of *Channa lucius* female kept in a concrete tank as much as four plots of each size 200×200×75 cm. *C. lucius* each concrete tank maintained by four females with an average weight of 300 ± 20 g/individual and four males with an average weight of 500 ± 50 g/individual. Treatment in this experiment consisted of four groups are control group (injection 0.9 NaCl) and exposed to 100 µg/kg body weight, 150 µg/kg body weight and 200 µg/kg body weight of LHRHa hormone preparations. Dose level best to increase the reproductive potential of *C. lucius* was 200 µg/kg body weight with time reaching a matured gonads 62 ± 12 days, fecundity was 2,617 ± 250 eggs/spaw, egg diameter 1.87 ± 0.02 mm, the survival embryo 85.92 ± 0.52% and hatching rate 82.41 ± 0.60%. Dose level of LHRHa was significantly ($p < 0.05$) with time mature gonadal, fecundity and eggs diameter. Whereas hatching rate and embryo survival not significantly ($p > 0.05$) between treatments, but significantly ($p < 0.05$) different from the control group.

Keywords: Domestication; *Channa Lucius*; Lhrha; Matured gonadal; Fecundity; Egg diameter; Survival embryo; Hatching rate

Introduction

Channa lucius is a family Channidae includes air breathing freshwater fishes, popularly known as snakeheads or murrels. This species has an important economic value at a high price in the local market ranged between 40.000 to 50.000 IDR/kg [1], These are economically important species having great potential for aquaculture and capture fisheries in Indonesia, especially in the province of West Sumatra, Riau, Jambi and South Sumatra [2,3]. In addition to economically valuable genera of Channa also has important value for human health [4-9]. Zuraini et al. [4] stated meat proximate composition of *C. lucius* is crude protein 19.9% (% DW), crude fat 11.9%, crude ash 1.2% and moisture 80.0% (% WW). This fish is one of the genus Channa are an important source of protein for people throughout the Asia Pacific region [10].

This fish has a chance to do domestication cause basic data on the biology and ecology in nature are known ie are carnivores eating fish, shrimp and frogs, male fish size larger than females. Maturation of female broodstock on the size of the standard length ranged from 245.0 to 400.0 mm and body weight between 156.0 to 560.0 g, total fecundity ranged from 1,152 to 3,746 grains/individual, egg diameter ranged between 1.35 to 1.70 mm. While the maturation of male broodstock on the size of the standard length ranged from 260.0 to 485.0 mm and body weight between 112.50 to 656.0 g [11]. In natural spawning of *C. lucius* putting the eggs around the water plants floating on the surface of the waters [2]. The size of egg *C. lucius* approached the size of the eggs *C. argus* ranged from 1.80 to1.85 mm (Soin, 1960) in [12], the size of egg *C. blehari* ranges from 0.9 to 1.1 mm [12], and size of egg *C. gachua* ranges from 2.1 to 2.6 mm [13]. Fecundity of Chevron snakehead (*C. striata*) in Malaysia ranges from 3,000 to 30,000 grains, egg diameter 1.25 mm [14] and fecundity of Chevron snakehead (*C. striata*) in the floodplain Musi River of South Sumatra Province Indonesia range from 1,141 to 16,468 grains [15]. Domestication of the family Channidae among other *C. punctatus* induced with Ovatide been successfully performed [16] and *C. striatus* with HCG [17]. Based on using LHRHa hormone implantation method was important to gonadal maturation of *C. lucius*.

Material and Methods

Broodstock

The broodstock of *C. lucius* females and males as much as 100 individuals (Figure 1) were collected from fishermen in Rangau River Bengkalis Regancy, Riau Province. Fish samples were transported to Laboratory of Freshwater Fisheries Development Faculty of Fisheries and Marine Sciences of Bung Hatta University Padang, using a plastic bag filled with oxygen with a stocking densities 4 individuals/10 liters of water. Transport equipment used is the car with the distance of five hours (± 200 km) of fish sampling sites. In the laboratory all the fish adapted for three months in the concrete tank size of 6.0×2.0×0.75 m

Figure 1: Female of *C. lucius* (A) Male of *C. lucius* (B).

*Corresponding author: Azrita, Department of Biology Education, Faculty of Education of Bung Hatta University Padang, Indonesia
E-mail: azrita31@yahoo.com

Figure 2: Maintenance concrete tank and spawning box.

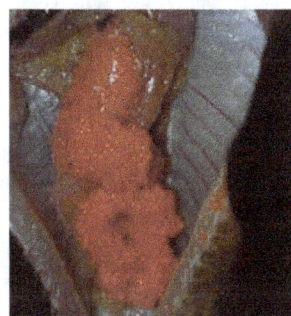

Figure 3: Spawning fish eggs of *C. lucius* stage IV.

with circulating water, volume of 6000 liters of water and fed of tilapia (*Orechromis nilaticus*) size 4 to 5 cm in adlibitum.

After three months maintained, female fish were selected as much as 16 tails with an average weight of 300 ± 20 grams which will be used for the four treatments and four replications with implantation of LHRHa hormone. *C. lucius* given treatment LHRHa with different doses twelve individuals and four individuals as a control (0.9% NaCl). All the fish are at stage 1 germinal vesicle in central position and distributed into four plots concrete tank size of 200×200×75 cm, the water height of 50 cm and in the each concrete tank maintenance supplied wooden boxes size of 60×40×40 cm two units are equipped with water plants functioning as a spawning naturally (Figure 2). The number of female fish every concrete tank is four individuals added with two male fish weighing 500 ± 50 g/individual. The concrete tank water comes from an artesian well to debit 0.1 m³/second, concrete tank temperature ranged from 26 to 28°C, pH 6.0 to 7.0, dissolved oxygen (mg/l) adalah 5,5 to 6,0. The food is given once each day of tilapia (*Orechromis nilaticus*) size 4 to 5 cm (10 g/individual) as many as 15% of the weight of biomass or equivalent to 33 individual/day/plot.

Checking the oocytes maturation

All fish were individually marked using floy-tags and weighed. Oocytes sampled in vivo were taken from females using the method described by Syandri [18], and were placed in Serra's solution (6:3:1, 70% ethanol, 40% formaldehyde and 99.5% acetic acid). for clarification of the cytoplasm. After 5 min, the position of oocytes nucleus was determined using a four-stage scale:

Stage 1 germinal vesicle in central position

Stage 2 early migration of germinal vesicle (less than half of radius)

Stage 3 late migration of germinal vesicle (more than half of radius)

Stage 4 periphery germinal vesicle or germinal vesicle breakdown, GVBD (Figure 3).

Hormonal treatment

Preparation of hormone pellets following the procedure as suggested by Lee et al. [19] is 1000 ug LHRHa hormone mixing with 70 mg cholesterol, 20 mg of cocoa butter (*unsalted butter*), 1 ml etanol 70%. The dough molded into pellets much as twenty grains and each pellet contains 50 µg of LHRHa. Experiments conducted on *C.lucius* female gonad maturity stage one with four groups and four replications. Control group (injection from 0.9% NaCl) and three experimental ones. Group two, three and four given a single dose LHRHa with 100 µg/kg, 150 µg/kg, 200 µg/kg body weight. LHRHa intra muscular implanted under the dorsal fin. Spawning conducted are naturally, without stimulation hormone. Egg samples from each treatment and replications. Gilson is preserved with a solution consisting of 100 ml, 60% alcohol, 880 ml of distilled water, 15 ml of nitric acid, 18 ml of glacial acetic acid and 20 grams of mercury chloride. Furthermore, the diameter of the eggs was measured with a microscope Olympus CX21 to 30 eggs of from each treatment and replications. Fertilized eggs were collected after ten hours and checked under a stereomicroscope. The eggs were randomly separated into four groups of aquarium (40x20x20 cm, water volume 8 liter), with four replicates for the control and four replicates for each exposure group. The total number of eggs in the control and the exposure groups was 100 eggs. The water temperature, dissolved oxygen, and pH levels of the test chambers were regularly monitored.

Time mature gonadal calculated from the time the fish began to LHRHa granted until the fish reaches a mature gonadal (days). Fecundity is the number of eggs produced per spawning, embryo survival (%) is the number of fertilized eggs compared to the number of eggs produced per spawning and s the hatching rate is number of larval compared to the number of fertilized eggs.

Statistical analyses

All data were analysed by analysis of variance (ANOVA), followed by comparisons of means by Duncan's multiple range test. In the case of time of mature gonadal, total fecundity, egg diameter, embryo survival and hatching rate. The mean of the duplicate samples was used for each fish. All statistical analyses were performed using the SPSS versi 13.

Results

The mean time to reach the gonadal mature of the control group was 122 days, whereas in LHRHa treatment group on average ranged from 92 to 112 days (Table 1). There were significant differences (p<0.05) between treatment groups LHRHa against time reaches mature gonadal. Irrespectively of applied hormonal dose (also in the control group) all females of the domestic stock ovulated.

The fast average time it reaches the gonadal mature female broodstok is 92 ± 12 days for 200 µg/kg body weight, whereas later in the control group is 122 ± 23 days. There is an indication of the higher-dose LHRHa, reached gonadal mature of *C.lucius* will be faster. There are significant differences between treatment groups to time reaches gonadal mature of *C.lucius* (p<0.05). Broodstock natural spawning 100% for each group in the treatment of LHRHa include control group.

The average number of eggs *C.lucius* from the natural spawning by LHRHa dose treatment and control are listed in Table 1. The number

Parameter	Dose hormone LHRHa			
	Group 1- control (0.9% NaCl)	Group 2- 100 µg/kg body weight	Group 3- 150 µg/kg body weight	Group 4- 200 µg/kg body weight
Avarage weight of female (g)	300 ± 25	300 ± 20	300 ± 28	300 ± 30
Time mature gonadal (days)	122 ± 23[aw]	112 ±15[b]	110 ± 20[c]	92 ± 12[d]
Percentace of natural spaw (%)	100[a]	100[a]	100[a]	100[a]
Fecundity (eggs/spaw)	1,778 ± 20[a]	2,106 ± 200[b]	2,685 ± 165[c]	3,600 ± 152[d]
Egg diameter (mm)	1.33 ± 0.05[a]	1.77 ± 0.02[b]	1.81 ± 0.01[c]	1.88 ± 0.01[d]
Embryo survival (%)	64.00 ± 3.77[a]	83.03 ± 0.51[b]	83.56 ± 0.14[b]	85.87 ± 0.16[b]
Hatching rate (%)	64.01 ± 3.35[a]	80.56 ± 0.38[b]	80.96 ± 0.63[b]	82.67 ± 0.30[b]
Survival rate of larvae 6 days (%)	100[a]	100[a]	100[a]	100[a]

[abcd]Values with the different superscript in each column are significantly different from each other ($p<0.05$).

Table 1: Average reproductive potential and quality of eggs *C. lucius*.

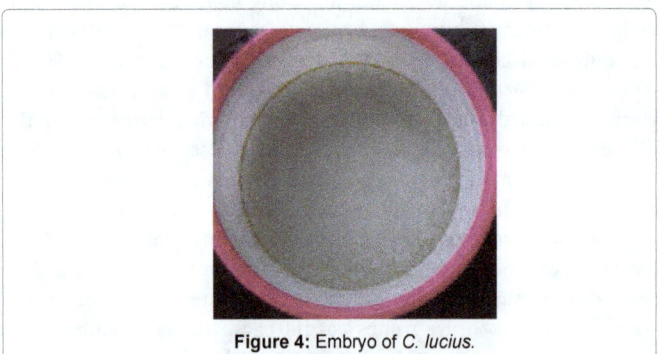

Figure 4: Embryo of *C. lucius*.

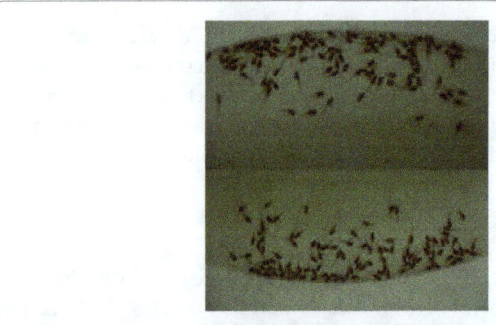

Figure 5: Larvae of *C. lucius*.

of eggs per spawning tendency to increase with increasing dose levels LHRHa. Average fecundity in the control group and the group that implanted with LHRHa preparations was significantly different (p <0.05). But every *C. lucius* broodstock that spawn at a dose level of LHRHa of 200 µg/kg body weight resulted in more number of eggs than the control group and at dose levels of 100 µg/kg and 150 µg/kg body weight.

The average egg diameter of *C.lucius* for each treatment dose of LHRHa hormone preparations and the control group was significantly different (p<0.05). The diameter of the largest egg of 1.87 ± 0.02 mm produced by *C.lucius* implanted LHRHa hormone preparations of dose level 200 µg/kg body weight and smallest size 1.33 ± 0.05 mm produced by the control group. The size of the diameter of the eggs was significantly different between treatments (p<0.05). Furthermore, the colour of egg *C.lucius* is dark orange (Figure 3), embryo is floating and do not stick to each other in the hatching medium (Figure 4). During the process of embryonic development not given the addition of oxygen and embryonic development to hatching at 28°C water temperature ranged from 30 to 36 hours. Embryo survival between treatment

groups were not significantly different (p>0.05), but significantly different from the control group (p<0.05). Average embryo survival in the treatment group LHRHa ranged between 83.03 ± 0.51% to 85.87 ± 0.16% and the highest embryo survival produced by female broodstock LHRHa preparations were implanted with a dose of 200 µg/kg body weight and 64.00 ± 3.77% lowest in the control group and maintained in the same environment with indicators of water quality parameters is important that is water temperatures ranged from 24 to 27°C, pH ranged from 6.5 to 7.0 and dissolved oxygen ranged from 6 to 7 mg/l.

Average hatching rate in the control group are 64.01 ± 3.35%, whereas in the group treated preparations LHRHa ranged between 80.56 ± 0.38% to 82.67 ± 0.30%. Hatching rate between LHRHa treatment group was not significantly different (P>0.05), and different from the control group (p<0.05). Egg yolk and oil bubble is a food reserve (first) after hatching the larvae of *C. lucius*. Absorbed yolk out for 92 hours (4 to 5 days) after hatching and for each treatment group and the control group survival rate of larvae on average 100% (Figure 4). *C.lucius* larvae at the beginning of exogenous feeding can directly take Moina sp, although in this research author has not been able to determine large mouth aperture larvae. Therefore, continued research on larval rearing is very important (Figure 5).

Discussion

Gonadal maturation process of *C. lucius* for increasing the reproductive potential and the best quality eggs needed LHRHa dose of 200 µg/kg body weight. Implantation of LHRH-a physiologically can release of GTH with slowly. Gonadotropin hormone will stimulate the development of granulosa cells and after reaching a certain development granulosa cells to release estradiol. Estradiol stimulates the liver to form the vitellogenin that will stimulate the vitelogenesis in ovaries. After reaching a certain cell targets, vitellogenesis process ended, the granulosa cells will secrete steroid hormones stimulating gonadal maturation. The optimal dose for accelerate the gonadal mature of each species of fish is different, Balanthiocheilus melanopterus Cyprinidae fish needed LHRHa 25 µg/kg body weight [20], *C. pleurothalmus* needed 150 µg/kg body weight [21], *Mystus nemurus* Bagridae 100 µg/kg body weight [22], *Epinephelus fuscoguttatus* Serranidae 150 µg/kg body weight [23]. Differences LHRHa dose every type of fish allegedly due to differences in fish species, thus providing different responses for each type of fish reaches a gonadal mature.

LHRHa dose effects on fecundity has been studied by some researchers but there are inconsistencies in the reported results. Fecundity produced by Botia *macracantha* with LHRHa dose of 150 µg/kg body weight a average of 19.615 eggs/spawning, whereas at a dose of 200 µg/kg body weight fecundity as much as 10,203 eggs/spawning

[24]. In this study with increasing doses of LHRHa hormone, then the increasing fecundity every spawning. Fecundity of *C.lucius* highest average of 3,600 ± 152 eggs / individuals, whereas the control group 1,778 ± 20 eggs/ individuals. Differences fecundity of each treatment group LHRH is suspected related with a dose of the hormone LHRHa in implantation. Whereas LHRH-a alone function to promote secretion of gonadotropin (GTH) from the pituitary gland, given fixed with a certain dose to ensure certainty of oocyte maturation. The success of the effect of the hormone on gonad development should be supported the maintenance of environmental conditions and adequate feeding [21]. *C. lucius* during maintained in the process of maturing the gonads were fed live ie tilapia juvenile (*Oreochromis nilaticus*).

The feed can sustain the life of *C. lucius* and assist in the maturation of the gonads because it allegedly contains compounds of fatty acids. Mokoginta [25] states that there is a tendency broodstock who gets feed to the linoleic fatty acid content of less or more will produce a low fecundity. The growth of the embryo very dependent on the quality and quantity of nutrients stored in the yolk, because the egg membranes are impermeable amino acids and nucleotides (Neyfaakh and Abromova, 1974) in [25]. Ontogenesis of success generated by the conversion of yolk material into embryonic tissues [18].

Increasing the diameter of the eggs *C. lucius* from 1.33 mm to 1.88 mm caused by the implantation of LHRHa hormone because it was bigger than the diameter of the eggs in the control group. According Subagja et al. [24] Botia fish (*Botia macrachanta*) eggs treated with LHRH-a can grow from the size of 0.35 mm to 0.80 mm and ready ovulated, while the parent is not given LHRH the godanal undeveloped. Grouper eggs which have diameters ranged from 60 to 380 μm required optimal dose of 100 ug LHRHa/kg body weight [26]. From the research it can be stated that egg size *C. lucius* relatively large and large eggs will larvae tend to have a large mouth openings making it easier to accept as natural food such as *Artemia salina* nauplii when the change of food from endogenous feeding to exogenous feeding. Therefore *C. lucius* can be recommended for performed as a candidate culture of farmed fish. This opinion refers to the results of research Syandri [27] to *Tor douronensis*, the size of the newly hatched larvae are 1.1 mm and at the age of four days can directly take *Artemia salina* nauplii as a natural feed so that larvae survival can reach 80%. Different from the larvae newly hatched *Mystacoleucus padangensis* Blkr have a size of 0.2 to 0.3 mm, very difficult to provide natural food and at the age of 13 days to take of *Artemia salina* nauplii, consequently larval survival to the age of 14 days only 10% [18]. During embryonic development and hatching eggs several water quality parameters monitored are as follows water temperature ranged from 27 to 29°C, pH ranged from 6.5 to 7.0 and dissolved oxygen ranged from 6 to 7 mg/l and ammonia 0.02 mg/l. According Boyd [28] good fish growth occurs at temperatures 25 to 32°C, pH 6.5 to 9.0 and dissolved oxygen >5 ppm.

Conclusion

Increasing doses of the hormone LHRH-a at broodstock females in general can increase the time mature gonadal, fecundity, eggs diameter, embryo survival and hatching rate. The best dose is 200 μg / kg body weight.

Acknowledgements

We are indebted to The Ministry of Education and Culture Directorate General of Higher Education to providing research funding Competitive Grants 2014 and Faculty of Fisheries and Marine Sciences of Bung Hatta University Padang for providing full laboratory facilities during the study period. There is no conflict of interest among the authors for the publication of this research article, and also thank to reviewers at JARD for valuable criticism of the manuscript.

References

1. Azrita (2011) Patterns of growth and condition factor of Channa lucius in Singkarak Lake.

2. Azrita, Syandri H (2013) Fecundity, egg diameter and food Channa lucius Cuvier in different waters habitats. J Fish Aqua Sci 4: 115-120.

3. Warsa A, Nastiti AS, Krismono, Nurfiarini A (2009) Fisheries resources in Koto Panjang Reservoir. Bawal 2: 93-97.

4. Zuraini A, Somchit MN, Solihah MH, Goh YM, Arifah AK, et al. (2006) Fatty acid and amino acid composition of three local Malaysian Channa spp. Fish. Food Chemistry 97 : 674-678.

5. Lu F, Ding Y, Ye X, Liu D (2010) Cinnamon and nisin in alginateecalcium coating maintain quality of fresh northern snakehead fish fillets. LWT- Food Science and Technology 43: 1331-1335.

6. Karapanagiotidis IT, Yakupitiyage A, Little D, Bell MV, Mente E (2010) The nutritional value of lipids in various tropical aquatic animals from rice–fish. J Food Com Analy 23: 1-8.

7. Sood N, Chaudhary DK, Rathore G, Singh A, Lakra WS (2011) Monoclonal antibodies to snakehead, Channa striata immunoglobulins: Detection and quantification of immunoglobulin-positive cells in blood and lymphoid organs. Fish Shellfish Immunol 30: 569-575.

8. Ghassem M, Arihara K, Babji AS, Mamot S, Ibrahim S (2011) Purification and identification of ACE inhibitory peptides from Haruan (Channa striatus) myofibrillar protein hydrolysate using HPLC-ESI-TOF MS/MS. Food Chemistry 129:1770-1777.

9. Xu FL, Wu WJ, Wang JJ, Qin N, Wang Y, et al. (2011) Residual levels and health risk of polycyclic aromatic hydrocarbons in freshwater fishes from Lake Small Bai-Yang-Dian, Northern China. Ecological Modelling 222: 275–286.

10. Mohsin AK, Ambak MA (1983) Freshwater fishes of arginine supplementation enhances diabetic wound healing: peninsular malaysia. Serdang, Malaysia: Universiti Pertanian involvement of the nitric oxide synthase and arginase pathways.

11. Azrita (2012) Genetic variation and reproductive biology of Channa lucius Cuvier [Actinopterygii: Channidae] on different aquatic habitats in an effort to domestication.

12. Courtenay WR, Williams JD (2004) Snakeheads (Pisces, Channidae) A biological synopsis and risk assessment.

13. Mishra SK (1991) Reproductive biology of freshwater teleost, Channa gachua (Ham): Proceedings of the National Symposium on New Horizons in Freswater Aquaculture.

14. Jhingran AG (1984) The fish genetic resources of India: Bureau of Fish Genetic Resources, Allahabad and Maya Press Pvt. Ltd, Allahabad.

15. Makmur S, Rahardjo MF, Sukimin S (2003) Biology reproductive of Channa striata Bloch in floodplain Musi River South Sumatra. J Iktiologi Indonesia 3: 57-62.

16. Marimuthu K, Haniffa MA, Rahman MA (2009) Spawning performance of native threatened spotted snakehead fish, Channa punctatus (Actinopterygii: Channidae: Perciformes), Induced with ovatide. Acta Ichthyologica Et Piscatoria 39: 1-5

17. Paray BA, Haniffa MA, Manikandaraja D (2012) Induced ovulation and spawning of a striped snakehead murrel, Channa striatus (Bloch) under captive conditions. J Res Ani Sci 1: 033-039.

18. Syandri H (1997) Development of fish embryos and larvae bilih, Mystacoleucus padangensis Blkr in Singkarak lake. J Garing 2: 28-38.

19. Lee CS, Tamaru K, Kelly CD (1986) Technique for Making Chronic-release LHRHa and 17 methyltestosterone pelets for intramuscular implantation in fishes. Aquaculture 59: 161-168.

20. Insan I, Satyani D, Munddriyanto H, Kusdiarti, Djajasewaka H (2001) Dose differences hormone LHRH for the maturation of gonads Balasark (Balanteocheilus melanopterus). J Biosfera 18: 13-19.

21. Syandri H, Basri Y, Maseriza (2008) The Use of LHRH-a and Vitamine E for eggs quality of Channa pleurothalmus). Journal of Sigmatek 1: 131-144.

22. Aryani N, Syawal H, Bukhari D (2002) The use of LHRH-a for gonadal

maturation of green catfish (Mystus nemurus). Torani 12: 163-168.

23. Purba R (1995) Supporting Role of LHRHa for Tiger Grouper (Epinephelus fuscoguttatus) breeding, Research News 17: 7-8.

24. Subagja J, Komarudin O, Effendi J (1997) Effects of LHRH-a hormone implantation in Botia (Botia macrachanta) on gonadal maturation variability. J Indonesian Fisheries Research 2: 10-17.

25. Mokoginta I (2000) Needs essential fatty acids, vitamins and minerals in feed broodstock (Pangasius sutchi) for reproduction. The final report VII/1-2 Competitive Grant Universities.

26. Setiadharma T, Prijono A, Giri NA (2003) LHRH-a hormone application for gonadal development and spawning grouper (Epinephelus microdon). Journal of Indonesian Fisheries Research 1: 9-16.

27. Syandri H (2004) The use of vitamin E for enhancing the reproductive potential of Garing fish (Tor douronensis Blk). J Agricultural Dynamics XIX: 141-151.

28. Boyd CE (1979) Water quality in warmwater fish ponds. Agricultural Auburn University. Alabama.

13

Caligus elongatus and *Photobacterium damselae* subsp *piscicida* Concomitant Infections Affecting Broodstock European Seabass, *Dicentrarchus labrax*, with Special Reference to Histopathological Responses

Elgendy MY[1]*, Abdelsalam M[2]*, Moustafa M[2], Kenawy AM[1] and Seida A[3]

[1]*Department of Hydrobiology, Veterinary Research Division, National Research Centre, 12622 Dokki, Giza, Egypt*
[2]*Department of Fish Diseases and Management, Faculty of Veterinary Medicine, Cairo University, 12211 Giza, Egypt*
[3]*a-Leibniz Research Institute for Environmental Medicine, Düsseldorf, Germany*
[3]*b-Department of Microbiology and Immunology, Faculty of Veterinary Medicine, Cairo University, Giza Egypt*

Abstract

Caligus elongatus and *Photobacterium damsela* subsp *piscicida* are pathogens of serious infections in European seabass, *Dicentrarchus labrax*. In this study, both agents were concomitantly isolated from moribund broodstock European seabass cultured within the hatchery unit at El-Max Research Station (NIOF), Alexandria governorate, Egypt. Externally, fish were heavily infested with *Caligus elongatus* ectoparasitic copepods. The overall prevalence, mean intensity and mean abundance of *C. elongatus* on examined fish were 92.3%, 23.3 and 21.5; respectively. Majority of samples noticed sever haemorrhages on the external body surface and fins. Internally, moribund fish showed characteristic whitish nodules and extensive adhesions of visceral organs. 88.46% of investigated fish were concurrently found to be infected with *P. damsela* subsp *piscicida*. No other bacterial species were detected. *P. damsela* subsp *piscicida* was also isolated from *C. elongatus* infesting clinically diseased fish. All *P. damsela* subsp *piscicida* isolates were confirmed by sequencing of the16S rRNA gene. Microscopically, multiple granulomas were regularly observed in haemopoietic organs. Our results as a whole indicate that *C. elongatus* may serve as a potential vector for *P. damsela* subsp *piscicida* and possibly enhance photobacteriosis dissemination among co-habitant fish, thus suggesting the desirability of redesigning the protocols presently used for microbial recognition during fish epidemiological studies to improve fish health.

Keywords: *Photobacterium damselae* subsp *piscicida*; *Caligus elongatus*; Vector; European seabass; Mortalities

Introduction

European seabass, *Dicentrarchus labrax*, is one of the most valued marine fish species used in fish farming worldwide [1]. Several pathogens can put the life of cultured seabass in jeopardy with consequent detrimental impacts on growth, fecundity and productivity [2]. Photobacteriosis caused by the halophilic bacterium *Photobacterium damselae* subsp. *piscicida*, has long been considered among the dominant limiting factors in mariculture all over the world [3]. The disease has caused substantial mortalities in seabass and many other marine fishes with colossal economic losses [3-8].

The pathogenesis of *P. damselae* subsp. *piscicida* is not completely elucidated. This pathogen possesses many virulence mechanisms that significantly contribute to the capacity of this bacterium to devastate and overcome fish immune defense mechanisms including a capsule and an iron uptake system [9,10]. Moreover, this pathogen produces variety of destructive extracellular products with phospholipase, cytotoxic, and hemolytic activities that contribute significantly to the development of the disease [8].

The ability of *P. damselae* subsp. *piscicida* to invade and replicate intracellularly is a critical issue in the pathogenesis of photobacteriosis enabling this pathogen to evade host defenses as well as decrease the need for adherence to fish surfaces to establish infection [11].

P. damselae subsp. *piscicida* infections are enhanced with fish ectoparasites infestations as they could facilitate the invasion and settlement of bacteria in fish blood stream [12]. Skin injuries, induced by fish ectoparasites, including sea lice, are effective portals of entry for diversity of opportunistic bacterial infections [13-15].

P. damselae subsp. *piscicida* is considered an obligate pathogen and

its survival is short lived outside the host [16,17]. The transmission of this fish pathogen is not fully understood. A symptomatic carrier and/ or reservoir of infection may coexist [18].

Sea lice, *Caligus elongatus*, feed on host mucus, tissues and blood thereby they could be potential vectors as well as transmitters for numerous pathogens among fish [19,20]. From this perspective, copepods, *Caligus elongatus* may be involved in dissemination of photobacteriosis infections.

Bacterial pathogens vectored by copepods pose serious threats to aquaculture and human health. Therefore, this study aimed to investigate the link between copepods infestation and *P. damselae* subsp. *piscicida* infections in European seabass broodstock with large scale mortalities, and to provide information about the histopathological alterations induced by host pathogens interactions. Additionally with the aim to improve its diagnosis, full phenotypic and molecular analyses were employed to identify *P. damselae* subsp. *piscicida*.

---

*Corresponding authors: Elgendy MY, Department of Hydrobiology, National Research Centre, 12622 Dokki, Giza, Egypt
E-mail: mamdouhyousif@yahoo.com

Abdelsalam M, Department of Fish Diseases and Management, Faculty of Veterinary Medicine, Cairo University, E-mail: m.abdelsalam@staff.cu.edu.eg

Materials and Methods

Fish sampling

On June 2014, twenty-six moribund broodstock European seabass *D. labrax* were obtained from El-Max Research Station, National Institute of Oceanography and Fisheries (NIOF) Alexandria, Egypt. This farm noticed mass mortalities approaching 70% among broodstock seabass reared within concrete ponds. Fish were fed on pelleted diet 45% protein. Samples were preserved in isothermal boxes within ice, to be transferred with the minimum time of delay to Fish diseases Lab. Cairo University. The average body weight of examined fish ranged from 1100 to 1300 gm. All fish were visually inspected for any lesions before the examination is adopted. The average recorded values for salinity, water temperature, pH, dissolved oxygen and un-ionized ammonia in the investigated farm were 32‰, 25°C, 8.8, 3.6 mg/L and 0.9 mg/L respectively.

Parasitic investigation

Detection of external parasites relied firstly upon visual inspection by naked eye. Furthermore specimens, including the gills, fins, body cavity and internal organs of fish, were also examined in petri-dishes under the dissecting microscope. Identification of copepods was based mainly on characteristic morphological features according to [21]. The prevalence, mean intensity and mean abundance of copepods infestations were calculated using quantitative Parasitology web version 3 [22].

Bacterial isolation

Loopfuls from lesions in liver, spleen and kidney of moribund fish were streaked onto marine agar (Difco), thiosulphate citrate bile salt sucrose agar (TCBS, Oxoid) and blood agar containing 2% NaCl. Cultures were incubated at 25°C for 48-72 h. Representative inocula of single colonies collected from the plates were re- streaked onto tryptic soy agar supplemented with 1.5% NaCl (v/v) (TSA, Difco) for purity and identification.

The ectoparasitic copepods were removed from infested fish by sterilized forceps then washed three times with saline (0.85% NaCl). Five randomly selected copepods from each fish were processed as one group and homogenized using a sterile plastic rod. The homogenates were serially 10-fold diluted with saline and inoculated onto marine agar and TCBS as described by [13] with minor modification. Identification of retrieved bacterial isolates from both moribund seabass and copepods was mainly performed by using API 20E systems (BioMerieux). Furthermore, sensitivity to 150 mg vibrio static agent (O/129) and motility on soft agar were also investigated [23].

Sequencing of *P. damsel* subsp *piscicida* 16s rRNA gene

Genomic DNA was extracted from cultivated *P. damsel* subsp *piscicida* strains using prepMan Ultra reagent (Applied biosystems, USA) according to protocol supplied. To amplify a 267-bp fragment of the target 16S rRNA gene; the specific primer pair of Car 1:5′-GCTTGAAGAGATTCGAGT-3′, and Car 2:5′-CACCTCGCGGTCTTGCTG-3′ was used [24]. The PCR product was directly sequenced using the BigDye Terminator v3.1 Cycle Sequencing Kit (Applied Biosystems, USA) with 310 Automated DNA Sequencer (Applied Biosystems, USA) using the same primers for annealing. The nucleotide sequences of the 16s rRNA genes of *P. damsel* subsp *piscicida* isolates retrieved from both *Caligus elongatus* and seabass were submitted to the DNA Data Bank of Japan (DDBJ) nucleotide sequence database.

Histopathological examination

Specimens from gills, liver, spleen and kidney of infected fishes were taken for histopathological studies. The trimmed samples were fixed in 10% phosphate buffered formalin for 24 hours, dehydrated by a series of upgraded ethanol solution and embedded in paraffin. Finally, sections were stained with Hematoxylin and Eosin (H & E) to be examined under light microscope [25].

Results

Clinical examination

The prominent clinical findings of moribund broodstock European seabass were lethargy and widespread haemorrhages on the external body surface and fins (Figure 1a). Majority of samples showed haemorrhagic and bleeding vent (Figure 1b). Internally, whitish nodules and extensive adhesions of visceral organs were characteristic.

Parasitological examination

Moribund seabass were heavily infested with *Caligus elongatus* copepods appeared as extensive focal brown spotted lesions on the skin, fins, head region and buccal cavity. No other parasitic infestations were recorded.

Taxonomic summary

Caligus elongatus Nordmann (1832)

Family: Caligidae Burmeister (1835)

Host: European seabass *Dicentrarchus labrax* Linnaeus (1758) (Perciformes: Serrandidae)

Locality: The hatchery unit at El-Max Research Station, the National Institute of Oceanography and Fisheries (NIOF), Alexandria City, Egypt

Site of infection: Skin, fins, head region and buccal cavity of infected fish

Prevalence, mean intensity and mean abundance of infection:

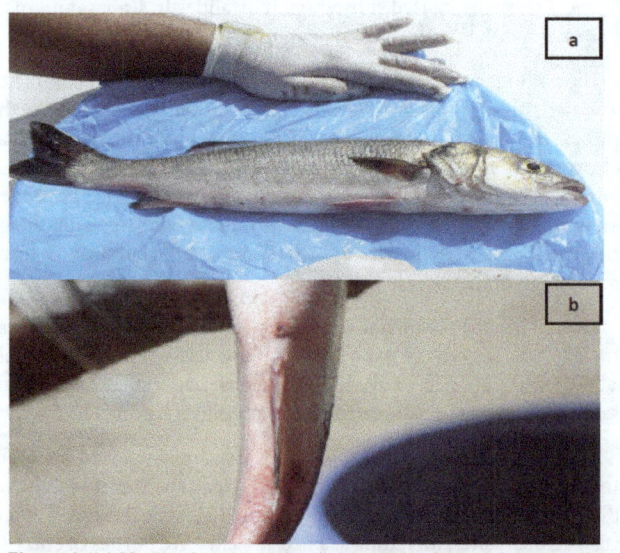

Figure 1: (a) Moribund seabass showing extensive haemorrhages on the external body surface and fins (b) Moribund seabass showing haemorrhagic and bleeding vent.

92.3% of examined fish were infected with 23.3% as mean intensity and 21.5% as mean abundance.

Bacteriological examination

88.46% of examined fish were found to be infected. *P. damselae* subsp. *piscicida* was the bacterial species solely obtained from investigated samples. *P. damselae* subsp. *piscicida* was retrieved also from all *C. elongatus* homogenates. No other bacterial species were detected in investigated fish specimens or copepod homogenates. Isolates were Gram negative, non-motile, pleomorphic rod-shaped with characteristic bipolar staining as well as sensitive to O/129 vibriostatic agent (150 mg). All *P. damselae* subsp. *piscicida* isolates obtained from both moribund seabass (23 isolates) and *C. elongatus* homogenates (24 isolates) had unique API 20 E profile, 2 005 004 (Table 1). All isolates reacted positively with the 16s rRNA gene specific primers and yielded the expected size of 267-bp. The accession numbers of sequenced 16s rRNA genes are LC017838 and LC017839 in this study. All sequences results revealed 100% homogenous with that of *P. damsel* subsp. *piscicida* ATCC29690 (GenBank accession number Y18496).

Histopathological findings

The tissues of infected fish revealed various proliferative, degeneration and circulatory changes. The circulatory changes included severe congestion, edema, hemorrhage in gills, kidney, liver and spleen. Moreover, diffuse hyperplasia of secondary lamellae is recorded. (Figure 2a).

Different sized granulomas were notcied in between hepatocytes, spleen and renal tubules together with vacuolar degeneration of

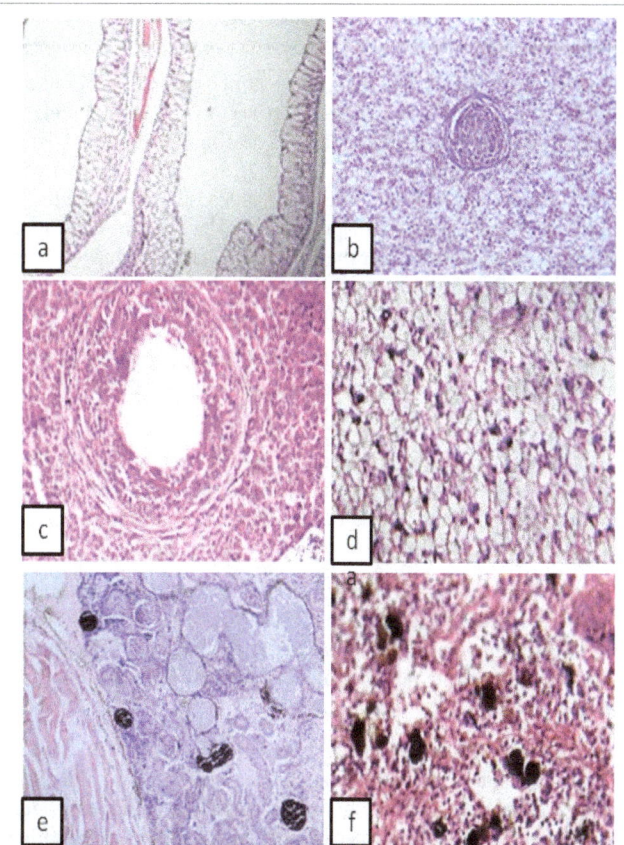

Figure 2: (a) Active goblet cell in gills of seabass (H&E).
(b) Granuloma in hepatic tissue (H&E).
(c) Large granuloma with necrotic area encapsulated with connective tissue (H&E).
(d) Sever vacuolar degeneration in hepatic tissue(H&E).
(e) Large granuloma in renal tissue (H&E).
(f) Activation of melanomacrophage centers in spleen (H&E).

Gram- staining	Gram-negative pleomorphic rod
Bipolar staining	+
Motility	Non motile
O/129 sensitivity (150 mg)	+
Cytochrome oxidase (OX)	+
Catalase	+
B–Galactosidase production (OPNG)	-
Arginine dihydrolase production (ADH)	+
Lysine decarboxylase production (LDC)	-
Ornithine decarboxylase production (ODC)	-
Citrate utilization (CIT)	-
H2S production (H2S)	-
Urease production (URE)	-
Tryptophane deaminase production (TDA)	-
Indole production (IND)	-
Acetoin production (VP)	+
Gelatinase production (CEL)	-
Acid from glucose (GLU)	+
Acid from manitol (MAN)	-
Acid from inositol (INO)	-
Acid from Sorbitol (SOR)	-
Acid from rhamnose (RHA)	-
Acid from sucrose (SAC)	-
Acid from from melibiose (MEL)	-
Acid from amygdalin (AMY)	-
Acid from arabinose (ARA)	-
O/F test	Fermentative without gas production

Table 1: Phenotypic and biochemical characteristics of retrieved *P. damselae* subsp. *piscicida* isolates.

hepatocytes and renal tubules. Moreover, inflammatory cell aggregation were notcied in pancreatic tissues. (Figure 2b, c and d).

Different granulomas were also notcied in kidney and spleen, with pronounced activation of melanomacrophage centers in spleen and activation of goblet cells in gills. (Figures 2e and 2f).

Discussion

In the aim to increase productivity per unit spaces, fish are intensively cultured in ponds and cages consequently they become feasible target for infectious diseases. At the top of the most critical pathogens limiting mariculture expansion, *P. damselae* subsp. *piscicida* and *C. elongatus* rank firstly [5,6,13]. These pathogenic agents are responsible for substantial economic losses among farmed marine fishes worldwide [26,27].

The majority of investigated moribund seabass, 88.46%, were found to be infected with *P. damselae* subsp. *piscicida*. No other bacterial infections were detected. *P. damselae* subsp. *piscicida* isolates were Gram negative, non-motile rod-shaped with bipolar staining and sensitive to O/129 vibrio static agent (150 mg). In addition, all *P. damselae* subsp. *piscicida* strains showed a unique API 20E profile, 2 005 004. All PCR reactions yielded definite amplicons of 267-bp fragment and the sequences of 16S RNA were 100% homogenous with that of *P. damsel* subsp. *piscicida* ATCC29690 (Gen Bank accession number Y18496).

The major part of investigated specimens were concomitantly found to be infested also with *C. elongatus* ectoparasitic copepods, the overall prevalence, mean intensity and mean abundance of *C. elongatus* on examined fish were 92.3%, 23.3 and 21.5; respectively. No other parasitic infestations were recorded [28] alleged that *C. elongatus* is a ubiquitous parasite among fish posing a significant problem for fish aquaculture operations all over the world by reaching high abundances that damage fish.

Strong evidence links potentiated bacterial infections to copepods infestations in fish [19,29,30], either providing portals of entry by damaging fish skin [31,32], or acting as a mechanical vector for numerous bacterial pathogens [19].

Sea lice also trigger diverse detrimental changes to host fish blood including, lymphopenia, anaemia, elevated cortisol level and ion imbalance [33,34]. These outcomes damage fish immunocompetence and predispose them to array of opportunistic pathogens.

Sea lice irritate and injury fish skin by their rasping piston-like mouthparts hence increase mucus secretions thereby provide a rich source of glycoproteins which is critical for bacterial adhesions, a significant step in the pathway of infectious diseases affecting fish [33-35].

Fish ectoparasitic copepods can transmit array of bacterial and viral pathogens a result of their feeding activities on fish blood and tissues [20,36]. Additionally, fish lice boost the spread of pathogenic agents as they infest diverse host species and switch between individuals consequently, transmit these agents to new hosts [37,38].

P. damselae subsp. *piscicida* was isolated from *C. elongatus* infesting broodstock seabass highlighting the potential role of sea lice in disseminating photobacteriosis among cultured fish. The indistinguishable biochemical and molecular profiles of recovered isolates irrespective of their source verified the previous hypothesis. This is in accordance with the findings of [39,40], who reported that sea louse, *Lepeophtheirus salmonis* is a potential vector for *Aeromonas salmonicida*.

The significance of environmental stress in the dynamics of fish diseases is renowned The pathways of fish disease are interrelated and variable factors relevant to invading pathogens, environment and fish should work together in synergism to define the nature of the triggered course of infection since the presence of pathogen alone not sufficient to produce disease [41-43]. The co-existence of unfavorable un-ionized ammonia levels, 0.9 mg/L, exacerbated the case and put more pressure on broodstock seabass [44] recommended 0.26 mg/L as a safe long-term limit for un-ionized ammonia in seawater. High ammonia levels enhance microbial infections through suppressing the immune capacity of fish. Phagocytic and clearance efficiency are diminished. As an ultimate fate for the staggering immumo-suppression of fishes inhabiting such conditions, parasitic and bacterial invasion will be the most probable event [45].

Outbreaks of bacterial diseases in fish are induced also by dissolved oxygen (DO) deficiency [46,47] recommended, 5- 8 mg/L, as optimal DO concentration for seabass which are far from that recorded in this study, 3.6 mg/L. The virulence of pathogens is exaggerated by exposure of farmed fish to reduced dissolved oxygen levels [48]. Moreover, oxygen consumption increased about twice in seabass when temperature increased from 15 to 25°C [49].

The severity and frequency of photobacteriosis infections boost at high water temperatures similar to conditions noticed in the investigated farm, 25°C, inducing fatal outbreaks in fish ultimately at water temperatures around 22°C [35]. There is also a significant seasonal variation in lice infestations with peak prevalence and intensity values during summer season [50,51].

The detected histopathological alterations were in conformity with our previous study in wild marine fishes [5,6]. The widespread multifocal granulomata in haemobiotec tissues as well as hyperactivity of melanomachrophage centers were frequently detected. These granulomas are thought to be crucial host-protective structures against virulent pathogens in attempt to restrict the expansion of infection by walling off bacteria. Pathogen proliferation and dissemination augment in infections without granulomas formation [52,53]. *P. damselae* subsp. *Piscicida* is also capable of intracellular growth in phagocytic cells. Accumulations of macrophages containing this bacterium block capillary flow resulting in local ischemia and focal necrotic changes [54].

The hyperactivity of goblet cells was also characteristic. This may indicate the dynamic involvement of these cells in the host responses [55]. Considerable evidence suggests that goblet cells have been found to act as an important barrier against many parasitic infections [56].

The recorded circulatory, degenerative and proliferative changes are attributed to the multifactorial virulence mechanisms of *P. damselae* subsp. *piscicida* including its toxic extra cellular products (ECP). These ECP were found to be lethal for different fish species including gilthead sea bream and sea bass. Phospholipases, cytotoxic, and hemolytic activities are among the (ECP) produced by this detrimental fish pathogen [9,57-59].

Conclusion

Our results as a whole indicate that *C. elongatus* may serve as a potential vector for *P. damsela* subsp *piscicida* and possibly enhance photobacteriosis dissemination among co-habitant fish, thus suggesting the desirability of redesigning the protocols presently used for microbial recognition during epidemiological studies not only focusing on diseased fish but should also include infesting ectoparasites to improve fish health.

References

1. Kousoulaki K, Saether BS, Albrektsen S, Noble C (2015) Review on European sea bass (*Dicentrarchus labrax*, Linnaeus, 1758) nutrition and feed management: a practical guide for optimizing feed formulation and farming protocols. Aquaculture Nutrition 21: 129-151.

2. Przybyla C, Fievet J, Callier M, Blancheton JP (2014) Effect of dietary water content on European sea bass (*Dicentrarchus labrax*) growth and disease resistance. Aquatic Living Resources 27: 73-81.

3. Romalde JL (2002) *Photobacterium damselae* subsp. *piscicida*: an integrated view of a bacterial fish pathogen. International Microbiology 5: 3-9

4. Toranzo AE, Magariños B, Romalde JL (2005) A review of the main bacterial fish diseases in mariculture system. Aquaculture 246: 37-61.

5. Elgendy MY (2007) Epizootiological studies on some bacterial infections in marine fishes. MVSc. Thesis. Cairo University, Cairo, Egypt.

6. Moustafa, M, Laila AM, Mahmoud MA, Soliman WS, Elgendy MY (2010) Bacterial Infections Affecting Marine Fishes in Egypt. Journal of American Science 6: 603 -612.

7. Wang R, Feng J, Su Y, Ye L, Wang J (2013) Studies on the isolation of Photobacterium damselae subsp. piscicida from diseased golden pompano (Trachinotus ovatus Linnaeus) and antibacterial agents sensitivity. Veterinary microbiology 162: 957-963.

8. Andreoni F, Magnani M (2014) Photobacteriosis: Prevention and Diagnosis. J Immunol Res.

9. Magarinios B, Romalde JJ, Lemos M, Barja J, Toranzo AE (1994) Iron uptake by *Pusteurella piscicida* and its role in pathogenicity for fish. Applied and Environmental Microbiology 60: 2990-2998.

10. Magarinos BA, Toranzo E, Romalde JL (1996) Phenotypic and pathobiological characteristics of *Pasteurella piscicida*. Annual Review of Fish Diseases 6: 41-64.

11. Elkamel A A, Thune R L (2003) Invasion and Replication of *Photobacterium damselae* subsp. *piscicida* in Fish Cell Lines. Journal of Aquatic Animal Health 15: 167-174.

12. Uzun E, Ogut H (2015) The isolation frequency of bacterial pathogens from seabass, *Dicentrarchus labrax*, in the Southeastern Black Sea. Aquaculture 437: 30-37.

13. Madinabeitia I, Ohtsuka S, Okuda J, Iwamoto E, Yoshida T, et al. (2009) Homogeneity among *Lactococcus garvieae* isolates from striped jack, *Pseudocaranx dentex* (Bloch & Schneider), and its ectoparasites. Journal of Fish Diseases 32: 901-905.

14. Jonsdottir H, Bron JE, Wootten R, Turnbull JF (1992) The histopathology associated with the pre-adult and adult stages of *Lepeophtheirus salmonis* on the Atlantic salmon, *Salmo salar* L. Journal of Fish Diseases 15: 521- 527.

15. Barker DE, Boyce B, Coombs MP, Braden LM (2009) Preliminary studies on the isolation of bacteria from sea lice, *Lepeophtheirus salmonis*, infecting farmed salmon in British Columbia, Canada. Parasitol Res 105: 1173-1177.

16. Janssen WA, Surgalla MJ (1968) Morphology, physiology, and serology of a Pasteurella species pathogenic for white perch, *Roccus americanus*. J Bacteriol 96: 1606-1610.

17. Plumb JA, Hanson LA (2011) Health maintenance and principal microbial diseases of cultured fishes. (3rdedn) John Wiley & Sons.

18. Robohm RA (1983) Pasteurella piscicida.

19. Cusack R, Cone DK (1986) A review of parasites as vectors of viral and bacterial diseases of fish. Journal of Fish Diseases 9: 169-171.

20. Gustafson LL, Ellis SK, Bartlett CA (2005) Using expert opinion to identify risk factors important to infectious salmon-anemia (ISA) outbreaks on salmon farms in Maine, USA and New Brunswick, Canada. Preventive Veterinary Medicine 70: 17-28.

21. Venmathi Maran BA, Leong TS, Susumu O, Kazuya N (2009) Records of Caligus (Crustacea: Copepoda: Caligidae) from marine fish cultured in floating cages in Malaysia with a redescription of the male of *Caligus longipedis* Bassett-Smith, 1898. Zoological Studies 48: 797-807.

22. Rozsa L, Reiczigel J, Majoros G (2000) Quantifying parasites in samples of hosts. J Parasitol 86: 228-232.

23. Buller NB (2004) Bacteria from Fish and Other Aquatic Animals: A Practical Identification Manual. CABI Publishing, Cambridge.

24. Osorio CR, Collins MD, Toranzo AE, Barja JL, Romalde JL (1999) 16S rRNA gene sequence analysis of *Photobacterium damselae* and nested PCR method for rapid detection of the causative agent of fish pasteurellosis. Appl Environ Microb 65: 2942-2946.

25. Bancroft JD, Gamble M (1996) Theory and practice of Histological techniques, 4th edition. Edinburgh, Chruchil living stone.

26. Muroga K (2001) Viral and bacterial diseases of marine fish and shellfish in Japanese hatcheries. Aquaculture, 202: 23-44.

27. Zorrilla M, Chabrillon AS, Rosales PD, Manzanares EM, Balebona MC, et al. (2003) Bacteria recovered from diseased cultured gilthead sea bream, *Sparus aurata* L. in southwestern Spain. Aquaculture 218: 11-20.

28. Jensen A (2013) Assessment of Sea Lice Infestations on Wild Fishes of Cobscook Bay. Honors College.

29. Bandilla M, Valtonen ET, Suomalainen LR, Aphalo PJ, Hakalahti T (2006) A link between ectoparasite infection and susceptibility to bacterial disease in rainbow trout. International J for Parasitology 36: 987-991.

30. Pylkko P, Suomalainen LR, Tiirola M, Valtonen ET (2006) Evidence of enhanced bacterial invasion during Diplostomum spathaceum. J Fish Dis 29: 79-86.

31. Kanno T, Nakai T, Muroga K (1990) Scanning electron microscopy on the skin surface of ayu *Plecoglossus altivelis* infected with *Vibrio anguillarum*. Diseases of Aquatic Organisms 8: 73-75.

32. Buchmann K, Bresciani J (1997) Parasitic infections in pond-reared rainbow trout Oncorhynchus mykiss in Denmark. Diseases of Aquatic Organisms 28: 125-138.

33. Tully O, Nolan DT (2002) A review of population biology and host-parasite interactions of the sea louse *Lepeophtheirus salmonis* (Copepoda: Caligidae). Parasitology 124: 165-182.

34. Johnson SC, Treasurer JW, Bravo S, Nagasawa NK, Kabata Z (2004) A review of the impacts of parasitic copepods on marine aquaculture. Zool Stud 43: 8-19.

35. Fouz B, Toranzo AE , Milan M, Amaro C (2000) Evidence that water transmits the disease caused by the fish pathogen *Photobacterium damselae* subsp. *Damsel*. J Appl Microbiol 88: 531-535.

36. Wagner GN, McKinley RS (2004) Anaemia and salmonid swimming performance: the potential effects of sub-lethal sea lice infection. J Fish Biol 64: 1027-1038.

37. Costello MJ (1993) Review of methods to control sea-lice (Caligidae, Crustacea) infestations on salmon farms. In Pathogens of Wild and Farmed Fish: Sea Lice (Boxshall, G.A. & Defaye, D., eds), Ellis Horwood 219- 252 p.

38. Pike AW, Wadsworth SL (1999) Sea lice on salmonids: Their biology and control. Advances in Parasitology 44: 233-337.

39. Nylund A, Wallace C, Hovland T (1993) The possible role of *Lepeophtheirus salmonis* (Krøyer) in the transmission of infectious salmon anemia.

40. Nylund A, Hovland T, Hodneland K, Nilsen F, Lovik P (1994) Mechanisms for transmission of infectious salmon anemia (ISA). Diseases of Aquatic Organisms 19: 95-100.

41. Moustafa M, Eissa AE, Laila AM, Gaafar AY, Abumourad IM, et al. (2014) Mass Mortalities in Mari-Cultured European Seabass, *Dicentrarchus labrax* at Northern Egypt. Research Journal of Pharmaceutical, Biological and Chemical Sciences 5: 95-109.

42. Elgendy MY, Moustafa M, Gaafar AY, Borhan T (2015) Impacts of extreme cold water conditions and some bacterial infections on earthen-pond cultured Nile tilapia, *Oreochromis niloticus*. .Research Journal of Pharmaceutical, Biological and Chemical Sciences 6: 136-145.

43. Moustafa M, Eissa AE, Laila AM, Gaafar AY, Abumourad IM, et al. (2015) Investigations into the Potential Causes of Mass Kills in Mari-Cultured Gilthead Sea Bream, *Sparus aurata*, at Northern Egypt. Research Journal of Pharmaceutical, Biological and Chemical Sciences 6: 466-477.

44. Dosdat A, Person-Le Ruyet J, Dutto G, Gasset E, Le Roux A, et al. (2003) Effect of chronic exposure to ammonia on growth, food utilisation and metabolism of the European seabass, *Dicentrarchus labrax*. Aquat Living Resour 16: 508-520.

45. Cheng W, Shan-Hsiao I, Jiann-Chu C (2004) Effect of ammonia on the immune response of Taiwan abalone Haliotis diversicolor supertexta and its susceptibility to Vibrio parahaemolyticus. Fish & Shellfish Immun 17: 193-202.

46. Department of Water Affairs and Forestry (1996) South African Water Quality Guidelines (second edition).

47. Mellergaard S, Nielsen E (1995) Impact of oxygen deficiency on the disease status of common dab *Limanda limanda*. Dis Aquat Org 22: 101-114.

48. Wedemeyer G (1981) The physiological responce of fishes to the stress of intensive aquaculture in recirculation systems.Proc World Symp on Aquaculture in Heated Effluents and Recirculation Systems 2: 3-18.

49. Person-Le Ruyet J, Mahé K, Le Bayon N, Le Delliou H (2004) Effects of temperature on growth and metabolism in a Mediterranean population of European seabass, *Dicentrarchus labrax*. Aquaculture 237: 269-280.

50. Schram TA, Knutsen JA, Heuch PA, Mo TA (1998) Seasonal occurrence of *Lepeophtheirus salmonis* and *Caligus elongatus* (Copepoda: Caligidae) on sea trout, *Salmo trutta*, off southern Norway. ICES Journal of Marine Science 55: 163-175.

51. Costello MJ (2006) Ecology of sea lice parasitic on farmed and wild fish. Trends in Parasitology 22: 475-483.

52. Andersen P (1997) Host responses and antigens involved in protective immunity to Mycobacterium tuberculosis. Scand J Immunol 45: 115-131.

53. Saunders BM, Cooper AM (2000) Restraining mycobacteria: role of granulomas in mycobacterial infections. Immunol Cell Biology 78: 334-341.

54. Kubota S, Kimura T, Egusa S (1970) Studies of bacterial tuberculosis of yellowtail, Symptomatology and histopathology. Fish Pathology 4: 11-18.

55. Wells PR, Cone DK (1990) Experimental studies on the effect of *Gyrodactylus colemanensis* and *G. salmonis* on density of mucous cells in the epidermis of fry of *Oncorhynchus myhss*. J Fish Biology 37: 599-603.

56. Ishikawa N, Horii Y, Oinuma T, Suganuma T, Nawa Y (1994) Goblet cell mucins as the selective barrier for the intestinal helminths: T-cell-independent alteration of goblet cell mucins by immunologically 'damaged' Nippostrongylus brasiliensis worms and its significance on the challenge infections with homologous and heterologous parasites. Immunology 81: 480-486.

57. Nakai T, Fujiie N, Muroga K, Arimoto M, Mizuta Y, et al. (1992) *Pasteurella piscicida* infection in hatchery-reared juvenile striped jack. Gyobyo Kenkyu 27: 103-108.

58. Rivas A, Balado M, Lemos ML, Osorio CR (2011) The *Photobacterium damselae* subsp. *damselae* Hemolysins Damselysin and HlyA Are Encoded within a New Virulence Plasmid .Infect Immun 79: 4617-4627.

59. Roberts RJ (2012) Fish Pathology.

14

Antigenic Characterisation of *Tenacibaculum maritimum* Isolates from Sea Bass (*Dicentrarchus labrax*, L.) Farmed on the Aegean Sea Coasts of Turkey

Yardimci RE* and Gülşen Timur

Istanbul University, Department of Aquaculture, Ordu Street, No: 200, Laleli, Turkey

Abstract

Tenacibaculosis, caused by *Tenacibaculum maritimum*, can result in severe mortalities of several marine fish species and thus represents a major challenge in Mediterranean aquaculture. Serological knowledge about this pathogen is required to develop effective preventive measures (vaccination). For this purpose, nineteen *T. maritimum* isolates, recovered between 2008 and 2010 from diseased European sea bass (*Dicentrarchus labrax*, L.) farmed at the Aegean Sea Coasts of Turkey, were characterised. All isolates produced flat, irregular, pale yellow colonies after incubation at 22-24°C for 48 hours, displayed pleomorphism with gliding motility with a size ranging between 4-20 × 0,5 µm and were otherwise biochemically identical to the *T. maritimum* NCIMB 2154[T] reference strain. The specific fluorescence appearance of the *T. maritimum* isolates were revealed by Indirect Fluorescent Antibody Technique (IFAT) which was also used to detect the bacterium in tissue samples. The presence of antibodies in the blood sera of the diseased fish against this pathogen was detected by using agglutination and Enzyme-Linked Immuno Sorbent Assay (ELISA). Dot-Blot testing identified all *T. maritimum* isolates as serotype O1. To our knowledge, this is the first report on O1 serotype *T. maritimum isolates from* sea bass farmed in Turkey.

Keywords: *Tenacibaculum maritimum;* Cultured sea bass; *Dicentrarchus labrax;* IFAT; ELISA; Serotyping; Dot-blot testing

Introduction

Tenacibaculum maritimum is the causative agent of tenacibaculosis in marine fish [1-3]. Since the first recognition of *T. maritimum* infection in farmed red and black sea bream (*Pagrus major* and *Acantopagrus schlagelli*) with high mortality in Japan, the presence of *T. maritimum* has become increasingly apparent in other marine fish species in Japan, USA, Canada, Australia, UK, France, Spain, Malta, Italy, Greece and Turkey [1,4-18].

This bacterium is difficult to distinguish from other phylogenetically and phenotypically similar species. In previous studies, serological methods such as slide agglutination, IFAT and ELISA were used for the identification of *T. maritimum* [5,7,13,19-21]. Although the bacterium is biochemically homogeneous, different O-serogroups, which seem to be related to the host species, could be detected by Avendano-Herrera et al., [21]. At least three groups of *T. maritimum isolates* from marine fish were distinguished [20]. These groups are associated with the host origin: group 1 comprises the strains isolated from sole (*Solea senegalensis and S. solea*), group 2 consists of the isolates from sea bream and sea bass, group 3 corresponds to the turbot isolates. These three groups of isolates could also be distinguished by randomly amplified polymorphic DNA-PCR [21]. By this methodology, the first group comprised all strains isolated from sole *and gilthead sea bream, the* second comprised the isolates from yellowtail (*Seriola quinqueradiata*), *Atlantic salmon* (*Salmo salar*) and turbot (*Scophthalmus maximus*) and the third group is formed by one isolate from *Pagrus major* and one from *Solea solea* [21].

It is important to determine the predominant *Tenacibaculum* serotype and different serotypes distribution to be able to develop effective preventive measurements like vaccines. In a serological characterization study of *T. maritimum* isolates, from farmed tub gurnard (*Chelidonichthys lucernus*) and wild turbot, carried out in Italy, it was determined that the isolates belonged to serotype O3 [16]. Castro et al., [22] reported that, by an old typing scheme, turbot isolated

strains of the bacterium belonged to serotype O2 in Spain. However, this needed revaluation according to the authors as they detected also serotype O3 in turbot and sole in the same study.

In Turkey, *T. maritimum was* isolated from farmed gilthead sea bream and sea bass at the Aegean sea coast [18-21,23-25] and from farmed rainbow trout in sea water of the Black Sea coast in a mixed infection case with other pathogen bacteria [26]. In our previous study, we described the isolation and identification of *T. maritimum* from infected sea bass by bacteriological, histopathological, and molecular methods [27]. *T. maritimum* isolates have also been detected in seven different fish species including sea bream, sea bass, meagre (*Argyrosomus regius*), turbot, corb (*Umbrina cirrosa*), sharpsnout sea bream (*Diplodus puntazzo*) and snappers (*Sparus pagrus*) in Turkey [28]. Until now, serological studies have never been carried out for this bacterial pathogen in Turkey. The aim of this study was to characterise isolates of *T. maritimum* from cultured sea bass (*Dicentrarchus labrax* L.) in Turkey.

Material and Methods

Bacterial strains

T. maritimum isolates were examined in this study. They were

---

*Corresponding author: Yardimci RE, Fisheries Faculty of Istanbul University, Department of Aquaculture, Ordu Street, No: 200, Laleli, Turkey E-mail: etepecik@istanbul.edu.tr

isolated from diseased European sea bass reared in five floating net cage farms at the Aegean Sea coasts of Turkey between 2008 and 2010. Conventional bacteriological tests and API ZYM test kits were used for biochemical identification of these strains. The T. *maritimum NCIBM 2154^T* reference strain was included as a positive control.

"O" antigen preparation and immunization of rabbits

"O" antigen preparation was performed as described by Toranzo et al., [29]. The density in each bacterial suspension was adjusted to 3 McFarland and boiled for 1 hour at 100°C for preparation of "O" antigens stored at 4°C until their use in Dot-Blot testing. Rabbits were immunized intravenously with 10^7 cell/ml formalin killed T. *maritimum* reference strain (NCIBM 2154^T) as well as representative strains (PC 503.1, PC424, ACC13.1). The polyclonal rabbit antiserums were later obtained according to Sorensen and Larsen [30] and stored at -20°C until used in the ELISA test as a positive control.

Indirect fluorescent antibody technique

A method described by Ainsworth et al., [31], for the diagnosis of *Edwardsiella ictulari,* was used to identify the T. *maritimum* isolates with minor modifications. PBS buffer was used as negative control. 20 μl test antigens were added in well of slides. After fixation, 10 μl of Rabbit anti-*Flexibacter maritimus* antisera (Microtek RFM01) diluted 400 × in PBS were added and incubated for 30 min at 37°C. Thereafter, slides were treated with 1:80 dilution of FITC labelled with goat anti-rabbit IgG for 30 min at 37°C, washed with PBS for three times and stained with 0.1% Evans blue for 30 min at 37°C. Finally, 100 μl of 25% glycerol solution (including 2.5 g DABCO) was added before slides were analysed under the fluorescent microscope. The IFAT procedure described by Lorenzen et al., [32] was also used for the detection of T. *maritimum* strains directly in fish tissues.

Slide agglutination test and enzyme linked immunosorbent assay

Blood samples were collected from the caudal artery of moribund fish. The antisera which had been stored at -20°C were used in slide agglutination test. This test was performed with a small amount of bacterial colonies mixed with several drops of serum obtained from fish samples. PBS buffer was again used as negative control [29] while immunized rabbit serum served as positive control. ELISA was performed as described by Knappskog et al., [33]. Monoclonal anti-European sea bass IgM marked with HRP (Aquatic diagnostics CO1) was used and PBS was included as a negative control.

Dot-blot analysis

The dot-blot analysis was performed as described by Cipriano et al., [34]. Rabbit sera were obtained from the University of Santiago de Compostela (Microbiology and Parasitology Department). Antisera against serotypes O1, O2 and O3 were prepared from representative strains PC 503.1, PC 424.1 and ACC 13.1, respectively as previously described by Avendano et al., [21].

Results

All bacterial isolates produced flat, irregular, pale yellow colonies after incubation at 22-24°C for 48 hours on MA and FMM. The bacteria showed pleomorphism with gliding motility within a size range 4-20 × 0.5 μm and reacted positive in the cytochrome oxidase and catalase tests, but did not produce flexirubine pigments. Morphological and phenotypical characteristics of the T. *maritimum* isolates are shown in Table 1. These isolates exhibited identical enzymatic profiles in API

ZYM tests to the reference strains in the test kit database.

Serologically, IFAT was used for the identification of T. *maritimum* strains and to show the specific fluorescence appearance of T. *maritimum* cells through microscopy. T. *maritimum* cells were detected in spleen, kidney and liver tissues of moribund fish samples using IFAT (Figure 1).

The slide agglutination test demonstrated positive reaction against fish antiserum (Figures 2a and 2b). Although these fish antiserums were cross absorbed with other *Tenacibaculum* sp.; specific monoclonal anti-European sea bass IgM marked HRP (Aquatic diagnostics CO1) was used in ELISA. The presence of antibodies in the blood sera of the diseased fish, against this pathogen, was also detected by ELISA and slide agglutination.

All isolates showed strong reaction only with the antiserum raised against the serotypes O1 (strains PC 503.1) in the dot-blot assays. It was therefore concluded that all T. *maritimum* isolates recovered from moribund sea bass samples were serotype O1 (Figures 3a and 3b).

Discussion

In this study, nineteen T. *maritimum* strains were isolated from

	T. *maritimum* (NCIMB 2154^T)	T. *maritimum* strains
Morphology	F	F
Motility	G	G
Gram staining	-	-
Oxidase	+	+
Catalase	+	+
Flexirubine pigment	-	-
Congo Red reduction	+	+
O/129 (150 μg) Resistance	S	S
Growth on TCBS	-	-
O/F	O	O
Indole	-	-
Methyl red	-	-
Voges Proskauer	-	-
Nitrate reduction	+	+
H$_2$S	-	-
Arginine dehydrolase	-	-
Lysine decarboxylase	-	-
Ornithine decarboxylase	-	-
Citrate	-	-
Aesculin degradation	-	-
Gelatinase	+	+
Urease	-	-
Acid production from		
Glucose	-	-
Maltose	-	-
Mannitol	-	-
Inositol	-	-
Sucrose	-	-
Lactose	-	-
Growth at		
4°C	-	-
37°C	-	-
44°C	-	-

Table 1: A summary of morphological and phenotypic characteristics of T. *maritimum* isolates examined in the study compared to T. *maritimum* reference strain (NCIMB 2154^T).

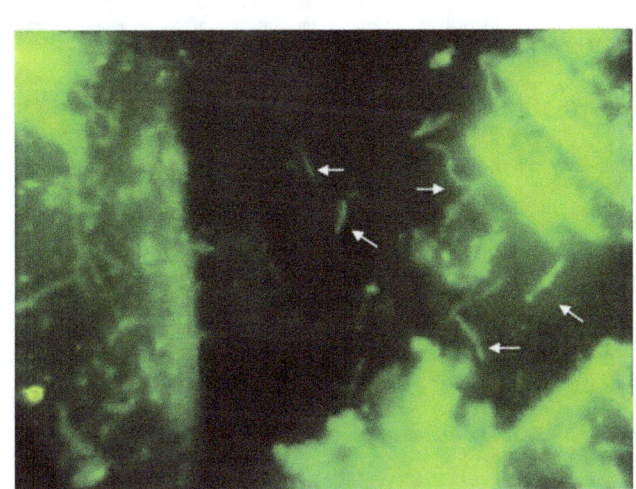

Figure 1: Filamentous T. maritimum cells (marked by arrows) located in spleen of moribund fish samples (IFAT, magnification X400).

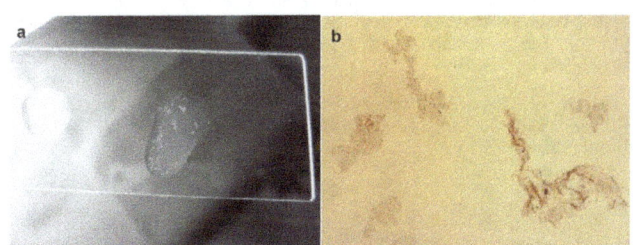

Figure 2: On slide agglutination with (a) white deposits composed of antigen-antibody complexes and (b) antigen-antibody complexes as observed under light microscope.

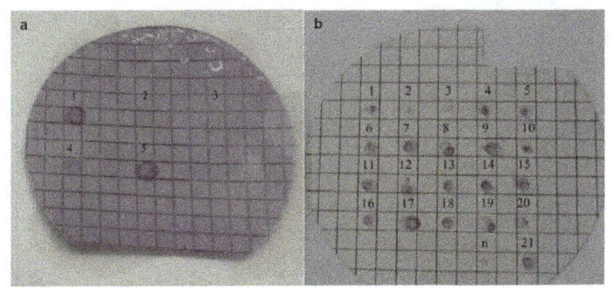

Figure 3: Dot blot assay using the antiserum obtained against Serotype O1 (a) 1: PC503.1, 2: PC424.1, 3: ACC13.1., 4: negative control, 5: T. maritimum isolate form.

diseased sea bass reared in five floating net cage farms at the Aegean Sea coasts of Turkey during 2008 and 2010. These filamentous, gram-negative, gliding bacteria are catalyse and cytochrome oxidase, Congo red absorption, nitrate reduction, gelatine hydrolysis positive, but were not producing flexirubine pigment. Biochemical homogeneity was determined among 19 T. maritimum isolates and when compared to other studies, a similarity was detected [1,2,4,11,12,21]. These isolates showed identical enzymatic profiles in API ZYM test kit with the previous records [6,11,12,21].

The nineteen T. maritimum isolates recovered from moribund

European sea bass were also serotyped using the Dot-blot method. All of our T. maritimum isolates from different farm locations reacted only with antiserums against the sole isolate PC 503.1 (serotype O1). Avendano et al., [20] originally noted that T. maritimum isolated from sea bream and sea bass from Spain reacted only with antiserum obtained against PC 424.1 (serotype O2). However, since then they have concluded that serotype O1 is the most predominant in Spanish reared sea bream and seabass, while serotype O2 is most common in turbot. In sole O1 and O2 serotypes are predominant but O2 isolates are increasing in numbers. In NW Spain the first isolates of this bacterium belonged to serotype O2, but recent isolates have also been found to be O3 [35].

In this study, IFAT was used for the detection of T. maritimum strains in fish tissues and identify bacterial cells. Specific fluorescence appearance of the bacteria cells of T. maritimum isolates were revealed as it was made previously by Powell et al., and Van-Gelderen et al., [9,23]. Baxa et al., [5] detected this pathogen in all tissues of black sea bream fry by using FAT technique, however this pathogen was only isolated from skin surface with culture methods. IFAT was used for the confirmation of recovery of this pathogen from gills following experimental inoculation by Powell et al., [9]. In this study, IFAT was also congruently used for the detection of T. maritimum in all tissue imprints of fish samples that T. maritimum was not isolated with culture methods. In this study, besides slide agglutination technique, the presence of antibodies in the blood sera of the moribund fish against this pathogen was also detected by using ELISA so recording false positive reactions was avoided. Both of these techniques have a short analysis time, and much less amounts of serum is used. Taken together, serological techniques proved to be more sensitive, rapid and efficient than the conventional bacteriological methods in the detection of T. maritimum in the moribund fish tissues and serum.

In conclusion, this first serotyping of Turkish T. maritimum isolates from sea bass revealed that they all belonged to serotype O1 and it suggest this serotype to be the predominant one in Turkey. However, further studies are needed to confirm this finding in order to produce effective vaccines against this pathogen in Turkey.

Acknowledgements

This study was supported by Istanbul University Scientific Research Projects (project number: 2618). All protocols were approved by the Ethics Committee for Animal Experiments of the University of Istanbul. We are grateful to Prof. Alicia E. Toranzo (Universidade de Santiago de Compostela, Spain) for generously providing antisera against T. maritimum serotypes' O1, O2 and O3 as well as for her helpful comments.

References

1. Wakabayashi H, Hikida H, Masumura K (1986) Flexibacter maritimus sp. nov., a pathogen of marine fishes. Int J Syst Bacteriol 36: 396-398.

2. Bernardet JF, Grlmont PAD (1989) Deoxyribonucleic acid relatedness and phenotypic characteristics of Flexibacter columnaris sp. nov., nom. rev., Flexibacter psychrophilus sp. nov., nom rev., and Flexibacter maritimus Wakabayashi, Hikida & Masamura, 1986. Int J Syst Bacteriol 39: 346-354.

3. Suzukiu M, Nakagawa Y, Harayama S, Yamamoto S (2001) Phylogenetic analysis and taxonomic study of marine Cytophaga-like bacteria: Proposal for Tenacibaculum gen. nov. with Tenacibaculum maritimum comb. nov. and Tenacibaculum ovolyticum comb. nov., and description of Tenacibaculum mesophilum sp. nov. and Tenacibaculum amylolyticum sp. nov. Int J Syst Evol Microbiol 51: 1639-1652.

4. Baxa DV, Kawai K, Kusuda R (1986) Characteristics of gliding bacteria isolated from diseased cultured flounder, Paralichthys oliveceous. Fish Pathology 21: 251-258.

5. Baxa DV, Kawai K, Kusuda R (1987) Experimental infection of Flexibacter

maritimus in black sea bream (Acanthopagrus schlegeli) fry. Fish Pathology 22: 105-109.

6. Chen ME, Henry-Ford D, Groff JM (1995) Isolation and Characterization of Flexibacter maritimus from Marine Fishes of California. Journal of Aquatic Animal Health 22: 7-11.

7. Ostland VE, La Trace C, Morrison D, Ferguson HW (1999) Flexibacter maritimus associated with a bacterial stomatitis in Atlantic salmon smolts reared in net-pens in British Columbia. J Aqua Anim Health 11: 35-44.

8. Mc Vicar AH, White PG (1982) The prevention and cure of an infectious disease in cultivated juvenile Dover sole Solea solea (L.). Aquaculture 26: 213-222.

9. Powell M, Carson J, Van-Gelderen R (2004) Experimental induction of gill disease in Atlantic salmon Salmo salar smolts with Tenacibaculum maritimum. Diseases of Aquatic Organisms 61: 179-185.

10. Schmidtkel L, Carson J, Howard T (1991) Marine Flexibacter infection in Atlantic salmon. Characterization of the putative pathogens. Proceedings of the Saltas Research Review Seminar.

11. Bernardet JF, Kerouault B, Michel C (1994) Comparative study on Flexibacter maritimus strains isolated from farmed seabass (Dicentrarchus labrax) in France. Fish Pathology 29: 105-111.

12. Pazos F, Santos Y, Nunez S, Toranzo AE (1993) Increasing occurence of Flexibacter maritimus in the marine aquaculture of Spain. FHS/AFS Newsletter 21: 1-2.

13. Pazos F (1997) Flexibacter maritimus: estudio fenotípico, inmunológico y molecular. Tesis doctoral, Universidad Santiago de Compostela, Spain.

14. Tabone J (1996) Isolation and characterization of the fish pathogen Flexibacter maritimus from cultured sea bass Dicentrarchus labrax L, B.Sc. Thesis, University of Malta, Malta.

15. Salati F, Cubadda C, Viale I, Kusuda R (2005) Immune response of sea bass Dicentrarchus labrax to Tenacibaculum maritimum antigens. Fisheries Science 71: 563-567.

16. Magi GE, Avendano-Herrera R, Magarinos B, Toranzo AE, Romalde JL (2007) First reports of flexibacteriosis in farmed tub gurnard (Chelidonichthys lucernus L.) and wild turbot (Scophthalmus maximus) in Italy Bull. Eur Ass Fish Pathol 27: 177-184.

17. Kolygas MN, Gourzioti E, Vatsos IN, Athanassopoulou F (2012) Identification of Tenacibaculum maritimum strains from marine farmed fish in Greece. Veterinary Record 170: 623.

18. Türk N (2006) Fish Disease, Flexibacteriosis. Aquaculture and Fisheries 2: 53-54.

19. Arenas J, Mata M, Santos Y (2003) Evaluation of an enzyme-linked immunosorbent assay for serological typing of Tenacibaculum maritimum. European Association of Fish Pathologist 11th International Conference of 'Disease of Fish and Shell Fish'.

20. Avendano-Herrera R, Magarinos B, Romalde JL, Toranzo AE (2003) An update on the antigenic diversity of Tenacibaculum maritimun strains isolated from marine fishes. FHS/AFS Newsletter 31: 24-26.

21. Avendano-Herrera R, Magarinos B, Lopez-Romalde S, Romalde JL, Toranzo AE (2004) Phenotyphic characterization and description of two major O-serotypes in Tenacibaculum maritimum strains from marine fishes. Diseases of Aquatic Organisms 58: 1-8.

22. Avendano-Herrera R, Magarinos B, Morinigo MA, Romalde JL, Toranzo AE (2005) A novel O-serotypes in Tenacibaculum maritimum strains isolated from cultured sole (Solea senegalensis). Bull Eur Ass Fish Pathol 25: 70-74.

23. Van-Gelderen R, Carson J, Nowak B (2009) Effect of extracellular products of Tenacibaculum maritimum in Atlantic salmon, Salmo salar L. Journal of Fish Diseases 32: 727-731.

24. Castro N, Magarinos B, Nunez S, Toranzo AE (2007) Reassessment of the Tenacibaculum maritimum serotypes causing mortalities in cultured marine fish. Bull Eur Ass Fish Pathol 27: 229-233.

25. Şen E (2007) Sea Bass (Dicentrarchus labrax) on a research Flexibacter maritimus infection in fish. M.Sc. T. C. Istanbul University, Institute of Science and Technology.

26. Timur G, Timur M, Akaylı T, Korun J (2007) Survey Study Of Pathologies Affecting Farmed Sea Bass (Dıcentrarchus labrax L. 1758) And Marine Cultured Rainbow Trout (Oncorhynchus Mykiss) in Turkey. 13th international conference of the EAFP "Diseases of fish and shellfish", Grado-Italy.

27. Yardımci RE, Timur G (2015) Detection of Tenacibaculum maritimum, the Causative Agent of Tenacibaculosis in Farmed Sea Bass (Dicentrarchus labrax) on the Aegean Sea Coast of Turkey. The Israeli Journal of Aquaculture - Bamidgeh 67: 1172-1182.

28. Avsever ML, Türk N, Ün C, Didinen BI, Tunalıgil S (2015) Detection of Tenacibaculum maritimum from Seven Different Cultured Marine Fish in Turkey. The Israeli Journal of Aquaculture-Bamidgeh, 6 pages, accepted manuscript.

29. Toranzo AE, Baya AM, Roberson BS, Barja JL, Grimes DJ et al. (1987) Specificity of slide agglutination test for detecting bacterial fish pathogens. Aquaculture 61: 81-97.

30. Sorensen UB, Larsen JL (1986) Serotyping of Vibrio anguillarum. Appl Environ Microbiol 51: 593-597.

31. Ainsworth AJ, Capley G, Waterstr Eet P, Munson D (1996) Use of monoclonal antibodies in the indirect fluorescent antibody technique (IFA) for diagnosis of Edwardsiella ictulari. Journal of Fish Diseases 9: 439-444.

32. Lorenzen E, Karas N (1992) Detection of Flexibacter pschrophilus by immunofluorescence in fish suffering from fry mortality syndrome: A rapid diagnostic method. Diseases of Aquatic Organisms 13: 231-234.

33. Knappskog DH, Rodseth OM, Slinde E, Endresen C (1993) Immunochemical analyses of Vibrio anguillarum strains isolated from cod, Gadus morhua L, suffering from vibriosis. Journal of Fish Diseases 16: 327-338.

34. Cipriano RC, Pyle JB, Starlıper CE, Pyle SW (1985) Detection of Vibrio anguillarum antigen by dot blot assay. Journal of Wildlife Diseaes 21: 211-218.

35. Toranzo AE (2015) Tenacibaculosis of farmed fish in Southern Europe. Tenacibaculum maritimum Workshop Maritime Centre, British Columbia, Canada.

15

Development of Cost Effective Nutritionally Balanced Food for Freshwater Ornamental Fish Black Molly (*Poecilia latipinna*)

Pai IK[1]*, Maryem Shaikh Altaf[1] and Mohanta KN[2]

[1]Department of Zoology, Goa University, Goa-403 206, India
[2]Fishery field Laboratory, ICAR Complex, Old Goa, Goa, India

Abstract

Fishery is one of the oldest professions of man and even today, it helps in food security and poverty alleviation, in many parts of the world. Fish culturing is major part of aquaculture. When natural fishery does not suffice the need of the society, world has gone for aquaculture to increase the productivity. On the same lines, in ornamental fishery too, man's interference in deciding the breed, formulating the feed, bringing changes in living conditions of the fishes, etc., is very much in vogue. To improve the health of the fishes, and their productivity of ornamental fisheries, nutrition plays a major role. In this direction, several attempts have been made by various workers, to alter the food and other environmental conditions. However, as there are hardly any attempt to develop low cost balanced food for ornamental fishes, present paper deals with formulating low cost nutritionally balanced food for Black Molly (*Poecilia latipinna*), a popular freshwater ornamental fishes has been attempted, by using locally available 'waste material' like Groundnut oil cake, Fish meal, Wheat bran, Snail, Marine Fish waste, Freshwater Fish waste, Chicken waste, Earthworms, Squids, Mussel, Chicken liver, Prawn meal, etc., The results obtained on growth and nutrition utilization indicated by weight gain and specific growth rate of Black Molly was highest, in fish fed with diet containing Snail meal, Prawn meal, Mussel meal and Chicken liver. The studies indicate that, these ornamental fishes can be reared well with animal based products, replacing agro based products, which can replace high cost commercial Fish meal.

Keywords: Ornamental fishery; Black molly; Fish meal

Introduction

The majority (>90%) of freshwater ornamental fish are captive bred, compared to only about 25 of a total of 8000, in the case of marine fish and the ornamental fish industry relies heavily on the export and import of introduced species [1]. In India too, like other countries, Guppy's, Swordtails, Platys and Molly's, due to their color, sturdiness, stability, ability to withstand considerable environmental variations, relatively easy maintenance etc., are most popular as aquarium fishes in ornamental fisheries.

Production of aquaculture species, in semi intensive pond culture system, demands the use of artificial feed, as supplementary source of nutrition. Using of such artificial diet for fishes can be traced as early as 1927 [2]. Most commonly used animal protein source in the diet is Fish meal, whose level could range between 10-50% of the operational costs [3]. Increasing cost and short supply of Fish meal has escalated the cost of Fish feed [4-6]. Thus, people have started concentrating on low cost Fish feed, by using plant protein sources [7]. Soybean products seem to be more promising, as a substitute, as they are almost half the price of Fish meal [8]. Though there is a tendency to use alternative plant and animal proteins, as a low cost substitute for Fish meal, such alternatives are known to have lower nutritional value, than Fish meal resulting in lower growth rates or a reduced performance of the cultured animals. In addition to the above, many of these protein sources containing substances that, might cause slight to severe effects on nutritional status of an animal. Insoluble fibers, Enzyme inhibitors, Saponins, Lectins, Tannis, Phytic acid and Gossypol are some of the important anti-nutrients, acting in the Gastro intestinal tract [9]. Despite the above, the search for alternative low cost substitutes is on throughout the world [10-13].

Ingested food nitrogen in fishes is either excreted as nitrogenous waste product or stored in growth products. The end product of protein catabolism in fishes is ammonia, which is generally considered as toxic, if allowed to accumulate. In many fishes, protein retention may be improved, by partly replacing dietary proteins by lipids. Such protein sparing effects have been demonstrated in Salmon [14,15], Trout [16], Carp [17], hybrid Striped Bass [18], Yellowtail [19], Red Sea Beam [20]. Improving digestibility of diet formulation and optimizing feeding regimens can improve feed utilization efficiency in farmed fishes [21]. It is also known that, fishes are able to regulate their daily food intake, based on their nutrient and energy requirements [22].

Fish meal is known for its high essential Amino acid a Fatty acids contents, high digestibility, low Carbohydrates, etc., Lipid is known to be one of the important nutrient next to protein, which plays a major role in optimum utilization of dietary Protein for growth. Lipids are almost completely digestible by fish and seem to be favored over Carbohydrates as an energy source [23]. Fishes are also known for utilizing protein preferentially over lipid or Carbohydrate as an energy source. Therefore, it is important from nutritional, environmental and economic point of view to improve protein utilization for tissue synthesis, rather than energy purposes.

In view of the above, locally available 'Waste material' like Groundnut oil cake, Fish meal, Wheat bran, Snail, marine Fish waste, freshwater Fish waste, Chicken waste, Earthworms, Squids, Mussel, Chicken liver, Prawn meal, etc., were analyzed to see their efficacy as alternative to Fish meal in one of the most popular aquarium fish species viz., Black Molly (*Poecilia latipinna*).

*Corresponding author: Pai IK, Department of Zoology, Goa University, Goa-403 206, India, E-mail: ikpai@unigoa.ac.in

Materials and Method

One of the most popular Aquarium Fish viz., Black Molly (*Poecilia latipinna*) was utilized in the present experiments. Uniform sized, healthy fish fingerlings were procured from Freshwater ornamental Fish hatchery of ICAR research complex, Old Goa, India and were treated with 0.05% Potassium Permanganate solution, for two minutes, to make them free from external parasites and pathogens present if any. Before initiating the feeding trials, they were kept in 1.0 ton capacity cement tanks for conditioning/acclimatize to laboratory condition. During this period, they were fed twice daily, with practical diet other than experimental diet.

All the necessary fish feed ingredients such as Snail, Freshwater Fish waste, Marine Fish waste, Chicken waste, Earthworm, Squid, Mussel, Chicken liver, Prawns, Fish meal, Groundnut oil cake, Wheat bran, Vitamin and Minerals, Oil were obtained from local market. The ingredients were dried at 105°C to make them moisture free. Each ingredient was powdered separately using domestic grinder. The powdered material was then sieved through a mesh size of 0.5mm dia. Later ingredients were analyzed, to determine their chemical composition (Table 1). Later, nine different feeds were formulated, based on the nutrient requirement (40% protein and 6% lipid, as determined earlier) (Table 2). The percent composition and proximate composition of the experimental diet is provided in Table 3. A weight of individual components of the diet was done, by using Sertorius Single Pan Balance

and the mixture was prepared by thoroughly mixing the required components in a plastic tray. To these mixtures, required quantity of lukewarm, distilled water was added to prepare the dough, which was then put into hand palletizer and the 2 mm diameter pallets were drawn. These pallets were sun dried for 3-4 days, till they contain 7-8% moisture level. Then, they were stored in air tight containers to be used later. Three each replicate of 10 fishes each, for each of the treatment, was placed in fish tank of 100 L capacity, in which 50 L of water of the quality as mentioned in Table 4 was used. All the batches were fed twice daily, with the experimental diet. The body weight gain after a month of treatment has been mentioned in Table 5. The data obtained was subjected for statistical analyses like ANOVA and the comparison among the groups, was done by Duncan multiple range test at $P > 0.05$.

Results

Table 1 provides the data on Composition (% of dry matter) of the food ingredients used for formulating diet for Molly. It can be seen that among the ingredients Fish meal has highest dry matter (93.12%) in comparison with lowest dry matter (15.12%) in Squids. Crude protein was observed to be highest in chicken liver (66.12%) in comparison with 10.94% of crude protein in wheat bran. Further, maximum and minimum of ether extract was recorded in Chicken waste (28.54%) and Wheat bran (2.24%) respectively. Analyses for total Ash, showed that the maximum ash was in Prawn meal (17.96%), compared to a minimum of 7.68% of ash in wheat bran.

Table 2 Exhibits the ingredients composition (% of dry matter) of the diet fed to Molly fishes. In all the experimental feeds fish meal and Vitamin and minerals was kept constant (10% and 5.0% respectively) while other ingredients were fed to the experimental batched in varied concentrations.

Proximate composition (% dry matter) of the experimental diet, which provides data on dry matters, crude protein, ether extract and ash, has been provided in Table 3. It can be seen that, dry matter ranged from 92.4 (in Snail) to 94.2% (in Squid meal and Prawn meal). Crude protein was maximum in Marine Fish waste meal (41.90%) and was minimum in Mussel meal (39.96%) though, the difference was statistically insignificant. Ether extract showed maximum in Chicken liver (13.07%) and a minimum of 7.23% in Snail meal. Similarly, Ash content was highest in Prawn meal (13.54%) in comparison with lowest Ash contents in Chicken waste (9.68%).

Ingredient	Dry matter	Crude protein	Ether extract	Total ash
Groundnut oil	92.86	43.75	9.48	7.84
Fish meal	93.12	56.00	8.80	14.12
Wheat bran	92.54	10.94	2.24	7.68
Snail	30.50	54.25	4.50	15.22
Marine fish waste	33.56	56.87	9.60	14.24
Freshwater fish waste	32.46	47.25	18.0	16.86
Chicken waste	30.25	51.62	28.54	13.46
Earthworm	17.16	50.75	15.00	12.66
Squids	15.12	69.12	6.20	11.88
Mussels	17.24	58.35	12.60	9.32
Chicken liver	23.22	66.12	24.00	10.16
Prawn meal	92.44	64.75	5.20	17.96

Table 1: Composition (% of dry matter) of food ingredient used for formulating diet for Black Molly (*Poecilia latippina*).

Ingredient	Experimental diet								
	T-1 (Snail Meal)	T-2 (Freshwater fish waste)	T-3 Marine fish waste)	T-4 (Chicken waste)	T-5 (Earthworm)	T-6 (Squid)	T-7 Mussel)	T-8 (Chicken liver)	T-9 (Prawn meal)
Snail meal	45.0	----	----	----	----	----	----	----	----
Freshwater fish waste	----	48.0	----	----	----	----	----	----	----
Marine fish waste	----	----	42.0	----	----	----	----	----	----
Chicken waste	----	----	----	46.0	----	----	----	----	----
Earthworm	----	----	----	----	54.0	----	----	----	----
Squid	----	----	----	----	----	36.0	----	----	----
Mussel	----	----	----	----	----	----	38.0	----	----
Chicken liver	----	----	----	----	-----	----	-----	40.0	----
Prawn meal	----	----	----	----	----	----	----	----	38.0
Fish meal	10.0	10.0	10.0	10.0	10.0	10.0	10.0	10.0	10.0
Groundnut cake	20.0	25.0	25.0	25.0	25.0	22.0	28.0	22.0	20.0
Wheat bran	18.0	12.0	14.0	15.0	15.0	25.0	29.0	23.0	25.0
Vit. and Min.	5.0.	5.0	5.0	5.0	5.0	5.0	5.0	5.0	5.0
Oil	2.0	----	----	----	----	2.0	----	----	2.0

Table 2: Ingredient composition (% of dry matter) of the diet fed to Black Molly (*Poecilia latippina*) Fingerlings.

Physico-chemical parameters of the water, in which experimental animals were maintained is provided in Table 4. The data reveals that, the water used for maintain the Fishes was very much normal and values obtained were acceptable for conducting such experiments.

Discussion

In aquaculture, diet cost accounts to over 50% of the operating cost, in intensive aquaculture, depending upon many factors, such as protein level, source, type of ingredient, and manufacturer's practices etc., the animal byproducts (ABM) meal, such as the one like Snails, Freshwater Fish waste, Marine Fish waste, Squid, Poultry industry waste can be used as substitute. Added advantage is that, these do not cause any deleterious effects on the fishes. Such attempts have been made earlier too [24], who used Earthworms, Poultry Egg Shell dust, Plant Rhizome, Wheat flour etc., as a substitute, for commercial Fish meal. The study revealed better that growth, length, gonad development, fertility in the Fishes fed with specially prepared feed, followed by dried Earthworm diet. Chong and Hashim observed higher man body length, in red sword tail *Xiphophorus helleri*, when they were fed several times. They reported that, more energy food is required, before spawning, for gonadal development and production of eggs. While, James and

Sampath [25] found that, females of *X. helleri* with frequent meals exhibited higher gonad weight, and Gonado-somatic index. In 1964, Hester reported that, scarcity of food will result in reduced fertility, in Guppy Fishes. He further reported that, female Fish needs adequate Protein, Fat, Vitamins and Minerals for egg development and spawning, as yolk is composed of Phospholipo proteins, an amalgam of Minerals (Phosphorus), a Protein and Lipid. It is also known that, Protein is also required for forming follicle in Embryo.

In the present studies, the replacement of high cost fish meal, with the animal by product meal (ABM) and agro-based products, are used to formulate practical diet, for Black Molly, to observe the growth and dietary performance. The study indicate that, the growth was better, in the fish fed with diet containing Squid meal, Prawn meal, Chicken liver and Mussel meal. Thus, it can be concluded that, such substitution could be used to replace the costly commercial Fish meal [26,27].

References

1. FAO (2015) World review of fisheries and aquaculture, FAO, United Nations.

2. Ida Mellen (1927) The Natural and Artificial Foods of Fishes. Transactions of the American Fisheries Society Volume 57: 120-142.

3. Akiyama DM, Dominy WG, Lawrence AL (1991) Penaeid shrimp nutrition. In: Marine Shrimp culture: Principles and Practices, Elsevier Publ. Amsterdam, The Netherlands.

4. HiGuera M, Gardenete G (1989) Fuentes alternativas de protean y energia en acuicultura. Alimentacion en Acquaculture FEUGA, Madrid.

5. McCoy HD (1990) Fishmeal-the critical ingredient in aquaculture feeds. Aquaculture magazine 16: 43-50.

6. Bimbo AP, Crowtber B (1992) Fish meal and oil: Current uses. Jl of Am Chem Soc 69: 221-227.

7. Tacon AGJ, Jaskson AJ (1985) Utilization of conventional and unconventional protein sources in practical fish feeds. Nutrition and Feeding in fish, Academic Press, London.

8. Dabrowski K, Poczynski P, Kock G, Berger B (1989) Effects of partially or totally replacing fish meal protein on growth, food utilization and proteolytic enzyme activities in rainbow trout (*Salmo gairdneri*): new *in vivo* test for exocrine secretion. Aquaculture 77: 29-49.

9. Krogdahl A (1986) Anti-nutrients effecting digestive functions and performance in poultry. Proc. Of the 7th European poultry conf., World Poultry Sci. Association, Paris.

10. Tacon AGJ (1993) Fish ingredient for carnivorous fish species: alternatives to fishmeal and other fishery resources, FAO, Fish Circ.

11. Stafford EA, Tacon AGJ (1985) The nutritional evaluation of dried earthworm meal (*Eisenia foetida* Savigny, 1826) included at low levels of production diets for rainbow trout, *Salmo Gairdneri* Richardson. Aquaculture and Fisheries Management 16: 213-222.

12. Pongmaneerat I, Watanabe T (1991) Nutritive value of protein of feed ingredient for carp. *Cyprinus carpio*, Nippon Suisan Gakkaishi 57: 503-510.

13. Rumsey GL (1993) Fish meal and alternate source of protein in fish feeds update. Fisheries 18: 14-19.

Treatment/diet	Parameters			
	Dry matter	Crude protein	Ether extract	Ash
T-1 (Snail)	92.4	40.73	7.23	12.72
T-2 (Freshwater fish waste)	92.6	40.53	12.16	12.88
T-3 (Marine fish waste)	93.2	41.90	7.68	13.96
T-4 (Chicken waste)	93.6	41.72	17.08	9.68
T-5 (Earthworm)	92.8	40.86	10.34	12.36
T-6 (Squid)	94.2	40.06	7.75	11.76
T-7 (Mussel)	93.4	39.96	8.03	11.42
T-8 (Chicken liver)	92.8	40.65	13.07	10.96
T-9 (Prawn meal)	94.2	40.28	7.32	13.54

Table 3: Proximate composition (% dry matter) of the experimental diet.

Parameter	Range
Temperature (Deg. C)	26.7-28.4
pH	7.2-7.7
Dissolved oxygen(mg/lit)	6.55-7.23
Hardness (CaCo$_3$/Lit)	88.0.-103.0
Alkalinity (CaCo$_3$/Lit)	97.0-107.0
Nitrate (mg/Lit)	12-20.0
Nitrite (mg/Lit)	0.06-0.09

Table 4: Water parameters measured for the experiment on formulation of practical diet for Black Molly (*Poecilia latipinna*).

Diet	Nutritional indices					
	Initial weight (g)	Final weight (g)	Weight gain (g)	FCR	Specific Growth Rate	PER
Snail meal	0.83 ± 0.02[a]	1.71 ± 0.06[c]	0.88 ± 0.04[d]	1.93 ± 0.03[a]	2.39 ± 0.04[c]	1.28 ± 0.01[c]
Freshwater fish waste	0.88 ± 0.02[a]	1.97 ± 0.07[b]	1.09 ± 0.05[b]	1.76 ± 0.02[c]	2.67 ± 0.06[b]	1.40 ± 0.02[b]
Marine fish waste	0.95 ± 0.03[a]	1.89 ± 0.02[bc]	0.94 ± 0.02[cd]	1.84 ± 0.03[b]	2.30 ± 0.08[c]	1.30 ± 0.01[c]
Chicken waste	0.92 ± 0.03[a]	2.07 ± 0.06[b]	1.15 ± 0.05[a]	1.70 ± 0.02[c]	2.69 0.09[b]	1.41 ± 0.01[b]
Earthworm meal	0.95 ± 0.03[a]	1.99 ± 0.04[b]	1.05 ± 0.01[bc]	1.75 ± 0.01[c]	2.48 ± 0.04[c]	1.41 ± 0.03[b]
Squid meal	0.91 ± 0.02[a]	2.53 ± 0.09[a]	1.62 ± 0.05[a]	1.47 ± 0.02[d]	3.42 ± 0.07[a]	1.69 ± 0.03[a]
Mussel meal	0.90 ± 0.02[a]	2.45 ± 0.09[a]	1.55 ± 0.05[a]	1.51 ± 0.02[d]	3.35 ± 0.07[a]	1.66 ± 0.04[a]
Chicken liver	0.86 ± 0.02[a]	2.37 ± 0.06[a]	1.51 ± 0.06[a]	1.53 ± 0.04[d]	3.37 ± 0.08[a]	1.62 ± 0.04[a]
Prawn meal	0.88 ± 0.02[a]	2.46 ± 0.08[a]	1.58 ± 0.06[a]	1.49 ± 0.02[a]	3.42 ± 0.04[a]	1.67± 0.02[a]

Table 5: Growth and dietary performance of Black Molly (*Poecilia latipinna*) fed with different diets.

14. Garcia M, Zamora S, Lopez MA (1981) The influence of partial replacement of protein by fat in the diet on the protein utilization by the rainbow trout (*Oncorhynchus mykiss*). Comp Biochem Physiol 68B: 457-460.

15. Johnson F, Hillested M, Austreng E (1991) High energy diets for Atlantic salmon. Effects on pollution, Fish nutrition in practice. Biarritz (France) Les Colloques 61: 391-401.

16. Beamish FWH, Mediandn TE (1986) Protein sparing effects in large rainbow trout, *Oncorhynhus mykiss* Aquaculture 55: 35-42.

17. Watanabe T (1987) Lipid nutrition in fish. Comp Biochem Physiol 73: 3-15.

18. Nematipour GR, Brown ML, Gatlin DM (1992) Effects of dietary energy: protein ratio on growth characteristics and body composition of hybrid striped bass *Morone chrysops* x *M. saxtilis.* Aquaculture, 107: 359-368.

19. Shimeno S, Hosokawa H, Takeda M (1979) The importance of carbohydrates in the diet of carnivores fish. Proc World Symp Finfish Nutr Fishfeed Technol 1: 20-23.

20. Takeuschi T, Shiina Y, Watanabe T (1991) Suitable protein and lipid levels in diet for fingerlings of red sea bream *Pagrus major.* Nipp Suis Gakk 57: 521-527.

21. Cho CY, Bureau DP (2001) A review of diet formation strategies and feeding systems to reduce excretory and feed waste in aquaculture. Aquacul Res 32: 349-360.

22. Kaushik SJ, Madela F (1994) Energy requirements, utilization and supply to salmonids. Aquaculture 124: 81-97.

23. Cowey CB, Sargent JR (1977) Lipid nutrition in fish. Comp Physiol 57B: 269-273.

24. Chakrabarty I, Gani Md.A, Misra A, Chaki KK, Sur R (1995) Digestive enzymes in 11 freshwater teleost fishes in relation to food habit and niche segregation. Comparative Biochemistry and physiology 112: 167-177.

25. James R, Sampath K (2004) Effect of animal and plant protein diets on growth and reproductive performances in an ornamental fish *Xiphophorus helleri.* Ind J Fish. 54: 75-86.

26. Hexter FJ (1964) Effects of food supply on the fecundity in the guppy, *Lebistes reticulates (*Peters). J Fish Res Board Can 21: 757-764.

27. Tacon AGJ, Stafford EA, Edwards CA (1983) A preliminary investigation on the nutritive value of three terrestrial lumbricid worms for rainbow trout. Aquaculture 35: 187-199.

16

A Study of *Clinostomum* (Trematode) and *Contracaecum* (Nematode) Parasites Affecting *Oreochromis Niloticus* in Small Abaya Lake, Silite Zone, Ethiopia

Mohammed Reshid[1], Marshet Adugna[2], Yisehak Tsegaye Redda[1]*, Nesibu Awol[1] and Awot Teklu[1]

[1]*Mekelle University College of Veterinary Medicine, P.O.Box 231 Mekelle, Ethiopia*
[2]*Ethiopian Institute of Agricultural Research, National Fisheries and other Aquatic Life Research Center, P.O. Box 64, Sebeta, Ethiopia*

Abstract

This study was conducted at Lake Small Abaya Ethiopia, to identify *Clinostomum and Contracaecum* parasites from a total of 384 *O. niloticus* species sampled during November 2013- April 2014. Of the 384 samples collected, 138 (35.9%) were infested with nematode of *Contracaecum* species and 72 (18.8%) were infected with trematode of *Clinostomum* species. The intensity of infestation by *Contracaecum* and *Clinostomum* was 1-19 worms per fish (mean intensity=4.47) and 1-12 worms per fish (mean intensity=3.56) respectively. There was no significant difference ($p > 0.05$) in the prevalence of infestation among host sex, host size and host weight. There was no any statistically significant ($p > 0.05$) correlation between the number of *Clinostomum*, *Contracaecum* and the mixed number of parasites and the fish's condition. In conclusion, the study show that fish parasite are prevalent in Small Abaya lake .Hence, further studies and appropriate control measure are recommended to reduce their effect on the fishery industry and public health.

Keywords: *Clinostomum; Contracaecum;* Condition factor; *O. niloticus;* Small abaya lake

Introduction

Ethiopia has large water resources, with an estimated surface area of 733k km² of major lakes and reservoirs, 275 km² of small water bodies and 7285 km long rivers within the country [1]. As a result of these ecological variations, Ethiopia has been the home of highly diversified flora and fauna. More than 200 species of fish are known to occur in lakes, rivers and reservoirs in Ethiopia [2]. The country depends on its inland water bodies for fish supply to its population.

Mostly, Nile tilapia (*Oreochromis niloticus*) has been introduced throughout the country because of its adaptive abilities and its suitability to match Ethiopian consumers' preferences. As a consequence of its natural occurrence plus its introduction into different water bodies, it is contributing about 40.9 % of the 13,253 tons of commercial fish catch in 2007/2008 [3].

One of the problems of the fishery sector in the wild populations are parasites and disease conditions of fish. Parasitic diseases reduce fish production by affecting the normal physiology and if left uncontrolled, it can result in mass mortalities or in some cases, can be served as a source of infection for human and other vertebrates that consumed fish [4]. Presence of a massive number of parasites on each fish might constitute a real threat to the fish population and require immediate action [5].

The digenea parasites are the main endo-parasites of fishes; the greater majority of fish are susceptible to infection with different stages of these parasites [6]. Clinostomum species, digenetic termatode, are common fish parasites throughout the world and the final hosts of this fluke are generally piscivorous birds, including herons and egrets [7]. The metacercariae embedded in the tissues of fish are freed in the host stomach and migrates up to the esophagus, and then attach to the throat or mouth cavity. Human infections are known to be resulting from eating raw freshwater fish. Metacercarial infection in fish is the main source of disease with subsequent economic loss. Metacercariae may affect growth and survival, or disfigure fish so that they loss their market value as a food or ornamental product [8].

Most adult nematodes are found in the intestine of fish but their larval stage, which is infective to human, has the greatest impact on the consumer acceptance of fish as a source of protein [9]. *Contracaecum* is an anisasakid nematode that infects fish-eating bird and marine mammals. Larval stage of *Contracaecum* usually occurs in the body cavity and mesenteries of fish while the adults occur in the gut of piscivorous birds, notably pelicans, cormorant's herons and darters [10].

So far, very few diseases have been described from fish of Ethiopia waters. Moreover there is no report of disease and prevalence of parasitic infection in fish in small Abaya Lake.

Therefore the objectives of the current study were: to determine the prevalence of *Clinostomum* and *Contracaecum* species in *Oreochromis niloticus* in Small Abaya Lake.

Materials and Study Methodology

The study area

The study was conducted at Lake Small Abaya (7°29 03' 65" N latitude and 38° 03' 17.79" E longitude), which is located at altitude of 1835 meter above the sea level. The Lake covers a total area of 1253ha and it is shallow lake with the maximum depth of 9 m. The mean monthly minimum and maximum temperatures varies between 10.8°C

*Corresponding author: Redda YT, Mekelle University College of Veterinary Medicine, P.O. Box 231 Mekelle, Ethiopia
E mail: yistseg@yahoo.com

to 14.1°C and 22.5°C to 28.7°C, respectively throughout sampling period. Before stocking, there were no commercially important fish species in the lake except for the naturally occurring *Barbus* species. In the 2005 National Fisheries and other Aquatic Life Research Center stocked *Oreochromis niloticus* or *Nile tilapia* and *Tilapia zilli* fry in to Lake small Abaya.

Study design

A cross sectional study was conducted from November 2013 to April 2014 at Small Abaya Lake to determine the prevalence of parasitic infestation of *O. niloticus* /Nile tilapia/. The desired sample size was calculated using the formula given by Thrusfield [11]. By considering 95% confidence interval, 5% desired absolute precision and 50% expected prevalence and the total number of sample found to be 384.

Study methodology

Sample collection: A total of 384 *O. niloticus* species of fish were sampled and examined. All the fish were caught using gill net with mesh size ranging from 6 to 12 cm. Harvested fishes were transported in ice to Ethiopian Institute of Agricultural Research, National Fisheries and other Aquatic Life Research Center, Sebeta, for analysis. The length (L) of fish was taken from the tip of the snout to the posterior tip of the caudal fin and was measured to the nearest ± 0.1 cm. The weight of the fish was measured to the nearest gram using an electric balance.

Laboratory examinations: Each fish was opened and its internal organs were fully examined for parasites. The entire digestive system was removed and placed in a Petri-dish with physiological saline, and the gut was divided into sections. The muscles, gonads, liver, and heart were examined with the aid of a dissection microscope and a phase contrast light microscope at 10 and 40 magnifications. Parasites were counted, their location recorded, and preserved in 70% ethanol. Identification of most parasites was made immediately following standard keys in literature [12-14].

Data analysis

The data obtained from the laboratory finding were summarized and then analyzed using SPSS version 16 analyzing software. Chi-square was applied to test association between sex, weight, and standard length with occurrence of the disease. The effect of the parasites on the health of their host was determined by calculating Fulton's condition factor (K), a measure of an individual fish's health that uses standard weight. Proposed by Fulton in 1904, it assumes that the standard weight of a fish is proportional to the cube of its length.

$$K = 100(W / L3)$$

Where W is the whole body wet weight in grams and L is the length in centimeters; the factor 100 is used to bring K close to a value of one. Pearson correlation was done to find the correlation between the body conditions of fish with the number of parasites.

Mean intensity was also calculated using the formula given below

$$Mean\ Intensity = \frac{Total\ number\ of\ parasites\ in\ a\ given\ host}{Total\ number\ of\ hosts\ infected}$$

Results

Monthly variations

Protein content significantly varied from 6.05 ± 0.45% with T_6 (banana leaf) at 6th month (September, 2010) to 31.20 ± 0.32% with treatment T_3 (mustard oilcake) at 2nd month (May, 2010). Lipid content

significantly varied from 2.95 ± 0.21% with treatment T_6 (banana leaf) at 5th month (August, 2010) to 13.72 ± 0.36% with treatment T_3 (mustard oilcake) at 4th month (July, 2010). Carbohydrate significantly varied from 32.85 ± 0.14% with treatment T_3 (mustard oilcake) at 4th month (July, 2010) to 66.35 ± 0.32% with T_2 (wheat bran) at 3rd month (June, 2010). In the same feed item no significant difference in the nutrient content was found during the study period (Tables 1-4).

Mean variations

The variations in the mean values of nutrient contents (protein, lipid and carbohydrate) with different treatments of feed items are presented in Table 3 and Figures 1-3. Protein content significantly varied from 6.18 ± 0.13% with treatment T_6 (banana leaf) to 30.53 ± 0.40% with treatment T_3 (mustard oilcake). Lipid content significantly varied from 3.06 ± 0.09% with treatment T_6 (banana leaf) to 13.33 ± 0.10% with treatment T_3 (mustard oilcake). Carbohydrate significantly varied from 32.95 ± 0.29% with treatment T_3 (mustard oilcake) to 66.12 ± 0.47% with treatment T_2 (wheat bran).

Discussion

Monthly variations of the nutrient contents

Protein content varied from 6.05 ± 0.45% with (T_6 at 6th month) to 31.20 ± 0.32% (T_3 at 2nd month). Lipid content ranged from 2.95 ± 0.21% (T_6 at 5th month) to 13.72 ± 0.36% (T_3 at 4th month). Carbohydrate

Parasites Taxonomy	No. of infected	Prevalence %	Total No. of parasites recovered	Mean intensity
Contracaecum	138	35.9%	617	4.47
Clinostomum	72	18.8%	257	3.56
Total	210	54.7%	874	4.0

Table 1: The prevalence and mean intensity of *Contracaecum* and *Clinostomum*.

Sex	Parasite genera		
	Clinostomum	*Contracaecum*	%
Female (n=144)	23 (16%)	56 (38.9%)	54.9%
Male (n=240)	49 (20.4%)	82 (34.2%)	54.6%
Total (n=384)	72 (18.8%)	138 (35.9%)	54.7%

χ^2=0.03; P>0.05; df=1

Table 2: The prevalence of parasites in relation to host sex.

	Parasite genera		
Length (n)	*Clinostomum*	*Contracaecum*	Prevalence %
11-17 (64)	8 (12.5%)	26 (40.6%)	8.85%
17.5-23 (242)	47 (19.4%)	86 (35.5%)	34.63%
23.5-29.5 (78)	17 (21.8%)	26 (33.3%)	11.19%
Total (384)	72 (18.8%)	138 (35.9%)	54.7%

χ^2=0.076; p>0.05; df=2

Table 3: Prevalence of parasites that infect fish in relation to the standard length of the host (cm).

	Parasite genera		
Weight (n)	*Clinostomum*	*Contracaecum*	Prevalence%
26-135 (n=125)	19 (11.2%)	54 (43.2%)	19.0%
136-178 (n=148)	28 (18.9%)	51 (34.5%)	20.5%
179-549 (n=111)	25 (22.5%)	33 (29.7%)	15.1%
Total (n=384)	72 (18.8%)	138 (35.9%)	54.7%

χ^2=1.063; P>0.05; df=2

Table 4: The Prevalence of parasite infections in fish of various host weight range.

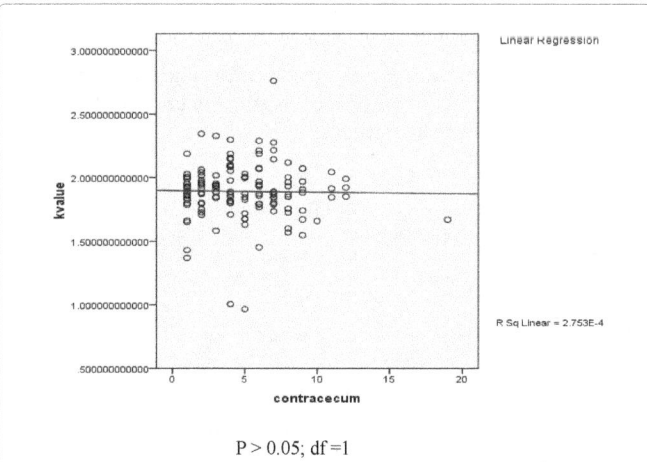

P > 0.05; df =1

Figure 1: Relationship between condition factor and the number of *Contracaecum* parasites.

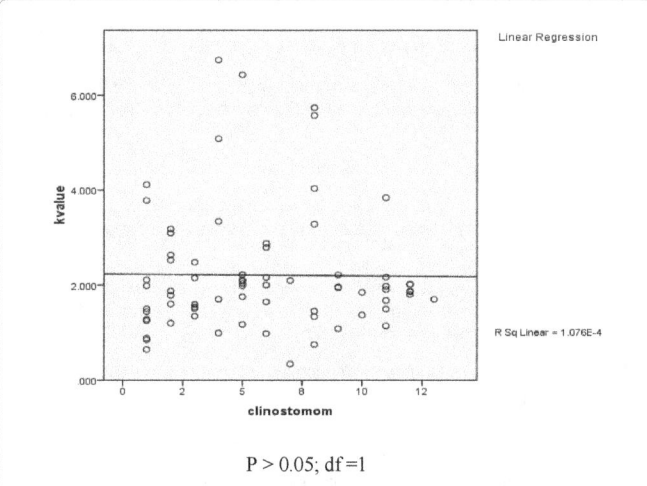

P > 0.05; df =1

Figure 2: Relationship between condition factor and the number of *Clinostomum* parasites.

content ranged from 32.85 ± 0.14% (T$_3$ at 4th month) to 66.35 ± 0.32% (T$_2$ at 3rd month). Suresh and Mandal worked on the determination of nutritive value of rice bran, mustard oil cake and Azolla for a period of 4 months from July to October. In rice bran they found crude protein and crude fibre as 12.6% and 21.9%, respectively. In mustard oilcake, crude protein and crude fibre was 38.6% and 6.8%, respectively and in Azolla, crude protein and crude fibred was 26.5% and 20.4%, respectively. Sithara and Kamalaveni worked on the formulation of low cost fish feed using Azolla as a protein supplement during September to March and reported 20-25.5% protein in Azolla. Ebrahim used Azolla as tilapia diet for a period of 90 days in summer season and reported 20% protein in Azolla. Fasakin and Balogan worked on the nutritional aspects of Azolla in August, 1997 and reported 20.9% protein in Azolla.

Present findings also indicated that in case of same feed item, no significant difference was found in the nutrient content at different months (Tables 1-4). This might be due to no major change in the temperature was found to affect the growth and composition of Azolla during the study period. This statement was almost agreed with Lumpkin and Plucknett who reported that change in Azolla composition was subjected to change in environment. Statement also agreed with Van-Hove and Ebrahim who reported that change in

Azolla composition was subjected to change in species.

Mean variation of the nutrient contents

In the present study the protein content varied from 6.18 ± 0.13% (T$_6$, banana leaf) to 30.53 ± 0.40% (T$_3$, mustard oilcake), lipid content varied from 3.06 ± 0.09% (T$_6$, banana leaf) to 13.33 ± 0.10% (T$_3$, mustard oilcake) and carbohydrate content varied from 32.95 ± 0.29% (T$_3$, mustard oilcake) to 66.12 ± 0.47% (T$_2$, wheat bran). The highest protein and lipid content was found in treatment T$_3$ (mustard oilcake) whereas the highest carbohydrate content was found in treatment T$_2$, wheat bran (66.12 ± 0.47%) followed by T$_4$, Azolla (50.21 ± 0.54%), T$_6$, banana leaf (48.50 ± 0.51%), T$_5$, grass (46.36 ± 0.16%), T$_1$, rice bran (44.09 ± 0.67%), T$_3$, mustard oilcake (32.95 ± 0.29%). Hepher reported the protein content of ricebran, wheat bran, oil cake and Azolla as 11.88%, 14.57%, 30-33% and 19.27%, respectively. Banerjee and Matai determined the nutritive status of *Azolla pinnata* and reported protein as 21.9% and Lipid as 3.8%. Gavina reported crude protein of 20.98%, crude fat of 5.17% and crude fiber of 19.30% in Azolla. Tavares observed 38.8% crude protein, 3.8% crude fat and 13.2% crude fiber in dried duck weed. They also reported that the protein content of duckweeds growing on nutrient poor and nutrient rich water varied between 15-25% and 35-45% (Dry matter basis), respectively. In case of conventional feed items the major nutrient like protein varied from 14.40 ± 0.32% (rice bran) to 30.53 ± 0.40% (mustard oilcake). Whereas in case of non-conventional feed items the protein varied from 6.18 ± 0.13% (banana leaf) to 18.58 ± 0.09% (Azolla). Being an omnivore, the fish can also feed on vegetation and may be able to assimilate Azolla in the diets.

The chemical composition of Azolla species varies with ecotypes and with the ecological conditions and the phase of growth. The crude protein content is about 19-30 percent dry matter basis during the optimum conditions for growth. The protein contents of Azolla species are comparable to or higher than that of most other aquatic macrophytes. Aquatic weeds' are highly nutritious with protein content of 20-30%, when cultivated in nutrient rich waters. Importantly, they

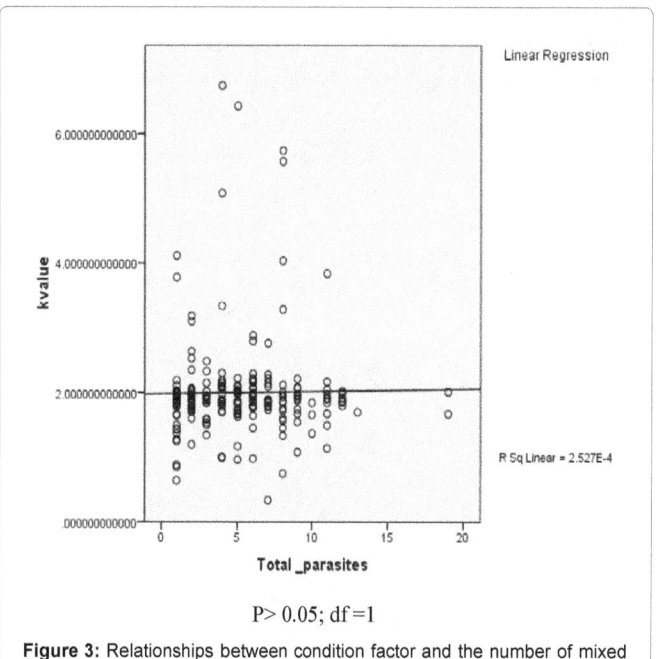

P> 0.05; df =1

Figure 3: Relationships between condition factor and the number of mixed parasites.

are preferred food of a wide range of herbivorous fish such as grass carp (*Ctenopharyngodon idella*), silver barb (*Barbonymus gonionotus*, *Puntius jerdoni*), tilapias (*Oreochromis niloticus*, *Tilapia rendalli*, *Tilapia zillii*) and rohu (*Labeo rohita*).

Overall findings indicated that inspite of having variations in nutrient contents, monthly supply of nutrients was almost same respective feed item under non-conventional feeds as with conventional feeds. Mean values of the nutrient contents under non-conventional feed items are found potentials for the development of low cost aquaculture.

Fish feed generally constitutes 60-70% of the operational cost in intensive and semi- intensive aquaculture system. The fish feed used in aquaculture is quite expensive, irregular and short in supply in many third world countries. These feeds are sometimes adulterated, contaminated with pathogen as well as containing harmful chemicals for human health. Naturally there is a need for the development of healthy, hygienic fish feed which influences the production as well as determines the quality of cultured fish. Considering the importance of nutritionally balanced and cost-effective alternative diets for fish, almost similar expression to evaluate the nutritive value of different non-conventional feed resources, including terrestrial and aquatic macrophytes was found with Wee and Wang and Mondal and Ray. However potentials roles of aquatic and terrestrial macrophytes as supplementary feeds in fish farming were also found to be expressed with Bardach and Edwards.

Conclusion

In case of conventional feed items, protein, lipid and carbohydrate varied from 14.40 ± 0.32% to 30.53 ± 0.40%, 6.69 ± 0.30% to 13.33 ± 0.10% and 32.95 ± 0.29% to 66.12 ± 0.47%. In case of non-conventional feed items, protein, lipid and carbohydrate varied from 6.18 ± 0.13% to 18.58 ± 0.09%, 3.06 ± 0.09% to 6.31 ± 0.13% and 46.36 ± 0.16% to 50.21 ± 0.54%. Inspite of variations weeds are moderately nutritive and low cost effective diets for fish. However, the present study did not evaluate the fish production and economy of feed and weed based systems.

Recommendation

Present findings explored the nutritive aspects of both conventional and non-conventional feed items and question raised about the response of utilizing the feed especially of aquatic weeds to fish growth and economy. Therefore, it is recommended to conduct further study on the evaluation of fish production and economy under different feed and weed based systems in polyculture ponds.

Acknowledgement

The research work was conducted under a financial support by the Ph. D. Fellowship Programme of Ministry of Science and Technology, Govt. of the People's Republic of Bangladesh which is gratefully acknowledged.

References

1. FAO (2005) Ethiopia Information on Fisheries management (From IFMC) water report No. 29, Fao, Rome, Italy.

2. JERBE (2007) Fish diversity in the main drainage system of Ethiopia.

3. MoARD (2008) Ministry of Agriculture and rural development annual report.

4. Ayotunde EO, Change ST, Okey IR (2007) Parasitological examination and food composition in the gut of feral Africa carp, LebeoCoubie in the cross river,South Estern Nigeria. Afr. Journal biotechnol 6: 625-630.

5. http://aqua.internet.com.

6. Mashengo SN (1989) Nematode Parasites of Barbus species in Lebowa and Venda, south Africa. South African J Wildlife Research 119: 35-37.

7. Aohagi Y, Shibahara T, Machida N, Yamaga Y, Kagota K, et al. (1992) Natural infections of Clinostomum complanatum (Trematoda: Clinostomatidae) in wild herons and egrets, Tottori Prefecture, Japan. J Wildl Dis 28: 470-471.

8. Paperna I (1980) Parasites, infection and disease of fish in Africa CIFA Tech.

9. Nagasawa K (1989) Note on parasites of aquatic organisms for visceral adhesion of high–seas sockeye salmon caused by the nematode Philonemaoncorhynchi. Aquabiology 11: 320-321.

10. Whitfield AK, Heeg J (1977) On the life cycle of the cestode ptychobothriumbelones and nematode of the genus *Contracaecum* from Lake St. Lucia, Zululand. SA J sci 73: 121-122.

11. Thrusfield M (1995) Veterinary epidemiology.

12. Roberts RJ (2001) Fish Pathology.

13. Pouder DB, Curtis EW and Roy PE (2005) Common fresh water fish parasites.

14. Klinger RE, Francis-floyd R (2002) Introduction to fresh water fish parasites.

17

Effect of Algal Oil Incorporated Diet on Growth Biochemical and Immunological Response in Ornamental Fish *Daniorerio rerio*

Blessy G, Ajan C, Citarasu T and Michael Babu M*

Planktology and Aquaculture Division, Centre for Marine Science and Technology, Manonmaniam Sundaranar University, Rajakkamangalam, Kanyakumari District, Tamil Nadu, India

Abstract

The study was undertaken to know the effect of algal oil as one of the diet ingredients for the betterment of growth, microbial identification and immunological parameters in the fish *Daniorerio rerio*. The oil was extracted from four micro algae such as *Tetraselmis* sp., *Dunaliella* sp., *Pavlova*sp sp., and *Chaetoceros* sp., The oil obtained from the four different algae were mixed with other feeding ingredients and fed to Zebra fish *Daniorerio rerio*. The water quality parameters like temperature, pH, dissolved oxygen, ammonia (NH_3), growth parameters such as weight (absolute growth rate, specific growth rate, food conversion ratio), food consumption and food conversion efficiency were studied biochemically in *Daniorerio rerio*. Bacterial clearance was evaluated and total viable count of bacteria in different parts of the fish such as gut, gill and body surface were enumerated. Among the oil incorporated diet prepared from the four species of microalgae, *Tetraselmis* sp., had the maximum growth that ranged from 1.34 g to 2.86 g and control had the minimum growth that ranged from 1.14 g to 2.16 g. The maximum food consumption rate recorded in *Pavlova* sp., was 0.27 g. Among the total protein estimated in 1st, 5th, 10th, 15th, 20th, 25th and 30th days, the maximum protein 6.147 mg/ml was noticed in the fishes which are fed with the feed incorporated with oil obtained from *Chaetoceros* sp. Among the total lipid estimation in 1st, 5th, 10th, 15th, 20th, 25th and 30th days of feeding, the maximum lipid 6.147 mg/ml was noticed in feed prepared from *Dunaliella* sp. Among the total carbohydrate estimation in 1st, 5th, 10th, 15th, 20th, 25th and 30th days, the maximum carbohydrate 2.751 mg/ml was noticed in feed *Pavlova* sp. Among the total carotenoid estimation in 1st, 5th, 10th, 15th, 20th, 25th and 30th days of feeding, the maximum carotenoid 0.70 mg/ml was noticed in the animal they are fed with the diet incorporated with oil obtained from *Chaetoceros* sp. Among the total bacterial clearance noticed in 1, 2, 3 and 4 hours, the maximum bacterial clearance was noticed in feed having *Pavlova* sp. oil incorporation after 4 hours. It was found that the Zebra fish fed with the diet incorporated with oil from *Tetraselmis* sp., gave good growth and pigment production.

Keywords: Feed ingredients; Food conversion ratio; Food consumption; Growth parameters

Abbreviations: PUFA: Polyunsaturated Fatty Acids; EFA: Essential Fatty Acids; DHA: Docosahexaenoic acid

Introduction

Globally the ornamental fish culture is a powerful income and employment generating industry. In the aquaculture sector, ornamental fish breeding, culture and trade provide excellent opportunities as a non-food fishery activity for employment and income generation. It is environment friendly, socially acceptable and involves low investment for adopting as a small-scale enterprise with high return. The attractive colouration and quiet disposition of ornamental fish provide a source of joy and peace for people irrespective of age group [1].

Polyunsaturated Fatty Acids (PUFA) are Essential Fatty Acids (EFA), which cannot be synthesized *de novo* by fish, nor in general by all animal, and it must be supplied through the diet. The exact dietary requirement of EFA in fish requires consideration not only for the relative and absolute amounts of individual fatty acids in the fish diets, but also the fish's innate abilities to metabolize these fatty acids, whether anabolically or catabolically [2].

DHA comprises about half of the fatty acids in the brain and is associated with the additional set of health benefits established for omega-3s, notably the protection of the retina, the development of the brain and the prevention of cognitive decline. DHA and EPA (eicosapentaenoic acid) are omega-3 fatty acids found in fatty fish such as salmon, tuna and mackerel. Both fatty acids are recommended for consumption, but recommendations are higher for those who are pregnant, lactating or at risk of CHD. Wild fish obtain these omega-3 fatty acids from the marine algae on which they feed. However, these fish populations are severely declining due to overfishing. Aquaculture (fish farms) had tried to fill the gap and provide an alternative source of fish, but there are environmental concerns surrounding its practice [3].

Studies have been conducted to extract the DHA and EPA directly from the microalgae. Michael Babu et al. in 2013 had enabled the study of the fatty acid profile of algal oil and provide the algal associated bacteria living along with the culture of microalgae enhanced the PUFA level in microalgae.

A few studies have been carried out using algal oil as a feed ingredient in the pellet diet of ornamental (or) any cultivable fishes and hence this study is important.

The major objective of the present study was to test the efficiency of algal oil as a feed ingredient in the pellet feed of Zebra fish (*Daniorerio rerio*).

*Corresponding author: Michael Babu M, Planktology and Aquaculture Division, Centre for Marine Science and Technology, Manonmaniam Sundaranar University, Rajakkamangalam, Kanyakumari District, Tamil Nadu, India
E-mail: michaelmsu@live.com

Materials and Methods

Extraction of algal oil from microalgae

The algal oil was extracted by following the method described by Browne et al. [4].

Extraction of algal oil

Hexane was heated within the miscilla tank, creating vapor rising to the condenser. The hexane then condenses and was released into the extraction chamber with the algae. The hexane begins to break down the cellular wall, releasing lipids into the extraction chamber. The hexane lipid mixture then reaches a critical height level within the extraction chamber. This initializes the siphoning process. Once siphoned back into the miscilla tank, the process starts over, turning the hexane into vapor under specified temperature and pressure while retaining the algae oil within the miscilla tank for roughly 2.5 hrs, until the cellular wall has been completely broken down. The hexane/lipid mixture was then heated once more, converting the liquid hexane into vapor. The hexane vapor was run through a condenser and released into the hexane chamber. All the hexane was released from the chamber leaving only algae oil.

Collection of experimental fish

The experimental fish *Daniorerio rerio* was collected from J.J. Fish farm Puthalam in Kanyakumari district, Tamil Nadu. The fish was purchased in polythene cover and placed by water until both temperatures comes equal. The pH level of water was 7.21. The collected fishes were taken to the laboratory and stocked in recirculation water tanks (5 litre capacity) and acclimatized to the ambient laboratory condition, prior to the experiment.

All the fish had size ranges from 3.5 cm to 5 cm. Water exchange was done in every alternative day. The oxygen level maintained was 5.3 mg/l to 6.2 mg/l.

Taxonomy of fish

Kingdom - Animalia

Phylum - Chordata

Class - Actinoptrygii

Order - Cypriniformes

Family - Cyprinidae

Genus - Danio

Species - rerio

Feeding ingredients

The following feed ingredients were used for the preparation of control and experimental diets [5] (Supplementary Table 1).

Fish meal: Fish meal was obtained from freshly dried and powdered anchovies. Fish meal carries large quantities of energy per unit weight and is an excellent source of protein, lipids (oils), minerals and vitamins. Fishmeal contains certain compounds that make the feed more acceptable and agreeable to the taste. This property allows for the feed to be ingested rapidly, and will reduce nutrient leaching.

Wheat flour: It contains more protein and lysine but a similar energy value to that of corn. Wheat has a tendency to flour and form small, fine particles. While wheat improves pellet quality, the non-starch polysaccharide decreases. For the present study Wheat flour obtained from the commercial market was used as an ingredient in fish feed preparation.

Rice bran: Rice bran is a good source of energy and is also a good source of starch, phosphorous, potassium, manganese and zinc, niacin, pantothenic acid and biotin. Ground whole rice flour is widely used in fish feeds. Gelatinization improved water stability. For the present study, rice bran obtained from the commercial market was used as an ingredient in fish feed preparation.

Topioca powder: Tapioca powder purchased from the commercial market was used as the ingredient as well as binder.

Chicken intestine: Chicken intestines were collected in raw from local market, washed with tap water and converted into meal after sun drying and grinding.

Vitamins and minerals: Vitamins are chemically diverse group of organic substances. It was used for the maintenance of normal metabolic and physiological functions, resulting in increase of growth and high survival rate of organisms.

Minerals are the important constituents of the structural compounds of tissues and skeleton in the regulations of osmotic pressure, nerve impulse transmission and in muscle contraction.

Collection of marine microalgae

The marine microalgae such as *Tetraselmis* sp., *Dunaliella* sp., *Chaetoceros* sp. and *Pavlova* sp were obtained from the Planktology division of Centre for Marine Science and Technology, Manonmaniam Sundaranar University, Rajakkamangalam, Tamil Nadu, India. The collected algal samples were then brought to the laboratory for stock culture and mass culture studies.

Stock culture maintenance of microalgae

The collected algal cells were multiplied in 5 ml test tube and then transferred to the test tubes and conical flasks. The culture media used in the stock culture was Walne's medium [6]. The stock culture was provided with 2000 lux fluorescent light and no aeration.

Tank preparation for mass culture of marine microalgae

The FRP tanks of two hundred and fifty liters capacity were used for mass culturing of algae. The tanks were rinsed with soap water and washed thoroughly with the tap water. The sea water enriched with Walne's medium was filled in the tank. Then, 10% to 20% of the inoculum of growing phase was added in to the respective mass culture tanks. Finally, the culture tanks were placed at the direct sunlight with continuous aeration. The growth rates of the algae were measured in every 6 hour interval by taking sample from the mass culture tank and counted the cells by using improved Neubauer chamber (haemocytometer).

Algal oil

It was obtained from the microalgae named *Tetraselmis* sp., *Dunaliella* sp., *Chaetoceros* sp. and *Pavlova* sp. This alga was isolated, mass cultured, harvested and made in to powder. The powder was used as the raw material for the oil extraction following the method described by Michael Babu et al. The oil possesses high HUFA and PUFA and also contains carotenoid and some minerals.

Control diet

The control diet was also prepared with all the above ingredients except algal oil.

Preparation of feeds

Fish feeds should have adequate energy for body maintenance and growth. It was contributed by three major nutrients, namely protein, fat and carbohydrate. The feeds should have vitamin and minerals to meet their deficiencies. Such attractants and flavors are needed to fish for quick consumption and effective utilization of feed.

The feed ingredients were weighed and mixed well in a container by adding sufficient quantity of distilled water and then the ingredients were made into dough. The dough was then placed in a container and boiled in a pressure cooker for 20 minutes. After boiling, the dough was taken out of the container and then vitamins and minerals mixture, and gelatin was added to the dough and mixed well.

The dough was then allowed to pass through a pelletizer having perforation diameter of 1.5 mm in the diet. Then the control as well as the experimental diets was dried in duration of 15 hours. Then the dried pellets were collected and stored in air tight plastic container.

Altogether, five diets were prepared, a control diet (C) and four experimental diets having 5 ml oil obtained from *Tetraselmis* sp., *Dunaliella* sp., *Chaetoceros* sp. and *Pavlova* sp. In control diet, there was no algal oil.

Water quality parameters

The water quality parameters were monitored and maintained at an optimum level during the entire experimental duration. Water samples were collected from respective tanks and analyzed. Temperature, pH, dissolved oxygen and ammonia [7] in culture tanks were measured and recorded.

Growth parameters

To find out the growth, the weight of the animal in each tank was measured once in 5 days. Weight was measured by using balance with least disturbance to the fish on wet weight basis.

Production (growth)

$$\text{Production (g)} = \text{final wet weight} - \text{initial wet weight}$$

Food consumption

$$\text{Food consumption (g dry weight)} = \text{Food provided} - \text{Unfed remains}$$

Food Conversion Efficiency (FCE)

$$FCE = \frac{\text{Wet weight of fish produced (g)}}{\text{Dry weight of the feed given (g)}} \times 100$$

Absolute Growth Rate (AGR)

$$AGR \text{ (g / body wt / day)} = \frac{\text{Final body weight - Initial body weight}}{\text{Total number of days}}$$

Specific Growth Rate (SGR)

$$SGR \text{ (\%)} = \frac{\text{ln final wet weight (g) - ln initial wet weight (g)}}{\text{Experimental period (days)}} \times 100$$

Food Conversion Ratio (FCR)

$$FCR = \frac{\text{Total amount of feed given (dry weight, g)}}{\text{Total production of fish (wet weight, g)}}$$

Biochemical analysis

The protein [8], carbohydrate [9], lipid [10] and carotenoid composition of fish were estimated.

Total viable count

The fish samples such as gut, gill, body surface and water sample were dissected out under aseptic conditions and samples was ground well using ethanol. To enumerate the bacterial load present in the fish sample, 1 ml of sample was taken and serially diluted (10^{-1} to 10^{-6}). From each dilution 0.1 ml of sample was taken and pour plated on nutrient agar. After this the plates were incubated at 37°C for 24 hours in an incubator and the total number of individual viable colonies was counted using cubic colony counter.

$$CFU / ml = \frac{\text{Average of CFU counted}}{\text{Volume of inoculums}} \times \text{dilution factor}$$

Bacterial clearance

To determine how rapidly bacteria are cleared from blood, fish were injected intraperitoneally near the caudal region with 0.1 units *Vibrio harveyi* suspension containing 1000 CFU/ml The *Vibrio harveyi* was obtained from Centre for Marine Science and Technology Microbiology Laboratory. After 1, 2, 3 and 4 hours, a 100 μl of blood was drawn using insulin syringe with sterile saline. Triplicates TCBS agar plates were prepared. The samples were immediately mixed with TCBS agar poured in to petridish and incubated at 37°C for 24 hours. Number of bacterial colonies per plates were counted and divided by the volume of blood extracted, to determine the number of colony forming units in milliliter of blood. Control was also maintained, but injected with only sterile saline without *Vibrio harveyi*.

Results

Water quality parameters

In the present study, water quality parameters were maintained at an optimum level in control (C) and experimental tanks during the culture period. The water quality parameters recorded in the culture tanks are given in Table 1.

In control tank (C), the range of temperature and pH recorded during the experimentation were ranged from 32°C to 34°C and 7.08 to 7.32 respectively. The dissolved oxygen content was ranged from 4.564 mg/l to 6.893 mg/l during the experimentation. The ammonia content of the water sample also varied between 0.623 μg/l and 0.837 μg/l during the culture period.

In experimental tank (*Tetraselmis* sp. oil incorporated diet fed group), the temperature and pH value recorded during the experiment ranged from 32°C to 34°C and 7.10 to 7.36 respectively. The dissolved oxygen content ranged from 4.345 mg/l to 6.861 mg/l during the experimentation. The ammonia content of the water sample also varied between 0.521 μg/l and 0.674 μg/l during the culture period.

In *Dunaliella* sp. oil group, the temperature and pH recorded during the experimentation ranged from 32°C to 34°C and 7.25 to 7.86 respectively. The dissolved oxygen ranged from 3.429 mg/l to 5.589 mg/l during the experimentation. The ammonia of the water also varied between 0.428 μg/l and 0.738 μg/l during culture period.

In *Chaetoceros* sp. oil group, the temperature and pH recorded during the experimentation ranged from 32°C to 34°C and 7.21 to 7.39 respectively. The dissolved oxygen ranged from 3.176 mg/l to 6.783 mg/l during the experimentation. The ammonia content of the water also varied between 0.378 μg/l and 0.842 μg/l during culture period.

In *Pavlova* sp. oil group, the temperature and pH recorded during the experimentation ranged from 32°C to 34°C and 7.14 to 7.33

Feed Type	Parameters	Water quality parameters of fish in five days intervals						
		1th day	5th day	10th day	15th day	20th day	25th day	30th day
Control feed	Temperature (°C)	34 ± 0.00	33 ± 0.00	32 ± 0.00	34 ± 0.00	32 ± 0.00	33 ± 0.00	32 ± 0.00
	pH	7.13 ± 0.13	7.32 ± 0.67	7.21 ± 0.05	7.24 ± 0.30	7.08 ± 0.08	7.21 ± 0.23	7.25 ± 0.67
	Dissolved O_2 (mg/ml)	6.761 ± 0.61	4.564 ± 0.015	6.893 ± 0.18	5.823 ± 0.31	5.213 ± 0.25	5.145 ± 0.21	4.976 ± 0.082
	Ammonia (mg/ml)	0.623 ± 0.03	0.642 ± 0.06	0.665 ± 0.08	0.687±0.14	0.723 ± 0.56	0.765 ± 0.54	0.837 ± 0.18
Tetraselmis sp.	Temperature (°C)	34 ± 0.00	33 ± 0.00	32 ± 0.00	34 ± 0.00	32 ± 0.00	33 ± 0.00	32 ± 0.00
	pH	7.23 ± 0.45	7.23 ± 0.07	7.34 ± 0.07	7.28 ± 0.46	7.36 ± 0.18	7.10 ± 0.16	7.32 ± 0.04
	Dissolved O_2 (mg/ml)	5.216 ± 0.13	6.861 ± 0.018	6.253 ± 0.14	5.341 ± 0.013	4.345 ± 0.16	4.717 ± 0.023	4.453 ± 0.051
	Ammonia (mg/ml)	0.576 ± 0.07	0.585 ± 0.09	0.545 ± 0.07	0.521 ± 0.18	0.538 ± 0.42	0.574 ± 0.19	0.674 ± 0.53
Dunaliella sp.	Temperature (°C)	34 ± 0.00	33 ± 0.00	32 ± 0.00	34 ± 0.00	32 ± 0.00	33 ± 0.00	32 ± 0.00
	pH	7.35 ± 0.08	7.45 ± 0.14	7.28 ± 0.09	7.86 ± 0.08	7.36 ± 0.08	7.25 ± 0.06	7.46 ± 0.08
	Dissolved O_2 (mg/ml)	5.561 ± 0.87	4.784 ± 0.09	4.853 ± 0.018	5.589 ± 0.021	4.537 ± 0.028	3.429 ± 0.023	3.827 ± 0.53
	Ammonia (mg/ml)	0.528 ± 0.05	0.428 ± 0.08	0.628 ± 0.09	0.687 ± 0.09	0.684 ± 0.08	0.738 ± 0.07	0.529 ± 0.09
Chaetoceros sp.	Temperature (°C)	34 ± 0.00	33 ± 0.00	32 ± 0.00	34 ± 0.00	32 ± 0.00	33 ± 0.00	32 ± 0.00
	pH	7.35 ± 0.07	7.27 ± 0.18	7.29 ± 0.08	7.21 ± 0.10	7.31 ± 0.09	7.35 ± 0.19	7.39 ± 0.08
	Dissolved O_2 (mg/ml)	4.780 ± 0.03	3.176 ± 0.016	6.783 ± 0.017	5.572 ± 0.08	5.131 ± 0.029	4.272 ± 0.25	3.761 ± 0.057
	Ammonia (mg/ml)	0.378 ± 0.09	0.387 ± 0.08	0.465 ± 0.08	0.598 ± 0.09	0.654 ± 0.07	0.842 ± 0.08	0.678 ± 0.867
Pavlova sp.	Temperature	34 ± 0.00	33 ± 0.00	32 ± 0.00	34 ± 0.00	32 ± 0.00	33 ± 0.00	32 ± 0.00
	pH	7.24 ± 0.09	7.14 ± 0.72	7.17 ± 0.14	7.22 ± 0.16	7.27 ± 0.15	7.28 ± 0.18	7.33 ± 0.09
	Dissolved O_2	4.239 ± 0.06	4.157 ± 0.019	5.782 ± 0.018	5.514 ± 0.12	5.324 ± 0.027	4.256 ± 0.28	3.345 ± 0.058
	Ammonia	0.325 ± 0.12	0.376 ± 0.56	0.451 ± 0.19	0.557 ± 0.08	0.678 ± 0.28	0.748 ± 0.12	0.645 ± 0.89

Table 1: Water quality parameters of fish when feeding different diet in different days of culture.

Days	Fish weight during five days intervals (g)				
	Control	Tetraselmis sp.	Dunaliellasp.	Chaetoceros sp.	Pavlova sp.
1	1.14 ± 0.19	1.34 ± 0.18	1.63 ± 0.17	1.54 ± 0.23	1.79 ± 0.32
5	1.25 ± 0.21	1.52 ± 0.14	1.84 ± 0.21	1.68 ± 0.28	1.81 ± 0.28
10	1.31 ± 0.28	1.81 ± 0.19	1.97 ± 0.38	1.94 ± 0.42	2.05 ± 0.15
15	1.58 ± 0.29	1.92 ± 0.17	2.15 ± 0.37	2.02 ± 0.34	2.25 ± 0.27
20	1.83 ± 0.18	2.23 ± 0.21	2.39 ± 0.28	2.16 ± 0.43	2.46 ± 0.42
25	1.91 ± 0.34	2.48 ± 0.20	2.57 ± 0.35	2.27 ± 0.21	2.68 ± 0.38
30	2.16 ± 0.26	2.86 ± 0.28	2.82 ± 0.14	2.38 ± 0.46	2.84 ± 0.51
The feed prepared from different algal oil significantly (P<0.01) enhanced the growth					

Table 2: Effect of different types of algal oil diet on fish weight during different days of culture.

Growth parameters	Fish growth parameter during five days of intervals (g)				
	Control	Tetraselmis sp.	Dunaliella sp.	Chaetoceros sp.	Pavlova sp.
Production growth (g)	0.31 ± 0.08	0.38 ± 0.027	0.46 ± 0.09	0.37 ± 0.051	0.52 ± 0.32
Food consumption (g/day)	0.19 ± 0.17	0.21 ± 0.08	0.23 ± 0.09	0.26 ± 0.08	0.27 ± 0.41
FCE (%)	4.12	5.34	12.38	5.26	13.78
AGR (g)	0.008 ± 0.002	0.011 ± 0.002	0.018 ± 0.008	0.014 ± 0.004	0.020 ± 0.007
SGR (%)	0.98 ± 0.34	1.03 ± 0.05	1.86 ± 0.17	1.64 ± 0.28	1.94 ± 0.04
FCR (%)	5:1	5:3	5:2	5:3	5:4
The feed prepared from different algal oil significantly (P<0.01) enhanced the growth					

Table 3: Effect of different types of algal oil diet on growth parameters during different days after feeding.

respectively. The dissolved oxygen content ranged from 3.345 mg/l to 5.782 mg/l during the experimentation. The ammonia content of the water also varied between 0.325 µg/l to 0.748 µg/l during culture period (Table 1).

Growth parameters

In all the experimental diet and control feed groups, minimum fish weight was 2.16 g during the experiment. In Tetraselmis sp., the maximum fish weight was 2.86 g during the experimentation (Table 2).

The maximum growth recorded in Pavlova sp., oil fed animal was 0.52 g. The food consumption rate recorded in control was 0.19 g and in Tetraselmis sp., Dunaliella sp., Chaetoceros sp. and Pavlova sp., oil incorporated diet fed animals were 0.21, 0.23, 0.26 and 0.27 respectively. The FCE value recorded in control was 4.12% and in Tetraselmis sp., Dunaliella sp., Chaetoceros sp. and Pavlova sp., oil incorporated diet fed animals were 5.34%, 12.38%, 5.26% and 13.78% respectively. The AGR value recorded in control was 0.008 g and in Tetraselmis sp., Dunaliella sp., Chaetoceros sp. and Pavlova sp. were 0.011 g, 0.018 g, 0.014 g and 0.020 g respectively. The SGR value recorded in control was 0.98% and in Tetraselmis sp., Dunaliella sp., Chaetoceros sp., and Pavlova sp. were 1.03%, 1.86%, 1.64% and 1.94% respectively. The FCR value recorded in control was 5:1 and in Tetraselmis sp., Dunaliella sp., Chaetoceros sp. and Pavlova sp. were 5:3, 5:2, 5:3 and 5:4 respectively (Table 3). The growth was significant increased by the (P<0.05) incorporation of algal oil in Zebra fish.

Protein estimation of fish

Total protein estimated in the first day of animal fed with diet made from algal oil of *Tetraselmis* sp., *Dunaliella* sp., *Chaetoceros* sp., *Pavlova* sp. and control diets were 3.192 mg/ml, 3.285 mg/ml, 3.294 mg/ml, 3.367 mg/ml and 3.513 mg/ml respectively.

Total protein estimated in the animal fed with diet made from algal oil of diet *Tetraselmis* sp., *Dunaliella* sp., *Chaetoceros* sp., *Pavlova* sp. and control in 5th day were 3.738 mg/ml, 3.973 mg/ml, 3.728 mg/ml, 3.812 mg/ml and 3.823 mg/ml respectively.

Total protein estimated in the animal fed with diet made from algal oil of diet *Tetraselmis* sp., *Dunaliella* sp., *Chaetoceros* sp., *Pavlova* sp. and control in 10th day were 3.948 mg/ml, 4.365 mg/ml, 4.676 mg/ml, 4.286 mg/ml and 3.918 mg/ml respectively.

Total protein estimated in the animal fed with diet made from algal oil of diet *Tetraselmis* sp., *Dunaliella* sp., *Chaetoceros* sp., *Pavlova* sp. and control in 15th day were 4.287 mg/ml, 4.843 mg/ml, 5.317 mg/ml, 4.723 mg/ml and 4.128 mg/ml respectively.

Total protein estimated in the animal fed with diet made from algal oil of diet *Tetraselmis* sp., *Dunaliella* sp., *Chaetoceros* sp., *Pavlova* sp. and control in 20th day were 4.638 mg/ml, 5.156 mg/ml, 5.567 mg/ml, 5.241 mg/ml and 4.320 mg/ml respectively.

Total protein estimated in the animal fed with diet made from algal oil of diet *Tetraselmis* sp., *Dunaliella* sp., *Chaetoceros* sp., *Pavlova* sp. and control in 25th day were 4.719 mg/ml, 5.487 mg/ml, 5.850 mg/ml, 5.579 mg/ml and 4.532 mg/ml respectively.

Total protein estimated in the animal fed with diet made from algal oil of diet *Tetraselmis* sp., *Dunaliella* sp., *Chaetoceros* sp., *Pavlova* sp. and control in 30th day were 4.913 mg/ml, 5.852 mg/ml, 6.147 mg/ml, 5.714 mg/ml and 4.756 mg/ml respectively.

Among the total protein estimated in 1st, 5th, 10th, 15th, 20th, 25th and 30th days, the maximum protein 6.147 mg/ml was noticed in the animals with the diet prepared using algal oil obtained from feed *Chaetoceros* sp., (Table 4). The growth due to algal oil feed was highly significant ($P<0.05$) in different microalgaloil feed of Zebra fish.

Lipid estimation of fish

Total lipid estimated in the first day of the animal fed with diet made from algal oil of *Tetraselmis* sp., *Dunaliella* sp., *Chaetoceros* sp., *Pavlova* sp. and control diet were 2.247 mg/ml, 2.076 mg/ml, 2.175 mg/ml, 2.268 mg/ml and 2.027 mg/ml respectively.

Total lipid estimated in the animal fed with diet made from algal oil of diet *Tetraselmis* sp., *Dunaliella* sp., *Chaetoceros* sp., *Pavlova* sp. and control in 5th day were 2.416 mg/ml, 2.314 mg/ml, 2.258 mg/ml, 2.382 mg/ml and 2.218 mg/ml respectively.

Total lipid estimated in the animal fed with diet made from algal oil of diet *Tetraselmis* sp., *Dunaliella* sp., *Chaetoceros* sp., *Pavlova* sp. and control in 10th day were 2.627 mg/ml, 2.516 mg/ml, 2.573 mg/ml, 2.576 mg/ml and 2.472 mg/ml respectively.

Total lipid estimated in the animal fed with diet made from algal oil of diet *Tetraselmis* sp., *Dunaliella* sp., *Chaetoceros* sp., *Pavlova* sp. and control in 15th day were 2.768 mg/ml, 2.824 mg/ml, 2.876 mg/ml, 2.741 mg/ml and 2.521 mg/ml respectively.

Total lipid estimated in the animal fed with diet made from algal oil of diet *Tetraselmis* sp., *Dunaliella* sp., *Chaetoceros* sp., *Pavlova* sp. and control in 20th day were 2.827 mg/ml, 2.976 mg/ml, 2.902 mg/ml, 2.934 mg/ml and 2.589 mg/ml respectively.

Total lipid estimated in the animal fed with diet made from algal oil of diet *Tetraselmis* sp., *Dunaliella* sp., *Chaetoceros* sp., *Pavlova* sp. and control in 25th day were 2.958 mg/ml, 3.016 mg/ml, 2.980 mg/ml, 3.047 mg/ml and 2.628 mg/ml respectively.

Total lipid estimated in the animal fed with diet made from algal oil of diet *Tetraselmis* sp., *Dunaliella* sp., *Chaetoceros* sp., *Pavlova* sp. and control in 30th day were 2.982 mg/ml, 3.176 mg/ml, 3.065 mg/ml, 3.123 mg/ml and 2.761 mg/ml respectively.

Among the total lipid estimated in 1st, 5th, 10th, 15th, 20th, 25th and 30th days, the maximum lipid 6.147 mg/ml was noticed in the animals fed with the diet prepared using algal oil obtained from *Dunaliella* sp., (Table 5). The growth of algal oil feed was highly significant ($P<0.05$) different microalgal oil feed of Zebra fish.

Days	Protein Estimation of five days of intervals (mg/dl)				
	Control	*Tetraselmis* sp.	*Dunaliella* sp.	*Chaetocero* ssp.	*Pavlova* sp.
1	3.513 ± 0.283	3.192 ± 0.189	3.285 ± 0.307	3.294 ± 0.193	3.367 ± 0.461
5	3.823 ± 0.184	3.738 ± 0.286	3.973 ± 0.735	3.728 ± 0.264	3.812 ± 0.587
10	3.918 ± 0.175	3.948 ± 0.321	4.365 ± 0.581	4.676 ± 0.391	4.286 ± 0.619
15	4.128 ± 0.063	4.287 ± 0.257	4.843 ± 0.839	5.317 ± 0.427	4.723 ± 0.783
20	4.320 ± 0.271	4.638 ± 0.981	5.156 ± 0.684	5.567 ± 0.371	5.241 ± 0.516
25	4.532 ± 0.056	4.719 ± 0.381	5.487 ± 0.421	5.850 ± 0.331	5.579 ± 0.718
30	4.756 ± 0.139	4.913 ± 0.273	5.852 ± 0.417	6.147 ± 0.428	5.714 ± 0.672

The feed prepared from different algal oil significantly ($P<0.01$) enhanced the body protein.

Table 4: Effect of different types of algal oil feed on total body protein during different days of culture.

Days	Lipid Estimation of five days of intervals (mg/dl)				
	Control	*Tetraselmis* sp.	*Dunaliella* sp.	*Chaetoceros* sp.	*Pavlova* sp.
1	2.027 ± 0.084	2.247 ± 0.068	2.076 ± 0.348	2.175 ± 0.281	2.268 ± 0.238
5	2.218 ± 0.093	2.416 ± 0.238	2.314 ± 0.183	2.258 ± 0.183	2.382 ± 0.176
10	2.472 ± 0.217	2.627 ± 0.163	2.516 ± 0.194	2.573 ± 0.249	2.576 ± 0.347
15	2.521 ± 0.035	2.768 ± 0.023	2.824 ± 0.098	2.876 ± 0.351	2.741 ± 0.150
20	2.589 ± 0.024	2.827 ± 0.034	2.976 ± 0.283	2.902 ± 0.278	2.934 ± 0.261
25	2.628 ± 0.134	2.958 ± 0.021	3.016 ± 0.094	2.980 ± 0.162	3.047 ± 0.374
30	2.761 ± 0.192	2.982 ± 0.026	3.176 ± 0.254	3.065 ± 0.078	3.123 ± 0.456

The feed prepared from different algal oil significantly ($P<0.01$) enhanced body lipid.

Table 5: Effect of different types of algal oil feed on total body lipid during different days of culture.

Days	Carbohydrate level in different intervals (mg/dl)				
	Control	*Tetraselmis* sp.	*Dunaliella* sp.	*Chaetoceros* sp.	*Pavlova* sp.
1	1.623 ± 0.085	1.818 ± 0.194	1.712 ± 0.027	1.763 ± 0.049	1.787 ± 0.037
5	1.835 ± 0.091	1.924 ± 0.045	1.839 ± 0.043	1.806 ± 0.036	1.826 ± 0.089
10	1.914 ± 0.039	2.250± 0.021	2.357 ± 0.091	1.939 ± 0.023	1.982 ± 0.054
15	1.720 ± 0.028	2.471 ± 0.094	2.418 ± 0.089	2.384 ± 0.035	2.178 ± 0.085
20	1.038 ± 0.037	2.543 ± 0.098	2.564 ± 0.035	2.585 ± 0.042	2.365 ± 0.057
25	1.145 ± 0.138	2.589 ± 0.029	2.575 ± 0.028	2.607 ± 0.039	2.578 ± 0.078
30	1.229 ± 0.085	2.638 ± 0.021	2.581 ± 0.087	2.728 ± 0.067	2.751 ± 0.084

The feed prepared from different algal oil significantly (P<0.01) enhanced the carbohydrates.

Table 6: Effect of different types of algal oil diet on total body carbohydrate during different days of culture.

Days	Carotenoid Estimation of five days of intervals (mg/dl)				
	Control	*Tetraselmis* sp.	*Dunaliella* sp.	*Chaetoceros* sp.	*Pavlova* sp.
1	0.54 ± 0.16	0.52 ± 0.16	0.53 ± 0.08	0.62 ± 0.19	0.50 ± 0.21
5	0.55 ± 0.23	0.53 ± 0.05	0.56 ± 0.19	0.65 ± 0.18	0.52 ± 0.16
10	0.53 ± 0.18	0.56 ± 0.19	0.59 ± 0.18	0.67 ± 0.16	0.56 ± 0.23
15	0.51 ± 0.09	0.60 ± 0.16	0.59 ± 0.19	0.68 ± 0.18	0.58 ± 0.02
20	0.53 ± 0.18	0.68 ± 0.18	0.61 ± 0.12	0.69 ± 0.13	0.60 ± 0.24
25	0.52 ± 0.29	0.69 ± 0.17	0.63 ± 0.17	0.71 ± 0.06	0.61 ± 0.15
30	0.53 ± 0.12	0.70 ± 0.34	0.65 ± 0.16	0.70 ± 0.1	0.63 ± 0.18

The feed prepared from different algal oil significantly (P<0.01) enhanced the carotenoid.

Table 7: Effect of different algal oil diet on total carotenoid in different days of culture.

Carbohydrate estimation of fish

Total carbohydrate estimated in the first day of the animal fed with diet made from algal oil obtained from *Tetraselmis* sp., *Dunaliella* sp., *Chaetoceros* sp., *Pavlova* sp., and control diet were 1.818 mg/ml, 1.712 mg/ml, 1.763 mg/ml, 1.787 mg/ml and 1.623 mg/ml respectively.

Total carbohydrate estimated in the animal fed with diet made from algal oil obtained from *Tetraselmis* sp., *Dunaliella* sp., *Chaetoceros* sp., *Pavlova* sp., and control in 5th day were 1.924 mg/ml, 1.839 mg/ml, 1.806 mg/ml, 1.826 mg/ml and 1.835 mg/ml respectively.

Total carbohydrate estimated in the animal fed with diet made from algal oil obtained from *Tetraselmis* sp., *Dunaliella* sp., *Chaetoceros* sp., *Pavlova* sp. and control in 10th day were 2.250 mg/ml, 2.357 mg/ml, 1.939 mg/ml, 1.982 mg/ml and 1.914 mg/ml respectively.

Total carbohydrate estimated in the animal fed with diet made from algal oil obtained from *Tetraselmis* sp., *Dunaliella* sp., *Chaetoceros* sp., *Pavlova* sp. and control in 15th day were 2.471 mg/ml, 2.418 mg/ml, 2.384 mg/ml, 2.178 mg/ml and 1.720 mg/ml respectively.

Total carbohydrate estimated in the animal fed with diet made from algal oil obtained from *Tetraselmis* sp., *Dunaliella* sp., *Chaetoceros* sp., *Pavlova* sp. and control in 20th day were 2.543 mg/ml, 2.564 mg/ml, 2.585 mg/ml, 2.365 mg/ml and 1.038 mg/ml respectively.

Total carbohydrate estimated in the animal fed with diet made from algal oil obtained from *Tetraselmis* sp., *Dunaliella* sp., *Chaetoceros* sp., *Pavlova* sp. and control in 25th day were 2.589 mg/ml, 2.575 mg/ml, 2.607 mg/ml, 2.578 mg/ml and 1.145 mg/ml respectively.

Total carbohydrate estimated in the animal fed with diet made from algal oil obtained from *Tetraselmis* sp., *Dunaliella* sp., *Chaetoceros* sp., *Pavlova* sp. and control in 30th day were 2.638 mg/ml, 2.581 mg/ml, 2.728 mg/ml, 2.751 mg/ml and 1.229 mg/ml respectively.

Among the total carbohydrate estimated in 1st, 5th, 10th, 15th, 20th, 25th and 30th days, the maximum carbohydrate 2.751 mg/ml was noticed in the animals with the diet prepared using algal oil obtained from feed *Pavlova* sp., (Table 6). The growth of algal oil feed was highly significant (P<0.05) different micro algal oil feed of Zebra fish.

Carotenoid estimation

The initial carotenoid level in the control fish was 0.54 µg/g. After 5th, 10th, 15th, 20th, 25th and 30th days, the carotenoids level estimated were 0.55 µg/g, 0.53 µg/g, 0.51 µg/g, 0.53 µg/g, 0.52 µg/g and 0.53 µg/g respectively.

The initial carotenoid level in the fish fed with feed prepared from *Tetraselmis* oil was 0.52 µg/g. After 5th, 10th, 15th, 20th, 25th and 30th days, the carotenoids level estimated were 0.53 µg/g, 0.56 µg/g, 0.60 µg/g, 0.68 µg/g, 0.69 µg/g and 0.70 µg/g respectively.

The initial carotenoid level in the fish fed with feed prepared from *Dunaliella* oil was 0.53 µg/g. After 5th, 10th, 15th, 20th, 25th and 30th days, the carotenoids level estimated were 0.56 µg/g, 0.59 µg/g, 0.59 µg/g, 0.61 µg/g, 0.63 µg/g and 0.65 µg/g respectively.

The initial carotenoid level in the fish fed with feed prepared from *Chaetoceros* oil was 0.62 µg/g. After 5th, 10th, 15th, 20th, 25th and 30th days, the carotenoids level estimated were 0.65 µg/g, 0.67 µg/g, 0.68 µg/g, 0.69 µg/g, 0.71 µg/g and 0.70 µg/g respectively.

The initial carotenoid level in the fish fed with feed prepared from *Pavlova* oil was 0.50 µg/g. After 5th, 10th, 15th, 20th, 25th and 30th days, the carotenoids level estimated were 0.52 µg/g, 0.56 µg/g, 0.58 µg/g, 0.60 µg/g, 0.61 µg/g and 0.63 µg/g respectively.

Among the total carotenoid estimated in 1st, 5th, 10th, 15th, 20th, 25th and 30th days, the maximum carotenoid 0.70 mg/dl was noticed in feed prepared from oil obtained from *Chaetoceros* sp. (Table 7).

Bacterial count

The initial bacterial count in the gut of control was TNTC, water sample was TNTC gut sample was 267 × 10⁻⁵, gill was TNTC and body surface was TNTC, whereas in *Tetraselmis* oil diet fed animal, the number of bacterial colony in water sample was TNTC, gut sample was TNTC, gill sample was TNTC and body surface was 245 × 10⁻⁵. In the *Dunaliella* sp. the bacterial count in water sample was TNTC, gut was TNTC, gill was TNTC and body surface was 218 × 10⁻⁵. In the *Chaetoceros* sp., the bacterial count in water sample was TNTC, gut was TNTC, gill was TNTC and body surface 225×10^{-5}. In the *Pavlova* sp., the bacterial count in water sample TNTC, gut was TNTC, gill was TNTC and body surface 226×10^{-5}.

After 30 days, the minimum bacterial count was found in fishes fed with diet prepared from *Pavlova* oil (water sample 112×10^{-5}, gut 64×10^{-5}, and gill 62×10^{-5} and body surface 76×10^{-5}). The bacterial count in fish fed with oil from *Chaetoceros* sp., was 124, 72, 68 and 81×10^{-5} in water sample, gut, gill and body surface respectively. In *Dunaliella* oil diets, 81, 108, 76 and 51×10^{-5} water sample, gut, gill and body surface respectively. In *Tetraselmis* oil diet, fed animal group, were found in water sample, gut, gill and body surface 138 CFU/ml, 136 CFU/ml, 107 CFU/ml and 96×10^{-5} CFU/ml (Table 8).

Bacterial clearance

The initial bacterial clearance in the control after 1 hour was 15 CFU/ml where as in *Tetraselmis* sp., *Dunaliella* sp., *Chaetoceros* sp. and *Pavlova* sp. were 18 CFU/ml, 17 CFU/ml, 17 CFU/ml and 16 CFU/ml respectively.

The bacterial clearance in the control after 2 hours was 12 CFU/ml whereas in *Tetraselmis* sp., *Dunaliella* sp., *Chaetoceros* sp., and *Pavlova* sp. were 11 CFU/ml, 10 CFU/ml, 12 CFU/ml and 9 CFU/ml respectively.

The bacterial clearance in the control after 3 hours was 7 CFU/ml where as in *Tetraselmis* sp., *Dunaliella* sp., *Chaetoceros* sp., and *Pavlova* sp. were 5 CFU/ml, 5 CFU/ml, 7 CFU/ml and 4 CFU/ml respectively.

The bacterial clearance in the control after 4 hours was 4 CFU/ml where as in *Tetraselmis* sp., *Dunaliella* sp., *Chaetoceros* sp., *Pavlova* sp. were 0 CFU/ml, 0 CFU/ml, 1 CFU/ml and 0 CFU/ml respectively.

Among the total bacterial clearance noticed in 1, 2, 3 and 4 hours, the minimum bacterial clearance CFU/ml was noticed in feed type *Pavlova* sp., after 4 hours (Table 9). The growth of algal oil feed was highly significant ($P<0.05$) different microalgal oil feed of bacterial clearance of Zebra fish.

Discussion

This work was carried out to find out the efficiency of the artificial diet prepared with algal oil on protein, lipid, carbohydrate, growth and immunological parameters in Zebra fish and to study the effect feed in changing the water quality parameters such as temperature, pH, dissolved oxygen and ammonia. The results obtained from the present study were compared with control to find out the efficiency of the experimental diet.

Artificial diet plays a major role in grow out culture system of fin and shell fishes. The efficiency of any feed can be determined by its FCR. When a feed is considered to be a best, it should have more self-life, more FCR, easy digestibility and more over it should not spoil the culture environment.

In our experiment, five diets were prepared and tested for pH, oxygen and ammonia in the culture water. Among the five diets tested, the level of ammonia in the culture water was minimum in *Pavlova* sp. oil mixing diet. It shows the fact that the particular diet did not spoil

Feed Type	Parameters	Bacterial count of fish in five days intervals ($\times 10^{-5}$)						
		1	5	10	15	20	25	30
Control feed	Water	TNTC	251 ± 4.73	194 ± 5.85	174 ±7.76	153 ± 6.82	135 ± 8.51	114 ± 5.61
	Gut	267 ± 2.48	292 ± 4.67	247 ± 8.28	238 ± 4.29	210 ± 6.13	195 ± 6.67	178 ± 7.30
	Gill	TNTC	139 ± 4.17	168 ± 6.17	135 ± 3.62	113 ± 8.78	97 ± 3.68	82 ± 1.83
	Body surface	TNTC	278 ± 7.18	274 ± 1.49	239 ± 8.65	212 ± 6.67	198 ± 4.56	145 ± 5.78
Tetraselmis sp.	Water	TNTC	TNTC	263 ± 4.76	214 ± 7.84	198 ± 5.28	174 ± 3.15	138 ± 3.49
	Gut	TNTC	245 ± 6.18	237 ± 5.63	216 ± 6.65	192 ± 4.56	145 ± 4.69	136 ± 4.71
	Gill	TNTC	187 ± 4.74	187 ± 2.98	167 ± 4.71	149 ± 3.12	123 ± 7.67	107 ± 3.65
	Body surface	245 ± 4.89	254 ± 8.87	216 ± 8.46	146 ± 4.74	123 ± 9.67	108 ± 7.15	96 ± 4.84
Dunaliella sp.	Water	TNTC	TNTC	179 ± 5.71	148 ± 4.76	124 ± 4.61	101 ±7.57	81 ± 4.85
	Gut	TNTC	251 ± 2.67	191 ± 6.37	172 ± 4.37	154 ± 8.45	124 ± 5.68	108 ± 3.49
	Gill	TNTC	185 ± 1.41	145 ± 3.74	137 ± 4.15	116 ± 4.56	91 ± 4.60	76 ± 5.38
	Body surface	218 ± 5.18	187 ± 5.54	123 ± 4.17	96 ± 7.56	78 ± 3.18	67 ± 4.72	51 ± 3.67
Chaetoceros sp.	Water	TNTC	268 ± 7.93	215 ± 8.04	193 ± 5.34	176 ± 4.76	141 ± 3.63	124 ± 4.62
	Gut	TNTC	136 ± 6.19	117 ± 4.65	106 ± 5.53	97 ± 4.52	81 ± 4.78	72 ± 2.65
	Gill	TNTC	148 ± 2.94	128 ± 8.28	108 ± 4.18	90 ± 7.34	76 ± 2.86	68 ± 2.50
	Body surface	225 ± 4.85	197 ± 7.34	167 ± 4.78	145 ± 3.67	126 ± 6.61	105 ± 8.71	81 ± 7.76
Pavlova sp.	Water	TNTC	231 ± 7.56	212 ± 8.67	175 ± 5.73	158 ± 4.78	134 ± 3.61	112 ± 4.78
	Gut	TNTC	123 ± 6.45	113 ± 4.34	108 ± 5.82	85 ± 4.34	76 ± 4.90	64 ± 2.62
	Gill	TNTC	135 ± 4.27	126 ± 8.67	101 ± 4.13	87 ± 7.37	78 ± 2.81	62 ± 2.48
	Body surface	226 ± 4.14	162 ± 7.38	156 ± 4.65	138 ± 3.18	112 ± 6.78	97 ± 8.89	76 ± 7.72

Table 8: Effect of different type of diet on Bacterial count during different days of culture.

Hours	Bacterial Clearance of Fish in 1 hour interval (100 μl)				
	Control	Tetraselmis sp.(I)	Dunaliella sp.(II)	Chaetoceros sp.(III)	Pavlova sp.(IV)
1	15.25 ± 0.75	18.20 ± 0.48	17.65 ± 0.82	17.52 ± 0.12	16.64 ± 0.42
2	12.54 ± 0.43	11.52 ± 0.14	10.94 ± 0.81	12.50 ± 0.65	9.54 ± 0.32
3	10.86 ± 0.60	5.34 ± 0.16	5.46 ± 0.8	7.34 ± 0.40	4.20 ± 0.80
4	7.42 ± 0.21	0 ± 0.50	0 ± 0.0	1 ± 0.4	0 ± 0.0

The diets prepared from oil obtained from different species of algae significantly ($P<0.05$) influenced in the bacterial clearance in different time intervals than the control.

Table 9: Efficiency of different types of algal oil diet on bacterial clearance.

the water quality. The nitrogen compounds such as ammonia dissolve in the fish rearing tank is the most important deteriorating chemicals in larval rearing system which affects the health of fish. Most of the nitrogen compounds enter the culture tank in the form of nitrogenous wastes of fish. Generally, ammonia is found in the water either as toxic unionized (NH_3) form or in non-toxic ionized form (NH_4). Ammonia containing nitrogenous organic matter is directly or indirectly toxic to many species of aquatic animals [11].

In this present study four types of diet were prepared with algal oil to study the effect on growth, food consumption, FCE, AGR, SGR and FCT. Among the 5 diets tested, the diet (II) had the maximum growth, food consumption, FCE, FCR, AGR and SGR. It shows that the particular combination of diet is most suitable for rearing Zebra fish. In this combination, the protein, lipid and carbohydrate were 6.147, 3.176 and 2.751 respectively, and in control the level of the said biochemical combination were protein 4.756, lipid 2.761 and carbohydrate 1.229.

Conclusion

From the above result, it can be concluded that the level of protein and lipid present in *Chaetoceros* oil was higher than the control. This may be the reason that the *Chaetoceros* oil only enhanced growth, food consumption, FCE, FCR, AGR and SGR. In the present study, it is clearly evident that Zebra fish showed increase in growth, SGR, FCR and protein, when *Chaetoceros* oil which is in agreement with the previous reports [12,13].

Protein is the main constituent of the fish body and so a sufficient dietary supply is needed for optimum growth. Protein is the most expensive macronutrient in fish diet [14]. So, the amount of protein in the diet should be enough for the fish growth whereas the excess protein in fish diets may be wasteful and cause diets to be unnecessarily expensive [15].

The protein compound of feed is responsible for its high cost [16] and most especially fishmeal [17]. Thus, efficient transformation of protein into tissue protein for growth is of immense significance [18]. Furthermore, metabolization of protein by fish should be directed towards body protein synthesis rather than energy supply [19,20]. Growth rates of fish may be highly variable and, in many cases, appear to be limited by food availability, quality and quantity of dietary non-protein to protein nutrients. In hybrid tilapia (*Oreochromis niloticus x O. aureus*), optimum dietary lipid for maximum growth has been reported to be about 12% [21]. However, Tilapia has been reported to utilize vegetable oil that is high in omega 6 (n-6) fatty acids better than fish oil that is rich in omega 3 (n-3) fatty acids for maximum growth [17]. Although the available dietary energy plays an important role in determining body lipid deposition, the dietary lipid content is regarded as the most important factor influencing carcass lipid in fish [2,22]. An increase in dietary lipid level elevates the body lipid level in *O. niloticus*. The increase in carcass lipids with increasing dietary lipids and the consequent reduction in carcass proteins have been reported for most species investigated [23-25].

In order to evaluate the quality of experimental diet in the culture system of Zebra fish, the bacterial growth was analyzed from the first day to 30th day of culture water, gut, gill and body surface of fish. Among the food diets tested for the bacterial growth on different parts as well as in water, the diet (IV) was found to be more effective in reducing the bacterial count in culture water, gut, gill and body surface of the animal. The fecal matter released by the animal after feeding the diet may promote the growth of bacteria. In our experiment, the diet (IV) had a less number of bacteria. The reason may be that the biochemical

components like protein, lipid and carbohydrate present in this type of diet might have been almost fully absorbed in to the animal than the other type of diets. This is the reason why the bacterial load is less in this type of diet than the other type.

The microbiology of fish skin and gastro intestinal tract has been subjected to many researches. The diet plays a major role in the existence of bacterial population in different organs of fishes. Fish can spoil from both outer surface and inner surfaces as fish stomach contain digested and partially digested food which can pass into the intestine. After fish is being caught and killed the immune system collapses and bacteria are allowed to proliferate freely on the skin surface and the stomach. The walls of intestines do break down sufficiently for bacteria to move into the flesh through the muscle fiber. It has been suggested that intestinal microflora is the causative agent for food spoilage [26]. Fish take a large number of bacteria into their gut from water sediment and food [27]. It has been well known that both fresh and brackish water fishes can harbor human pathogenic bacteria particularly the coliform group [28]. Fecal coliform in fish demonstrates the level of pollution in their environment because coliform are not named flora of bacteria in fish [29].

The colouring pigments by carotenoid play a major role in the development of colour in ornamental fishes. The diet prepared with algal oil claimed to the best producer of carotenoids because it enhanced the colouration in ornamental fishes.

In the present study, the diet (IV) offered the maximum carotenoid production in ornamental fish than the other three types of diets. The effect of the diet on production of a carotenoid had been declining from the first day to the 30th day of culture. The reason may be that the animals might have been under feeding or the water quality might not have been maintained well from first to the 30th day.

Pigments are responsible for the wide spectrum of colours in fishes which is an essential prerequisite for the quality as they fetch higher price in the commercial market. As fishes cannot synthesize their own colouring pigments *de novo*, the colouring agents which are synthesized by some plants, algae and microorganisms, need to be incorporated in their diet [30,31]. Varieties of colouring agents are used in aqua industry to impart colour for the muscle and skin of fishes. Thus, pigmentation is an important criterion for fishes, since their colour affect commercial acceptability.

One of the greatest challenges in the ornamental fish industry is appearance of the accurate natural colour of the fish in the captive environment. Various products have been introduced to alleviate this problem, but none has performed so effectively and consistently as carotenoid pigment. Varieties of carotenoids pigments are used in fish diet for colouring enhancement. The most promising carotenoids proved to be successful in enhancing colour is astaxanthin that shows marked improvement in colour on most species of brightly coloured ornamental fishes like Tetras, Cichlids, Gouramis, Goldfish, Koi, Danios and many other species [32].

Effect of the diet may also indirectly have determined by its effect in clearing or reducing the pathogenic bacteria injected in to its body if the diet is considered to be more efficient, then it should control the injected pathogenic bacteria in short time than the animal fed with other type of diet. In our experiment, the diet (IV) was found to be more effective than the other three diets. In the diet (IV) fed animal, the pathogenic bacteria injected were completely cleared or destroyed by the immune system of the fishes within 2 hours after 24 hours incubation. It shows that the diet (IV) fed animal was found to be healthier to fight against the invaded pathogens.

Acknowledgment

Authors are thankful to the UGC SAP for the financed support to conduct this study.

References

1. Swain SK, Mallik D, Mishra S, Sarkar B, Soutray P (2007) Ornamental fish as model animals for biotechnological research. Environ Biotechnol pp: 293-328.

2. Sargent JR, Tocher DR, Bell JG (2002) The lipids. Fish Nutrition. Elsevier Science, USA.

3. Bernstein AM (2012) A meta-analysis shows that docosahexaenoic acid from algal oil reduces Serum triglycerides and increases hdl-cholesterol and ldl-cholesterol in persons without coronary heart disease. J Nutr 142: 99-104.

4. Browne B, Gibbs R, McLeod J, Parker M, Schwanda W, et al. (2009) Oil extraction from microalgae.

5. Fiji (2007) Fish feed formulation. Ministry of Agriculture Fisheries and Forestry.

6. Walne PR (1970) Studies on the food value of nineteen genera of algae to junenile bivalves of the genera Ostea, Crassostrea and Mytilis. Fishery Invest 26: 162.

7. Solorzano I (1969) Determination of ammonia in natural waters by the phenolhypochlorite method, Limnol. Oceanogr 14: 799-801.

8. Lowry LOH, Rosebrough NJ, Farr AL, Randall RJ (1951) Protein measurement with the folin-phenol reagents. J Biol Chem 193:265-275.

9. Folch J, Lees M, Stanley GHS (1957) A simple method for the isolation and purification of total lipids from animal's tissues. J Biol Chem 226: 497.

10. Seifter S, Dayton S, Novic V, Muntwyler E (1950) The estimation of glycogn with the anthrone reagent. Arch Biochem Phys pp: 186-200.

11. Prema LY, Lipton AP (2007) Water quality management in gold fish reading tanks using different filter materials. Indian Hydrobiology 10: 301-302.

12. Eric CH (2012) The use of algal in fish feeds as alternatives to fishmeal. J International Aqua feed 10: 3-27.

13. Zahira Y, Ehsan A, Mohd ST (2014) Biomolecules from microalgae for animal feed and aquaculture. J Biol Res (Thessalon) p: 21.

14. Pillay TVR (1990) Aquaculture: Principles and practices. Fishing News Book. Blackwell Scientific Publications Ltd., Oxford, UK.

15. Ahmad MH (2000) Improve productive performance in fish. Dissertation, Animal Prof. Department, Faculty of Agriculture, Zagazig University.

16. Shiau SY, Lin SF (1993) Effect of supplementary dietary chromium and vanadium on the utilization of different carbohydrates in tilapia, Oreochromis niloticus X O. areus. Aquacul 110: 321-330.

17. National Research Council (NRC) (1993) Nutrient requirement of fish. Committee on animal nutrition, Board on agriculture, National Academy of Sciences. National Academy Press, Washington DC., USA.

18. Weatherley AH, Gill HS (1987) Recovery growth following periods of restricted rations and starvation in rainbow trout, Salmon gairdneri Richardson. J Fish Bio 18: 195-207.

19. Shiau SY (1997) Utilization of carbohydrates in warm water fish–with particular reference to tilapia, Oreochromis niloticus X O. aureus. Aquacul 151: 79-96.

20. Borba DM, Francalossi IE, Pezzato LE (2006) Dietary energy requirement on Piracan juba fingerlings, Bryconorbi gynanus, and relative utilization of dietary carbohydrate and lipid. Aqua Nutr 12: 183-191.

21. Chou BS, Shiau SY (1999) Optimal dietary lipid for growth of juvenile hybrid tilapia, Oreochromis niloticus X O. aureus. Aquacul 43: 185-195.

22. Hanley F (1991) Effects of feeding supplementary diets containing varying levels of lipid on growth, food conversion, body composition of Nile tilapia, Oreochromis niloticus (L.). Aquacul 93: 323-334.

23. Refstie T, Austreng E (1981) Carbohydrate in rainbow trout diets, III. Growth and chemical composition of fish from different families fed four levels of carbohydrate in the diet. Aquacul 25: 35-49.

24. Sargeant J, Henderson RJ, Tocher DR (1989) The lipids. Fish Nutrition, Academic Press, New York, USA.

25. De-Silva SS, Gunasekara RM, Shim KF (1991) Interactions of varying dietary protein and lipid levels in young red tilapia: Evidence of protein sparing. Aquacul 95: 305-318.

26. Kaneko S (1971) Microbiological study of fresh fish. New Food Industries 13: 176-180.

27. Sugita H, Tsunohara M, Ohkoshi T, Deguchi (1988) The establishment of an intestinal microflora in developing goldfish (Carassius auratus) of culture ponds. Microbiology Ecol 15: 333-344.

28. Leung CY, Huang TT, Pancardo O (1990) Journal of Agriculture in the Tropics 5: 87-90.

29. Cohen J, Shuval HI (1973) Water Soil Pollution 2: 85-95.

30. Johnson EA, An GH (1991) Astaxanthin from microbial sources. Crit Rev Biotechnology 11: 297-326.

31. Davis BH (1985) Carotenoid metabolism in animals, a biochemist's view. Pure Appl Chem 57: 679-684.

32. Gupta SK, Jha AK, Venkateshwarlu G (2007) Use of natural carotenoids for pigmentation in fishes. Natural Product Radiance 6: 46-49.

18

Morphometric Parameters and Allometric Growth in Paradise Threadfin *Polynemus paradiseus* (Linnaeus, 1758) from a Coastal River of Bangladesh

Chaklader MR[1], Siddik MAB[1]*, Ashfaqun Nahar[2], Hanif MA[1], Alam MJ[3] and Sultan Mahmud[4]

[1]*Department of Fisheries Biology and Genetics, Patuakhali Science and Technology University, Bangladesh*
[2]*Department of Marine Fisheries and Oceanography, Patuakhali Science and Technology University, Bangladesh*
[3]*Department of Fisheries Management, Patuakhali Science and Technology University, Bangladesh*
[4]*Department of Aquaculture, Patuakhali Science and Technology University, Bangladesh*

Abstract

Morphological parameters comprising length-weight relationships (LWRs), sex ratio, condition factor (KF) and allometric growth of paradise threadfin, *Polynemus paradiseus* from the southern coast of Bangladesh were estimated. A total of 221 specimens were collected with the help of local fishermen ranging size 8.30-13.70 cm standard length (SL) and 11.64-50.67 g body weight (BW) during the period of January to October, 2104. The overall sex ratio of the samples did not reveal significant variation from the expected value of 1:1 (male: female=1:0.99, χ^2=0.004, P < 0.05). Length-frequency distribution revealed a size predominance of males over females, where mean variation of males consistently exceeded that of females throughout the year. The allometric coefficient b of LWR significantly deviated from 3 indicating allometric growth in males and females. The analysis of covariance (ANCOVA) showed significant differences in slope and intercept between the sexes (P < 0.001). KF by month in both sexes indicating the fishes were thriving very well in the coast of Bangladesh. This study reports the first findings regarding the LWRs, sex ratio and KF of *Polynemus paradiseus* in the coastal waters of Bangladesh.

Keywords: Length-weight; Size-frequency; Allometric growth; *Polynemus paradiseus*; Bangladesh

Introduction

The coastal rivers and estuaries of the southern Bangladesh are categorized by high level of commercial fish catch have direct contribution to the national economy of the country [1-3]. Like other commercial species, paradise threadfin *Polynemus paradiseus* is one the vital component of estuaries and Bay of Bengal fishery in Bangladesh commonly known as threadfin fish [4,5]. This species is available in the Indian subcontinent including Bangladesh, India, Pakistan, Sri Lanka and also reported in Indo-pacific ocean confluence of the Bay of Bengal [5,6]. It has an increasingly commercial importance in southern coastal regions of Bangladesh because of its nutritional value and placed as important food item next to importance of Hilsa fishery [5]. In spite of having great economic value, the abundance of fish is declining in Bangladesh due to over exploitation, pollutions, habitat loss and other ecological changes to its habitat [7-11].

Length-weight relationship plays a vital role in the field of fish biology, physiology, ecology, fisheries assessment and fish conservation [3,12]. It is considered an essential tool in the studies of fish stock assessment and management of fisheries resources [13,14]. Life-history traits as sex ratio, length–frequency distribution (LFD), length weight relationships (LWRs), and relative-condition factor (*K*) are generally less known for tropical and subtropical finfish species [15] although this knowledge can be utilized to convert growth-in-length equations to growth-in weight in stock assessment models [16-18], population biomass estimation [16] and fish condition [19]. Few studies have been carried out in different aspects as biology, breeding and nutritional quality of this species but knowledge on length-weight relationships and condition factor of the *P. paradiseus* is not available of this species. Therefore, this article is the first complete and comprehensive description of the LFD, LWRs, sex ratio, K_F and allometric growth of *P. paradiseus* from the coastal waters of Bangladesh in order to provide sufficient data and information for the sustainable management of the species.

Materials and Methods

The specimens were collected from eight stations of the Payra River located in the southern coastal region of Bangladesh (Figure 1) with the help of fishermen during January-October 2014. Water temperature was also recorded monthly at each sampling. The collected specimens were preserved with ice and transferred to the laboratory prior to analysis. The standard length (SL) of this species was measured to the nearest 1 cm using a measuring scale, while body weight (BW) was recorded using a digital balance (Shimadzu, EB-430DW, Japan) to 0.01 g accuracy.

Calculations of length-weight data was done separately and also combined using the conventional formula described by as $W = a\, L^b$ where, W is the total weight (expressed in g), L is the standard length (expressed in cm), a is a coefficient related to body form and b is an exponent indicating isometric growth when equal to 3 and indicating allometric growth when significantly different from 3 [20]. The parameters a and b were calculated using linear regression analysis equation based on natural log equations of the relationship between BW and SL as follows: $ln\,(W) = ln\,(a) + b\,ln\,(L)$. Following formula was used for the calculation of condition factors (K) of individual fish species for each month using the expression as $K = (W/L^3) \times 100$ for

*Corresponding author: Muhammad Abu Bakar Siddik, Department of Fisheries Biology and Genetics, Patuakhali Science and Technology University, Patuakhali-8602, Bangladesh, E-mail: siddik@pstu.ac.bd

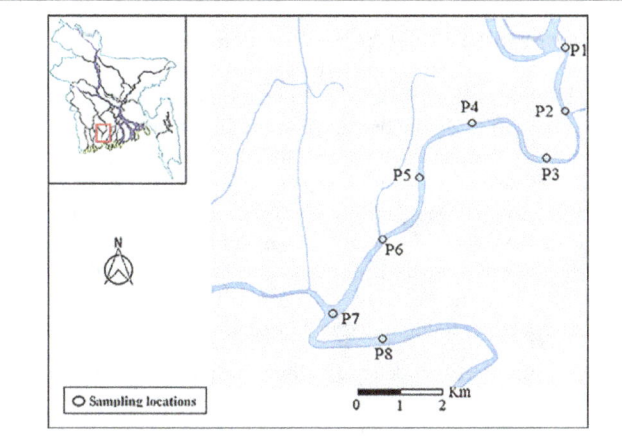

Figure 1: Sampling stations of *Polynemus paradiseus* in the Payrariver located in the southern coastal region of Bangladesh.

both monthly and in terms of size (SL in cm) class.

The form factor ($a_{3.0}$) which was calculated for *P. paradiseus* using the equation given by Froese [21] as: $a_{3.0} = 10^{\log a - s(b-3)}$, where a and b are regression parameters of LWR, and S is the regression slope of log a vs. b. A mean slope $S = -1.358$ Froese [21] was used for estimating the form factor during this study because information on LWRs is not available for this species for estimation of the regression (S) of $\ln a$ vs b. Statistical analysis was performed with SPSS Version 15.0 software package SPSS Inc. 2004.

Results

A total of 221 specimens were collected of which 50.22% were males and 49.77% were females. The overall sex ratio did not show significant difference from the expected value of 1:1 (male: female = 1:0.99, $\chi^2 = 0.004$) (Figure 2).

The SL of the specimen varied between 8.90-13.70 cm for males and 8.30-13.60 cm for females while BW was ranged between 15.00-50.67 g and 11.64-44.73 g for males and females, respectively. The SL-frequency distribution showed that the males and females of *P. paradiseus* were not normality distributed ($P < 0.001$) in the southern coastal waters of Bangladesh (Figure 3). The calculated regression parameter b based on relationships between SL and BW indicated negative allometric growth both in males and females for all sample size. There was significant difference observed for both slope (b) and intercept (a) between sexes ($P < 0.001$) (Figures 4 and 5). The estimated allometric coefficients monthly varied between 2.07-3.38 in males and between 2.22-3.54 in females of *P. paradiseus* during the entire study period. Values of determination coefficients were less than 0.9, therefore, it was highly insignificant (P < 0.05).

Condition factor (K_F) of *P. paradiseus* in different months is shown in (Table 1). The K_F value of males was varied from 1.82 to 2.23 and 1.80 to 2.05 in males and females, respectively. The monthly highest K_F was found in April at 9.40 to 12.70 cm length group with a mean of 2.225 ± 0.053 and lowest in July at 8.90 to 11.80 cm length group with a mean of 1.830 ± 0.073 while the highest K_F value for females was found in May at 8.30 to 10.30 cm length group with a mean of 2.054 ± 0.066 and lowest in August at 9.20 to 12.70 cm length group with a mean of 1.802 ± 0.057. The highest K_F value for males is higher than that in females in majority of the months except July. (Figure 6) revealed that the overall K_F value of male population was comparatively higher than the female population.

Monthly variations of water temperature along the southern coast of Bangladesh are shown in (Figure 7). The low temperature was observed in January with increasing gradually at beginning of February and remained high from April to October and thereafter started to decrease gradually until December.

Discussion

Taxonomic studies of this fish have been carried out by Chaklader et al. [5] presented the diversity and standard measurements of male and female fishes. The length–weight relationship, condition factor and

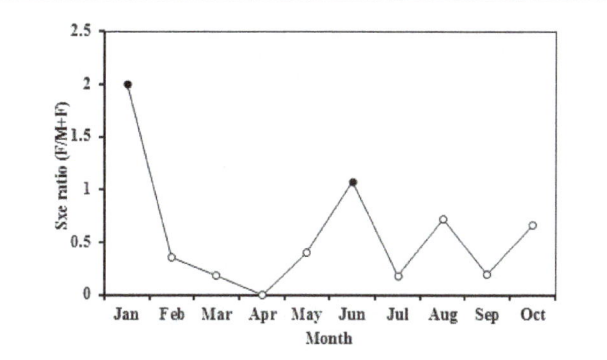

Figure 2: Temporal variation in sex ratio of *P. paradiseus* (● indicates statistically significant difference from 1:1 ratio, non-significant) collected from the Payra river, southern Bangladesh.

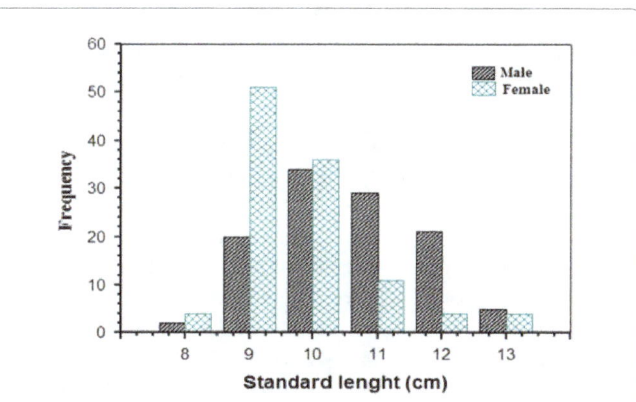

Figure 3: Size-frequency distribution of male and female *P. paradiseus* (Linnaeus, 1758) collected from the Payrariver, southern Bangladesh.

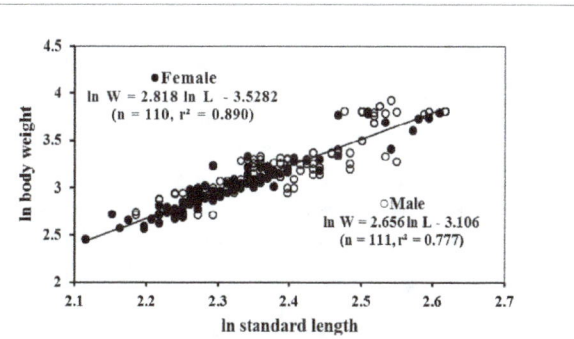

Figure 4: Relationship between log-transformed SL and log-transformed BW for male and female *P. paradiseus* (Linnaeus, 1758) collected from the Payrariver, southern Bangladesh.

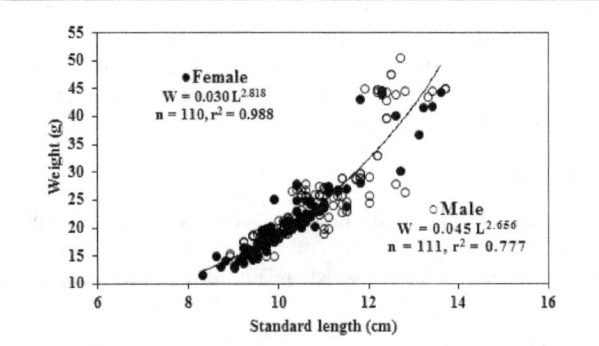

Figure 5: Relationships between SL and BWof male and female *P. paradiseus* (Linnaeus, 1758) collected from the Payrariver, southern Bangladesh.

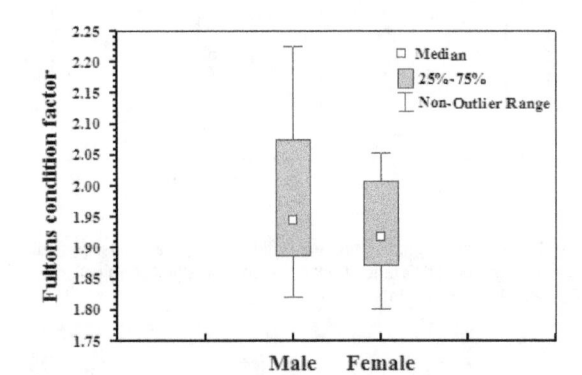

Figure 6: Fulton's condition factor for combined male and female *P. paradiseus* (Linnaeus, 1758) collected from the Payrariver, southern Bangladesh.

form factor are recorded for the first time. Female to male ratio vary between 1:1 and 1:1.3 in a typical population reported by Hossain et al. [22]. For most aquatic animals, deviation from a 1:1 sex ratio is not expected but some finfish and prawn populations may shows a strong bias in this ratio [23]. The overall monthly sex ratio was found to be slightly in favor of male and it was revealed that male and female equally presented in most of the months except January, February and August when male was predominant over female but female was predominant male in June and October. In the present study, the overall male and female sex ratio was 1:0.99 in the coastal waters which did not differ significantly from the expected value of 1:1. Sex ratio variations may be influenced by reproduction, growth and longevity of a species [24].

The monthly size-frequency distributions indicated that there were more than one size groups found to be present in each month for both sexes. The size predominance in female is a common feature reported

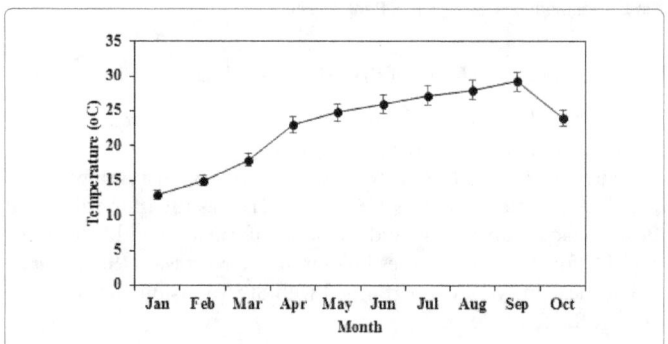

Figure 7: Monthly changes of water temperature in the Payrariver, southern Bangladesh.

Month	Sex	n	Length Characteristics		Weight Characteristics		Parameters of the LWR					GT	$a_{3.0}$	K_F
			MSL ± SE	SL_{min}-SL_{max}	MW ± SE	W_{min}-W_{max}	a	LCL – UCL	b	LCL – UCL	r^2			
Jan-14	M	12	11.52 ± 0.28	10.00-12.80	30.97 ± 3.10	18.81-44.76	0.008	0.003-0.156	3.38	2.15-4.60	0.79	A +	0.026	1.969
	F	6	10.17 ± 0.25	9.50-10.90	20.38 ± 1.54	14.92-20.38	0.022	0.0002-2.09	2.95	0.98-4.92	0.812	A +	0.016	1.924
Feb	M	14	11.86 ± 0.19	10.00-11.00	32.51 ± 2.54	13.40-23.87	0.012	0.0001-2.44	3.18	1.03-5.33	0.465	A +	0.021	1.921
	F	11	10.00 ± 0.20	8.80-10.80	18.74 ± 0.93	13.39-22.66	0.071	0.015-0.251	2.42	1.87-3.10	0.903	A-	0.011	1.864
Mar	M	10	11.48 ± 0.22	10.40-12.40	29.39 ± 2.17	21.03-43.00	0.014	0.005-0.395	3.13	1.76-4.49	0.776	A +	0.021	1.919
	F	12	10.08 ± 0.23	8.70-11.50	19.42 ± 1.29	13.06-19.42	0.026	0.008-0.089	2.86	2.33-3.39	0.935	A-	0.017	1.872
Apr	M	11	10.68 ± 0.32	9.40-12.70	28.18 ± 3.26	18.76-50.67	0.012	0.003-0.044	3.28	2.71-3.85	0.949	A +	0.029	2.225
	F	11	9.95 ± 0.17	9.00-11.00	18.91 ± 0.89	12.90-23.66	0.049	0.007-0.341	2.59	1.74-3.44	0.841	A +	0.013	1.913
May	M	13	10.06 ± 0.19	8.90-11.00	21.32 ± 1.14	15.00-27.50	0.042	0.013-0.132	2.7	2.20-3.20	0.927	A +	0.017	2.075
	F	10	9.51 ± 0.19	8.30-10.30	17.92 ± 1.28	11.64-25.25	0.01	0.0007-0.139	3.32	2.15-4.50	0.842	A +	0.027	2.054
Jun	M	9	9.84 ± 0.22	8.90-10.90	20.42 ± 1.12	15.50-27.00	0.126	0.014-1.076	2.23	1.28-3.17	0.817	A-	0.011	2.133
	F	14	9.62 ± 0.10	9.10-10.30	17.95 ± 0.73	14.30-25.25	0.021	0.0008-0.496	2.98	1.58-4.38	0.642	A +	0.019	2.007
Jul	M	10	10.30 ± 0.22	9.70-11.40	19.97 ± 1.21	15.00-27.00	0.083	0.003-2.609	2.35	0.87-3.82	0.626	A-	0.011	1.82
	F	12	10.44 ± 0.24	9.40-11.80	23.63 ± 2.24	14.87-43.17	0.006	0.0005-0.071	3.54	2.46-4.62	0.843	A +	0.032	2.018
Aug	M	13	11.20 ± 0.33	9.80-13.70	26.70 ± 2.33	19.00-45.00	0.116	0.009-1.437	2.25	1.20-3.30	0.688	A-	0.011	1.882
	F	9	10.41 ± 0.38	9.20-12.70	20.48 ± 1.69	14.95-30.19	0.113	0.036-0.343	2.22	1.74-2.70	0.944	A-	0.01	1.802
Sept	M	9	11.38 ± 0.41	10.20-13.70	27.88 ± 2.66	19.00-45.00	0.182	0.006-5.223	2.07	0.68-3.47	0.638	A-	0.01	1.888
	F	11	11.25 ± 0.45	9.20-13.60	28.33 ± 3.58	13.78-44.36	0.011	0.003-0.034	3.25	2.76-3.74	0.961	A +	0.024	1.879
Oct	M	10	11.44 ± 0.45	9.20-13.70	31.84 ± 3.70	17.67-47.65	0.031	0.004-0.225	2.85	2.01-3.69	0.885	A +	0.019	2.054
	F	14	10.73 ± 0.37	8.60-13.40	25.07 ± 2.52	15.14-44.73	0.019	0.014-0.162	2.63	2.11-3.15	0.91	A-	0.006	1.969
Overall	M	111	11.00 ± 0.11	8.90-13.70	27.09 ± 0.87	15.00-50.67	0.045	0.023-0.086	2.67	2.39-2.93	0.777	A-	0.016	1.987
	F	110	10.21 ± 0.10	8.30-13.60	21.19 ± 0.67	11.64-44.73	0.03	0.019-0.045	2.82	2.63-3.01	0.89	A-	0.017	1.935
	B	221	10.61 ± 0.07	8.30-13.70	24.16 ± 0.58	11.64-50.67	0.032	0.022-0.045	2.8	2.64-2.96	0.849	A-	0.017	1.961

Table 1: Monthly descriptive statistics, estimated parameters of length-weight relationships, Fulton Condition Factor (KF) and Form factor (a3.0) for *P. paradiseus* (Linnaeus, 1758) along the southern coast of Bangladesh.

by several studies [25]. Numerous factors, such as physiological changes influenced by temperature change, feeding regime and reproductive cycle might be responsible for size difference between male and female [26]. Besides, traditional fishers used numerous types of fishing gear with various mesh sizes leading to the selectivity of target species which may greatly influence the size distribution of the individuals caught resulting in highly biased estimations of the various population parameters including the maximum size [22].

In this study, the b values for male, female and combined gender of *P. paradiseus* were found as 2.66, 2.82 and 2.80, respectively indicating negative allometric growth of *P. paradiseus* in the coastal waters i.e., fish gets slimmer with increasing length [27]. It is widely recognized that a number of factors including growth phase, sex, seasons, food (quantity, quality and size), stage of maturity and health and general fish condition, preservation techniques and differences in the observed length ranges of the specimen caught are known to influence the length-weight relationship in fish [28] all of which were not accounted in the present study.

Condition factor is an index reflecting interactions between biotic and abiotic factors in the physiological condition of the fishes indicating the well-being of the population during various life cycle stages [21]. In the present study, the K_F values of *P. paradiseus* was highest in April for males and May for females respectively, whilst the lowest was in July for males and August for females (Table 1). The monthly mean K_F values were always above 1 and no major variation was found in different length groups. Le Cren [29], reported that K_F values greater than 1 indicates good general condition of the fish whereas values less than 1 denotes the reverse condition. It is clearly indicated that the growth pattern of the fish was ideal in the study areas. In general, the seasonal cycle is related with the condition of fishes suggested a relationship with gonadal development. Since the paradise threadfin spawns in March to June [5], the highest K_F was found that month for both male and female [30] pointed out that the condition factor was constant during the pre-spawning period, decreased in the period of spawning and was lowest immediately after spawning in case of silver hatchet chela, *Chela cachius* (Hamilton 1822). K_F values are generally influenced by the changing of seasons, amount of food supply, and maturity of gonads [31]. Fluctuation of water temperature occurred in all month and optimum range (20-25°C) found from April to July during the experimental period shown in (Figure 6). K_F is strongly influence by both biotic and a biotic environmental conditions and can be used as an index to assess the status of the aquatic ecosystem in which fish live [32]. Several studies stated that the water temperature has considerable importance for controlling spawning of fish [30]. However, no references dealing with the condition factors of the *P. paradiseus* are available in the coastal waters, preventing the comparison with previous results from the same population.

There are few studies dealing with form factor of fishes available in Bangladeshi waters [33]. Nevertheless, reference concerning the form factor of this species is not available as it is the first study of this species. Therefore, the study provides basic information on some morphological parameters including the sex ratio, LFD, LWRs, condition factor and growth type of *P. paradiseus* in the southern coastal waters which should be useful for the sustainable management of finfish fishery not only in southern Bangladesh but its whole geographic locations. Yet, more detailed studies should be conducted to answer several questions about body-size range and the spawning periodicity of this finfish fishery.

References

1. Hanif M, Siddik M, Chaklader M (2015) Fish diversity in the southern coastal waters of Bangladesh: present status, threats and conservation perspectives. Croatian Journal of Fisheries 73: 251-274.

2. Sharker M, Mahmud S, Siddik M, Alam M, Alam M, et al. (2015) Livelihood status of hilsha fishers around Mohipur fish landing site, Bangladesh. World Journal of Fish and Marine Sciences 7: 77-81.

3. Siddik M, Hanif M, Chaklader M, Nahar A, Mahmud S, et al. (2016) Fishery biology of gangetic whiting *Sillaginopsis panijus* (Hamilton, 1822) endemic to Ganges delta, Bangladesh. Egyptian Journal of Aquatic Research 41: 307-313.

4. Hanif M, Siddik M, Chaklader M, Mahmud S, Nahar A (2015) Biodiversity and conservation of threatened freshwater fishes in Sandha River, South West Bangladesh. World Applied Sciences Journal 33: 1497-1510.

5. Chaklader M, Siddik M, Nahar A (2015) Taxonomic diversity of paradise threadfin *Polynemus paradiseus* (Linnaeus, 1758) inhabiting southern coastal rivers in Bangladesh. Sains Malaysiana 44: 1241-1248.

6. Siddik M, Nahar A, Ahamed F, Hossain M (2014) Over-wintering growth performance of mixed-sex and mono-sex Nile tilapia *Oreochromis niloticus* in north-eastern Bangladesh. Croatian Journal of Fisheries 72: 70-76.

7. Hoq M (2007) An analysis of fisheries exploitation and management practices in Sundarbans mangrove ecosystem, Bangladesh. Ocean and Coastal Management 50: 411-427.

8. IUCN (2000) The world conservation union, Red book of threatened fishes of Bangladesh, Dhaka, Bangladesh, IUCN.

9. Siddik M, Nahar A, Ahamed F, Masood Z, Hossain M (2013) Conservation of critically endangered olive barb *Puntius sarana* (Hamilton, 1822) through artificial propagation. Our nature 11: 96-104.

10. Chaklader M, Nahar A, Siddik M, Sharker R (2014) Feeding habits and diet composition of Asian Catfish *Mystus vittatus* (Bloch, 1794) in shallow water of an impacted coastal habitat. World Journal of Fish and Marine Sciences 6: 551-556.

11. Sharker M, Siddik M, Nahar A, Shahjahan M, Faroque A (2015) Genetic differentiation of wild and hatchery populations of Indian major carp *Cirrhinus cirrhosis* in Bangladesh. Journal of Environmental Biology 36:1223-1227.

12. Mustac B, Sinovcic G (2010) Morphometric and meristic parameters of sardine (*Sardina pilchardus*, Walbaum, 1792) in the Zadar fishing area. Croatian Journal of Fisheries 68: 27-43.

13. Al-Beak A, Ghoneim, El-Dakar A, Salem M (2015) Population Dynamic and Stock Assesment of White Seabream *Diplodus sargus* (Linnaeus, 1758) in the Coast of North Sinai. Fish Aquac J 6:152. doi:10.4172/2150-3508.1000152.

14. Rodríguez-Romero J, Palacios-Salgado D, López-Martínez J, Hernández Vázquez S, Velázquez-Abunader J, et al (2009) The length-weight relationship parameters of demersal fish species off the western coast of Baja California Sur, Mexico. J Appl Ichthyol 25: 114-116.

15. Ecoutin J, Albaret J, Trape S (2005) Length-weight relationships for fish populations of a relatively undisturbed tropical estuary: The Gambia. Fisheries Research 72: 347-351.

16. Qambrani G, Soomro A, Palh Z, Baloch W, Tabasum S, et al.. (2016) Reproductive Biology of *Glossogobius giuris* (Hamilton), in Manchar Lake Sindh, Pakistan. J Aquac Res Development 7: 392.

17. Morato T, Afonso P, Lourinho P, Barreiros J, Santos R, et al. (2001) Length-weight relationships for 21 coastal fish species of the Azores, north-eastern Atlantic. Fisheries Research 50: 297-302.

18. Ozaydin O, Taskavak E (2006) Length-weight relationships for 47 fish species from Izmir Bay (eastern Aegean Sea, Turkey). Acta Adriatica 47: 211.

19. Chatterji A, Pati S, Dash B (2015) A Study on the Growth of Juveniles of Tiger Prawn, *Penaeus monodon* (Fabricius) Under Different Photoperiods. J Aquac Res Development 6: 385.

20. Simon K, Mazlan A (2008) Length-weight and length-length relationships of archer and puffer fish species. The Open Fish Science Journal 1:19-22.

21. Froese R (2006) Cube law, condition factor and weight–length relationships: history, meta-analysis and recommendations. J Appl Ichthyol 22: 241-253.

22. Hossain M, Khatun M, Jasmine S, Rahman M, Jewel M, et al. (2013) Life

history traits of the threatened freshwater fish *Cirrhinus reba* (Hamilton 1822) in the Ganges River, Bangladesh. Sains Malaysiana 42: 1219-1229.

23. Hossain M, Sharmin J, Rahman M (2015) Biological Aspects of the critically endangered fish, *Labeo boga* in the Ganges River, Northwestern Bangladesh. Sains Malaysiana 44: 31-40.

24. Chilari A, Thessalou-Legaki M, Petrakis G (2005) Population structure and reproduction of the deep-water shrimp *Plesionika martia* (Decapoda: Pandalidae) from the eastern Ionian Sea. J Crustacean Biol 25: 233-241.

25. Ohtomi J, Yamamoto S, Koshio S (1998) Ovarian maturation and spawning of the deep-water mud shrimp Solenocera melantho De Man, (Decapoda, Penaeoidea, Solenoceridae) in Kagoshima Bay, southern Japan. Crustaceana 71: 672- 685.

26. Newman S, Cappo M, Williams D (2000) Age, growth, mortality rates and corresponding yield estimates using otoliths of the tropical red snappers, *Lutjanus erythropterus*, *L. malabaricus* and *L. sebae*, from the central Great Barrier Reef. Fisheries Research 48:1-14.

27. Jobling M (2008) Environmental factors and rates of development and growth, Oxford: Blackwell Publishing, England.

28. Koutrakis E, Tsikliras A (2003) Length-weight relationships of fishes from three northern Aegean estuarine systems (Greece). J Appl Ichthyol 19: 258-260.

29. Le Cren E (1951) The length-weight relationship and seasonal cycle in gonad weight and condition in the perch (*Perca fluviatilis*). The Journal of Animal Ecology 20: 201-219.

30. Ahamed F, Ohtomi J (2012) Growth patterns and longevity of the pandalid shrimp *Plesionika izumiae* (Decapoda: Caridea). J Crustacean Biol 32: 733-740.

31. Iqbal K, Masuda Y, Suzuki H, Shinomiya A (2006) Age and growth of the Japanese silver-biddy, *Gerres equulus*, in western Kyushu, Japan. Fisheries research 77: 45-52.

32. Anene A (2005) Condition factor of four Cichlid species of a man-made lake in Imo State, Southeastern Nigeria. Turkish Journal of Fisheries and Aquatic Sciences 5: 43-47.

33. Sani R, Gupta B, Sarkar U, Pandey A, Dubey V, et al. (2010) Length-weight relationships of 14 Indian freshwater fish species from the Betwa (Yamuna River tributary) and Gomti (Ganga River tributary) rivers. J Appl Ichthyol 26: 456-459.

19

Effect of Use Fresh Macro Algae (Seaweed) *Ulva fasciata* and *Enteromorpha flaxusa* With or Without Artificial Feed on Growth Performance and Feed Utilization of Rabbitfish (*Siganus rivulatus*) fry

Mohamed FA Abdel-Aziz* and Mohammed A Ragab

Fish Rearing Lab, Aquaculture Division, National Institute of Oceanography and Fisheries (NIOF), Egypt

Abstract

This study a low-cost aquaculture diet with fresh seaweed was tested with the herbivorous rabbitfish (*Siganus rivulatus*) fry. Two fresh seaweed was genera, *Ulva* and *Enteromorpha* (belong to family Ulvaceae) were used to replace the artificial feed by 0, 50 and 100 percent regardless the protein percentage. Initial average weight of fry was 0.18 g. This trial consisted of sixth treatments, the first treatment (T1) fish fed on artificial feed only, the (T2) fish fed of half feeding rate on artificial feed and other fresh *Ulva*, the (T3) fish fed of half feeding rate on artificial feed and other fresh *Enteromorph*, the (T4) fish fed on fresh *Ulva* only, the (T5) fish fed on fresh *Enteromorpha* only and the (T6) fish fed of half feeding rate on fresh *Ulva* and other fresh *Enteromorph*. Feeding rat was 7% of biomass and this trial continued for 70 days. There were significant differences between the treatments in all the growth performance parameters. The T3 was the highest in final weight (W_2), total weight gain (TG), average daily gain (ADG), relative growth rate (RGR) and specific growth rate (SGR), followed by both the T2 and T1. And the best feed conversion ratio (FCR) was achieved with T3, T1 and T2 followed by T6 and T5 but the T4 had the worst FCR.

Keywords: Rabbitfish; Macro algae; Feeding rate; Growth performance; Feed utilization

Introduction

Marbled spinefoot rabbitfish *Siganus rivulatus* is a potential for warm water marine aquaculture diversification [1,2]. Rabbitfishes belong to the genus *Siganus* of the family siganidae [3]. Siganids are herbivorous marine and brackish water fishes that are found throughout the indo west pacific [4], and the more common species are the objects of traditional subsistence and commercial fisheries throughout this region. There has been interest in the culture of these fishes in ponds or cages in several areas [5].

Being herbivorous, the *siganus* species need a big quantity of algae feed to assure their biological activities. The stomach of these fishes is an acid medium able to digest marine plants before entering in the digestive tract for complete digestion and thus excreting feces. In addition to algae feed, they can feed accidentally on some non-digestible substances such as mollusk shells and other invertebrates attached to algae [6,7]. Egypt production of rabbitfish was about 1363 ton in 2014, Mediterranean Sea took part in 822 ton production, Red Sea (466 ton) and lakes (75 ton) according to GAFRD [8].

Marine macro algae, or seaweed, are plantlike organisms that generally live attached to rock or other hard substrata in coastal areas. They belong to three different groups, empirically distinguished since the mid-nineteenth century on the basis of thallus color: - red algae (phylum Rhodophyta), brown algae (phylum Heterokontophyta (also known as the Ochrophyta), class Phaeophyceae), and green algae (phylum Chlorophyta, classes Bryopsidophyceae, Chlorophyceae, Dasycladophyceae, Prasinophyceae, and Ulvophyceae). There are about 8,000 species of macro algae (seaweeds) along the world's coast live and they may extend as deep as 270 m [9]. A total of 25 species of green sea weeds, 90 species of brown and 350 species of red seaweeds are found in the world sea area that are commercially important because of their protein, amino acids and mineral contents [10]. The protein content of seaweed differs according to species and seasonal period. Generally, the protein fraction of brown seaweeds is low (3% to 5% of the dry weight) compared with that of the green or red seaweed (10% to 47% of the dry weight). The content of crude protein, crud lipid, ash and fiber in green meals from 7% to 29%, 0.5% to 4%, 13% to 36% and from 3% to 6% respectively [11,12].

Seaweeds are rarely promoted for the nutritional value of their proteins [13]. Considering the importance of seaweeds, it can be said that, sea weeds can play a vital role in various aspects compared to other aquatic resources. Much attention should be given on seaweed to compensate the food problem to some extent and fulfill the deficiency of nutrition for erecting the economy of several countries [14].

All green seaweed belong to the classes Ulvophyceae, Bryopsidophyceae, and Dasycladophyceae, which include approximately 1,500 species currently referred to eight orders [15]. The genus *Ulva* (Phylum: Chlorophyta, Class: Ulvophyceae, Order: Ulvales, Family: Ulvaceae) was first identified by Linnaeus in 1753 [16]. Since then many taxonomists and phycologists have been involved in the identification of *Ulva* species which are notoriously difficult to classify due to the morphological plasticity expressed by many members as well as the few reliable characters available for differentiating taxa [17].

Ulva is a good source of protein, pigments, minerals and vitamins, and is especially rich in vitamin C [18,19] and, in recent years, *Ulva* species have become important macro algae, which have been investigated as a dietary ingredient for a wide range of fish species.

*Corresponding author: Mohamed FA Abdel-Aziz, Fish Rearing Lab, Aquaculture Division, National Institute of Oceanography and Fisheries (NIOF), Egypt
E-mail: m_fathy8789@yahoo.com

Enteromorpha (Phylum: Chlorophyta, Class: Ulvophyceae, Order: Ulvales, Family: Ulvaceae) have also been used as a source of bioactive compounds similar to those which cause an inhibitory effect against the bacterium *Xanthomonas oryzae*, which causes leaf blight disease in paddy crops [20]. People in the Philippines and Japan also use *Enteromorpha* spp. as food [21,22].

The study aimed to evaluate the effect of use fresh macro algae including feeding rate with artificial feed on growth performance, feed efficiency and feed cost of rabbitfish fry. Whereas the most of the cultured marine fish are carnivorous species, and seldom are herbivorous or omnivorous species, and the higher prices and uncertainty of availability of fishmeal (FM) is limiting the development of mariculture industry, especially the culture of carnivorous fish. So FAO (food and agriculture organization) quite canonizes the culture of herbivorous or omnivorous species [23]. From this point, the present study presented some information for the development of low-cost diets with fresh macro algae as dietary ingredient for the culture of marine fish such as herbivorous rabbitfish fry.

Materials and Methods

The present study was conducted using the research facilities of Shakshouk Fish Research Station, Fayoum Governorate, National Institute of Oceanography and Fisheries (NIOF), Egypt. Rabbitfish (*Siganus rivulatus*) fry were obtained from (Mediterranean Sea) National Institute of Oceanography and Fisheries (NIOF), Alexandria Governorate, Egypt, initial average weight for this fry was 0.18 ± 0.012 g (SE standard error) and initial average length was 2.76 ± 0.057 cm.

Fish acclimatization and diet preparation

Fry were acclimatized to be adapted to water salinity of Lake Qaroun 33 part per thousand (ppt) for one week. An artificial diet was formulated by hand, the diet formulated to be almost containing 36% crude protein (Table 1) and two genuses of macro algae belong to family Ulvaceae, the first genus *Ulva* (*Ulva fasciata*) was collected from Mediterranean sea, Alexandria Governorate, Egypt and the second genus *Enteromorpha* (*Enteromorpha flaxusa*) was collected from Qaroun lake, Fayoum Governorate, Egypt. After operation collection,

Ingredients	(g/100 g)
Fish meal (72% CP)	22
Extruded full fat Soybean meal (37% CP)	43
Wheat bran fine	28
Fish oil	4
Super yeast	1
Starch	1.7
Vit. & Min. & premix	0.3
Total	100
Chemical analysis % on dry matter basis	
Moisture (M)	6.94
Dry matter (DM)	93.06
Crude protein (CP)	36.44
Ether extract (EE)	13.78
Crude fiber (CF)	3.10
Nitrogen free extract (NFE)	39.02
Ash	7.66
Gross energy (GE, Kcal/g)*	5.09
Chemical analysis was determined according to AOAC [26] and NFE was calculated by difference. *Calculated according to NRC [27].	

Table 1: Ingredients and a proximate chemical analysis of the experimental diet.

Items	Ulva (*Ulva fasciata*)	Enteromorpha (*Enteromorpha flaxusa*)
Moisture (M)	76.10	78.60
Dry matter (DM)	23.90	21.40
Crude protein (CP)	27	25.03
Ether extract (EE)	0.57	1.74
Crude fiber (CF)	9.81	4.61
Nitrogen free extract (NFE)	42.56	38.43
Ash	20.06	30.19
Gross energy (GE, Kcal/g)*	3.73	3.00
Chemical analysis was determined according to AOAC [26] and NFE was calculated by difference *Calculated according to NRC [27].		

Table 2: A proximate chemical analysis % on dry matter basis of the fresh macro algae (seaweed).

the seaweed were kept in colder and used freshly in fish feeding (Table 2). Feed was offered by hand.

The trial began at 25/7/2015 and ended 3/10/2015, (70 days). An average initial weight (W_1) of fry was 0.18 ± 0.012 g and initial average length (L_1) was 2.76 ± 0.05 cm.

Trial design and distribution of fish in ponds

The indoor ponds laboratory were made of concrete, this trial consists of twelve concrete ponds. The dimensions of each pond were 1 m length × 1 m width × 1 m height and the water volume of each pond was 0.95 m³.

This trial consists of six treatments, the first treatment, fish fed on artificial feed only. The second treatment, fish were fed of half feeding rate on artificial feed and other fresh *Ulva fasciata*, the third treatment, fish were fed of half feeding rate on artificial feed and other fresh *Enteromorpha flaxusa*, the fourth treatment, fish fed of feeding rate on fresh *Ulva fasciata* only, the fifth treatment, fish fed on fresh *Enteromorpha flaxusa* only and sixth treatment fish fed of half feeding rate on fresh *Ulva fasciata* and other fresh *Enteromorpha flaxusa*. It did not take into consideration the percentage of protein diet, but was taking the variety feed. Feeding rate was 7% of fish body weight, 40 fish were stocked in each pond (n = 80 fry for each treatment) and the feeding was twice daily. A random sample was taken every three week for change of feed quantity without any mortality for fish.

The system of running water in experimental units

The system contained on water pump, sand filter unit and two large tanks (10000 liter/tank) used to storage the water at a point between the water source (Lake Qaroun water) and experimental units. The water pump was raising the water from water source to the sand filter unit then to the large tanks and hence to experimental units.

The system of aeration in experimental units.

The system contained on Blower connected to a network of plastic pipes this pipes transport the air to each experimental unit, the air was controlled by tap of each pond or tank and the air diffusers was used to distribute of air in all experimental unit trends.

Water quality

Some water quality parameters were measured of each treatment, temperature, pH and salinity were measured daily at 1 pm by centigrade thermometer, Orion digital pH meter model 201 and Refractometer (VITAL Sine SR-6, China) respectively. Dissolved oxygen (DO) was

Items	Treatments					
	T1	T2	T3	T4	T5	T6
	Artificial Feed Only	Artificial feed with *Ulva*	Artificial feed with *Enteromorpha*	*Ulva* Only	*Enteromorpha* Only	*Ulva* with *Enteromorpha*
Temperature (ºC)	28.502 ± 0.159	28.369 ± 0.174	28.466 ± 0.162	28.488 ± 0.163	28.487 ± 0.164	28.341 ± 0.179
pH	8.418 ± 0.079	8.447 ± 0.0746	8.436 ± 0.0782	8.478 ± 0.0497	8.455 ± 0.068	8.523 ± 0.039
Salinity (‰)	32.772 ± 0.401	32.710 ± 0.421	32.622 ± 0.443	32.791 ± 0.685	32.523 ± 0.396	32.635 ± 0.415
DO (mg/L)	7.300 ± 0.129	7.425 ± 0.165	7.100 ± 0.208	6.700 ± 0.147	7.400 ± 0.168	8.000 ± 0.258
Nitrite (mg/L)	0.078 ± 0.001	0.107 ± 0.003	0.081 ± 0.002	0.082 ± 0.002	0.074 ± 0.002	0.066 ± 0.001
Nitrate (mg/L)	0.210 ± 0.002	0.143 ± 0.002	0.116 ± 0.007	0.135 ± 0.006	0.154 ± 0.004	0.136 ± 0.005
Total ammonia (mg/L)	0.418 ± 0.008	0.537 ± 0.007	0.362 ± 0.004	0.382 ± 0.002	0.741 ± 0.008	0.467 ± 0.004

Table 3: Mean (± SE) of water quality parameters.

Items	Treatments						SED*
	T1	T2	T3	T4	T5	T6	
	Artificial Feed Only	Artificial feed with *Ulva*	Artificial feed with *Enteromorpha*	*Ulva* Only	*Enteromorpha* Only	*Ulva* with *Enteromorpha*	
Initial weight (w₁), (g)	0.18	0.18	0.18	0.18	0.18	0.18	-
Final length (L₂), (cm)	6.20ᵃ	6.26ᵃ	6.40ᵃ	3.83ᶜ	3.93ᶜ	4.21ᵇ	0.090
Final weight (W₂), (g)	2.49ᵇ	2.95ᵇ	3.41ᵃ	0.73ᵈ	0.67ᵈ	0.92ᶜ	0.070
Total weight gain (TG, g)	2.31ᵇ	2.41ᵇ	3.23ᵃ	0.55ᵈ	0.49ᵈ	0.73ᶜ	0.070
Average daily gain (ADG), g/day)	0.0330ᵇ	0.0340ᵇ	0.0460ᵃ	0.0075ᶜ	0.0070ᶜ	0.0100ᶜ	0.001
Relative growth rate (RGR), (%)	1283.30ᵇ	1338.80ᵇ	1794.40ᵃ	315.55ᶜᵈ	272.2ᵈ	408.33ᶜ	41.950
Specific growth rate (SGR/day, %)	3.75ᵇ	3.80ᵇ	4.20ᵃ	1.99ᵈ	1.87ᵈ	2.31ᶜ	0.118
Survival rate (SR, %)	75ᵃ	65ᵃᵇ	55ᵇ	25ᶜ	28.75ᶜ	26.20ᶜ	5.460

(a, b, c and d) Average in the same row having different superscripts significantly different at (P ≤ 0.05).
*, SED is the standard error of difference ($\sqrt{2\,mean\,square\,of\,error\,/\,replicates}$).

Table 4: Effect of use fresh macro algae with or without artificial feed on growth performance of rabbitfish (*Siganus rivulatus*) fry.

measured every week by oxygen meter (Cole Parmer model 5946). Nitrite, nitrate, total ammonia were measured every two week by the chemical methods according to [24,25].

Measurements of growth performance and some of the internal organs

Total weight gain (TG), average daily gain (ADG), Relative growth rate (RGR), specific growth rate (SGR) and survival rate (SR).

These parameters were calculated according the following equations:

$$TG, g = Final\ Weight\ (W_2) - Initial\ Weight\ (W_1)$$

$ADG, g/day = Average\ Weight\ Gain, g\ /\ Experimental\ Period, day$

$$RGR, \% = [(W_2 - W_1)/W_1] \times 100$$

$$SGR, \%/day = [(ln\ W_2 - ln\ W_1)/t] \times 100$$

whereas ln: is the natural log. and t: is the time in days,

$SR\% = (Number\ of\ fish\ at\ end\ /\ Number\ of\ fish\ at\ start) \times 100$

Measurements of feed utilization efficiency

Feed intake g/fish (FI), feed conversion ratio (FCR), feed conversion efficiency (FCE), protein efficiency ratio (PER), protein productive value (PPV), energy efficiency ratio (EER) and energy productive value (EPV).

These parameters were calculated according the following equations:

FI g/fish feed intake during the trial period/ the final number of fish for this trial

$$FCR = Feed\ Intake, g\ /\ Weight\ Gain, g$$

$FCE\% = (Weight\ Gain, g\ /\ Feed\ Intake, g) \times 100$

$$PER = Weight\ Gain, g\ /\ Protein\ Intake, g$$

$PPV\% = (Retained\ Protein, g\ /\ Protein\ Intake, g) \times 100$

$$EER = Weight\ Gain, g\ /\ Energy\ Intake, Kcal$$

$EPV\% = (Retained\ Energy, Kcal\ /\ Energy\ Intake, Kcal) \times 100$

Chemical analysis of feeds and whole body fish

The conversional chemical analysis of diet and whole body fish samples were carried out as described by AOAC [26] and Gross energy (GE) was estimated for formulated diets the factors 5.64 Kcal/g, 9.44 Kcal/g and 4.11 Kcal/g for CP, EE and carbohydrates respectively were used [27], for fish 5.5 Kcal/g and 9.5 Kcal/g for protein and fat respectively [28].

Statistical analysis

The analysis of variance and LSD of Duncan Waller were used to compare treatment means. Data were analysed using stratigraphic package software [29] SPSS Inc. Released 2007. SPSS for Windows, Version 16.0. Level of significant was 0.05.

Results and Discussion

Water quality

Some water quality parameters recorded in this trial were shown in Table 3. The averages of water temperature, water pH, water salinity, dissolved oxygen (DO), and nitrite, nitrate and total ammonia values in all treatments were within the acceptable limits for rabbitfish (*Siganus rivulatus*) fry as reported by [2,30-37].

Growth performance

As shown in Table 4, there were significant differences between the treatments in all the growth performance parameters. The differences between the T1 (fry fed at artificial feed only), the T2 (fry fed at artificial feed with fresh *Ulva*) and the T3 (fry fed at artificial feed with fresh *Enteromorpha*) in final length (L_2) were not significant followed by the T6 (fry fed at fresh *Ulva* and fresh *Enteromorpha*). While the lowest L_2 were obtained by T4 (fry fed at fresh *Ulva* only) and T5 (fry fed at fresh *Enteromorpha* only). The T3 was the highest in W_2, TG, ADG, RGR and SGR. Followed by both the T2 and T1.

The T6 was the lower in these parameters than both the T2 and T1. But both the T4 and T5 were the lowest in all treatments in these parameters. The highest Survival rate (SR) was achieved with the T1 followed by the T2 and the T3. While insignificant differences between T4, T5 and T6.

From these results, it can be observed that, the total replacement of artificial feed with fresh macro algae had negative effect on growth performance of rabbitfish fry, however, use of the macro algae as half of the feeding rate with artificial feed had positive effect on growth performance of rabbitfish fry and reduce of the feed cost. The positive effects of macro algae on growth performance may be due to the algae are a strongly appreciated source of protein, essential amino acids [38] and vitamins [39].

As well as the positive effect of the used additive algae decrease the cholesterol and fat level and improved lipid metabolism in fish too [40]. And, Seaweeds cannot be considered as a main source of energy but they have nutritional value regarding vitamin, protein and mineral contents [41]. According to Chapman and Chapman [42], 100 g seaweed provides more than the daily requirement of Vitamin A, B1 and B12 and two thirds of Vitamin C. Also seaweeds are natural sources of hydrosoluble and liposoluble vitamins, such as thiamine and riboflavin, *b*-carotene and tocopherols, as well as of long-chain polyunsaturated essential fatty acids from the omega-3 (ω-3) (family such as eicosapentaenoic acid [43]. Moreover, Some dietary macroalge meals are improved the growth, lipid metabolism, physiological activity, stress response, disease resistance and carcass quality of various fish species [44-46]. So gut weed can be used as a direct feed or as ingredient in diets for herbivorous fish according to Teimouri et al. [47].

Over and above, rabbitfish are herbivorous fish and mainly graze on seaweeds. However, they can also feed on formulated feed or trash fishes when seaweeds are absent [48].

Due to seasonal availability of seaweeds used for feed, it is necessary to develop a formulated feed in order to promote the development of an intensive aquaculture industry for rabbitfish. Studies of stomach contents and food preference revealed that rabbitfish feed on fresh green or red algae or benthic marine plants. Studies revealed that among the many different algal species, *Gracilaria* is the preferred species of food [49].

As well as most herbivorous fishes prefer fleshy seaweeds over calcareous coralline and encrusting seaweeds [50].

In view of the economic side, the fishmeal and fish oil remain the main protein and lipid sources of feed for marine fish, and the development and application of formulated feed are restricted by the high cost of these ingredients. Therefore, efforts should focus on producing commercial feed for rabbitfish at low cost and high efficiency [51]. Such as the use of seaweeds in the development of low-cost, highly nutritive diets for animal nutrition, especially animal nutrition since sea vegetables are able to accelerate the growth of oysters, tilapia, salmon, trout, etc., all of great commercial interest [52,53]. As well as marine macro algae could be a potential low-cost source of protein for fish [54]. Moreover, the economic comparison of feed cost indicated that increasing level of fresh and dried gut weed in alternative feeding treatments, commercial feed used for fish growth was reduced leading to significantly reduction of feed cost. Compare to commercial feed, treatments [55].

Likewise, Xu et al. [51] reported that, the *Gracilaria lemaneiformis* GL is the preferred food for *S. canaliculatus*, it is logical to utilize Dried *Gracilaria lemaneiformis* DGL as a dietary ingredient for developing low-cost feeds for *S. canaliculatus*. Utilization of DGL can perhaps partially replace protein from fish meal in feeds and reduce fishmeal inclusion, and secondly, the high level of carbohydrates in GL can reduce the amount of supplemental starch in formulated feeds and minimize competition with human food sources. Furthermore, application of GL in feed can promote the coordinated development of aquaculture industries for GL and rabbitfish and simultaneously increase incomes of fisherman. Previous studies [56] investigating inclusion of other plant ingredients in the diets of *S. canaliculatus* showed different results, not influencing growth.

In the light of the results were shown in Table 4 it can be observed that, the (T3) fish fed at artificial feed with fresh *Enteromorpha* was better in growth performance than the (T2) fish fed at artificial feed with fresh *Ulva*, this may be due to the fresh *Ulva* contained the higher percentage of fiber than fresh *Enteromorpha* as shown in Table 2 this was supported by Leary and Lovell [57] high levels of fiber in the diets of many finfish species, have been shown to reduce growth. For tilapia, Anderson et al. [58] concluded that dietary fiber levels above 5% reduce food utilization and digestibility and protein utilization were reduced with excess fiber levels in the diets. As well as Ortiz et al. [18] reported that *U. lactuca* contains about 60% fiber this might reduce its value in aqua feeds. Early studies of stomach contents and food preference revealed that among the many different algal species and vascular plants eaten, the presence of *Enteromorpha* was high and was the preferred species. This preference for *Enteromorpha* by siganids is not directly related to the calorific value of the algae but is related to the texture of its thallic which is crispy and thin [59,60]. Moreover, the results confirmed that, the replacement at 50% of artificial diet with fresh macro algal (*Enteromorpha* or *Ulva*) from rabbitfish fry feed led to an increase in the growth performance parameters compared with 100% artificial diet. But the total replacement of artificial feed with fresh macro algal (*Enteromorpha* or *Ulva*) led to a decrease in the growth performance parameters, because protein percentage of the artificial feed was high level (35% CP) in addition to artificial feed contained on fish meal (FM) which is high protein content and balanced essential amino acids EAA profile. Fish meal FM is also an excellent source of essential fatty acids (EFA), digestible energy, minerals and vitamins and it is well known as being highly palatable and digestible to fish [56,61]. The T6 (fish fed at *Ulva* with *Enteromorpha*) was the better in the growth performance and survival rate than T5 or T4, this may be due to both the *Ulva* with *Enteromorpha* has more a balanced amino acid composition than *Ulva* alone or *Enteromorpha* alone, in addition to the varying and mixture between *Ulva* and *Enteromorpha* resulted in highly palatable and digestible to fish.

In general, these results are in agreement with Costa et al. [62]

Items	Treatments						SED*
	T1	T2	T3	T4	T5	T6	
	Artificial Feed Only	Artificial feed with *Ulva*	Artificial feed with *Enteromorpha*	*Ulva* Only	*Enteromorpha* Only	*Ulva* with *Enteromorpha*	
Feed intake (FI), g/ fish	3.54[f]	4.05[e]	4.70[c]	6.1[a]	4.34[d]	5.04[b]	0. 083
Feed conversion ratio (FCR)	1.53[c]	1.68[c]	1.45[c]	11.22[a]	8.85[b]	7.74[b]	0.752
Feed conversion efficiency (FCE, %)	65.25[b]	59.50[c]	68.72[a]	9.01[e]	11.29[e]	14.55[d]	1.260
Protein utilization							
Protein efficiency ratio (PER)	1.79[c]	1.87[b]	2.23[a]	0.33[f]	0.46[e]	0.56[d]	0.0116
Protein productive value (PPV, %)	86.32[c]	97.29[b]	107.56[a]	20.46[f]	25.32[e]	32.34[d]	0.0141
Energy utilization							
Energy efficiency ratio (EER, g/ Kcal)	0.128[b]	0.134[b]	0.170[a]	0.24[d]	0.38[cd]	0. 43[c]	0.0059
Energy productive value (EPV, %)	77.44[b]	75.32[c]	94.27[a]	12.86[f]	18.94[e]	22.30[d]	0.0141
(a, b, c, d, e and f) Average in the same row having different superscripts significantly different at (P ≤ 0.05).							
*, SED is the standard error of difference $(\sqrt{2\,mean\,square\,of\,error\,/\,replicates})$.							

Table 5: Effect of use fresh macro algae with or without artificial feed on feed utilization efficiency of rabbitfish (*Siganus rivulatus*) fry.

reported that fresh and dried gut weed can be used as a feed to substitute commercial feed for herbivorous fish such as spotted scat, (*Scatophagus argus*), red tilapia (*Oreochromis* sp.) and giant gourami (*Osphronemus goramy*) juveniles.

In the same trend, Siddik et al. [55] found that, the lowest final body weight and SGR were observed in treatments feeding fresh and dried gut weed as single feeds.

Moreover, herbivorous fish like tilapia generally accept plant originated ingredients better than animal originated ingredients in their diet. For example, Swain and Padhi [63] reported that tilapia grew better fed diet replacing 75% fish meal with okra meal (a by-product of soybean meal) than fed diet containing 100% fish meal. Moreover, the rabbitfish were fed a control diet with addition of a known weight of fresh *Enteromorpha* placed in plastic baskets at the bottom of the rearing tanks were the higher in final body weight and SGR than fish fed at the control diet or the other treatments [56].

The results were in partial disagree with El-Tawil [64] who reported that, specific growth rate of red tilapia (*Oreochromis* sp) improved significantly (P<0.05) with increasing *Ulva* level in the diet up to 15%. Increasing *Ulva* level beyond 15% had no significant effects on growth. Elmorshedy [65] showed that final body weight, weight gain and specific growth rate of gray mullet *Liza ramada* 0.094 g initial body weight were increased significantly with increasing seaweeds level (*Ulva sp.*) up to 28% in the fish diet. And Xu et al. [51] recommend a level of less than 33% DGL in the diet for *S. canaliculatus* and the optimum level require to be investigated in future studies.

On the other hand, carnivorous fish, such as African catfish, would tend to prefer diets with animal ingredients rather than plant feedstuff, and such a conclusion is in accordance with our results showing a decrease in feed intake by fish at the levels 20% and 30% of *U. lactuca* in their diets was increased [66].

The highest survival rate was achieved by the T1 (fish fed at the artificial diet only) and decreased with T2 and T3 while the lowest survival rate was achieved by the T4, T5 and T6. Hence, it can be said that the total replacement or partial of artificial diet with macro algae negatively affected on survival rate of rabbitfish fry this may be due to the important role of fish meal. Whereas, fishmeal has a balanced amino acid composition [56].

But Siddik et al. [55] found that, the equal survival (P>0.05) of tilapia juvenile in all dietary treatments was in agreement with the study of Rahman and Meyer [67] who observed similar survival of fish fed diet with seaweed and without seaweed.

Feed utilization efficiency

As shown in Table 5 there were significant differences between the treatments in all the feed utilization parameters the highest feed intake (FI, g/fish) was achieved by T4 followed by T6, T3, T5 and T2 respectively. While the lowest FI achieved by T1. The best feed conversion ratio (FCR) was achieved with T1, T2 and T3 followed by T6 and T5 but the T4 had the worst FCR. The highest protein efficiency ratio (EPR) and protein productive value (PPV) were achieved with the T3 Followed by T2, T1, T6, T5 and T4 respectively. In the same trend was energy efficiency ratio (EER), whereas the highest EER was obtained by T3 followed by T2, T1, T6, T5 and T4 respectively. The highest energy productive value (EPV) was obtained by T3 followed by T1, T2, T6, T5 and T4 respectively.

From these results, it can be observed that, the T3 (fish fed at the artificial diet with fresh *Enteromorpha*) was the highest in FCE, PER, PPV, EER and EPV. The best FCR was achieved by T3, T1 and T2 whereas the statistical analysis disappeared any significant difference between these the treatments. This confirmed that, the total replacement of artificial diet by fresh macro algae negatively affected on the feed utilization of rabbitfish fry while the replacement of artificial diet at 50% of feeding rate by fresh macro algae did not only negatively effect on the feed utilization of rabbitfish fry but also the replacement of artificial diet at 50% of feeding rate by fresh macro algae excelled in the feed utilization parameters compared with the use of the artificial diet only in particular the replacement of artificial diet at 50% of feeding rate by *Enteromorpha*.

This may be returned to macro algae improved rabbitfish fry palatable and digestible at the artificial feed as mentioned by Yone et al. [68] who interpreted the effect on growth as due to an acceleration of nutrient absorption of dietary algae. These results are in agreement with Yousif et al. [56] who found that, the fish was fed a control diet with addition of a known weight of fresh *Enteromorpha* placed in plastic baskets at the bottom of the rearing tanks was the best in FCR, PER and PPV than the other treatments, also they added that, Protein productive value (PPV) affected significantly (P<0.05) by different levels of *Ulva sp*. It increased significantly with increasing *Ulva* level in the diet Siddik et al. [55] who found that, tilapia fed alternative1 day

Items	Start	Treatments						SED*
		T1 Artificial Feed Only	**T2** Artificial feed with *Ulva*	**T3** Artificial feed with *Enteromorpha*	**T4** *Ulva* Only	**T5** *Enteromorpha* Only	**T6** *Ulva* with *Enteromorpha*	
Moisture (M, %)	80.70	70[c]	73.83[abc]	72.71[bc]	80.43[a]	79.71[ab]	80.20[ab]	2.88
Dry matter (DM, %)	19.30	30[a]	26.17[abc]	27.29[ab]	19.57[c]	20.29[bc]	19.80[c]	2.88
Crude protein (CP, %)	50.17	48.35[b]	51.55[ab]	48.07[b]	58.63[a]	54.54[ab]	56.03[ab]	3.48
Ether extract (EE, %)	9.75	33.84[a]	27.50[b]	29.48[ab]	17.29[c]	17.63[c]	18.44[c]	2.27
Ash, %	34.57	13.35[b]	15.26[b]	18.04[ab]	20.36[ab]	18.44[ab]	27.35[a]	4.08
Gross energy (GE, Kcal/g)	3.68	5.87[a]	5.45[ab]	5.44[ab]	4.86[c]	4.67[c]	4.84[c]	0.32
(a, b and c) Average in the same row having different superscripts significantly different at (P ≤ 0.05). *, SED is the standard error of difference $(\sqrt{2\,mean\,square\,of\,error\,/\,replicates})$.								

Table 6: Effect of use fresh macro algae with or without artificial feed on whole body chemical composition and energy content of rabbitfish (*Siganus rivulatus*) fry.

commercial feed and 1 consecutive day fresh or dried gut weed showed similar feed utilization to tilapia feed the commercial feed.

These results clearly indicated that gut weed can be used 1 day after using 1 day commercial feed without affecting feed utilization of tilapia. Also these results are partial in agreement with El-Tawil [64] reported that, supplementation of *Ulva sp.* to the prepared fish diet had a positive effect on FCR except fish fed the diet containing 25% *Ulva* level with the poorest FCR value. As well as, Diler et al. [69] stated that PPV improved significantly with increasing dietary *Ulva* inclusion rate up to 15%. Also, similar results were found by Elmorshedy [65] with gray mullet.

On the other hand, Abdel-Warith et al. [66] reported that, a decrease in feed intake by African catfish at the levels 20% and 30% of *U. lactuca* in their diets was improved FCR and added that, the fish fed the control diet displaying a superior PER, Protein productive values (PPV%) values also showed a decrease when fishmeal was replaced by the *U. lactuca* meal source. Also, Ergün et al. [44] suggested that low-level inclusion of *Ulva* meal can significantly improve growth performance and nutrient utilization of tilapia fed high-lipid diets.

Body chemical composition and energy content

Whole body chemical composition and energy content of rabbitfish fry (*siganus rivulatus*) at the beginning and the end of the experimental period are shown in Table 6. There were significant differences between the treatments at the end of the experimental period in moisture (M, %), crude protein (CP, %), ether extract (EE, %), ash (%) and gross energy (GE, kcal/g).

The highest moisture (M, %) was achieved by T4 followed by T6, T5, T2 and T3 but the lowest (M) was achieved by T1. the highest value of (CP) was obtained by the T4 and insignificant differences between T6, T5 and T2 while the lowest value of (CP) was obtained by T1 and T3. The highest value of (EE) was achieved by the T1 and T3 followed by T2 while there are not significant differences in EE between the T6, T5 and T4. The highest ash was achieved with the T6 and there are not significant differences in ash between the T5, T4 and T3 while the lowest ash was achieved T1 and T2. The highest value of GE was recorded by T1 followed T2 and T3 while there were insignificant differences in ash between the T4, T6 and T5.

It can be said that, an increase of EE and GE level of whole body rabbitfish fry at the end of the experimental period with the fish fed at the artificial feed or the fish fed at the artificial feed with fresh macro algae, may be due to the artificial feed contained on high level of EE (13.78%) this resulted in increment of lipid level in body fish, this are in agreement with Siddik et al. [55] who found that, the fish which fed at the commercial feed had the highest EE followed by the fish which fed at the fresh and dried gut weed with the commercial feed while the

lowest EE content was noticed with the fish were fed at fresh or dried gut weed alone.

The highest (M, %) was achieved by the fish were fed at fresh macro algae but the lowest (M, %) was achieved by the fish were fed at the artificial feed, this was agree with Siddik et al. [55] found that, The highest moisture content in tilapia carcass was observed in treatments fresh gut weed and dried gut weed while lowest was observed in treatments commercial feed.

While the highest ash was achieved by the fish fed at the diet without artificial feed (fresh macro algae only). This may be due to the fresh macro algae contained on high level of ash content in *Ulva* (20.06%) and *Enteromorpha* (30.19%) this resulted in increment of lipid level in body fish.

Conclusion

The total replacement of artificial feed with fresh macro algae had negative effect on growth performance of rabbitfish fry; however, use of the macro algae as half of the feeding rate with artificial feed had positive effect on growth performance of rabbitfish fry and reduce of the feed cost. Whereas, the T3 was the highest in the final weight (W_2) total weight gain (TG), average daily gain (ADG), relative growth rate (RGR) and specific growth rate (SGR). Followed by both the T2 and T1. Moreover, There are significant differences between the treatments in all the feed utilization parameters. The best feed conversion ratio (FCR) was achieved with T1, T2 and T3 followed by T6 and T5 but the T4 had the worst FCR.

References

1. Lam TJ (1974) Siganids: Their biology and mariculture potential. Aquacult 3: 325-354.

2. Saoud IP, Mohanna C, Ghanawi J (2008) Effects of temperature on survival and growth of juvenile spinefoot rabbitfish (*Siganus rivulatus*). Aquacult Res 39: 491-497.

3. Woodland DJ (1990) Revision of the fish family Siganidae with descriptions of two new species and comments on distribution and biology. Indo-Pacific Fishes, No. 19 Bernice Pauahi Bishop Museum, Honolulu, Hawaii.

4. Woodland DJ (1983) Zoogeography of the Siganidae (Pisces) and interpretation of distribution and richness pattern. Bulletin of Mar Sci 33: 713-717.

5. Duray MN (1990) Biology and culture of siganids. Aquaculture Department, Southeast Asian Fisheries Development Center, Tigbauan, Iloilo, Philippines.

6. Choat JH (1991) The biology of herbivorous fishes on coral reefs. The ecology of fishes on coral reefs. Academic Press, New York, USA. pp. 120-155.

7. Bariche M (2002) Biologie et Ecologie de deux especes lessepsiennes (*Siganus rivulatus* et *Siganus luridus*) Siganidae sur les cotes du Liban. These de Doctorat de l'Universite d'Aix Marseille II.

8. GAFRD (2014) Fish statistics year book. (24th edn), General Authority for Fish Resource Development, Agriculture ministry, Egypt.

9. Luning K (1990) Seaweeds, their environment, biogeography and eco-physiology. Willey Inter-science Publications 3; 370.

10. Santhanam RN, Remananthan N, Jagathusan G (1990) Coastal aquaculture in India. CBS Publishers & distributors pp.159-162.

11. Wong KH, Cheung PC (2000) Nutritional evaluation of some subtropical red and green seaweeds I. proximate composition, amino acid profiles and some physico-chemical properties. Food Chem 71: 475-482.

12. Marsham S, Scott GW, Tobin ML (2007) Comparison of nutritive chemistry of a range of temperate seaweeds. Food Chem 100: 1331-1336.

13. Fleurence J (1999) Seaweed proteins: biochemical, nutritional aspects and potential uses. Trends in Food Sci Technol 10: 25-28.

14. Satpati GG, Pal R (2011) Biochemical composition and lipid characterization of marine green alga Ulva rigida- a nutritional approach. J Algal Biomass Utln 2: 10-13.

15. Mine I, Menzel D, Okuda K (2008) Morphogenesis in giant-celled algae. Int. Review of Cell and Molecular Biol 266: 37-83.

16. Kong F, Mao Y, Cui F, Zhang X, Gao Z (2011) Morphology and molecular identification of Ulva forming green tides in Qingdao China. J Ocean China Uni 10: 73-79.

17. Wolf MA, Sciuto K, Andreoli C, Moro I (2012) Ulva (Chlorophyta, Ulvales) Biodiversity in the North Adriatic Sea (Mediterranean Italy), Cryptic species and new introductions. J Phycol 48: 1510-1521.

18. Ortiz J, Romero N, Robert P, Araya J, Lopez-Herna´ndez J, et al. (2006) Dietary fiber, amino acid, fatty acid and tocopherol contents of the edible seaweeds (Ulva lactuca) and (Durvillaea antarctica). Food Chem 99: 98-104.

19. Garcia-Casal MN, Pereira AC, Leets I, Ramirez J, Quiroga ME (2007) High iron content and bioavailability in humans from four species of marine algae. J Nutr 137: 2691-2695.

20. Manimala K, Rengasamy R (1993) Effect of bioactive compounds of seaweeds on the Phytopathogen (Xanthomonas oryzae). Phykos 32: 77-83.

21. Hoppe HA (1966) Nahrungsmittel aus Meeresalgen. Bot Mar 9: 18-40.

22. Tamura T (1970) Marine aquaculture, Nat Sci Foundation PB 194 OSIT, Part II, Washington, D.C.

23. Tolentino-Pablico G, Bailly N, Froese R, Elloran C (2008) Seaweeds preferred by herbivorous fishes. J Appl Phycol 20: 933-938.

24. Mullin JB, Riley JP (1955) The spectrophotometric determination of nitrate in natural waters, with particular references to see water. Analytica Chemica Acta 12: 464-480.

25. APHA (1992) Standard methods for the examination of water and waste water. (18th edn) Amer Public Health Association, Washington DC.

26. AOAC (1984) Official Methods of Analysis. Association of Official Analytic Chemists, Inc. Arlington, Virginia, USA.

27. NRC (1993) National Research Council, Nutrient requirements of fish. National Academy Press, Washington, D.C, USA.

28. Viola S, Malady S, Rappaport U (1981) Partial and complete replacement of fish meal by soybean meal in feeds for Intensive culture of carp. Aquacult 26: 223-236.

29. SPSS (2007) Statistical package for social science (for Windows). Release 16 Copyright ©, SPSS Inc., Chicago, USA.

30. Westernhagen HM, Rosenthal H (1975) Rearing and spawning siganids (Pisces: Teleostei) in a closed sea water system. Helgol Wiss Meeresunters 27: 1-18.

31. Huguenin JE, Colt J (1989) Design and operating guide for aquaculture seawater systems. Dev Aquacult Fish Sci 20: 264.

32. Meade JW (1989) Aquaculture management. New Branch. York: Van Nostrand Reinhold, NY.

33. Davis J (1993) Survey of aquaculture effluents permitting and standards in the South. Southern Regional Aquaculture Centre, SRAC publications, USA.

34. Lawson TB (1995) Fundamentals of Aquacultura Engineering. New York: Chapman & Hall, USA.

35. ANZECC (2000) (Australian and New Zealand Environment and Conservation Council) and ARMCANZ (Agriculture and Resource Management Council of Australia and New Zealand), Australian Guidelines for Water Quality Monitoring and Reporting. National Water Quality Management Strategy Paper No. 7, ANZECC and ARMCANZ, Canberra.

36. EPA (2003) The Environment Protection (Water Quality) Policyían overview, both the overview and a copy of the Water Quality Policy with an accompanying explanatory report are available on the EPA.

37. Saoud IP, Kreydiyyeh S, Chalfoun A, Fakih M (2007) Influence of salinity on survival, growth, plasma osmolality and gill Na-K-ATPase activity in the rabbitfish (Siganus rivulatus). J Exp Mar Biol Ecol 348: 183-190.

38. Becker E (1994) Microalgae biotechnology and microbiology. Cambridge University Press, Cambridge, Great Britain.

39. Becker E (2004) Microalgae in human and animal nutrition. Handbook of microalgal culture, Blackwell, Oxford, UK.

40. Sirakov I, Velichkova K, Nikolov G (2012) The effect of algae meal (Spirulina) on the growth performance and carcass parameters of rainbow trout (Oncorhynchus mykiss). J Biol Sci Biotechnol SE/Online: pp. 151-156.

41. Norziah MH, Ching CHY (2000) Nutritional composition of edible seaweeds (Gracilaria changgi). Food Chem 68: 69-76.

42. Chapman VJ, Chapman DJ (1980) Seaweeds and their uses. Chapman and Hall, London and New York, USA.

43. Khotimchenko SV, Vaskovsky VE, Titlyanova TV (2002) Fatty acids of marine algae from the Pacific coast of north California. Botanica Mar 45: 17-22.

44. Ergün S, Soyuturk M, Guroy B, Guroy D, Merrifield D (2009) Influence of Ulva meal on growth, feed utilization, and body composition of juvenile Nile tilapia (Oreochromis niloticus) at two levels of dietary lipid. Aquacult Int 17: 355-361.

45. Güroy D, Güroy B, Merrifield DL, Ergun S, Tekinay AA, et al. (2011) Effect of dietary Ulva and Spirulina on weight loss and body composition of rainbow trout, Oncorhynchus mykiss (Walbaum), during a starvation period. J Anim Physiol & Anim Nutr 95: 320-327.

46. Güroy B, Ergün S, Merrifield DL, Güroy D (2013) Effect of autoclaved Ulva meal on growth performance, nutrient utilization and fatty acid profile of rainbow trout (Oncorhynchus mykiss). Aquacult Int 21: 605-615.

47. Teimouri M, Amirkolaie AK, Yeganeh S (2013) Effect of Spirulina platensis meal as a feed supplement on growth performance and pigmentation of rainbow trout (Oncorhynchus mykiss). World J Fish and Mar Sci 5: 194-202.

48. Li YY, Hu CB, Zheng YJ, Xia XA, Xu WJ, et al. (2008) The effects of dietary fatty acids on liver fatty acid composition and D6-desaturase expression differ with ambient salinities in (Siganus canaliculatus). Comp Biochem Physiol 151: 183-190.

49. Jackson AJ, Capper BS, Matty AJ (1982) Evaluation of some plant proteins in complete diets for tilapia (Sarotherodon mossumbicus). Aquacult 27: 97-109.

50. Ojeda FP, Muñoz AA (1999) Feeding selectivity of the herbivorous fish Scartichthys viridis: Effects on macroalgal community structure in a temperate rocky intertidal coastal zone. Mar Ecol Prog Ser 184: 219-229.

51. Xu S, Zhang L, Wu Q, Liu X, Wang S, et al. (2011) Evaluation of dried seaweed Gracilaria lemaneiformis as an ingredient in diets for teleost fish (Siganus canaliculatus). Aquacult Int 19: 1007-1018.

52. Horn M (1989) Biology of marine herbivorous fishes. Oceano Mar Biol Annu Rev 27: 167-272.

53. Fleming AE, Barneveld RJ, Hone PW (1996) The development of artificial diet for abalone. Aquacult 140: 5-53.

54. Vinoj KV, Kaladharan P (2007) Amino acids in the seaweeds as an alternate source of protein for animal feed. J Mar Biol Ass Indian 49: 35-40.

55. Siddik MAB, Nahar A, Rahman MM (2014) Bossier gut weed, Enteromorpha sp. as a partial replacement for commercial feed in Nile Tilapia (Oreochromis niloticus). Culture World J Fish Mar Sci 6: 267-274.

56. Yousif OM, Osman MF, Anwahi AR, Zarouni MA, Cherian T (2004) Growth response and carcass composition of rabbitfish (Siganus canaliculatus Park) fed diets supplemented with dehydrated seaweed (Enteromorpha sp.). Emir J Agric Sci 16: 18-26.

57. Leary DF, Lovell RT (1975) Value of fiber in production type diets for channel catfish. Trans Amer Fish Soc 104: 328-332.

58. Anderson J, Jackson AJ, Matty AJ, Capper BS (1984) Effects of dietary carbohydrate and fiber on tilapia (*Oreochromis niloticus* Linn.). Aquacult 37: 303-314.

59. Von Westernhagen H (1973) The natural food of the rabbitfish (*Siganus oramin* and *S. striolata*). Mar Biol 22: 367-370.

60. Von Westernhagen H (1974) Food preference in cultured rabbitfishes (Siganidae). Aquacult 3: 109-117.

61. El-Sayed AM (2006) Tilapia culture. CAB International, Wallingford, UK.

62. Costa MM, Oliveira STL, Balen RE, Bueno JG, Baldan LT, et al. (2013) Brown seaweed meal to Nile Tilapia fingerlings. Arch Zootec 62: 101-109.

63. Swain PK, Padhi SB (2011) Utilization of seaweeds as fish feed in aquacult. J Biol Sci 2: 35-46.

64. El-Tawil NE (2010) Effects of green seaweeds (*Ulva sp.*) as feed supplements in Red Tilapia (*Oreochromis sp.*) diet on growth performance, feed utilization and body composition. J Arabian Aquacult Soc 5: 179-193.

65. Elmorshedy I (2010) Using of algae and seaweeds in the diets of marine fish larvae. Fac Agri Saba Bacha, Alexandria University, Egypt.

66. Abdel-Warith AWA, Younis EMI, Al-Asgah NA (2015) Potential use of green macroalgae (*Ulva lactuca*) as a feed supplement in diets on growth performance, feed utilization and body composition of the African catfish (*Clarias gariepinus*). Saudi J Biol Sci 23: 404-409.

67. Rahman MM, Meyer CG (2009) Effects of food type on diel behaviours of common carp (*Cyprinus carpio* L.) in simulated aquaculture pond conditions. J Fish Biol 74: 2269-2278.

68. Yone Y, Furuichi M, Urano K (1986) Effects of wakame *Undaria pinnatifida* and *Ascophyllum nodosum* on absorption of dietary nutrients, and blood sugar and plasma free amino-N levels of red sea bream. Nippon Suisan Gakkaishi 52: 1817-1819.

69. Diler IA, Tekinay A, Guroy D, Guroy BK, Soyuturk M (2007) Effects of (*Ulva rigida*) on the growth, feed intake and body composition of common carp (*Cyprinus carpio* L). J Biol Sci 7: 305-308.

20

Effects of Fish Meal Substitution with Poultry By-product Meal on Growth Performance, Nutrients Utilization and Blood Contents of Juvenile Nile Tilapia (*Oreochromis niloticus*)

Yones AMM* and Metwalli AA

National Institute of Oceanography and Fisheries, Shakshouk Fish Research Station, El-Fayoum, Egypt

Abstract

Four isonitrogenous and isocaloric diets (30.22 ± 0.02% CP and 19.007 ± 0.015 MJ kg^{-1} diet) were formulated to represent four dietary treatments. The first treatment (control) without poultry by-product (PM0), while the second, third and fourth diets formulated with substitution of fish meal by poultry by-product meal as 50, 75 and 100%, respectively. Each diet was fed to 100 juvenile tilapia (1.5 ± 0.05 g), in triplicate cement ponds of 2 m^3. Fish were fed diets at a rate of 3% of its biomass daily divided into two equal portions. The highest ($P<0.05$) growth performance parameters (finial weight, weight gain, daily gain and specific growth rate) and best nutrient utilization (feed conversion ratio, protein efficiency ratio and net protein utilization) were recorded with PBM0 and PBM100% groups. The applied treatments showed insignificant effects on nutrients digestibility coefficient among dietary groups for Dry matter, Energy, CP, Fat and nitrogen free extract. No significant effects on dry matter, crude protein, fat and ash contents were recorded between dietary treatments. Juvenile tilapia fed the experimental diets showed insignificant different in blood contents ($P<0.05$) between groups. The present study recommended substitution of 100% fish meal with poultry by-product meal in juvenile Nile tilapia diets.

Keywords: *Oreochromis niloticus*; Poultry by-product meal; Digestibility coefficient; Growth performance; Blood contents

Introduction

The global production of tilapia drastically increased from 124 thousand metric tons (Mt) in 1997 to 2.5 million Mt in 2010 [1]. This trend suggests that there will be even greater increases in the future. Among the cichlid species, it is the Nile tilapia (*Oreochromis niloticus*) that has dominated in different culture system. The tilapia market has expanded from a subsistence level to meet the protein needs of the middle class because of the year-round supply, delicious flavour and reasonable price of that fish [2]. Golbal tilapia production was recorded 3.500.000 metric tons in 2011, 3.8000.000 metric ton in 2012,4.850.000 metric ton in 2014 and by 2015, world production tilapia is forecast to reach 4.6-5.0 million metric ton [3,4].

Traditionally, fish meal has provided a major part of protein sources in formulated feeds because of its suitable protein quality. Science the recent scarcity and uncertain consistency of supply encourage its replacement by alternative protein sources that are of high quality, but less expensive has been investigated in many fish species. The limitations on the world's food supply provide additional motivation [5,6]. Therefore, numerous studies have undertaken to examine the effects of replacing fish meal by another source of protein such as animal by-product or plant-based protein in diets that can be fed to tilapia [7,8].

Animal by-products such as meat meal, bone meal and poultry by-product meal have considerable potential as feed ingredients in fish production system [9-12] and comparatively less expensive than fish meal [13]. These animal protein ingredients are good sources of amino acids with high protein content, total digestible dry matter and energy similar to fish meal [9]. Therefore, poultry by-product meal is considered a probable replacement for fish meal [14-21]. Many studies have also shown that animal protein ingredients can be useful for fish feed formulation and they are comparatively much less expensive than fish meal [13,22,23].

Some studies have shown that poultry by-product meal cannot replace more than 50% of fish meal in fish diets [24], but other studies have shown that with the recent improvement of the quality of poultry by-product meal it could replace 75% or 100% of fish meal without significant decrease in fish growth [25-28].

The present study aims to evaluate the effects of use poultry by-product meal as a alternative protein source to replace fish meal on growth performance, nutrient utilization, digestibility coefficient and some blood constituents of Nile tilapia (*Oreochromis niloticus*) reared in cement tanks.

Materials and Methods

Fish culture and experimental diets

The present study was conducted using the research facilities of the experimental station at Shakshouk, Fayoum Governorate, National Institute of Oceanography and Fisheries (NIOF). The system contained two water pumps and upstream sandy filter units at a point between the water source and tanks. Each pump was drowning the water from the lake Qaroun to collection cement pond and forced it through storage units and then to the rearing tanks in open system. Physicochemical characteristics of water tanks were examined every week, (Table 1) according to APHA [29].

*Corresponding author: Yones AMM, National Institute of Oceanography and Fisheries, Shakshouk Fish Research Station, El -Fayoum, Egypt
E-mail: yones_552000@yahoo.com

Diets				
Parameters	PBM0	PBM50	PBM75	PBM100
Temperature °C	27.0 ± 0.12	27.1 ± 0.12	27.4 ± 0.1	27.1 ± 0.15
pH	7.6 ± 0.1	7.8 ± 0.12	7.8 ± 0.11	7.6 ± 0.1
Dissolved oxygen (mg l⁻¹)	6.3 ± 0.11	6.2 ± 0.12	6.2 ± 0.11	6.1 ± 0.12
Salinity (g l⁻¹)	2.22 ± 0.1	2.23 ± 0.1	2.22 ± 0.1	2.22 ± 0.1
Unionized ammonia (mg l⁻¹)	0.03 ± 0.01	0.028 ± 0.01	0.026 ± 0.002	0.028 ± 0.001

Table 1: Averages of water physicochemical characteristics parameters during experimental period.

Ingredientes	DM	CP	EE	NFE	CF	ASH
Fish meal	916	700	128	-	-	172
Poultry by-product meal	921	560	134	128	24	154
Soybean meal	915	480	44	368	42	66
Gluten meal	906	350	48	532	22	48
Wheat bran	904	144	34	666	92	68
Yellow maize	896	90	24	760	72	54

DM: Dry Matter, CP: Crude Protein, EE: Ether Extract, NFE: Nitrogen Free Extract, CF: Crude Fiber.

Table 2: Proximate composition of feed ingredients (g kg⁻¹ d.m) n=3.

The fry of Nile tilapia (*Oreochromis niloticus*) used in the present study were obtained after brood stock hatching in the station. The fry were acclimatized for two weeks in rearing tanks and fed on prepared powder diet contain 30% crude protein, formulated from the same ingredients use in the growth trial. Juvenile tilapia with an initial average weight of 1.5 ± 0.05 g were randomly distributed and stocked at 100 juvenile per tank in 12 cement tanks, each with a water volume of (2 m³) and the treatments were performed in triplicates. The diets were given at 3% of live body weight (BW) and offered in two equal portions at 10.00 a.m and 16.00 p.m. The experiment lasted 120 days after start.

Four isonitrogenous diets were formulated to contain an average of 30.11 ± 0.07% crude protein for meeting the recommended nutritional requirements of tilapia [30]. The ingredients were obtained from Zoocontrol fish Co at 6 October city, Egypt. Ingredients, diets formulation and chemical composition analysis are presented in Tables 2 and 3. The first diet was formulated without poultry by-product meal and considered as a control diet (PM0), Diets 2 (PM50), 3 (PM75) and 4 (PM100) were formulated to be comprised with partial and total inclusion levels of 50, 75 and 100% poultry by-product meal, respectively. All diets were processed into dry sinking pellet form, using California pelleting machine with 1 mm diameter.

Apparent digestibility coefficient

The experimental test diets with addition of 0.5% chromic oxide (Cr_2O_3) were fed to fishes after the growth trial lasted for a period of two week in order to study the apparent digestibility coefficient (ADC %) of nutrients. Any uneaten and fecal residues were siphoned out from the tank bottom after two hours of first feeding (10.00 a.m.) and discarded. Fish fecal samples were collected every afternoon before the second feeding, new fecal materials were carefully siphoned and collected using the filtration system developed by Choubert et al. [31]. After freeze-drying of 20 g fecal samples in each replicate tank, the feces were analyzed. Dry matter was calculated by gravimetric analysis at 105°C for 24 hrs. Chromic oxide levels were determined spectrometry (Spectra, AA220FSNarin) based on the method described by Bolin et al. [32].

The apparent digestibility coefficients ADC for test diets were calculated according to the equation described by Cho [33].

$$ADC\ (n) = 100 - \{100\ (\%Cr_2O_3d)/\%Cr_2O_3f) \times (\%\ Nf/\%\ Nd)\}.$$

Where ADC (n)=apparent digestibility coefficients of a nutrient in the test diets; Cr_2O_3d=% chromic oxide of the diet; Cr_2O_3f=% chromic oxide of the feces; Nd=nutrient in the test diet; Nf=nutrients in feces.

Chemical analysis

The chemical composition of the experimental diets, feces and whole fish samples was performed via proximate composition analysis according to standard methods [34]. Briefly dry matter was determined gravimetrically in an oven dried samples at 105°C for 24h, protein (N×6.25) content was determined using Kjeldhal method and crude fat by chloroform-methanol extraction (2:1) using Soxhlet system. Ash was determined by incinerating samples in a Germany muffle furnace at 550°C for 18h. Nitrogen free extract (NFE) was calculated by the difference.

Diets samples were hydrolyzed in 6 N (HCL) at 106°C over 24 h in nitrogen-flushed glass vials before amino acid analysis. Total amino acids were analysed by high pressure liquid chromatography (HPLC) in a Pico-Tag amino acid analysis system (Water, Bedford, MA, (USA), using norleucine as internal standard and according to the procedure described by Cohen et al. [35].

Gross energy (MJ Kg⁻¹ diet) was calculated according to Schulz et al. [36] using the following calorific values: 23.9, 39.8 and 17.6 MJ g⁻¹ diet for protein, ether extract and nitrogen free extract, respectively. The metabolizable energy contents of the experimental diets were calculated as 18.9, 35.7 and 14.7 MJ g⁻¹ diet for protein, lipid and nitrogen free extract, respectively according to Jobling [37].

Diets				
Ingredients	PBM0	PBM50	PBM75	PBM100
Fish meal	200	100	50	-
Poultry by-product meal	-	150	200	260
Soybean meal	150	130	150	150
Corn gluten meal	110	100	100	100
Wheat brain	200	200	200	2100
Yellow corn maize	260	240	220	200
Fish oil	30	30	30	30
Sunflower oil	30	30	30	30
Vitamin and mineral mix¹	15	15	15	15
Chromic oxide	5	5	5	5
Proximate analysis (*n*=3)				
Dry matter	928	922	924	926
Crude protein	301.9	301	301.8	300
Ether extract	110.4	115.9	116.6	118.8
Nitrogen free extract	453.9	451.3	449.2	447.8
Crude fiber	45.8	46.8	50.4	49.4
Crude ash	88	85	82	84
Gross energy (MJ kg⁻¹ diet)²	19.58	19.74	19.75	19.77
ME (MJ kg⁻¹diet)³	16.31	16.44	16.46	16.49

¹Vitamin-mineral premix, mg Kg⁻¹ dry diets: vitamin A (as acetate), 7500 Iu kg⁻¹ dry diet, Vitamin D3 (as cholecalcipherol); 6000IU kg⁻¹ dry diet, vitamine E (as DL-L-tocopheryl-acetate); 150 IU kg⁻¹ dry diet, vitamin k (as menadione Na-bisulphate); 0.06 ascorbic acid (as ascorbyle polyphosphate), 150 D-biotin, 240 choline (as chloride) 3000; folic acid, 3 niacin (as nicotinic acid), 30 pantothenic acid, 60 pyridoxine, 15; ribflavine, 0.06; manganese sulphate, 0.18; potassium iodide, 0.02 zinc sulphate.

²Schulz et al. [36].

³Jobling [37].

Table 3: Formulation and approximate composition of experimental diets (g kg⁻¹).

Growth performance and feed utilization

Standard formulae were used to assess growth-feed utilization and other relevant parameters during the growth trial and these included, initial average weight, final average weight, total feed consumed, weight gain (g), average daily gain (g fish day⁻¹), specific growth rate (SGR% day⁻¹), feed conversion ratio (FCR), protein efficiency ratio (PER) and net protein utilization (NPU%).

Blood assays

Blood samples were collected using heparinized syringes from the caudal vein of the experimental fish at the end of the growth trial. Blood samples were centrifuged at 3000 rpm ×15 min at 4°C to allow separation of plasma which use to determine the blood parameters. Total plasma protein were carried out using Colorimeteric method, (Roch Diagnostics, GmbH, Monnheim, Germany) as recorded by Ruane et al. [38]. Creatinine was determined according to Pincus [39]. The activities of aspartate aminotransferase (AST) and alanine aminotransferase (ALT) were carried out using Colorimeteric method, Roch Diagnostics, GmbH, Monnheim, Germany kits according to Reitman and Frankel [40].

Statistical analysis

One way Analysis of Variance (ANOVA) was applied to test the effects of partially and totally replacement of poultry-by product meal on various growth parameters, chemical composition, blood constituents and apparent digestibility coefficients according to Snedecore and Cochran [41]. Duncan Multiple Range test was used to detect the significant differences between the means of treatments [42]. All analysis were performed using SAS (version 9.1 2004 SAS Institute, Cary, NC, USA) [43]. The level of significance was chosen at $p \leq 0.05$ and the results are presented as a group means (n=3 per tanks in each treatments ±S.E.M.).

Results

Physicochemical characteristics

Water physicochemical characteristics (Table 1) revealed that temperature, pH, dissolved oxygen, salinity and unionized ammonia are within the optimum ranges for rearing Nile tilapia according to Wangead et al., El-Shafai et al. and Ferreira et al. [44-46]. Similar physicochemical conditions were found in all tanks.

Chemical composition of diets

As can be seen in Table 3, the four experimental diets were almost similar in protein content (30.0-30.19%) and gross energy (19.58-19.77 MJ kg⁻¹ diets). However, they differed in their amino acids contents (Table 4). The all essential amino acids met the requirements of this species as recommended by NRC.

Growth performance

As presented in Table 5 averages of initial weights ranged between 1.5 to 1.6 g/fish with insignificant differences among the dietary groups indicating the random distribution of the experimental fish among treatment groups. Concerning finial weights the fish fed the tested treatments recorded an insignificant values (P<0.05) between each other. The same trend was observed with total gain in weight, the daily gain and specific growth rate of fish fed the different inclusion levels (50, 75 and 100% PM) of poultry by-product meal.

As shown in Table 5 average amounts of feed intake were found

Diets					
Amino acids	RE*	PBM0	PBM50	PBM75	PBM100
Essential amino acids					
Arginine	11.8	16.8	17.4	18.2	18.3
Hisitdine	8.4	9.2	8.8	8.9	9.1
Lysine	14.3	14.6	14.8	14.5	14.8
Leucine	9.5	20.4	21.6	22	22.4
Isoleucine	8.7	10.8	11.3	11.7	12
Valine	7.8	13.8	14.8	15.3	15.8
Methionine +Cystine	9	9.8	9.6	10.4	12.1
Threonine	10.5	10.3	11.1	11.6	11.9
Phenyl alinine+Tyrosine	15.5	20.7	20.4	24.9	24.5
Tryptophan	2.8	2.9	3.1	3	3.1
Non essential amino acids					
Glutamic acid	-	43	42.9	43.8	43.9
Aspartic acid	-	24.3	24.2	24.9	24.5
Glycine	-	18.8	19.2	18.9	17.9
Serine	-	11.7	15	16.4	17.7
Proline	-	18.4	20.4	21.1	21.8

*Requirements according to, NRC (2011).

Table 4: Amino acids contents of the experimental diets (g kg⁻¹) n=3.

Diets				
Parameters	PBM0	PBM50	PBM75	PBM100
Initial average weight (g fish⁻¹)	1.5 ± 0.05	1.6 ± 0.06	1.6 ± 0.05	1.5 ± 0.06
Final average weight (g fish⁻¹)	54.6 ± 2.5	53.8 ± 1.8	54.1 ± 2.0	54.3 ± 1.5
Gain in weight (g fish⁻¹)	53.1ᵃ ± 1.1	52.2ᵃ ± 1.4	52.5ᵃ ± 1.5	52.8ᵃ ± 1.6
Average daily gain (g fish day⁻¹)	0.44 ± 0.15	0.43 ± 0.2	0.43 ± 0.3	0.44 ± 0.1
Specific growth rate (% day⁻¹)¹	3.0 ± 0.5	2.92 ± 0.4	2.93 ± 0.2	2.99 ± 0.3
Feed consumed (g fish⁻¹)	70.0 ± 3.0	71.0 ± 2.0	71.0 ± 4.0	71.0 ± 3.0
Feed conversion ratio²	1.31 ± 0.15	1.36 ± 0.2	1.35 ± 0.2	1.34 ± 0.1
Protein efficiency ratio³	2.51 ± 0.2	2.44 ± 0.15	2.45 ± 0.18	2.46 ± 0.16
Net Protein Utilization (NPU%) ⁴	39.75 ± 4.0	37.10 ± 3.0	37.98 ± 3.0	38.51 ± 4.0

Values are the mean ± S.E. of triplicate groups of each treatment.

¹Specific growth rate = 100 × (Ln final weight- Ln initial weight)/ 120 day.

²Feed conversion= (feed given per fish)/ (weight gain per fish).

³Protein efficiency ratio = (weight gain per fish)/ (protein intake per fish).

⁴Net protein utilization (%) = (final body protein - initial body protein)/(protein intake) ×100.

Table 5: Growth performance of Nile tilapia fed the experimental diets.

to be 70.0, 71.0, 71.0 and 71.0 g for the PBMO, PBM50, PBM75 and PBM100% groups, respectively. On the other hand, the best FCR (lowest) values were obtained by the PBMO group without significant differences (P<0.05) were registered among treatments. As presented in the same table the PER and NPU values showed an insignificant values between the different groups.

Apparent digestibility coefficient

Data on apparent digestibility coefficients in present study for dry matter (DM), energy (E), crude protein (CP), fat and nitrogen free extract (NFE) are presented in Table 6. Results revealed that apparent digestibility coefficients DM, E, CP, fat and NFE were not significantly affected with the inclusion levels of poultry by-product meal.

Blood characteristics

Blood characteristics revealed that insignificant differences

Diets				
Nutrients	PBM0	PBM50	PBM75	PBM100
Dry matter	85.8 ± 3.0	85.5 ± 2.0	85.6 ± 1.0	85.4 ± 2.0
Energy	90.4 ± 4.0	89.8 ± 2.0	89.6 ± 3.0	90.2 ± 1.0
Protein	94.4 ± 3.0	94.5 ± 4.0	94.3 ± 2.0	94.6 ± 3.0
Fat	95.5 ± 2.0	95.3 ± 2.0	95.4 ± 3.0	95.6 ± 3.0
Nitrogen free extract	71.8 ± 1.0	71.9 ± 3.0	71.6 ± 2.0	71.5 ± 4.0

Table 6: Apparent digestibility coefficients (%) for the experimental diets (Mean ± S.E *n=3*).

Diets				
Parameters	PBM0	PBM50	PBM75	PBM100
Total protein g/dl	6.4 ± 1.11	6.4 ± 0.8	6.42 ± 0.9	6.3 ± 0.6
Albumin g/dl	3.22 ± 0.33	3.24 ± 0.22	3.3 ± 0.32	3.26 ± 0.2
Urea mg/dl	7.2 ± 0.14	7.13 ± 0.12	7.3 ± 0.16	7.5 ± 0.15
Creatinine	1.12 ± 0.3	1.14 ± 0.34	1.16 ± 0.2	1.18 ± 0.26
Ast[1]	116.0 ± 0.65	117.0 ± 0.48	124.0 ± 0.56	126.0 ± 0.52
Alt[2]	44.0 ± 0.41	44.0 ± 0.32	45.0 ± 0.22	44.5 ± 0.38

Values are the mean ± S.E. of triplicate groups of 20 fishes

1-Aspartate aminotransferase 2-Alanine aminotransferase

Table 7: An average of blood characteristics parameters of tilapia fed the experimental diets.

Diets					
Items	Initial	PBM0	PBM50	PBM75	PBM100
Dry matter	242 ± 1.6	271 ± 3.18	270 ± 3.6	276 ± 3.14	274 ± 1.16
Crude protein	152 ± 1.8	158 ± 2.16	152 ± 2.2	156 ± 2.2	154 ± 2.11
Crude lipid	32 ± 1.5	55 ± 1.8	62 ± 1.6	61 ± 1.4	58 ± 1.16
Crude ash	58 ± 1.2	58 ± 1.4	56 ± 1.6	59 ± 1.2	54 ± 1.6

Table 8: Carcass analysis of Nile tilapia fed the experimental diets (g kg^{-1} wet basis) mean ± S.E. n=3.

(P<0.05) were detected between the experimental groups (Table 7) for total protein, albumin, creatinine, Aspartate aminotransferase (Ast) and Alanine aminotransferase (Alt) of tilapia, which indicate that substitution of fish meal by poultry by-product had no hazardous effects on blood parameters tested.

Carcass analysis

Results of whole fish body composition in terms of wet weight, were not significantly affected (P<0.05) by the increment of PBM levels. The chemical analysis showed that the applied dietary treatments had no significant effects (P<0.05) on whole body dry mater, crude protein, fat and ash content between groups (Table 8).

Discussion

The replacement of dietary FM in aqua feeds with readily available and more economical alternatives sources, such as poultry by-product meal is an important aim for each aquaculture industry and feed-manufacture company. This by-product of poultry processing industry is high in protein, low price and contains a favorable profile of indispensable amino acids for fish production.

The results of the present study indicated that PBM is a suitable replacement of fish meal in practical formulation diets for juvenile tilapia.

The growth performance (final weight gain, daily gain and specific growth rate) and nutrient utilization (feed conversion ratio, protein efficiency ratio and net protein utilization) of tilapia has shown enhancement for dietary PBM without significant reduction when the replacement level of fish meal up to 100%. These findings are in agreement with other studies in tilapia [47]. However, PBM can be used to replace 75% of the fish meal in diets without amino acid supplementation for gilthead sea bream [26] up to 100%, red sea bream [27], sunshine bass [48]; grouper [49] and gible carp [50]. In contrast, fish meal could only be replaced with PBM at a level which did not exceed 50% for some marine fish species [51-53]. In the present study the growth performance and feed utilization recorded comparable results with fish meal diets. In contrast, Rawles et al. [16], recorded that PBM had lower growth than fish fed the control diet with FM in sunshine bass. They attributed that imbalance of some limiting amino acid content may have caused reduced growth performance in sunshine bass. They also speculated that reduced growth observed in this species may be due to reduced palatability for PBM. However, in the current study the balance in limiting amino acid, all diets were consumed similarly and different species may be enhance the growth performance in juvenile tilapia.

As can be seen from Table 5, the FCR, PER and NPU were not significantly differed (P<0.05) between groups and the best value was recorded by PBM0. Similar and comparable results of FCR, PER and NPU were recorded with tilapia [47,54,55].

Amongst the experimental diets in the present trial, fish receiving the PM0 and PM100 diets showed high growth performance and net protein utilization. This result is in agreement with the other several warm water finfish species obtained by Hernandez et al. and Pine et al. [47,56]. They recorded that poultry by-product has been shown to enhanced growth performance of Nile tilapia and sunshine bass.

The major reasons for different results may be due to the different fish species and the varying quality of tested PBM, which are significantly influenced by their processing methods [49].

The experimental diets in the present study showed a good digestibility coefficient for the tested diets. These results are in agreement with the results of Hernandez et al. [47]. Similar and comparable ADC values of feed dry matter, protein, lipid and energy were also observed by several authors in digestibility studies with tilapia fed the conventional commercial ingredient [55,57-60]. Similar results were recorded in rainbow trout by Burea et al. [9,23].

The carcass proximate composition of tilapia indicated that the dry matter, protein, lipid and ash were not affected by incorporation of 100% poultry by-product meal. Similar results have been reported for tilapia by Hernandez et al. [47] and sunshine bass [56], grouper [53] and gibel carp [50].

Measures of blood parameters not significantly different between treatments, where the total protein, albumin, urea, creatinine Ast and Alt are comparable with the control diet. These results are in agreement with the previous study in tilapia recorded by Metwalli, Yue et al. [61,62] and other species such as sturgeon [63] and sunshine bass [48].

Results of the present study suggest that potential replacing 100% of fish meal with poultry by-product meal in the feed of tilapia *Oreochromis niloticcus*, without compromising growth performance, nutrients utilization and some blood contents. It is important to establish that alternative dietary sources to fish meal are not only supplied in the correct quantities and balance for optimal growth and feed efficiency, but can maintain optimal whole body composition and blood contents.

References

1. FAO (2010) FAO Fisheries Department Fishery Information Data and Statistics Units DISHSTAT Plus: Universal software for fishery statistical time series, Version, 23 2000 Data sets: Aquaculture production: quantities and values 1950-2009, capture production 1950-2009.

2. Vechklang K, Boonanuntanasarn S, Ponchunchoovong S, Pirarat N, Wanapu C (2011) The potential for rice wine residual as alternative protein source in a practical diet for Nile tilapia (Oreochromis niloticus) at the juvenile stage Aquaculture Nutrition 17: 685-694.

3. Burden D (2012) Tilapia profile International and special projects Iowa State university.

4. Fitzsimmons (2014) World tilapia production 4 international trade and technical conference and exposition on tilapia, Kuala Lumpur Malaysia.

5. Naylor RL, Goldburg RB, Primavera JH, Kautsky N, Beveridge MCM, et al. (2000) Effect of aquaculture on world fish supplies Nature 405: 1017-1024.

6. New MB, Wijkstrom UN (2002) Use of fish meal and fish oil in aqua feed: further thoughts on the fish meal trap FAO fish circular No 975, FAO, Rome Italy.

7. Cavalheiro JMO, Souza EO, Bora PS (2007) Utilization of shrimp industry waste in the formulation of tilapia (Oreochroms niloticus Linnaeus) feed. Bioresour Technol 98: 602-606.

8. Nguyen TN, Davis DA (2009) Evaluation of alternative protein sources to replace fish meal in practical diets for Juvenile Tilapia, Oreochromis spp. J World Aquaculture Soc 40: 113-121.

9. Bureau DP, Harris AM, Bevan DJ, Simmons LA, Azevedo PA, et al. (2000) Feather meals and meat bone meals from different origins as protein sources in rainbow trout (Oncorhynchus mykiss) diets. Aquaculture 181: 281-291.

10. Millamena OM (2002) Replacement of fish meal by animal by-product meals in a practical diets for grow-out culture of grouper (Epinephelus coioides). Aquaculture 204: 75-84.

11. Wei Z, Kangsen M, Baigang Z, Fuzhen W, Yu U (2004) A study on the meat and bone meal and poultry by-product meal as protein substitutes of fish meal in practical diets for (Lipopenaeus vannami) juveniles. J Ocean Univ China (Ocean, Coast Sea Res) 3: 157-160.

12. Fasakin EA, Serwata RD, Davies SJ (2005) Comparative utilization of rendered animal derived products with or without composite mixture of soybean meal in hybrid tilapia (Oreochromis niloticcus×mossambicus) diets. Aquaculture 249: 324-338.

13. Abdel-Warith A, Davies SJ, Russell P (2001) Inclusion of commercial poultry by-product meal as a protein replacement of fish meal in practical diets for the African cat fish (Claris gariepinus). Aquacult Res 32: 296-306.

14. Gaylord TG, Rawles SD (2005) The modification of poultry by-product meal for use in hybrid striped bass (Morone chrysops×Msaxatilis) diets. J World Aquac Soc 36: 363-374.

15. Muzinic LA, Thompson KR, Metts LS, Dascupta S, Webster CW (2006) Use of turkey meal as partial and total replacement of fish meal in practical diets for sunshine bass (Morone chrysops × Morone saxatilis) grown in tanks Aquaculture 12: 71-81.

16. Rawles SD, Richie M, Gaylord TG, Webb J, Freeman DW, et al. (2006) Evaluation of poultry by-product meal in commercial diets for hybrid striped bass (Morine chrysops × Morone saxatilis) in recirculated tank production Aquaculture 259: 377-389.

17. Thompson KR, Metts LS, Muzinic LA, Dascupta S, Webster CD, et al. (2007) Use of turkey meal as a replacement for menhaden fish meal in practical diets for sunshine bass grown in cages N Am J Aquacult 69: 351-359.

18. Soltan MA (2009) Effect of dietary fish meal replacement by poultry by-product meal with different grain source and Enzyme supplementation on performance, feces recovery, body composition and nutrient balance of Nile tilapia. Pakistan Journal of Nutrition 8: 395-407.

19. El-Sayed AF (1998) Total replacement of fish meal with animal protein sources in Nile tilapia (Oreochromis niloticus L) feeds. Aquaculture 29: 275-280.

20. Yang Y, Xie SQ, Cui YB, Zhu XM, Lei W, et al. (2006) Partial and total replacement of fish meal with poultry by-product meal in diets for gibel carp, Carassius auratus gibelio Bloch. Aquaculture Research 37: 40-48.

21. Soaud IP, Rodgers LJ, Davis DA, Rouse DB (2008) Replacement of fish meal with poultry by-product meal in practical diets for red claw crayfish (Cherax quadrinatus). AquacultNutr 14: 139-142.

22. Rodriguez-Serna M, Olvera-Novoa MA, Carmona-Osalda C (1996) Nutritional value of animal by-product meal in practical diets for Nile tilapia (Oreochromis niloticus L) fry. AquacultRes 27: 67-73.

23. Bureau DP, Harris AM, Cho CY (1999) Apparent digestibility of rendered animal protein ingredients for rainbow trout (Oncorhynchus mykiss). Aquaculture 180: 345-358.

24. Fowler LG (1991) Poultry by-product meal as a dietary protein source in fall chinook salmon diets. Aquaculture 99: 309-321.

25. Alexis MN, Paparaskeva-Papoutsoglou E, Theochri V (1985) Formulation of practical diets for rainbow trout (Salmo gairdneri) made by partial or complete substitutes for fish meal by poultry by-product and certain plant by-products. Aquaculture 50: 61-73.

26. Nengas L, Alexis MN, Davies SJ (1999) High inclusion levels of poultry meal and related byproducts in diets for gilthead sea bream (Sparus aurata L). Aquaculture 179: 13-23.

27. Takag S, Hosokawa H, Shimeno S, Kawa MU (2000) Utilization of poultry by-product meal in a diet for red sea bream Pagrus major. Nippon Suisan Gakkaishi 66: 428-438.

28. Thompson KR, Rawles SD, Metts LS, Gannam AL, Brady YJ, et al. (2008) Digestibility of dry matter, protein, lipid and organic matter of two fish meals, two poultry by-product meals, soybean meal and distillers dried grains with solubles in practical diets for sunshine bass (Morone chrysops×M saxatilis). J World Aquac Soc 309: 352-363.

29. APHA (1992) Standard methods for the examination of water and waste water American Public Health Association, Washington, DC.

30. Jobling M (2011) National Research Council(NRC): Nutrient Requirements of fish and shrimp National Academy Press, Washington, DC, USA 20: 601-602.

31. Choubert G, De la Noue J, Luquet P (1982) Digestibility in fish: improved device for the automatic collection of faeces. Aquaculture 29: 185-189.

32. Bolin DW, King RP, Klosterman EW (1952) A simplified method for the determination of chromic oxide (Cr_2O_3) when used as an index substance. Science 116: 634-635.

33. Cho CY (1993) Digestibility of feedstuffs as a major factor in aquaculture waste management. Fish nutrition in practice. INRA, France.

34. AOAC (2006) Official methods of analysis of AOAC International (18 EDN) AOAC International Maryland, USA.

35. Cohen SA, Mamd M, Tavin TL (1989) The pico-Tag Method-A method of advanced techniques for amino acid analysis Bedford, MA, USA.

36. Schulz C, Wickert M, Kijora C, Ogunji J, Rennert B (2007) Evaluation of pea protein isolate as alternative protein source in diets for juvenile tilapia (Oreochromis niloticus). Aquaculture Research 38: 537-545.

37. Jobling M (1994) Fish bioenergetics, Fish and fisheries Series, Chapman & Hall-2-6 Boundary, London.

38. Ruane NM, Huisman EA, Comen J (2001) Plasma cortisol and metabolite level profiles in two isogenic strains of common carp. J Fish Biol 59: 1-12.

39. Pincus MR (1996) Interpreting laboratory results reference values and decision making, Clinical Diagnosis and Management by laboratory Methods Saunders, Philadelphia, USA.

40. Reitman A, Franks S (1957) Determination of aspartate glutamic amino transferase and alanin aminotransferase. American Journal of Clinical Pathology.

41. Snedecore, Gand W, Cochran WC (1987) Statistical Methods Iowa state Univ, USA.

42. Duncan DB (1955) Multiple ranges and multiple F tests. Biometric 11: 1-42.

43. SAS (1986) SAS User's Guide Version 6 Edition SAS Institute, Cary, NCUSA.

44. Wangead C, Greater A, Tansakul R (1988) Effects of acid water on survival and growth rate of Nile tilapia (Oreochromis niloticus) In: RSV Pulin, T Bhukaswan, K Tonguthai and J Maclean, Proceedings of the Second International Symposium on Tilapia in Aquaculture ICLARM Conference Proceedings No15, Deparment of Fisheries, Bangkok, Thailand ICLARM, Manila, Philippines.

45. El-Shafai SA, El-Gohary FA, Nasr FA, Der Steen NP, Gijzen HJ (2004) Chronic ammonia toxicity to duckweed-fed tilapia (*Oreochromis niloticcus*). Aquaculture 232: 117-127.

46. Ferreira MW, Araujo FG, Costa DV, Rosa PV, Figueiredo HCP, et al. (2011) Influence of dietary oil sources on muscle composition and plasma lipoprotein concentration in Nile tilapia (*Oreochromis niloticus*). Journal of World Aquaculture Society 42: 24-33.

47. Hernandez C, Olvera-Novoa AM, Hardy RW, Hemosillo A, Reyes C, et al. (2010) Complete replacement of fish meal by poultry by-product meals in practical diets for fingerlings Nile tilapia(*Oreochroms niloticus*) digestibility and growth performance. Aquaculture Nutrition 16: 44-53.

48. Rawles SD, Thompson KR, Brady YJ, Metts LS Aksoy MY, et al. (2011) Effects of replacing fish meal with poultry by-product meal and soybean meal and reduced protein level on the performance and immune status of pond-grown sunshine bass (*Morine chrysops* × *Morone saxatilis*). Aquacult Nutr 17: e708-e721.

49. Shapawi R, Ng WK, Mostafa S (2007) Replacement of fish meal with poultry by-product meal in diets formulated for the humpback grouper (*Cromileptes altivelis*). Aquaculture 273: 118-389.

50. Hu M, Wang Y, Wang Q, Zhao M, Xiong B, et al. (2008) Replacement of fish meal by rendered animal protein ingredients with lysine and methionine supplementation to practical diets for gibel carp (*Carassius auratus* gibelio). Aquaculture 275: 260-265.

51. El-Sayed AF (1994) Evaluation of soybean meal, spirulina meal and chicken offal meal as protein sources for silver sea bream (*Rhabdosargus sarba*) fingerlings. Aquaculture127: 169-175.

52. Rawles SD, Gaylord TG, McEntire ME, Freeman DW (2009) Evaluation of poultry by- product meal in commercial diets for hybrid striped bass (*Morine chrysops* × *Morone saxatilis*) in pond production. J World Aquac Soc 40: 141-156.

53. Wang Y, Li K, Han H, Zheng Z, Bureu DP (2008) Potential using a blend of rendered animal protein ingredients to replace fish meal in practical diets for Malabar grouper (*Epinephelus malabricus*). Aquaculture 281: 133-117.

54. Gao W, Liu YGJ, Tian LX, Mai KS, Liang GY, et al. (2011) Protein sparing capability of dietary lipid in herbivorous and omnivoraus freshwater finfish: a comparative case study on grass carp (*Ctenopharyngodon idella*) and tilapia(*Oreochromis niloticus* × *Oaureus*). Aquaculture Nutrition 17: 2-12.

55. Yones A, Abdel-Hakim NF (2010) Study on growth performance and apparent digestibility coefficients on some common plant protein ingredients used in formulated diets of Nile tilapia (*Oreochromis niloticus*). Egyptian Journal Nutrition and Feeds 13: 589-606.

56. Pine HJ, Daniels WH, Davis DA, Jiang M, Webster CD (2008) Replacement of fish meal with poultry by-product meal as a protein source in pond-raised sunshine bass (*Morine chrysops* × *Morone saxatilis*). Journal World Aquaculture Society 39: 586-597.

57. Shiau SY, Chuang JL, Sun CL (1987) Inclusion of soybean meal in tilapia (*Oreochromis niloticus* × *Oreochromis aureus*) diets at two protein levels. Aquaculture 65: 251-261.

58. Koprucu K, Ozdemir Y (2005) Apparent digestibility of selected feed ingredients for Nile tilapia (*Oreochromus niloticus*). Aquaculture Research 250: 308-316.

59. Gaye-Sliessegger J, Focken U, Abel HJ, Becker K (2005) Improving estimates of tropic shift in Nile tilapia (*Oreochromus niloticus* L) Comp Biochem Physiol A Mol Integr Physiol 140: 117-124.

60. Yones AM (2010) Effect of lupin kernel meal as plant protein sources in diets of red hybrid tilapia (*Oreochromis niloticus* × *O mossambicus*), on growth performance and nutrients utilization. African Journal Biological Science 6: 1-16.

61. Metwalli AA (2013) Effects of partial and total substitution of fish meal with corn gluten meal on growth performance, nutrients utilization and some blood constituents of the Nile tilapia (*Oreochromus niloticus*). Egypt J Aquat Biol & fish17: 91-100.

62. Yue YR, Zhou QC (2008) Effect of replacing soybean meal with cottonseed meal on growth, feed utilization and hematological indexes for juvenile hybrid tilapia (*Oreochromis niloticus*× *Oaureus*). Aquaculture 284: 185-189

63. Zhu H, Gong G, Wang J, Wu X, Xue H, et al. (2011) Replacement of fish meal with blend of rendered animal protein in diets for Siberian sturgeon (*Acipenser baerii*), results in performance equal to fish meal fed fish. Aquacult Nutr 17: e389-e395.

21

Histopathology and Wound Healing in Oxytetracycline Treated *Oreochromis niloticus* (L.) Against *Aeromonas hydrophila* Intramuscular Challenge

Julinta RB¹*, Abraham TJ¹, Anwesha Roy¹, Jasmine Singha¹, Gadadhar Dash¹, NageshTS² and Patil PK³

¹*Department of Aquatic Animal Health, Faculty of Fishery Sciences, West Bengal University of Animal and Fishery Sciences, Chakgaria, Kolkata, West Bengal, India*
²*Department of Fisheries Resource Management, Faculty of Fishery Sciences, West Bengal University of Animal and Fishery Sciences, Chakgaria, Kolkata, West Bengal, India*
³*Central Institute of Brackishwater Aquaculture, Indian Council of Agricultural Research, Raja Annamalai Puram, Chennai, Tamil Nadu, India*

Abstract

Antibiotics are very important tools for the control of fish bacterial diseases. Yet, there are strict regulations controlling the use of antibiotics in aquaculture. This study assessed the efficacy oxytetracycline dihydrate (OTC) at a dose of 2 g, 4 g, 6 g and 8 g/100 pounds fish/day against *Aeromonas hydrophila* challenge in Nile tilapia *Oreochromis niloticus* as well as the histopathological alterations in kidney and muscle, and wound healing. The commercial pellet feed was top dressed with OTC using 5 ml vegetable oil as a binder. The fish were injected intramuscularly with *A. hydrophila* at ≈1 × 10⁸ cells/fish and then fed OTC feeds at 2% of their body weight for 10 days. The fish fed with 8g OTC/100 pounds fish/day recorded the lowest mortality (3.33%). The untreated fish recorded 8.33% mortality. Histologically, the kidney tissues of *O. niloticus* exhibited nephropathy and glomerulopathy. The kidney of OTC fed groups had improved organization of nephritic tubules and glomerulus. The muscle tissues exhibited haemocyte infiltration with mild necrosis initially, followed by melanization and disrupted muscle bundles. The results demonstrated that the OTC treatment for 10 days could bring out improved functioning of fish kidneys that carry infectious agents. Within 3 days of OTC therapy, tissue reddening and inflammation subsided with the formation of black scar. Full recovery of normal skin architecture was reached within 26-31 days post-injection. Based on the results, prudent use of 4-8 g OTC/100 pounds fish/day is recommended for the control of *A. hydrophila* infection in *O. niloticus*.

Keywords: *Aeromonas hydrophila; Oreochromis niloticus;* Oxytetracycline; Oral therapy; Nephropathy; Necrotic lesion

Introduction

The tilapias are freshwater fish that belong to the family Cichlidae, and they are exclusively associated with Africa and Middle East [1]. The Nile tilapia (*Oreochromis niloticus* L.) is one of the first fish species to be cultured in the world. Illustrations from Egyptian tombs suggested that the Nile tilapias had been cultured more than 4,000 years ago [2]. In 2010, more than 73% of total tilapia production were represented by *O. niloticus* [3]. The farmed tilapia production statistics from 135 countries and territories on all continents are available currently [4]. Considering the demand for more fish, it has become an important species for aquaculture in India [5]. Initially, tilapias were considered to be more resistant to microbial diseases. But, in recent times tilapias have been found to be susceptible to various diseases. Bacterial infections are the most serious problem in tilapia production causing 80% of fish mortalities [2,6-7]. *Aeromonas hydrophila* is considered as a persuasive pathogen that causes mortalities in tilapia and other freshwater fish [8,9]. Fish exposed to poor water quality such as high nitrite levels, low levels of dissolved oxygen, or high levels of carbon dioxide are more susceptible to infection by bacterial pathogens [2].

Most bacterial infections can be treated effectively with antibiotics. However, the farmers use a variety of aquadrugs for the control of fish diseases [10]. Treatment of *A. hydrophila* is currently limited to two antibiotics, Terramycin˚, an oxytetracycline (OTC), and Remet-30˚, a potentiated sulfonamide. Oxytetracycline is one of the USFDA (United States Food and Drug Administration) approved chemotherapeutic as an oral antibacterial to treat specific bacterial diseases in temperate and warm water finfish [11-13]. The preparation of medicated feed for use with fish intended for human consumption is, however, regulated by the FDA [13]. The stability of OTC in feed premixes has been evaluated and is well established at >90% retained potency after 24 months of storage under ambient conditions [14]. Though the effectiveness and safety levels of the FDA approved antibiotics, including OTC

on temperate fish have been established, such studies on fish species cultured in topical condition are not attempted. This study was, therefore, aimed at to evaluate the effectiveness of feeds containing different concentrations of OTC on Nile tilapia *O. niloticus* against *A. hydrophila* infection with particular reference to histopatological alterations in kidney and muscle, and wound healing.

Materials and Methods

Bacterial strain

The β-haemolytic and oxytetracycline sensitive bacterial strain *Aeromonas hydrophila* BBT₄K₃ (NCBI accession number KY484791) used in this study was from the collections of the Department of Aquatic Animal Health, Faculty of Fishery Sciences, West Bengal University of Animal and Fishery Sciences, Kolkata, India. It was isolated from the kidney of haemorrhagic septicemic *O. niloticus*. The isolation, identification and preparation of bacterial cell suspension were as described in Abraham et al. [15].

Feed top dressing

As per the FDA standards [12], the medicated feeds for feeding

*Corresponding author: R. Beryl Julinta, Department of Aquatic Animal Health, Faculty of Fishery Sciences, West Bengal University of Animal and Fishery Sciences, Chakgaria, Kolkata-700094, West Bengal, India
E-mail: julintaa@gmail.com

the experimental fish at 2% of the body weight (BW) were prepared. In brief, the medicated feeds to feed fish at a dose of 2 g, 4 g, 6 g and 8 g OTC/100 pounds fish/day were prepared by mixing appropriate quantities of oxytetracycline dihydrate (HiMedia, India) in 5 ml vegetable oil and then admixed with 1 kg basal feed. The feeds containing OTC were mixed thoroughly, air dried at room temperature for 24 h, and stored separately in air tight plastic containers. Control feed was prepared as above without OTC. All feeds were prepared freshly and used immediately.

Efficacy of OTC

Plastic tanks of size L58 cm × H45 cm × B45 cm, and healthy *Oreochromis niloticus* (L.) juveniles of 8.25 ± 0.25 cm and 7.91 ± 0.31 g were used for the experiments in triplicate. Thirty tilapias were introduced into each of the tank and fed with basal pellet feed containing 30% crude protein at 2% BW. About 50% water was exchanged, and waste feed and faeces were removed periodically. The water quality parameters were maintained optimally (Water temperature: 22.0-29.0°C; pH: 8.0 - 8.7; dissolved oxygen: 4.2-4.9 ppm; nitrate and ammonia: <0.02 ppm). The experimental fish were divided into 7 groups, namely group 1) negative control (unchallenged and fed with basal feed), group 2) positive control (saline injected, unchallenged and fed with basal feed), group 3) 0 g OTC/100 pounds fish/day, group 4) 2 g OTC/100 pounds fish/day, group 5) 4 g OTC/100 pounds fish/day, group 6) 6 g OTC/100 pounds fish/day, and group 7) 8 g OTC/100 pounds fish/day. After acclimatization for 5 days, the fish were injected intramuscularly at the base of the dorsal fin. Prior to challenge, the fish were starved for a day and anesthetized using clove oil at 50 μl/litre water. The fish of groups 2-7 were injected intramuscularly with aliquots (0.1 ml each) of *A. hydrophila* BBT_4K_3 cell suspension at a predetermined dose of 1×10^8 cells/fish. After injection, the fish were transferred to the respective tanks. The fish of groups 1 and 2 were fed with basal feed and the group 3 was fed with control feed during the entire study period. The fish of groups 4, 5, 6 and 7 were fed with control feed during the pre-treatment (1-5 days) and disease progression (6-8 days) periods. During the treatment period for 10 days (9-18 days), they were fed with respective OTC feeds at 2% BW twice daily. During the post-treatment period for 21 days (19-39 days), the fish were fed with control feed. The unconsumed feed, if any, in each tank was removed daily, air dried and weighed carefully. Observations on feeding behaviour, behavioural changes, external signs of infections and mortality were recorded daily. Depending on the feed consumption, the feeding behavior of tilapia was rated using a scale ranging from 0 to 4, i.e., 4) 100% feed consumption, 3) 75% feed consumption, 2) 50% feed consumption, 1) 25% feed consumption and 0: No feed consumption.

Histopathology

The muscle and kidney samples of *A. hydrophila* BBT_4K_3 challenged and subsequently OTC treated *O. niloticus* were fixed in Bouin's solution for 24 h. The fixed samples were processed by standard techniques and embedded in paraffin wax. Thin (5 μm) sections were prepared and stained with haematoxyline and eosin [16].

Wound healing

The wounds at the site of intramuscular injection were first digitally photographed during the treatment regime. Tissue damages were assessed using a score ranging from 0 to 6, depending on the degree and extent of damage based on the scale proposed by Bernet et al. [17]. The extent of wound progression and healing was qualitatively classified as

0: undamaged; 2: mild damage; 4: moderate damage; 6: severe damage. Intermediate values were also considered.

Statistical analysis

The results of the experiment are expressed as mean ± standard deviation and analyzed by one-way analysis of variance (ANOVA) using Microsoft excel version 2010 to test the significance of differences among the experimental fish groups. Comparison of mean values was done by Duncan's Multiple Range Test [18]. A probability level of 0.05 was used to find out the significance in all cases.

Results

Efficacy of OTC therapies

The mortalities and cumulative mortalities recorded during the treatment regime in *A. hydrophila* challenged and OTC fed *O. niloticus* juveniles are presented in Figures 1 and 2, respectively. During the disease progression period, i.e., on day 8, 3.33-5.00% mortalities were noticed in treatment groups. At the end of 10 day OTC treatment, i.e., on day 18, the mortalities observed in challenged and OTC fed groups were 3.33-6.66%. The highest mortality (8.33%) was observed in *A. hydrophila* challenged and control feed fed fish. Significant differences existed between the positive control (saline injected), and challenged and control feed fed as well as OTC fed (2 g, 4 g and 6 g OTC/100 pounds fish/day) groups (P<0.05). Significant differences also existed among 0 g OTC/100 pounds fish/day group and the OTC feeds fed (4 g, 6 g and 8 g OTC/100 pounds fish/day) group (P<0.05). The difference in the mortality between 2 g and 8 g OTC/100 pounds fish/day groups was significant (P<0.05).

The rating of feeding behaviour of *A. hydrophila* challenged *O. niloticus* juveniles during the treatment regime is presented in Table 1. The feed intake of challenged and control feed fed tilapia was low till 18 dpi, which became normal subsequently. In OTC fed groups,

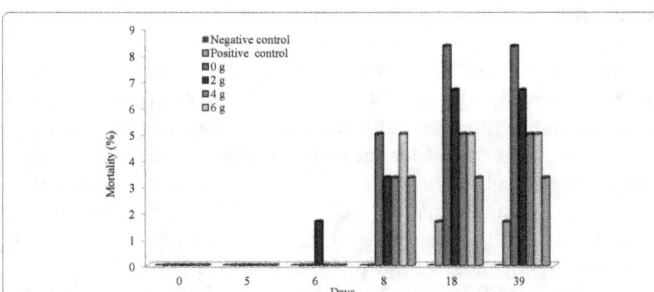

Figure 1: Mortalities in *Oreochromis niloticus* juveniles when challenged intramuscularly with *Aeromonas hydrophila* and subsequently fed oxytetracycline feeds (g OTC/100 pounds fish/day) for 10 days.

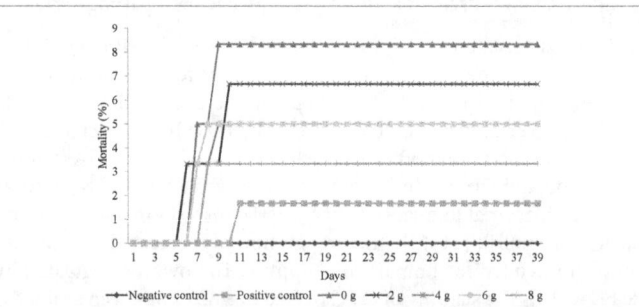

Figure 2: Cumulative mortalities in *Oreochromis niloticus* juveniles when challenged intramuscularly with *Aeromonas hydrophila* and subsequently fed oxytetracycline feeds (g OTC/100 pounds fish/day) for 10 days.

reduced feed intake was noted during the early OTC treatment period. The freshly dead fish were subjected to necropsy and bacteriology. Internally, discoloration and liquefaction of the internal organs, viz., kidney and liver were observed in the intramuscularly challenged fish. Bacterial inocula from the kidney of fresh dead fish on Rimler-Shotts agar yielded exclusive growth of yellow colour colonies at 35°C in 24-48 h. The standard biochemical test results confirmed *Aeromonas* infection (data not shown).

Histopathology

Histologically, the kidney tissues of unchallenged tilapia showed

Treatments	Feeding behaviour score# in Mean ± standard deviation			
	Pre-treatment (1-5 days)	Disease progression (6-8 days)	OTC treatment (9-18 days)	Post-treatment (19-39 days)
Negative control	4.00 ± 0.00	4.00 ± 0.00	4.00 ± 0.00	4.00 ± 0.00
0 g OTC*	4.00 ± 0.00	3.64 ± 0.67	2.70 ± 0.83	4.00 ± 0.00
2 g OTC*	4.00 ± 0.00	3.64 ± 0.67	3.77 ± 0.43	4.00 ± 0.00
4 g OTC*	4.00 ± 0.00	3.64 ± 0.67	3.77 ± 0.43	4.00 ± 0.00
6 g OTC*	4.00 ± 0.00	3.68 ± 0.57	3.74 ± 0.45	4.00 ± 0.00
8 g OTC*	4.00 ± 0.00	3.76 ± 0.54	3.85 ± 0.37	4.00 ± 0.00

OTC: oxytetracycline; *: Fish groups were fed with OTC at an appropriate dose/100 pounds fish/day for 10 days after challenging intramuscularly with *Aeromonas hydrophila*. Negative Control: Unchallenged, saline injected and fed basal feed. #: Feeding behaviour scale: 4: 100% feed consumption, 3: 75% feed consumption, 2: 50% feed consumption, 1: 25% feed consumption and 0: No feed consumption.

Table 1: Feeding behaviour of *Oreochromis niloticus* juveniles challenged with *Aeromonas hydrophila* and fed oxytetracycline (g OTC/100 pounds fish/day) at 2% of body weight for 10 days.

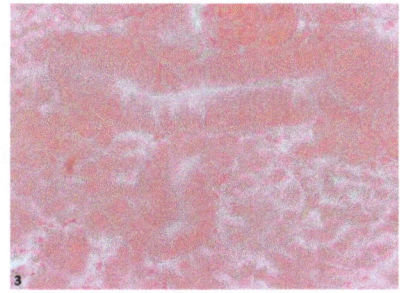

Figure 3: The normal histology of the kidney tissues of *Oreochromis niloticus* juveniles showing the typical structural organization of nephritic tubules with well-defined glomerulus, X200 H&E staining.

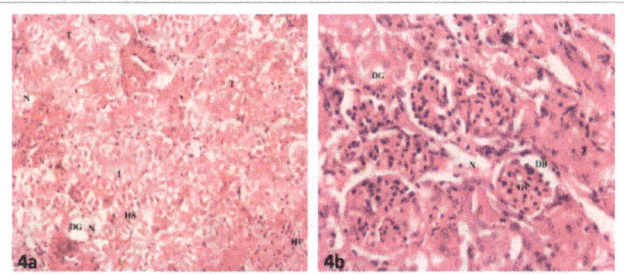

Figure 4: Histopathological changes in the kidney tissues of control feed (0 g OTC/100 pounds fish/day group) fed *Oreochromis niloticus* juveniles intramuscularly injected with *Aeromonas hydrophila* BBT$_4$K$_3$ on 10th day treatment showing (a) inflamed (I) nephritic tubules, haemocyte infiltration (HI), hydropic swelling (HS) with necrotized areas (N), degeneration of nephritic tubules (DG), deeply stained nucleus (DS) and thickening of nephritic tubule (T), X200 H&E staining; (b) degeneration of nephritic tubules (DG), glomerulopathy (GP) with dilated Bowman's space (BS) and necrosis (N), X400 H&E staining.

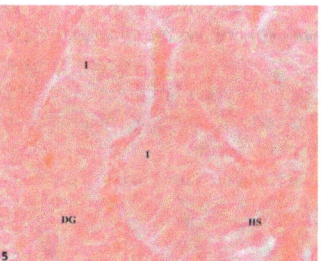

Figure 5: Histopathological changes in the kidney tissues of OTC fed (8 g OTC/100 pounds fish/day group) *Oreochromis niloticus* juveniles intramuscularly injected with *Aeromonas hydrophila* BBT$_4$K$_3$ on the 10th day OTC treatment showing hydropic swelling (HS), inflammation (I) and degeneration of nephritic tubules (DG), X400 H&E staining.

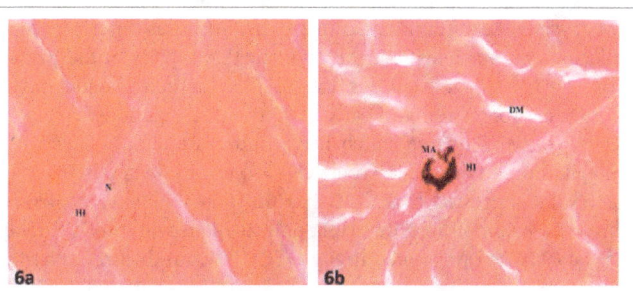

Figure 6: Histopathological changes in the muscle tissues of *Oreochromis niloticus* juveniles intramuscularly injected with *Aeromonas hydrophila* BBT$_4$K$_3$ (a) on 2 day post-injection showing haemocyte infiltration (HI) with mild necrosis (N), X200 H&E staining; (b) on 10th day OTC (8 g OTC/100 pounds fish/day) treatment showing melanized area (MA) with haemocyte infiltration (HI) and disrupted muscle bundles (DM), X400 H&E staining.

the typical structural organization of the nephritic tubules with well-organized glomerulus (Figure 3). While the kidney tissues of challenged and control feed fed *O. niloticus* juveniles recorded nephritic tubules inflammation, haemocyte infiltration, thickening of nephritic tubule, hydropic swelling with necrotized areas (Figure 4a), and glomerulopathy with dilated Bowman's space and degeneration of nephritic tubules (Figure 4b). The histopathological changes noticed in the kidney tissues of OTC (8 g OTC/100 pounds fish/day) fed *O. niloticus* on 10th day OTC feeding were inflammation of nephritic tubules, glomerulus with no pathological changes and thin epithelial layer, hydropic swelling, degeneration of nephritic tubules (Figure 5). The muscle tissues of challenged *O. niloticus* on 2 dpi had haemocyte infiltration with mild necrosis (Figure 6a), while the OTC fed (8 g OTC/100 pounds fish/day) *O. niloticus* on 10th day OTC feeding exhibited melanization and disrupted muscle bundles (Figure 6b).

Wound healing

The qualitative rating of wound progression and healing in *O. niloticus* challenged intramuscularly with *A. hydrophila* and fed OTC feed (8 g OTC/100 pounds fish/day group) for 10 days is depicted in Figure 7. The challenged fish were weak and lethargic initially. Tissue reddening, inflammation and skin peeling at the site of injection, and open sub-epithelial wounds started to become obvious within 24 and 48 h of challenge with *A. hydrophila,* respectively. A membrane over the wound was observed on 3 day post-injection (dpi). Within 3 days of OTC therapy, the reddening and inflammation subsided with the formation of black scar in the ulcerated area. The areas surrounding the wound become very dark in 3-6 days of therapy. Though there were depressions at the site of injection, all wounds examined were closed within 9 days of wounding or 6 days of OTC therapy (dot).

Aquaculture Production Systems

Disappearance of black scar, onset of dermal fibrous tissue re-growth and the development of skin at the ulcerated scar region were seen on 13 dpi (10 dot). The epidermis gradually became thinner with the development of scales in the underlying tissues and appeared normal on 20 dpi (7 days after OTC therapy). Full recovery of normal skin architecture was reached within 26-31 dpi (Figure 8). The wounds of OTC fed groups healed faster than the control.

Discussion

Disease is a component of the overall welfare of fish [19]. A perusal of the literature revealed the use of a wide range of antibiotics in attempts to control bacterial diseases in aquaculture [9-11,13]. The dosage rate used in medicated feed may vary according to the specific antibiotic used, but usually the rate is based on a number of grams/100 pounds fish/day [13]. The experimental transmission route used in the present study was intramuscular injection, which was reportedly more replicable and efficient than the other methods [20]. In the present study A. hydrophila challenge caused fish mortalities only during the disease progression and early OTC treatment periods. Following OTC sensitive A. hydrophila challenge and oral therapy, a reduction in tilapia mortality with increasing OTC concentrations was observed. The results suggested that OTC has the potentiality to control A. hydrophila infection in Nile tilapia only at higher doses. The highest survival (96.67%) was noted in 8 g OTC/100 pounds fish/day group,

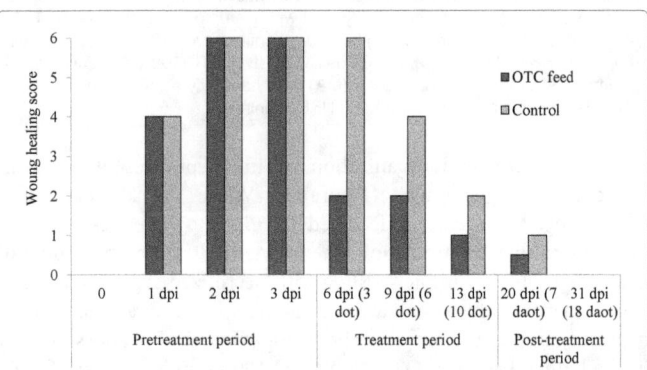

Figure 7: The qualitative rating of wound progression and healing in *Aeromonas hydrophila* challenged *Oreochromis niloticus* when fed with oxytetracycline (OTC) (8 g OTC/100 pounds fish/day) and control feeds during treatment regime.

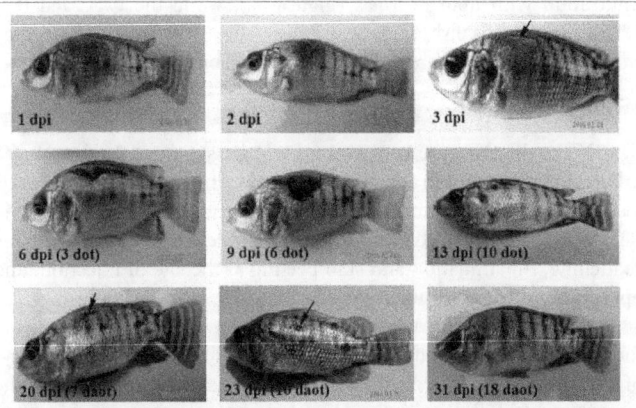

Figure 8: Digital images showing the wound progression and healing in *Aeromonas hydrophila* challenged, and oxytetracycline feed fed *Oreochromis niloticus* juveniles during the treatment regime. dpi: day post injection; dot: day OTC treatment; daot: day after OTC treatment.

followed by 95% each in 4 g and 6 g OTC/100 pounds fish/day groups. Statistically, the dose of 2g OTC/100 pounds fish/day was found to be ineffective. These results further suggested that the dose of 4-8 g OTC/100 pounds fish/day, equivalent to 2-4 g OTC/kg feed, is ideally used to control A. hydrophila infection in Nile tilapia under tropical Indian condition. Our results provided supportive evidences to Plumb [6], who recommended a dose of 2-4 g OTC/kg feed (50-100 mg/kg fish) for 14 days for bacterial disease treatment. The present study, thus, demonstrated the effectiveness of OTC oral therapy in reducing the A. hydrophila induced mortalities in Nile tilapia, which was found to be dose-dependent.

Histopathology is a well-established tool to figure out the qualitative changes in the affected organs and the patterns of recovery. Histopathologically, the main worth mentioning lesions is located in the kidney [21] and, therefore, the kidney samples were examined for the qualitative tissue level changes. The muscle samples were also examined as the fish were challenged intramuscularly. In unchallenged control fish, normal structure and systematic arrangement of kidney tissues with well-defined glomerulus were observed histologically. On the other hand, the changes observed in the kidney tissues of O. niloticus challenged with A. hydrophila on 10th day feeding with control feed and OTC feed (8 g OTC/100 pounds fish/day) during the treatment period were variable. The pathological changes observed in the kidney of A. hydrophila challenged and control feed fed O. niloticus are suggestive of nephropathy with severe cellular and tissue level alterations. The hydropic swelling, i.e., marked cell swelling of nephritic tubular epithelial cells is indicative of severe cellular oedema. Besides, there were glomerulopathy with dilated Bowman's space, necrotized areas, severe inflammation, degeneration and thickening of nephritic tubules, which are in accordance with the previous studies [22,23]. Though indeed there was inflammation, the 10 days OTC feeding controlled the tissue level changes in the kidney of A. hydrophila challenged O. niloticus, which exhibited glomerulus with no pathological changes, thin epithelial layer, and improved architecture and organization of nephritic tubules. The results demonstrated that the OTC treatment for 10 days could bring out improved functioning of fish kidneys that carry infectious agents. Nonetheless, several earlier studies demonstrated negative effects of OTC administration at high concentrations; for example, tissue damages in the kidneys when OTC was administered intramuscularly [21], congestion, severe fatty changes and vacuolations in the hepatocytes and periglomerular lymphocytic aggregation in the kidney of O. niloticus when fed with OTC at 100 mg/kg diet [24], diffuse cytoplasmic vacuolization of the renal duct epithelium in the kidney of C. carpio when fed OTC feed at 15 g/kg live weight [25]. Necrosis was observed only in challenging and non-treated fish, while it was absent in unchallenged control fish as well as challenged and OTC treated fish. It is clear from the present study that the normal kidney tissue exhibited well-defined nephrons, which when challenged with A. hydrophila showed nephropathy and glomerulopathy and finally, when treated with OTC, the kidney tissue showed improvement in nephritic tubular organization.

Melanin containing macrophages within the epidermal layer plays an important role in the healing of the injected area [21]. The haemocyte infiltration with mild necrosis at the intramuscularly injected site of O. niloticus on 2 dpi is an indication of a severe cellular and inflammatory reaction together with the disorganization of the muscular tissue, which seemed to be in agreement with the observations of Soler et al. [21]. The muscle tissue of A. hydrophila infected fish on the 10th day OTC feeding showed melanized area as dermomuscular necrotic lesion at the site of injection, which conform the previous reports [26,27]. These melanized areas were identified as aggregates of cells containing dark pigments. The melanin within the dermis of tilapia indicated mature healing ulcer at the site of injection. All these histopathological findings suggested marked improvements in the kidney and muscle tissues of *A. hydrophila* infected *O. niloticus* up on OTC treatment. The reduction in bacterial population and its consequences may probably be the underlying mechanism for the beneficial effect of OTC as was observed with antibiotics [9].

The rate of wound progression and recovery in *A. hydrophila* challenged *O. niloticus* juveniles upon oral OTC therapy was also investigated in the present study. The tissue reddening, inflammation and formation of a membrane over the wound within 3 dpi is an indication of initial protective reactions of fish to ward off the bacterial challenge. With OTC therapy, the reddening and inflammation were subsided relatively faster than control. The formation of black scar in the ulcerated area within 3 days of OTC therapy indicated the closure of the wound by the layers of epidermal cells. Darkened areas surrounding the wound in 3-6 days of therapy may be due to increased number of melanocytes, thereby, suggesting an increased melanocytes activity after the injury [28,29]. Within 6-10 days of OTC therapy, regeneration of muscle tissue with the closure of wounds, disappearance of black scar, onset of dermal fibrous tissue growth and the development of skin at the ulcerated scar region was obvious. The epidermis appeared normal with the development of scales in the underlying tissues after 7 days of completion of OTC therapy. Ohira et al. [30] also observed the growth of new scales with the size and characteristics of mature scales within a few weeks. The repair of dermal and muscle structure took much longer time in comparison with the epidermis. These results corroborate the findings of Ashley et al. [31], who described a temporal precedence of epidermal over dermal repair. Quilhac and Sire [32] also observed a rapid differentiation of the epidermal basal layer cells during the re-epithelialization process after wounding of a cichlid fish. The depression at the site of injection during the recovery period is an indication that the tissue re-growth or dermal repair had not reached steady state levels. Full recovery of normal skin architecture was reached within 26-31 dpi. Though the rate of wound healing was initially faster in OTC treated fish, the wounds were healed completely within 31 days even in untreated survivors. The results confirmed that the degree of wound healing was promoted by OTC feed, which was more prominent during the treatment period. Unlike mammals, no scar or scab was left behind on the wounded region, thereby representing a more advanced healing progression in teleost.

Conclusion

In general, concerns about the consequences of antibiotic use on public health have encouraged the development of strict regulations controlling the use of antibiotics and have led to only a few antibiotics being licensed for use in aquaculture. In this context, the present study demonstrated the effectiveness of dose-dependent OTC oral therapy in reducing the *A. hydrophila* induced mortalities in Nile tilapia and improving the kidney tissue architecture. The results suggested the prudent use of 4-8 g OTC/100 pounds fish/day at 2% body weight ration to control *A. hydrophila* infection in Nile tilapia under tropical Indian conditions. Since OTC is one of the FDA approved antibiotics for treating bacterial diseases of temperate and warm water finfish, responsible use of OTC as outlined by FAO [11,13] is recommended for the control of *Aeromonas* infection in Nile tilapia. Further, our results would provide the baseline data on the efficacy of OTC to the policy makers and regulatory authorities of Indian and global aquaculture.

Conflict of Interest

The authors declare that there is no conflict of interest.

Acknowledgements

The research work was supported by the Indian Council of Agricultural Research, Government of India, New Delhi under the All India Network Project on Fish Health (Grant F. No. CIBA/AINP-FH/2015-16 dated 02.06.2015). The authors thank the Vice-Chancellor, West Bengal University of Animal and Fishery Sciences, Kolkata for providing necessary infrastructure facility to carry out the work.

References

1. Trewaves E (1983) Tilapia fishes of the Genera Sarotherodon, *Oreochromis. Danakilia*. British Museum of Natural History.

2. Amal MNA, Zamri-Saad M (2011) Streptococcosis in tilapia (*Oreochromis niloticus*): A review. Pertanika Journal of Tropical Agricultural Science 34: 195-206.

3. FAO (2012) The state of world fisheries and aquaculture. FAO Fisheries and Aquaculture Department.

4. FAO (2014) The state of world fisheries and aquaculture: Opportunities and challenges. Food and Aquaculture Organization of the United Nations, Rome.

5. Lende SR, Mahida PJ, Mapwesera H, Chavda G, Rana G, et al. (2014) Prospects of tilapia farming in Gujarat region. Indian Farmer 1: 33-37.

6. Plumb JA (1999) Health maintenance and principal microbial diseases of cultured fishes. Iowa State University Press, Ames, Iowa.

7. Pradeepa PJ, Suebsing R, Sirthammajak S, Kampeera J, Jitrakorna S, et al. (2016) Evidence of vertical transmission and tissue tropism of Streptococcosis from naturally infected red tilapia (*Oreochromis* spp.). Aquaculture Reports 3: 58-66.

8. Aoki T (1999) Motile aeromonads (*Aeromonas hydrophila*). Fish Diseases and Disorders. CABI Publishing, Wallingford, UK.

9. Austin B, Austin DA (2007) Bacterial fish pathogens: Diseases of farmed and wild fish. (3rd edn), Springer-Praxis, Godalming, UK.

10. Romero J, Feijoo CG, Navarrete P (2012) Antibiotics in aquaculture – Use, abuse and alternatives. Health and Environment in Aquaculture, Agricultural and Biological Sciences, InTech.

11. Serrano HP (2005) Responsible use of antibiotics in aquaculture. FAO Fisheries Technical Paper. No. 469. FAO, Rome.

12. FDA-CVM (2007) Aquaculture website. FDA-CVM.

13. Bondad-Reantaso MG, Arthur JR, Subasinghe RP (2012) Improving biosecurity through prudent and responsible use of veterinary medicines in aquatic food production. FAO Fisheries and Aquaculture Technical Paper. No. 547. FAO, Rome.

14. FAO (1996) Residues of some veterinary drugs in animals and foods. Monographs prepared by the forty-fifth meeting of the Joint FAO/WHO Expert Committee on Food Additives, FAO Food and Nutrition Paper 41/8, FAO, Rome.

15. Abraham TJ, Sarker S, Dash D, Patra A, Adikesavalu H (2017) *Chryseobacterium* sp. PLI$_2$ and *Aeromonas hydrophila* co-infection in pacu, *Piaractus brachypomus* (Cuvier, 1817) fries cultured in West Bengal, India. Aquaculture 473: 223-227.

16. Roberts RJ (2012) Fish pathology (4th edn), Wiley-Blackwell, UK.

17. Bernet D, Schmidt H, Meier W, Burkhardt-Holm P, Wahli T (1999) Histopathology in fish: Proposal for a protocol to assess aquatic pollution. Journal of Fish Diseases 22: 25-34.

18. Duncan DB (1955) Multiple range and multiple 'F' test. Biometrics 11: 1-42.

19. Bergh O (2007) The dual myths of the healthy wild fish and the unhealthy farmed fish. Diseases of Aquatic Organisms 75: 159-164.

20. Perera RP, Johnson SK, Lewis DH (1997) Epizootiological aspects of Streptococcus iniae affecting tilapia in Texas. Aquaculture 152: 25-33.

21. Soler F, Reja A, Garcia-Rubio L, Miguez MDP, Roncero V (1996) Anatomo-pathological effect of OTC in tench (Tinca tinca). Toxicology Letters 88: 104.

22. Ghosh R, Homechaudhuri S (2012) Transmission electron microscopic study of renal haemopoietic tissues of *Channa punctatus* (Bloch) experimentally infected with two species of *Aeromonas*. Turkish Journal of Zoology 36: 767-774.

23. Laith AR, Najiah M (2013) *Aeromonas hydrophila*: Antimicrobial susceptibility and histopathology of isolates from diseased catfish, *Clarias gariepinus* (Burchell). Journal of Aquaculture Research and Development 5: 215.

24. Reda RM, Ibrahim RE, Ahmed EG, El-Bouhy ZM (2013) Effect of oxytetracycline and florfenicol as growth promoters on the health status of cultured *Oreochromis niloticus*. Journal of Fisheries and Aquatic Sciences 39: 241-248.

25. Svobodova Z, Sudova E, Nepejchalova L, Aervinka S, Vykusova B, et al. (2006) Effects of OTC containing feed on pond ecosystem and health of carp (*Cyprinus carpio* L.). Acta Veterinaria Brno 75: 571-577.

26. Lio-Po GD, Albright LJ, Alapide-Tendencia EV (1992) *Aeromonas hydrophila* in the Epizootic Ulcerative Syndrome (EUS) of snakehead, *Ophicephalus striatus* and catfish, *Clarias batrachus*: quantitative estimation in natural infection and experimental induction of dermo-muscular necrotic lesion. Diseases in Asian Aquaculture I. Asian Fisheries Society, Manila, Philippines.

27. Agius C, Roberts RJ (2003) Melanomarcophage centers in fish. Journal of Fish Diseases 26: 499-509.

28. Guerra RR, Santos NP, Cecarelli P, Silva JRMC, Hernandez-Blazquez FJ (2008) Healing of skin wounds in the African catfish *Clarias gariepinus*. Journal of Fish Biology 73: 572-583.

29. Ottesen OH, Amin AB (2011) Mortality and cellular response in the skin and gills of plaice (*Pleuronectes platessa* L) to parasite and bacteria infection. Bulletin-European Association of Fish Pathologist 31: 16-22.

30. Ohira Y, Shimizu M, Ura K, Takagi Y (2007) Scale regeneration and calcification in goldfish *Carassius auratus*: quantitative and morphological processes. Fisheries science 73: 46-54.

31. Ashley LM, Halver JE, Smith RR (1975) Ascorbic acid deficiency in rainbow trout and coho salmon and effects on wound healing. The Pathology of Fishes. University of Wisconsin Press, Madison.

32. Quilhac A, Sire JY (1999) Spreading, proliferation and differentiation of the epidermis after wounding a cichlid fish, *Hemichromis bimaculatus*. Anatomical Record 254: 435-451.

22

Effect of Dietary Salt (Sodium Chloride) Supplementation on Growth, Survival and Feed Utilization of *Oreochromis shiranus* (Trewavas, 1941)

Mzengereza K[1]* and Kang'ombe J[2]

[1]*Department of Fisheries Science, Mzuzu University, Private Bag 201, Mzuzu 2, Malawi*
[2]*Department of Aquaculture and Fisheries Science, Lilongwe University of Agriculture and Natural Resources, Bunda College, P.O. Box 219, Lilongwe, Malawi*

Abstract

A study was conducted to determine growth response, survival and feed utilization efficiency of *Oreochromis shiranus*. Fingerings of 12.32 ± 0.34 g were fed diets containing different levels of sodium chloride. (Diet 1= 0%, diet 2=1%, diet 3=1.5% and diet 4=2%). Treatments were replicated three times and 25 fish were stocked in each of the 100 L grass tanks. Fish were fed 5% body weight twice a day. Weight measurements were recorded fortnightly. Fish fed diet 3 and diet 1 had final a weight gain (6.45 g) and (5.25 g) respectively, higher than other diets. In diet 3 fish grew from 12.24 g to 18.69 g and in diet 2 fish grew from 12.34 g to 17.19 g on average. Feed Conversion ratio (FCR) was best in diet 2 (1.51) and diet 3 (1.44), highest was in diet 4 with FCR of 1.87. Diet 1 had an average weight gain of 3.86 g which was statistically not ($P>0.05$) different from that of diet 4. % Survival was higher in diet 1=0% NaCl level *(97.7%) and lowest in diet 2 (94.8%)*. The study indicates that salt can be incorporated in fish diets enhance growth, but can be used up to a limit beyond which growth is compromised.

Keywords: *Oreachromis shiranus*; Dietary salt; Feed utilization; Growth; Survival

Introduction

Malawi's fish production is dwindling and the development is not in tandem with the human population increase. In view of this, it is imperative to increase fish production to tally with the high market demand. To achieve this, there is a need to increase aquaculture production in Malawi through, among other important ways, fish feed supplementation. Thus, nutrition is critical in an effort to produce more fish. FAO [1] indicated that supplementation of fish diet increase yield over and above what would have been achieved without it.

A supplement is defined as a thing added to supply deficiencies. In aquaculture nutritional context, it will be tantamount to the supply of feed to meet one or more nutrient deficiencies of the system for the well-being of the stock (FAO) [1]. Supplementing the naturally available food in a culture system is the most simplistic functional interpretation of supplementary feeds. Supplementary feeding can affect feeding habits and food selection; the fish tend to select narrower range of food on supplementary feed regimes (FAO) [1]. According to Zaugg et al. [2] adding salt to the diet of fish has several advantages some of which are it increases appetite and also acts as humectants by reducing water activity.

Additives like sodium chloride are essentially ideal to enhance growth if incorporated in artificial feed as supplements. The use of salt (sodium chloride) is not a new advent. Salt is one of the essential mineral elements required by the animal and plant bodies for their normal functioning namely; making food taste better, regulating osmotic pressure of the body, form acid in mucous membrane of the stomach (activation of pepsin and enzymes of the salivary glands of the throat and keeping digestive processes normal). Elsewhere, research on salt supplementation has been conducted and yielded significant results. According to Nandeesha et al. [3] addition of salt to the diet of freshwater carp at 1.5% inclusion level resulted in better growth and is widely used in India. Therefore, knowing the optimal levels of different level of salt to be incorporated as a supplement in fish diet would play a vital role in enhancing feed intake and ultimately promote growth.

In freshwater-adapted fish, the passive outward flux of ions such as Na and CI from the fish to the external medium, via the gills, faeces and renal system must be overcome by active uptake of ion (e.g. Na+, CL- K+ and Ca2+) from the water and/or from the diet. Therefore, the diet constitutes an important source of salts that can satisfy the osmoregulatory requirements of fish kept in freshwater and thus spare energy used for osmoregulation leaving more energy available for somatic growth.

Addition of 2% NaCl and 2% potassium chloride to practical diets has shown to have a positive effect on growth of red drum in freshwater and brackish (6 ppt) water but no positive effect were in full strength artificial seawater. The beneficial effect of dietary salt supplementation for red drum in dilute water appears to be due to provision of ions, which were relatively scarce in this hypotonic environment [4].

The beneficial effects of dietary salt supplementation on growth of freshwater and euryhaline fish have not been consistently observed; however, significant improvements in seawater have been noted [4]. A high-salt diet prior to transfer to sea water has also been found to reduce osmoregulatory stress and increase the survival rates in sea water of some African Tilapia (*Oreochromis* species).

Oreochromis shiranus being a freshwater is fish hyper osmotic to its surrounding medium, and encounter the physiological problem of solute loss and in order to compensate this, they resort to active uptake of salt ions from the medium. It is reasonable to expect that diet is an important source of salts that would satisfy the osmolegulatory requirement of *Oreochromis shiranus* and that supplemental salt

*Corresponding author: Mzengereza K, Department of Aquaculture and Fisheries Science, Lilongwe University of Agriculture and Natural Resources, Bunda College, P.O. Box 219, Lilongwe, Malawi
E-mail: kumbumzenge@yahoo.com

spares energy used in osmoregulation, thereby leaving more energy available for growth [5]. Study by Nandeesha et al. [3], affirms that the addition of salt to the diet of freshwater carp at 1.5% levels resulted in better growth and is widely used in India. This corresponds with an experiment conducted on juvenile red drum, where it was found out that addition of 2% NaCl to the diet resulted in greater feed efficiency and greater weight gain [5].

Commercial aquaculture in Malawi is stagnating due to costly protein source especially fishmeal. Therefore, it is worthwhile finding ways of reducing protein intake without realizing significant decrease in growth. Salt act as a feed attractant can therefore help to minimize protein intake if incorporated into a diet. Freshwater fish, being hyper osmotic to the surrounding medium, encounter the physiological problem of solute loss and in order to compensate this, they resort to active uptake of salt ions from the medium. It is reasonable to expect that diet is an important source of salt that would satisfy the osmoregulatory requirement of the fish and that supplement salt spare energy used in osmoregulation, thereby leaving more energy available for growth. Until the recent survey, there had been no documentation that common salt was used as an additive to supplementary feed(s) (FAO) [1]. Therefore, results of this study are expected to provide information of salt as a fish diets supplement impact on growth performance; survival and feed convention ration *Oreochromis shiranus*.

Material and Methods

Experimental site and fish

The experiment was conducted inside the wet laboratory at Bunda fish farm using 100 litre tanks each stocked with 25 fish. It used water from the reservoir which is first pumped to the biofitration tank before it goes to the elevated storage tanks from where it is channeled through pipes to the hatchery and the laboratory. Seed (fingerings) were procured from earthen ponds at the farm with an average weight and total length of 12 g and 93 mm respectively. This was done to achieve uniform growth and minimize competition over space and food as recommended by Webster and Lim [6].

Acclimatization

Fingerings destined for the experiment were first acclimatized in concrete tanks for a fortnight.

Diet preparation

Soybean meal, maize bran and cassava flour were bought from areas in the vicinity of Bunda College, Lilongwe, Malawi. Dry fish, rice bran, unionized salt (sodium chloride) mineral and vitamin premixes were bought from Lilongwe market, Malawi. Soy bean was first roasted before grinding in line with Jauncey [7] who reported that raw bean contains a number of anti-nutritional factors, Soybean trypsin inhibitor which is destroyed by heating during processing. Samples of the ingredients were subjected to proximate analysis following the (AOAC) [8]. Mixed ingredients were used to formulate the diets as illustrated in Table 2. Finally, pellets of approximately 2-2.5 mm were produced using a meat mincer.

Proximate analysis of feed ingredients

The ingredients were assayed for in the aquaculture laboratory using proximate analysis using appropriate methods (AOAC) [8] before feed formulation (Tables 1 and 2).

Experimental design

The experiment was layed out in a Completely Randomized Design (CRD) using 100 litre tanks each stocked with 25 fingerings in triplicate totaling to 12 tanks. Fingerings were stocked at mean weight of 12 g to minimize competition over space and food as recommended by Webster and Lim [6]. Each of the diets was fed at 5% body weight twice a day (8:00 – 9.00 am and 2:00 - 3:00 pm) for 90 days and the quantity of feed was adjusted after a fortnight. Mortality was monitored every day to calculate survival rate. Data on water quality parameters was recorded twice daily before feeding, temperature at (6:00 - 7:00 and 1:00 – 1:30) using a water checker.

Sampling techniques and growth monitoring

Sampling was done every fortnight. 15 fish from each tank were weighed (g) using an electronic scale while measuring board was used to record standard length (mm) Mortality was recorded every day. The following growth parameters were calculated using the following formulae:

$$G_R = \frac{WG}{IW} \; ;$$

$$SGR = \frac{Log_{W_2}}{T_2} - \frac{Log_{w_1}}{T_1}$$

$$FCR = \frac{FI}{WG} \; ;$$

$$Survival = \frac{N_{H}-N_S}{N_S}$$

Statistical analyses

Data analysis was done using SPSS-Statistical Package for Social Scientists (16.0 versions). One-way analysis of variance (ANOVA) was

Proximate Component	Maize bran	Fish meal	Rice bran	Soy bean
Dry Matter %	91.3[a]	91.9[a]	91.4[a]	92.09[b]
Crude Protein %	10.2[a]	62.14[b]	9.53[b]	42.1[d]
Ash %	3.41[a]	11.64[b]	5.48[c]	4.13[d]
Fat %	0.17[a]	13.87[b]	1.65[c]	18.90[d]

Means in the row with the same superscript letter are not significantly different (P>0.05).

Table 1: Proximate analysis of feed ingredients used in feed formulation.

Ingredient	Diet 1 (0% salt)	Diet 2 (1% salt)	Diet 3 (1.5% salt)	Diet 4 (2% salt)
Fish meal	21.90	21.65	25.52	21.40
Soybean meal	21.90	21.65	25.52	21.40
Maize bran	25.60	25.35	25.23	25.10
Rice bran	25.60	25.35	25.23	25.10
Cassava flour	2.0	2.0	2.0	2.0
Salt	0	1.	1.5	2.0
Vitamins 1*	1.5	1.5	1.5	1.5
Minerals 2*	1.5	1.5	1.5	1.5
Total	100	100	100	100

1*.Vitamin premixes (as active matter g/100): A, 2000, 000 UI; D3 245, 000 UI; E, 1, 000 UI; C, 500g; B12, 3 mg; d-Pantothenic, (B6), 100 mg; pyridoxine, 100 mg

2*. Mineral premixes (as active matter g/100): Calcium hydrogen orthophosphate, 10.29; Magnesium sulphate, 50.00; Sodium chloride, 30.00; Potassium chloride, 5.00; Zinc sulphate, 2.75; Manganese sulphate, 1.27; Copper sulphate, 0.39; Cobalt sulphate, 0.24; Chromic chloride, 0.064. Premixes contained also 1.06 g Potassium and 21.2 g, Sodium salts.

Table 2: Composition of the experimental diets (Kg) fed to *Oreochromis shiranus*

used to test significant differences among treatment means at 0.05% alpha level. Least Significant difference (LSD) test was employed to separate significantly different means.

Results

Fish growth

There was no significant difference (P>0.05) among treatments in the first 2 weeks in terms of weight gain. From week 2 to week 10, growth was significantly different (P<0.05) among treatments. Treatment 3 (1.5% salt level) and treatment 2 (1% salt level) had higher mean weight gains recording 18.69 g and 17.19 respectively while treatment 4 and treatment1 registered 15.93 g and16.32 mean weights at harvest (Figure 1 and Tables 3-5).

Survival of fish

Survival was not significantly different (P>0.05) among treatments (Table 4). Survival rate was higher in treatments 1 and 4 while treatment 3 had the lowest survival.

Discussion

Results show that higher fish growth was observed at 1% and 1.5% salt inclusion level (Table 4). Thus, consistent to a large dimension with other researchers in that significant growth can be observed when fish are fed on salty diet than salt free diet. The addition of salt to the diet of freshwater carp at a level of 1.5% resulted in significantly better growth and is in widespread use in India [3]. The final weight gain of Rohu, *Labeo rotiha* fed 0.5% and 1% NaCI –incorporated diets were significantly (P<0.005) higher than that of the control (0% NaCI) [9]. Freshwater fish, being osmotic to the surrounding medium, encounter the physiological problem of solute loss and in order to compensate this, they resort to active uptake of salt ions from the medium [5]. Therefore, it is reasonable to speculate that dietary salt satisfied the osmolegulatory requirement of *Oreochromis shiranus* as it is a freshwater fish and that supplemental salt spared energy used in osmoregulation, thereby leaving more energy for somatic growth. Smith et al. [10] reported that the dietary sodium intake of salmonids kept in fresh water increased by eightfold from winter to summer. This corresponds to the increase in feeding and shows that almost all the sodium required can be derived from dietary salt. This can therefore be used as a source of salts for fish kept in freshwater, providing ions which the fish cannot obtain from the hypotonic environment.

In addition, Asian sea bass *Lates calcirifer* reared in freshwater recirculating fed 1% NaCI diet exhibited a better feed conversion ratio and a higher (yet not significant) growth, not only that, but also enhanced the activity of brush body enzymes e.g. leucine amino peptidase and were pronounced in pyloric caeca [11]. The better enzymatic activity can be explained by the absorption mechanism of end products –glucose and amino acids. Since the glucose and most of amino acids is dependent on the Na+/K+ ATPse pump [12], a higher concentration of Na+ in the lumen might lead to a better absorption of carbohydrates and amino acids. Since the activity might be inhibited by its end products (amino acids and carbohydrates) [13], reduction of the end products concentration can lead to better enzyme activity in the lumen of fish fed feed enhanced with NaCl. Therefore, the previous work corroborates with this study in that *Oreochromis shiranus* had a better enzyme activity owing to dietary NaCl, thus registering a better growth 1% and 1.5% inclusion level.

Salt appears to affect growth rate inversely when the level of supplementation interferers with the balance of other essential dietary

components. Feeding supplementary NaCI in freshwater has strongly stimulated gill Na+ -K+ - ATPsc activity, which has been found to be accompanied by an increase in the number of chloride cells. This is ascribed to the need for increased gill NaCI extrusion capacity because the salty diet imposed a salt load that exceeded the normal capacity for diffusion loss across the gills and the excretion through urine [14]. The decrease in weight gain and SGR of *Oreochromis shiranus* in the present study could be at 2% could be because of excess salt level that hindered other metabolic processes responsible for growth.

However, there was slow fish growth in the first two weeks among all treatments but increased later. This might be because the fish were still in the process of being accustomed to the new feed and the environment of the experimental unit. According to Likongwe [15], and Lovell [16], there is a specific period of rapid growth in the growth curve of fish.

The experiment also indicates that fish fed on 1% NaCI and 1.5% NaCI levels had a better utilization of feed as evidenced in the lower Feed Conversion Ratio of 1.51 and 1.47 respectively (Table 4). This is an indication that feed was appropriately utilized and hence a superior growth rate. The results agrees favorably with Gangadhara et al. [9] reported that 0.5% NaCI and 1% NaCI produced better FCRs at 1.56 and 1.44 respectively and subsequently higher growth rate. Fish feeds constitute one of the most expensive components in rearing of fish, and high protein levels required for these fish are also a major source of nitrogenous products harmful to fish in closed systems. Therefore, the fact that the addition salt to the feed resulted in a better FCR is of great importance as this is an indication of good feed conversion into fresh. 2% NaCI had the highest FCR clearly showing that feed utilization was poor. Low digestibility and faster evaluation of food have been associated with high levels of NaCI in diets [17]. This in turn could affect assimilation and conversion efficiency. It is, therefore, reasonable to suggest that 2% salt level led to excessive salt loading and adversely affected feed intake, digestion and/or absorption, because of

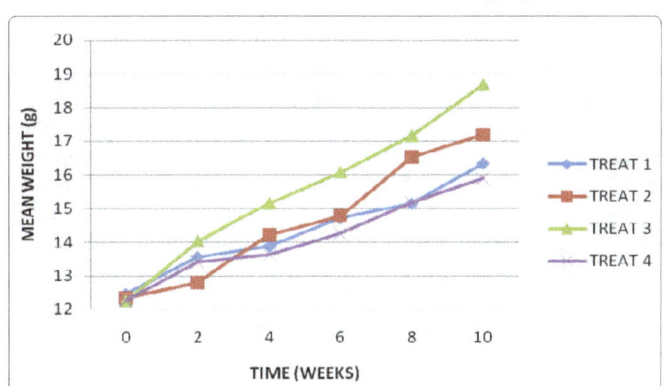

Figure 1: Growth trend in weight (g) of *Oreochromis shiranus* fed on different salt (sodium chloride) levels in the diets.

Proximate Component	Diet 1 (0% salt)	Diet 2 (1.0% salt)	Diet 3 (1.5% salt)	Diet 4 (2.0% salt)
Dry mater	92.28 ± 0.27[a]	92.22 ± 0.12[a]	92.44 ± 0.27[a]	90.48 ± 0.6[b]
Fat (%)	14.21 ± 0.05[a]	16.01 ± 0.0[b]	19.87 ± 0.06[c]	15.14 ± 0.04[a]
C P (%)	30.70 ± 0.02[a]	30.63 ± 0.12[a]	30.48 ± 0.05[a]	30.41 ± 0.02[a]
Ash	14 ± 0.04[a]	12.42 ± 0.2[b]	10 .65 ± 0.03[c]	13.70 ± 0.23[a]
Energy(kJ/g)	21.04 ± 0.2[a]	21.35 ± 0.10[a]	21.56 ± 0.23[a]	20.21 ± 0.15[b]

Means in the row with the same superscript letter are not significantly different (P>0.05).

Table 3: Proximate composition of experimental diets (Mean ± SE).

the changed gastric/intestinal environment and may even have had pathological effects on *Oreochromis shiranus* in this experiment. That is probably the reason for a depressed weight gain and SGR.

Water quality parameters were within limit throughout the experiment (Table 5). Temperature ranged from 20°C to 24°C which is within the recommended levels. Claude reported that the optimal temperature range for warm water fish is 20°C to 28°C but these are typically encountered in the natural regimes [18]. Dissolved oxygen ranged from 7.02 to 8.71 mg/L and this was within tolerable limit. Claude reported that dissolved oxygen of above 5 mg/L augurs well for the survival of fish [18]. In the present study, pH levels ranged from 7.78 to 8.21 and are considered desirable and could in no way negatively affect growth, survival and feed utilization of *Oreochromis shiranus*. According to FAO [19], the optimal PH levels are 4 to 10 of which below or above become detrimental to fish survival. NH_3 and salinity were also not significantly different (P>0.05) among treatments. However, NH_3 were sometimes slightly higher with a maximum of 0.38 mg/L (Table 4) but this did not have adverse effects on the survival of *Oreochromis shiranus*. According to Lovell, Tilapia can tolerate NH_3 levels up to 2.4 mg/L [16]. Survival was higher in diet1, 0% NaCl level (97.7%) and lowest in diet 2, 1%NaCl (94.8%). This meant that survival did not depend on sodium chloride level.

Conclusion

Different salt (NaCl) levels in a diet of *Oreochromis shiranus* exhibit different growth responses. Salt level of 1.5% (treatment 3) is optimum for incorporation in diets of *Oreochromis shiranus*. 1% inclusion level plays second fiddle. Feed utilization was better in both 1% and 1.5% salt inclusion levels with 1.54 and 1.44 FCRs respectively. At 2% SGR, FCR and weight gain diminished, thus an increase in salt level beyond a certain (optimum) level make the feed less appropriate for consumption and affect growth negatively.

Recommendation

The intensive rearing of fish in closed systems is costly due to high cost of feed like fish meal and calls for methods that will enhance food utilization and growth while generating economic returns. Utilizing a diet enhanced with 1% to 1.5% salt has an advantage on Tilapia species (*Oreochromis shiranus*), as observed from the results of the present study. It can lead to better feed utilization under intensive production conditions and can reduce the cost of feed since there is dilution of expensive component with a cheap mineral. This simple method can be used by small-scale fish farmers and does not require special means. Optimum salt content must be taken into account to prevent growth reduction owing to excess salt incorporation.

PARAMETER	Treatment 1 (0% salt)	Treatment 2 (1.0% salt)	Treatment 3 (1.5% salt)	Treatment 4 (2.0% salt)
Initial Weight(g)	12.46 ± 0.19ᵃ	12.34 ± 0.09ᵃ	12.24 ± 0.24ᵃ	12.26 ± 0.11ᵃ
Final Weight(g)	16.32 ± 0.99ᵃ	17.19 ± 0.19ᵇ	18.69 ± 0.22ᶜ	15.93 ± 0.19ᵃ
Weight gain (g)	3.86 ± 0.86ᵃ	4.84 ± 0.41ᵇ	6.45 ± 0.38ᶜ	3.67 ± 0.15ᵃ
Weight gain/day(g)	0.05 ± 0.01ᵃ	0.06 ± 0.01ᵇ	0.07 ± 0.003ᶜ	0.05 ± 0.003ᵃ
SGR (%)	0.33 ± 0.06ᵃ	0.41 ± 0.04ᵇ	0.52 ± 0.03ᶜ	0.32 ± 0.01ᵃ
FCR	1.77 ± 0.06ᵃ	1.51 ± 0.06ᵇ	1.44 ± 0.08ᶜ	1.87 ± 0.09ᵃ
Survival (%)	97.7 ± 0.5	94.8 ± 0.8	96.2 ± 0.9	95.5 ± 0.7

Means in the row sharing the same superscript letter are not significantly different (P>0.05).

Table 4: Mean initial weights, daily weight gain, Feed conversion ratio (FCR), Specific growth rate (SGR %/day) of *Oreochromis shiranus* raised on different salt (NaCl) diets in 100 L tanks: (Mean ± SE).

Parameter	Treatment 1 (0% salt)	Treatment 2 (1.0% salt)	Treatment 3 (1.5% salt)	Treatment 4 (2.0% salt)
Temperature morning afternoon	20.53 ± 0.13ᵃ 22.57 ± 0.11a	20.57 ± 0.12ᵃ 23.27 ± 0.10ᵃ	21.53 ± 0.13ᵃ 23.31 ± 0.10ᵃ	20.08 ± 0.13ᵃ 23.24 ± 0.10ᵃ
PH morning afternoon	8.10 ± 0.03ᵃ 7.96 ± 0.05ᵃ	7.9 ± 0.04ᵃ 8.09 ± 0.06ᵃ	8.13 ± 0.04ᵃ 8.09 ± 0.06ᵃ	8.27 ± 0.03ᵃ 8.17 ± 0.06ᵃ
Ammonia (NH_3)	0.389 ± 0.02ᵃ	0.227 ± 0.03ᵇ	0.327 ± 0.04ᵃ	0.343 ± 0.04ᵃ
Dissolved oxygen (DO)	7.87 ± 0.07ᵃ	8.06 ± 0.08ᵇ	8.23 ± 0.08ᵇ	7.02 ± 0.07ᵃ

Means in the row sharing the same superscript letter are not significantly different (P>0.05).

Table 5: Water quality parameters observed throughout the experimental period.

Acknowledgement

The authors wish to thank the crew at Lilongwe University of Agriculture and Natural Resources, department of aquaculture and fisheries science laboratory for technical assistance. Special mention is Mr Elton Nyali, chief technician for his tireless effort.

References

1. FAO (1995) Farm-made aqua feeds, Viale delle Terme di Cracalla, 00100 Rome, Italy.

2. Zaugg WS, RoD D, Prencice EF, Gores KX, Waknizt FW (1983) Increased Seawater survival and contribution to the fishery of Chinook salmon (*Oncorhynchus Tshawytscha*) by supplemental dietary salt. Aquaculture 32: 183-188.

3. Nandeesha MC, Gangadhar B, Keshavanath P, Varghese TJ (2000) Effect of dietary salt supplementation on growth, biochemical composition and digestive enzyme activity of young *Cyprinus carpio* (Linn) *Cirrhinus mrigala* (Ham). Journal of aquaculture in the tropics 15: 135-144.

4. Lim C, David JS (1994) Nutrition and Utilization Technology in Aquaculture, AOUS press, Chapman and Hall, Illinois, USA.

5. Gatlin DM, Mackenzie DS, Craig SR, Neill WAH (2000) Effects of dietary sodium chloride on red drum juveniles in waters of various salinities. Progressive fish Culturist.

6. Webster CD, Lim CE (2002) Nutrient requirement and feeding of finfish for Aquaculture, CABI publishing, Wallington, U.K.

7. Jauncey K, Ross B (1998) A guilde to Tilapia feeds and feeding. University of Stirling.

8. AOAC (2003) Official Methods of Analysis.(15edn), Association of Official Analytical Chemists. Washington, DC, USA.

9. Gangadhara B, Keshvaanar P, Varghese TJ (2004) Growth performance, Feed Utilization and Body Composition of Rohu (*Lebo rotiha*) fed salt incorporated diets. Canadian Journal of Fisheries Science 23: 98-99.

10. Smith AA, Craig SR, Neil WAH (1989) Effect of supplemental dietary sodium chloride on growth rate of Tilapia, *Oreochromis Shiranus*. Canadian Journal of Fisheries and Aquatic science 11: 719-726.

11. Harpaz M (2005) Marine fish Biology. Elsevier Company, Porstmouth, UK.

12. Klein S, Cohn SM, Alpers DH (1998) The alimentary canal in nutrition. (9thedn), William and Wilkins, Baltimore, MD.

13. De la Fuente JL, Rumbero A, Martin JF, Liras P (1997) Delta-1-piperideine-6-carboxylate dehydrogenase, a new enzyme that forms alpha-aminoadipase in *stretomyces clavuligerus* and other cephamycin C-producing actinomycetes. Journal of Biochemistry 327: 59-64.

14. Staurnaes M, Asgurd T, Griffiths D, Husbay J, Einarsdottir I, et al. (1990) Effects of dietary NaCl supplementation on Smoltification and sea water tolerance in Atlantic salmon reared to smolt development at time release. Aquaculture 118: 327-337.

15. Likongwe JS (1998) Effect of varying dietary energy level on growth and feed conversion of juvenile Tilapia , *Oreochromis niloticus* at salinity level. Proceeding of the first regional workshop in Aquaculture, Bunda college of Agriculture, Lilongwe, Malawi.

16. Lovell T (1989) Nutrition and Feeding of Fish. Van Nostrad Reinhold, New York.

17. Salman NA, Eddy FB (1987) Effect Of different sodium chloride on growth food intake and conversion efficiency in rainball trout (*salmo qaidneri* Richardson) Aquaculture 70: 131-144.

18. Claude EB (1990) Water quality in pond for Aquaculture, Birmingham publishing company. Birmigham, United Kingdom.

19. FAO (1988) Inland fish enhancement, Toowimba, Queensland, Australia.

23

Non-Selectivity of R-S Media for *Aeromonas hydrophila* and TCBS Media for *Vibrio* Species Isolated from Diseased *Oreochromis niloticus*

Ibrahim M Aboyadak[1]*, Nadia GM Ali[1], Ashraf MAS Goda[2], Walaa Saad[3] and Asmaa ME Salam[4]

[1]*Fish Disease Lab, National Institute of Oceanography and Fishery (NIOF), Egypt*
[2]*Aquaculture Division, National Institute of Oceanography and Fishery (NIOF), Egypt*
[3]*Central Diagnostic and Research Lab, Faculty of Veterinary Medicine, Kafrelsheikh University, Egypt*
[4]*Faculty of Aquatic and Fisheries Sciences, Kafrelsheikh University, Egypt*

Abstract

The current study was conducted to determine the bacterial pathogens incorporated in mass mortality observed in cultured *Oreochromis niloticus* farms at Kafrelsheikh province, Egypt, during the summer season of 2015. Moribund fish samples were collected from six infected farms. General signs of septicemia were dominant in the clinical and gross internal examination of diseased fish. The pathogenic bacteria were isolated on specific media then confirmed by polymerase chain reaction (PCR). Out of thirty isolates, nineteen *Aeromonas hydrophila*, seven *Vibrio cholera* and three *Vibrio alginolyticus* isolates were recovered and identified using PCR. Current study indicated non-selectivity of Rimler-Shotts media for selective isolation of *Aeromonas hydrophila* also non-selectivity of TCBS media for *Vibrio* spp. isolated from diseased *Oreochromis niloticus*.

Keywords: *Aeromonas*; *Vibrio alginolyticus*; *Vibrio cholera*; PCR.

Introduction

Fish is cheap delicious animal protein of high nutritive value; it represents a good substitute to red meet especially in developing countries. Expanding in global aquaculture help to provide the excessive demand by increased human population. Finfish production comprised 49.8 million tons, (99.2 billion USD), Egypt ranked as the 8th aquaculture producing country in 2014 [1], it is producing about 1.130.000 ton of finfish that representing 2.26% of global aquaculture of these species. Egypt also, ranked as the second global tilapia producer after China, it produces about 867557 tons in 2014 [2].

Economic losses from diseases outbreaks in global aquaculture estimated by 3 Billion USD in 1997 [3]. Bacterial diseases reported to be the main etiological agents responsible for severe economic losses in cultured fish, including infection with *Aeromonas* spp., *Pseudomonas* spp., *Vibrio* spp., *Streptococcus* spp. and *Enterococcus* spp. as recorded by Aboyadak et al. [4], Zhang et al. [5], Olugbojo and Ayoola [6], Darak and Barde [7], Thune et al. [8].

Bacterial fish diseases resulted in various clinical finding including skin hemorrhage, ascites and exophthalmia [9-11]. Enlarged congested internal organs including liver spleen and kidney, and intestinal hemorrhage with or without presence of ascetic fluid in abdominal cavity are the main internal lesions [12-14].

The genus *Vibrio* is a member of the family *Vibrio*naceae, it is a Gram negative non-spore forming comma shape rods. Thiosulphate citrate bile salt sucrose (TCBS) agar is highly selective for the isolation of *V. cholerae* and another *Vibrio* spp. The acidification of the medium resulting from the fermentation of sucrose by *Vibrio* makes bromthymol blue turns yellow.

The genus *Aeromonas* is belongs to the family Aeromonadaceae, it is a Gram negative facultative anaerobic cocco-bacilli. Rimler-Shotts (RS) agar used for selective isolation and identification of *Aeromonas hydrophila*, the organisms that ferment maltose are seen as yellow colonies.

Isolation and biochemical identification of pathogenic bacteria are laborious and time consuming, application of molecular methods such as the polymerase chain reaction (PCR), is an easy, less expensive, and more rapid means for diagnosis of such diseases [15,16].

The present work was conducted to determine the bacterial pathogens responsible for mass mortality in cultured tilapia farms at Kafrelsheikh province using selective media and polymerase chain reaction.

Materials and Methods

Study area

Samples were taken from six affected tilapia farms at Torombat seven, Alreiad district, Kafrelsheikh governorate northern Egypt. The affected farms complain increased mortality during the study period from March to August 2015.

Samples

Thirty live diseased Nile tilapias (*Oreochromis niloticus*) with obvious signs of septicemia were collected; five fish were taken from each farm. Diseased fish weight was ranged between 100-250 g and 16-24 cm in total body length. Each fish sample was packed alive in a separate sterile labeled plastic bag and transported in ice pox to lab.

Clinical examination

The clinical examination including observation of any external lesions was performed according to the method described by Noga [13].

Post mortem examination

The post mortem examination was performed according to the method described by Heil [17].

*Corresponding author: Ibrahim M. Aboyadak, Fish Disease Lab, National Institute of Oceanography and Fishery (NIOF), Egypt
E-mail: i.aboyadak@gmail.com

Isolation and identification of the causative agent

Under complete aseptic condition, a small tissue pieces from heart, hepatopancreas, spleen, posterior kidney and from lesions in musculature were taken from each fish to a labeled test tube containing 10 ml sterile peptone water, after that the sample was homogenized at 3000 rounds per minute (rpm) for 1 min using homogenizer pro® USA. Test tubes were centrifuged at 500 rpm for 30 sec and one ml from supernatant was added to another test tube containing sterile tryptic soy broth then incubated at 33°C for 24 h. Rimler-Shotts media with novobiocin selective supplement (HiMedia, India), tryptic soy agar (Oxoid, England), *Pseudomonas* selective agar with CFC (Cetrimide – Fucidin – Cephaloridine) and glycerol supplement, (Lab M, United Kingdom), TCBS agar media (Oxiod, England) and Edwards media modified (Oxoid, England) were streaked from each sample then incubated at 33°C for 24h. Few morphologically similar colonies from isolates grown on each media were picked up and inoculated to trypticase soy broth, incubated at 33°C for 24 h then 0.5 ml from this broth was preserved on sterile 50% glycerol (biotechnology grade) one volume from overnight broth : one volume from 50 % glycerol, after that it was stored on - 80°C for - PCR identification.

Identification of collected strains using PCR

DNA extraction: Bacterial DNA extraction was performed by thermolysis based on destruction of bacterial cells using dry heat as described by Ahmed et al. [18]. Two hundred microliter of overnight cultured broth was mixed with 800 µl of nuclease free double deionized distilled water in eppendorf tube, after that the eppendorf tubes were placed in heat block for 5 min at 95°C followed by cool centrifugation for 2 min at 15000 rpm and 4°C, the supernatant was used as DNA template and stored at -20°C for PCR study.

Primers

Primers used for identification of recovered isolates shown in (table 1).

PCR procedures

PCR mixture for amplification of targeted DNA: The reaction volume for all performed PCR reaction was 25 µl, each reaction volum consists of 5 µl of 5X master mix (taql/high yield- Jena Bioscience, Jena, Germany, consists of DNA polymerase, dNTPs mixture, (NH4) SO4, MgCl2, Tween 20, Nonidet P-40, stabilizers) + 1.25 µl of each primer (forward and reverse) (20 pmol/µl), + 5 µl of extracted bacterial DNA + 12.5 µl of nuclease free double deionized distilled water. Amplification of targeted DNA gens was carried out in thermal cycler (T100TM Thermal cycler, BIORAD, USA).

A- Thermal cycle adjustment for amplification of genus *Aeromonas* and *Aeromonas hydrophyla* target DNA: Amplification of genus *Aeromonas* targeted DNA was performed according to the method described by Lee et al. [19], with an initial denaturation at 94°C for 4 min followed by 35 cycles of denaturation each at 94°C for 1 min, annealing at 68°C for 30 sec and extension steps at 72°C for 45 secs, then final extension step at 72°C for 10 min.

Amplification of *Aeromonas hydrophila* targeted DNA (specific-16S rRNA gene) was done using the same primer described by Trankhan et al. [20] but with different PCR condition. The thermal cycler condition was adjusted as following: initial denaturation at 94°C for 2 min, followed by 35 cycles of denaturation, at 94°C for 30 secs, annealing at 55.5°C for 30 sec and extension steps at 72°C for 30 secs, then one final extension step at 72°C for 10 min.

B- Thermal cycle adjustment for amplification of genus Vibrio, Vibrio cholera, Vibrio parahaemolyticus and Vibrio alginolyti-

Table 1: Primers used for detection of Genus *Aeromonas* and *Aeromonas hydrophila*, Genus *Vibrio*, *Vibrio cholera*, *Vibrio parahaemolyticus* and *Vibrio alginolyticus*.

Target bacteria	Primer Name	Oligonucleotide Sequence (5´-3´)	Size (Bp)
Genus Aeromonas	AER-F	CTA CTT TTG CCG GCG AGC GG	953 bp
	AER-R	TGA TTC CCG AAG GCA CTC CC	
A. hydrophila	16SrRNA-F	GGC CTT GCG CGA TTG TAT AT	103 bp
	16SrRNA-R	GTG GCG GAT CAT CTT CTC AGA	
Genus Vibrio	16SrRNA-F	CCT GGT AGT CCA CGC CGT AA	168 bp
	16SrRNA-R	CGA ATT AAA CCA CAT GCT CCA	
V. cholera	OmpW-F	CAC CAA GAA GGT GAC TTT ATT GTG	427 bp
	OmpW-R	CGT TAG CAG CAA GTC CCC AT	
V. parahaemolyticus	Collagenase-F	GAA AGT TGA ACA TCA TCA GCA CGA	271 bp
	Collagenase-R	GGT CAG AAT CAA ACG CCG	
V. alginolyticus	GyrB-F	GAG AAC CCG ACA GAA GCG AAG	337 bp
	GyrB-R	CCT AGT GCG GTG ATC AGT GTT G	

cus **target DNA:** Thermal cycler adjustment for amplification of genus *Vibrio* and mentioned *Vibrio* species targeted DNA were the same, it performed according to the method described by Wei et al. [15] except, reaction for each species was performed separately (not multiplex). Amplification starts with an initial denaturation at 94°C for 3 min, followed by 30 cycles of (denaturation at 94°C for 30s, annealing at 60°C for 30s, and extension step at 72°C for 120 s), finishing with a final extension step at 72°C for 10 min then storage at 4°C.

DNA assay

DNA was assayed by agarose gel-electrophoresis, 5 µl from the PCR products were loaded in 1.5% agarose 0.6 µg/ml ethidium bromide in 1X tris acetate EDTA buffer using gel electrophoresis apparatus (SCIE-PLAS, UK). DNA bands were visualized using UV transilluminator (Winpact Scientific, USA).

Results and Discussion

Bacterial infections represent the major cause of economic losses in fish farms [21]. The recorded mortality in affected farms was 8% to 15% during 2 weeks after disease onset (appearance of clinical signs as daily mortality was 0.5% to 1%) which considered direct economic losses due to bacterial infection.

General signs of septicemia including presence of hemorrhagic patches on the skin and at the base of pectoral fin together with scale desquamation and skin ulceration were the most common observed signs (Figure 1). Increase the abdomen size (ascites), (Figure 1) and exophthalmia was occasionally observed in diseased fish. Congestion of internal organs with enlarged liver and enlarged gall bladder (Figure 1), congested stomach walls and intestine was the dominant postmortem lesion (Figure 1). Similar clinical and post mortem finding was observed by Asaad [22], Zakaria [23], Laith and Najiah [24] and Zhang et al. [5] in naturally infected fish with *Aeromonas hydrophyla*. The same clinical finding was also observed in naturally infected *Oreochromis niloticus* with different *Vibrio* species as described by Chen et al. [25] and Okasha et al. [26].

Isolated bacterial pathogens express their effect through different virulence factors. *Aeromonas hydrophila* release a variety of virulence factors which are important for their pathogenicity [27], these virulence factors including outer membrane protein, extracellular products (cytotoxins and proteases) and enterotoxins [12,28]. Pathogenic *Vibrio* species produce a wide variety of proteases and extracellular enzymes that capable of causing tissue and cell damage in infected fish, purified *Vibrio* proteinases are toxic to fish [8], *Vibrio cholera* is a highly virulent fish pathogen, [29], it has the ability to bind collagen, fibrinogen, gelatin, and fibronectin and have specific surface receptors for connective tissue [30]. Haemolysins and toxic proteases produced by *Vibrio alginolyticus* play great role in pathogenicity [31].

Out of thirty sample twenty-two isolate grow on R-S media producing yellow colonies which is specific for *Aeromonas hydrophyla* (Figure 1) while on TCBS media fourteen isolate grow giving yellow colonies (Figure 1) specific for genus *Vibrio* and none of the thirty isolates growth over *Pseudomonas* selective agar or modified Edwards media. Only one sample not grows on any of previously mentioned specific media.

Results of growth on specific media was controversial, as 7 bacterial isolates grow on both R-S and TCBS media in the same time that means either presence of mixed infection between *Aeromonas* and *Vibrio* in the same sample or ability of some strains to grow on both media in the same time, and to determine any of these possibilities is true all the recovered yellow colonies grow on Rimler-Shotts media were subjected to PCR identification for genus *Aeromonas* and for genus *Vibrio*, and so for all isolates recovered on TCBS were subjected to PCR identification for genus *Vibrio* and for genus *Aeromonas*.

PCR identification results become more obvious, out of twenty-two isolate grow on Rimler-Shotts media nineteen isolate was positive for genus *Aeromonas* and all of them are *Aeromonas hydrophyla*, and three isolates were positive for genus *Vibrio* (Table 2) after that these three-*Vibrio* species were identified by PCR as two *Vibrio alginolyticus* isolates and one *Vibrio cholera* isolates.

Table 2: Number of clinically diseased fish samples and recovered bacterial isolates.

Number of affected farms	Number of clinically diseased fish samples	Number of recovered bacterial isolates
6	30	30
No. of isolates grown on R-S media	No. of isolates grown on R-S, identified as genus *Aeromonas* by PCR	No. of isolates grown on R-S, identified as genus *Vibrio* by PCR
22	19	3
No. of isolates grown on TCBS media	No. of isolates grown on TCBS, identified as genus *Vibrio* by PCR	No. of isolates grown on TCBS, identified as genus *Aeromonas* by PCR
14	10	4

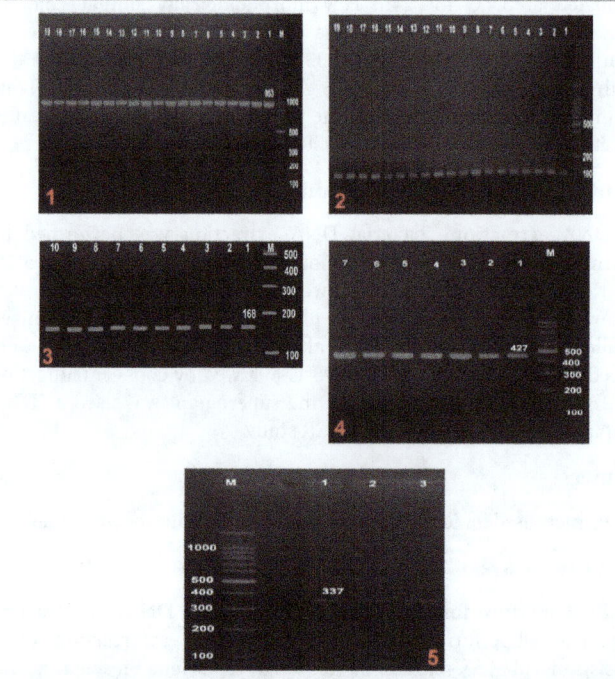

Figure 2: (1) Agarose gel electrophoresis of PCR product showing specific genus *Aeromonas* bands at 953 bp. (2) Agarose gel electrophoresis of PCR product showing specific *Aeromonas hydrophyla* bands at 103 bp. (3) Agarose gel electrophoresis of PCR product showing specific genus vibrio bands at 168 bp. (4) Agarose gel electrophoresis of PCR product showing specific *Vibrio cholere* bands at 427 bp. (5) Agarose gel electrophoresis of PCR product showing specific *Vibrio alginolyticus* bands at 337 bp.

Figure 1: (1) Naturally infected *Oreochromis niloticus* showing scale desquamation, skin ulceration and hemorrhage. (2) Naturally infected *Oreochromis niloticus* showing abdominal distention with presence of hemorrhagic patches at caudal peduncle. (3) Naturally infected *Oreochromis niloticus* showing inflamed and severely congested hepatopancreas with presence of hemorrhagic patches. (4) Naturally infected *Oreochromis niloticus* showing inflamed and congested stomach and intestine. (5) R-S media with novobiocin selective supplement showing characteristic yellow colonies of *Aeromonas hydrophyla*. (6) TCBS media showing yellow colonies of *Vibrio cholera* and *Vibrio alginolyticus*.

PCR identification for fourteen isolates grown on TCBS indicating presence of only ten isolates were positive for genus *Vibrio*, from them seven isolates were *Vibrio cholera*, and three was *Vibrio alginolyticus* and *Vibrio parahaemolyticus*. While the remaining four isolates identified as *Aeromonas hydrophyla* (Table 2).

PCR is a rapid and specific molecular technique provide a highly accurate results, amplification of 16S rRNA gene fragment indicated presence of nineteen *Aeromonas* sp. Isolate that characterized by appearance of specific bands at 953 bp, all these isolates were further identified by PCR as *Aeromonas hydrophyla* as characteristic bands appeard at 103 bp (Figure 2). Other researches indicating identification of such genus and species using the same primers design including Lee et al. [19] and Zakaria [23] for genus areomonas and Trankhan et al. [20] and Aboyadak et al. [4] for *Aeromonas hydrophyla*.

Agar gel electrophoresis of PCR product for detection genus *Vibrio* and its different species revealed detection of ten isolates related to genus *Vibrio* at 168 bp by amplification of 16S rRNA gene (Figure 2), seven isolates were identified as *Vibrio cholera* at 427 bp through amplification of OmpW gene (Figure 2) and three isolates were identified as *Vibrio alginolyticus* at 337 bp through amplification of GyrB gene (Figure 2). In current research, no isolates were identified as *Vibrio parahaemolyticus*. Wei et al. [32] and Wei et al. [15] recorded the same results using the same primers but different thermal cycler conditions.

PCR results indicated the ability of 4 *Aeromonas hydrophila* isolates to grow on TCBS producing yellow colonies which was in complete agreement with Buller [33], Bridson [34] and Dworkin et al. [35], they indicated the ability of some *Aeromonas hydrophila* strains to grow on TCBS producing small yellow colonies. The PCR results also assert the ability of three *Vibrio* isolates to grow on Rimler-Shotts media, the three isolates were PCR negative for genus aeromonas and positive for genus *Vibrio*, one of them identified as *Vibrio cholera* and other two identified as *Vibrio alginolyticus*. Davis and Sizemore [36] and Arcos et al. [37] reported the ability of *Vibrio alginolyticus* to grow on Rimler-Shotts media. The current research results indicated non-selectivity of both Rimler-Shotts and TCBS media for isolation *Aeromonas* and *Vibrio* species from diseased fish and proved the inevitable need for other diagnostic techniques as PCR.

Current work refered the similarity between different isolated bacteria colonies color on specific media (R-S and TCBS) to the biochemical characters of each recovered bacterial species. On R-S media *Aeromonas hydrophila* ferment maltose that increase the pH to acidic side that subsequently turns bromthymol blue to yellow color, *Vibrio cholera* and *Vibrio alginolyticus* also ferment maltose producing similar change in media color to yellow, but the role of antibiotic substance (novobiocin) present in R-S media is to prevent the growth of any microorganisms other than aeromonas, but if *Vibrio cholera* and *Vibrio alginolyticus* resist novobiocin they can grow producing yellow color colonies similar to that of *Aeromonas hydrophila*, Rahim and Aziz [38] recorded resistance of 82% of toxogenic and 75% of non toxogenic *Vibrio cholera* strains isolated from water to novobiocin, and this support our point of view. On the other hand, *Vibrio cholera* and *Vibrio alginolyticus* can grow on TCBS giving yellow colonies as they ferment sucrose producing acid that decrease pH which makes bromthymol blue turn yellow color, this media depends on the presence of ox bile in inhibition of another bacteria growth. Current research documents the ability of some strains of *Aeromonas hydrophila* to grow on TCBS media giving yellow colonies, this can be explained by the ability of such microorganism to grow in the presence of bile and this suspect is supported by the ability of *Aeromonas hydrophila* to induce enteritis, moreover *Aeromonas hydrophila* has the ability to ferment sucrose as mentioned by Cipriano et al. [9] the produced acid shifts the pH to acidic side developing yellow color on TCBS media.

Thirty bacterial isolates were recovered from diseased fish samples (Table 3), *Aeromonas hydrophyla* represent 63.3% of total isolated pathogens, it considered one of the most important bacterial pathogens affecting cultured freshwater fish particularly tilapia inducing motile

Table 3: Total number and percent of each recovered bacterial isolates.

Recovered bacteria	Number of isolates	Percent (%)
Aeromonas hydrophila	19	63.33
Vibrio angiolyticus	3	10
Vibrio cholera	7	23.33
Other bacteria (not identified)	1	3.33

Aeromonas septicemia [9,39]. Ten *Vibrio* isolates represent 33.3% of isolated bacterial pathogens, 23.3% of them were identified as *Vibrio cholera* and 10% were *Vibrio alginolyticus*. *Vibrio* spp, which are considered a significant problem affecting aquaculture worldwide [40,41]. *Vibrio alginolyticus* and *Vibrio cholerae* are major pathogens in Chinese aquatic products [15]. Senderovich et al. [42] isolated *Vibrio cholera* from *Oreochromis aureus*; Rehulka et al. [43] isolated pathogenic *Vibrio cholerae* non-O1/non-O139 from moribund freshwater fish, and proved its ability to induce disease condition through experimental infection of common carp, infected carb developed typical signs of *Vibrio*sis. Unidentified pathogenic bacteria represent 3.3% (one out of 30 isolates).

Conclusion

In conclusion, the disease condition affecting studied farms was induced by Gram negative bacterial pathogens including *Aeromonas hydrophila*, *Vibrio angiolyticus* and *Vibrio cholera* that represent (96.6%) of total isolates. Isolation and identification of bacterial fish pathogens depending on culture characters on specific media is inadequate as current research proves non-selectivity of R-S to *Aeromonas hydrophyla* and TCBS media for *Vibrio* sp. Isolated from fish, so polymerase chain reaction identification is inevitable.

References

1. FAO (2016) The State of World Fisheries and Aquaculture. Rome, Italy.

2. GAFRD (2016) Fish Statistics yearbook 2014, General Authority for Fish Resources and Development, Ministry of Agriculture, Cairo, Egypt.

3. Subasinghe RP, Bondad-Reantaso MG, McGladdery SE (2001) Aquaculture development, health and wealth. Aquaculture in the third millennium. Technical Proceedings of the Conference on Aquaculture in the Third Millennium, Rome, Italy.

4. Aboyadak IM, Ali NGM, Goda AMAS, Aboelgalagel WH, Salam AME (2015) Molecular detection of *Aeromonas hydrophila* as the main cause of outbreak in tilapia farms in Egypt. J Aquacul Mar Biol 2: 5.

5. Zhang D, Xu DH, Shoemaker C (2016) Experimental induction of motile *Aeromonas septicemia* in channel catfish (*Ictalurus punctatus*) by waterborne challenge with virulent *Aeromonas hydrophila*. Aquaculture Reports 3: 18-23.

6. Olugbojo JA, Ayoola SO (2015) Comparative studies of bacteria loads in fish species of commercial importance at aquaculture unit and lagoon front of the University of Lagos. J of Fish and Aqua 7: 37-46.

7. Darak O, Barde RD (2015) *Pseudomonas fluorescens* associated with bacterial disease in *Catla catla* in Marathwada region of Maharashtra. Int J Adv Biotechnol Res 6: 189-195.

8. Thune RL, Stanley LA, Copper RK (1993) Pathogenesis of gram-negative bacterial infection in warm water fish. Ann Rev Fish Dis 3: 37-68.

9. Cipriano RC, Bullock GL, Pyle SW (2001) *Aeromonas hydrophila* and motile *Aeromonad septicemias* of Fish (Fish Disease Leaflet 68), US Department of the Interior Fish and Wildlife Service Division of Fishery Research Washington DC.

10. AOAD (2005) Study on fish diseases in Arab countries (in Arabic). Arab Organization for Agricultural Development publication. Khartoum, Sudan.

11. Lewbart GA (2008) Bacterial diseases of pet fish. Proceedings of Michigan Veterinary Conference, Michigan, USA.

12. Austin B, Austin DA (2012) Bacterial fish pathogens, Disease of farmed and wild fish (5th edn). Springer Dordrecht Heidelberg, New York, London.

13. Noga EJ (2010) Fish disease diagnosis and treatment. (2nd edn), Blackwell Publishing, USA.

14. Yanong RPE, Floyd FR (2002) Streptococcal infections of fish. Report from University of Florida (Report No. 57). Series from the Department of Fisheries and Aquatic Sciences, Florida Cooperative Extension Service. Institute of Food and Agricultural Sciences, University of Florida, Florida.

15. Wei S, Zhao H, Xian Y, Hussain M, Wu X (2014) Multiplex PCR assays for the detection of *Vibrio alginolyticus*, *Vibrio parahaemolyticus*, *Vibrio vulnificus*, and *Vibrio cholerae* with an internal amplification control. Diagn Microbiol Infect Dis 9: 115-118.

16. Sebastião FA, Furlan LR, Hashimoto DT, Pilarski F (2015) Identification of bacterial fish pathogens in Brazil by direct colony PCR and 16S rRNA Gene Sequencing. Adv Microbiol 5: 409-424.

17. Heil N (2009) National wild fish health survey- laboratory procedures manual, (5th edn). US Fish and Wildlife Service, Warm springs, GA, USA.

18. Ahmed AM, Motoi Y, Sato M, Maruyama A, Watanabe H (2007) Zoo animals as a reservoir of gram-negative Bacteria Harboring Inte-grones and Antimicrobial Resistance Genes. Appl Environ Microbiol 73: 6686-6690.

19. Lee C, Cho JC, Lee SH, Lee DG, Kim SJ (2002) Distribution of Aeromonas spp. as identified by 16S rDNA restriction fragment length polymorphism analysis in a trout farm. J Appl Microbiol 93: 976-985.

20. Trakhna F, Harf-Monteil C, Abdelnour A, Maaroufi A, Gadonna-Widehem P (2009) Rapid Aeromonas hydrophila identification by TaqMan PCR assay: comparison with a phenotypic method. Lett Appl Microbiol 49: 186-190.

21. De Ocenda VR, Almeida-Prieto S, Luzardo-Álvarez A, Barja JL, Otero-Espinar FJ, et al. (2016) Pharmacokinetic model of florfenicol in turbot (Scophthalmus maximus): establishment of optimal dosage and administration in medicated feed. J Fish Dis 40: 411-424.

22. Asaad TMA (2011) Further studies on Aeromonas hydrophila isolates recovered from fish. Alexandria University, Alexandria, Egypt.

23. Zakaria AA (2014) Molecular characterization of Aeromonas hydrophila strains isolated from diseased marine and freshwater fish. Kafrelsheikh University, Kafrelsheikh, Egypt.

24. Laith AR, Najiah M (2013) Aeromonas hydrophila: Antimicrobial susceptibility and histopathology of isolates from diseased catfish, Clarias gariepinus (Burchell). J Aqua Res Dev 5: 215.

25. Chen CY, Chao CB, Bowser PR (2006) Infection of tilapia Oreochromis species by Vibrio vulnificus. J Worl Aquacult Soc 37: 82-88.

26. Okasha LA, Ammar A, El-Hady MA, Samir A, Samy AA, et al. (2016) Identification of common fish bacterial pathogens in Kafr El Sheikh governorate, Egypt using PCR. Int J Biol Pharm Allied Sci 5: 522-537.

27. Janda JM, Abbott SL (2010) The genus Aeromonas: Taxonomy, pathogenicity, and infection. Clin Microbiol Rev 23: 35-73.

28. Aboyadak IMI (2011) A study on the role of antibiotics in controlling Aeromonas hydrophila infection in cultured freshwater fish. Kafrelsheikh University, Kafrelsheikh, Egypt.

29. Yamanoi H, Muroga K, Takahashi S (1980) Physiological characteristics and pathogenicity of NAG vibrio isolated from diseased ayu. Fish Pathol 15: 69-73.

30. Ascencio F, Aleljung P, Wadstrom T (1990) Particle agglutination assays to identify fibronectin and collagen cell surface receptors and lectins in Aeromonas and Vibrio species. Appl Environ Microbiol 56: 1926-1931.

31. Li J, Zhou L, Woo NYS (2003) Invasion routes and pathogenicity mechanisms of Vibrio alginolyticus to silver sea bream Sparus sarba. J Aquat Anim Health 15: 302-313.

32. Wei S, Xian Y, Zhao H, Wu X (2013) Simultaneous Detection of Vibrio alginolyticus, Vibrio parahaemolyticus, Vibrio vulnificus and Vibrio Cholera using multiplex PCR. Scientia Agricultura Sinica 46: 1682-1686.

33. Buller NB (2004) Bacteria from fish and other aquatic animals, A practical identification manual. CABI Publishing, UK.

34. Bridson EY (2006) The OXOID Manual. Published by OXOID Limited, Wade Road, Basingstoke, Hampshire, England.

35. Dworkin M, Falkoe S, Rosenberg E, Schleifer KH, Stackebrandt E (2006) A Handbook on the Biology of Bacteria, Volume 6: Proteobacteria: Gamma Subclass, third Edition. Springer Science and Business Media, LLC, 233 Spring Street, New York, USA.

36. Davis JW, Sizemore RK (1981) Non-selectivity of Rimler-Shotts Medium for Aeromonas hydrophila in Estuarine Environments. Appl Environ Microbiol 42: 544-545.

37. Arcos ML, De Vicente A, Morinigo MA, Romero P, Borrego JJ (1988) Evaluation of several selective media for recovery of Aeromonas hydrophila from polluted waters. Appl Environ Microbiol 54: 2786-2792.

38. Rahim Z, Aziz KMS (1992) Isolation of enterotoxogenic Vibrio cholera nonO1 from the Buriganga river and two ponds of Dhaka, Bangladesh. J Diarr Diseas Res 10: 227-230.

39. Furmanek-Blaszk B (2014) Phenotypic and molecular characteristics of an Aeromonas hydrophila strain isolated from the River Nile. Microbiol Res 169: 547-552.

40. Chatterjee S, Haldar S (2012) Vibrio related diseases in aquaculture and development of rapid and accurate identification methods. J Mar Sci Res Develop S1:002.

41. Haenen OLM, Fouz B, Amaro C, Isern MM, Mikkelsen H, et al. (2014) Vibriosis in aquaculture (Workshop report). Bulletin European Association of Fish Pathologists 34: 138.

42. Senderovich Y, Lzhaki I, Haloperm M (2010) Fish as reservoirs and vectors of Vibrio cholera. PLoS ONE 5: e8607.

43. Rehulka J, Petras P, Marejkova M, Aldova E (2015) Vibrio cholerae non-O1/non-O139 infection in fish in the Czech Republic. Veterinarni Medicina 60: 16-22.

24

Enhancement of Antioxidant Activity, Non-specific Immunity and Growth Performance of Nile Tilapia, *Oreochromis Niloticus* by Dietary Fructooligosaccharide

Eman A Abd El-Gawad[1*], **Ashraf M Abd El-latif**[1] **and Ramy M Shourbela**[2]

[1]*Department of Fish Diseases and Management, Faculty of Veterinary Medicine, Benha University, Egypt*

[2]*Department of Animal Husbandry and Animal Wealth Development, Faculty of Veterinary Medicine, Alexandria University, Egypt*

[*]**Corresponding author:** Eman A Abd El-Gawad, Department of Fish Diseases and Management, Faculty of Veterinary Medicine, Benha University, Egypt
E-mail: eman.mahmoud@fvtm.bu.edu.eg

Abstract

In this study, Nile tilapia was fed experimental diets containing different levels (0, 1, 2, and 3%) of fructooligosaccharide (FOS) for 6 weeks to investigate its effect on the antioxidant activity, non-specific immunity and growth performance of Nile tilapia. Liver and serum samples were taken after 3 and 6 weeks feeding. Results showed that malondialdehyde level and superoxide dismutase activity decreased significantly ($P<0.05$) with dietary FOS supplementation after 3 and 6 weeks feeding compared to the control. Catalase and glutathione peroxidase activities were significantly decreased in groups fed 1 and 2% FOS for 3 weeks. Serum immunoglobulin M and lysozyme activity were significantly increased with dietary FOS after 3 and 6 weeks feeding. Nitric oxide revealed significant increase with 2% dietary FOS after 3 weeks feeding and there were no significant difference ($P>0.05$) in other treated groups fed for 6 weeks compared to the control. Weight gain also recorded significant increase in group fed 2% FOS for 6 weeks. These results indicated that dietary FOS supplementation could significantly enhance the antioxidant activities, non-specific immune response and growth performance of *Oreochromis niloticus*. It could be conclude that 2% dietary FOS was the most suitable and beneficial dose for Nile tilapia.

Keywords: Antioxidant enzymes activity; Growth performance; Fructooligosaccharide; Nile tilapia; Non-specific immunity

Introduction

Nile tilapia, *Oreochromis niloticus (O. niloticus)* is considered as one of the most important freshwater species for commercial aquaculture in Egypt, due to its high nutritional values, rapid growth rate and resistance to diseases [1]. However, bacterial diseases outbreak continues to occur among cultured *O. niloticus* due to high intensification, causing considerable economic losses in fish farms. Several approaches such as vaccination and chemotherapy have been carried out to increase fish immunocompetence and prevent aquatic diseases [2,3].

On the other hand, there are strict regulations on the application of antibiotics and chemotherapeutics in aquaculture because of its negative impacts which includes the development of antibiotic resistant bacteria, suppression of host's immune system, destruction of the microbial population in the aquatic environment and bioaccumulation [4,5]. Therefore, increasing attention is being paid to the dietary supplementation of an alternative friendly probiotics, prebiotics and immunostimulants which help to improve fish immune response and hence reduce the susceptibility of fish to diseases [6-8].

Prebiotics are defined as non-digestible feed ingredients that stimulating the growth and metabolism of beneficial bacteria in the host gastrointestinal tract [9]. Fructooligosaccharide (FOS or oligofructose) is one of the most common prebiotics used in human, terrestrial animals and has received great attention as dietary supplement for different finfish species during the past years [10-13] as well as shellfish [14]. Beneficial effects of dietary FOS supplementation on growth performance, immune response and disease resistance in several fish species have been demonstrated in previous studies [12,15-19].

Furthermore, FOS has shown positive effects on the fish antioxidant activity [12,13,20,21]. However, other studies have reported that FOS has no beneficial effects on fish growth or immunity [7,10,22-24] or even has adverse effect on fish performance [25]. Therefore, prebiotics evaluations are recommended before suggesting specific prebiotic strategies for certain fish species [26]. The aim of the present study was to evaluate the effect of different dietary levels of FOS on the antioxidant capability, non-specific immunity and growth performance of Nile tilapia.

Materials and Methods

Experimental diets and fish

Fructoligosaccharides powder (FOS) (Nutraflora)® obtained from GTC Nutrition company (Westchester, USA) was added at a different concentrations on the commercial basal diet (Joe Trade Company, Egypt) containing 30% crude protein. The first diet was kept as a control without additive. The second, third and fourth diets were contained FOS at a concentration of 1, 2% and 3% (w/w) respectively. The experimental diets were repelleted using a hand pelletizer with a die of 3 mm and left air-dried at room temperature. After drying, the diets were broken up into appropriate size and stored in tight plastic bags at 4ºC. The proximate analysis of basal diet was carried out according to the AOAC standard [27]. Ingredients and proximate analysis of the basal diet and Gross energy was calculated by Brett [28] in Table 1.

Ingredients	(%)	Proximate analysis	%
Fish meal	15	Dry matter	89.07
Yellow Corn meal	28	Crude protein	30
Soybean meal	41	Crude fat	3.17
Wheat bran meal	10	Ash	10.76
Corn oil	4	Crude fiber	4.94
Vitamin mix a Mineral mix b	1 1	Nitrogen free extract (NFE)c	40.2
Total	100	Gross energy (MJ/Kg)d	15.24

aVitamin (each 3 kg) supply the following: vitamin A, 1200.000 IU; vitamin D3, 300.000 IU; Vitamin E, 700 mg; vitamin K3, 500 mg; vitamin B1, 500 mg; vitamin B2, 200 mg; vitamin B6, 600 mg; vitamin B12, 3 mg; ascorbic acid, 450 mg; Biotin, 6 mg; panthonic acid, 670 mg; methionine, 3.000 mg; folic acid, 300 mg; chlolin chloride, 10.000 mg.

bMineral (each kg) contain the following: zinc sulphate, 50.000 mg; copper, 4.000 mg; ferrous sulphate, 20.000 mg; Manganous sulphate, 60.000 mg; sodium, 2.000 mg; selenium, 100 mg; cobalt chloride, 100 mg; iodine, 500 mg; c NFE = 100 - (% crude protein + % crude fat + % ash + % crude fiber + % moisture); d: Gross energy was calculated according to Brett [28].

Table 1: Ingredients and proximate analysis of basal diet.

Nile tilapia was obtained from private fish farm at Kafr El Sheikh Governorate, Egypt. The fish were transported to the Lab of Fish Diseases and Management at Fac. of Vet. Med, Benha University, Egypt and stocked in fiberglass tanks (750 L capacity). Fish were randomly distributed into eight tanks (35 fish per tank) and allowed to acclimate to lab conditions for 2 weeks before starting the feeding experiment. Each treatment was carried out in duplicate. During the acclimation period, fish were fed basal diet two times a day.

Experimental design

After acclimation, the first control group was fed on free FOS diet. The second, third and fourth groups have been received diets containing 1%, 2%, and 3% FOS respectively. All experimental groups were fed by hand, twice daily at 08:00 and 16:00 at a rate of 3% of their body weight for 6 weeks. Nearly half of tank water was exchanged every day to maintain water quality. The water quality parameters were determined according to the guidelines of APHA [29]. The water temperature was adjusted at 28 ± 2°C using electrical heaters (Xilong, China), pH was 7.1 ± 0.2 and the dissolved oxygen was maintained at 5.2 ± 0.7 mg L^{-1} along the experimental period. Fish were routinely checked for health and any mortality throughout the experiment.

Sampling

Two sampling points were taken after 3 and 6 weeks feeding. Three fish per tank (Total = 6 fish per treatment) were taken randomly and anaesthetized with benzocaine (80 mg L^{-1}). The blood was collected without anticoagulant from the caudal vessels of each individual fish using 3 ml sterile syringe (23-gauge needle) and allowed to clot at room temperature for 2 h. After that, all the samples were centrifuged at 3,000 rpm for 15 min at 4°C. The separated sera were pooled together (n=3) and stored at -80°C for later analysis of immunological parameters. Immediately after blood collection, the liver was quickly removed, weighted and stored at -80°C for lipid peroxidation (LPO) and antioxidant enzymes assay.

Lipid peroxidation and antioxidant enzymes assay

The liver was homogenized in cooled phosphate buffer saline pH 7.2 at a ratio 1:10 (w/v) using electrical homogenizer (Heidolph, Germany). The procedure was performed on ice. The homogenates were centrifuged at 13,000 rpm at 4°C for 15 min and the resultant supernatants were separated in aliquots and stored at -80°C until the determination of malondialdehyde (MDA) level and antioxidant enzymes which performed within one week after extraction. The liver malondialdehyde (MDA), which used as marker of LPO was measured according to Ohkawa et al. [30]. Superoxide dismutase (SOD) activity was determined using commercial kits (Biodiagnostic Company, Egypt) following the method described by Nishikimi et al. [31]. Catalase (CAT) activity was determined according to Sinha [32] and Glutathion peroxidase (GPX) activity was assayed according to Paglia and Valentine [33].

Immunological parameters assay

Serum lysozyme activity was measured using lysoplate assay [34]. Serum Immunoglobulin M (IgM) was measured spectrophotometrically following the manufacture protocol of Enzyme-Linked Immunosorbent Assay (ELISA) kits obtained from (Cusabio Biotech Co. Ltd, USA). Nitric oxide was determined using Griess reagent according to the method described by Granger et al. [35].

Growth performance

At the end of the experiment, growth performance and physiological indices were assessed according to the following formula: Weight gain (WG) = W2-W1, Feed conversion ratio (FCR) = F/WG; Condition (CF) factor = 100 × (body weight (g)/body length (cm)^3; Spleensomatic index (SSI) = weight of spleen (g)/total body weight (g)) ×100; Hepatosomatic Index (HSI) = weight of liver (g)/total body weight (g)) ×100. Where, W1 is initial weight (g), W2 is final weight (g), and F is amount of feed intake (g).

Statistical analysis

Data were statistically analyzed by one-way analysis of variance (ANOVA) and Duncan's multiple range test within each sampling time to determine the significant difference (P<0.05) between means using SPSS version 16.0 software package. All data were presented as means ± SE (standard error). Normality and homogeneity of variance for all obtained data were confirmed with Shapiro–Wilk and Levene's test respectively. Two-way analysis was carried out to evaluate the interaction between dietary FOS and feeding duration.

Results

Liver MDA content and SOD activity of Nile tilapia fed diets containing different levels of FOS were significantly (P<0.05) decreased after 3 and 6 weeks feeding compared to the control (Figures 1 and 2). The lowest SOD activity was observed in group fed 3% (361.67 ± 2.03 U/g) for 6 weeks compared to the control (460.67 ± 3.18 U/g). CAT activity exhibited significant decrease in groups fed 1 and 2% FOS for 3 weeks (14.0 ± 0.83 U/g) and (16.20 ± 0.97 U/g) respectively compared to the control (36.69 ± 2.03 U/g). However, after

6 weeks feeding, CAT activity recorded no significant difference with 2 and 3% dietary FOS (Figure 3). Moreover, GPX activity revealed significant decrease in group fed 1 and 2% dietary FOS for 3 weeks (201.0 ± 2.31 and 341.0 ± 5.77 U/g) respectively. While, after 6 weeks feeding, GPX activity exhibited significant decrease in all FOS treated groups (Figure 4). There were highly significant (P<0.001) interaction between dietary FOS and feeding duration on liver SOD and GPX activities.

decrease in serum IgM (117.61 ± 2.62 µg/ml) compared to the group fed 1 and 2 % FOS (133.36 ± 2.75 and 129.89 ± 2.39 µg/ml) respectively. Nitric oxide was significantly (P<0.05) increased in group fed with 2% dietary FOS (24.15 ± 3.36 µmole/l) for 3 weeks feeding compared to the control (5.86 ±1.42 µmole/l). However, after 6 weeks, nitric oxide has not significantly affected in all treated groups in compare with the control (Figure 7). Interaction between dietary FOS and feeding duration was significantly (P<0.01) observed on serum nitric oxide level.

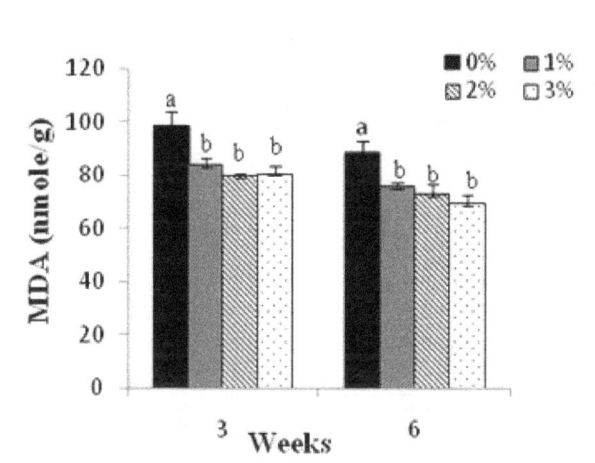

Figure 1: Lipid peroxidation as malondialdehyde (MDA) level in liver of Nile tilapia fed different levels of dietary FOS. Data represent the mean ± S.E. Values with different superscript letters are significantly different among group at each sampling point.

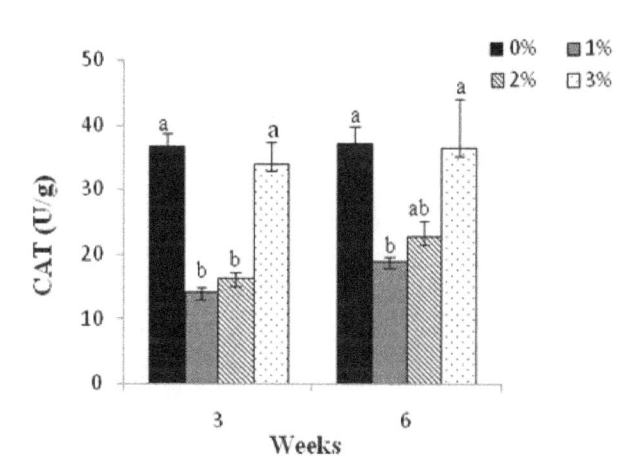

Figure 3: Catalase (CAT) activity in liver of Nile tilapia fed different levels of dietary FOS. Data represent the mean ± S.E. Values with different superscript letters are significantly different among group at each sampling point.

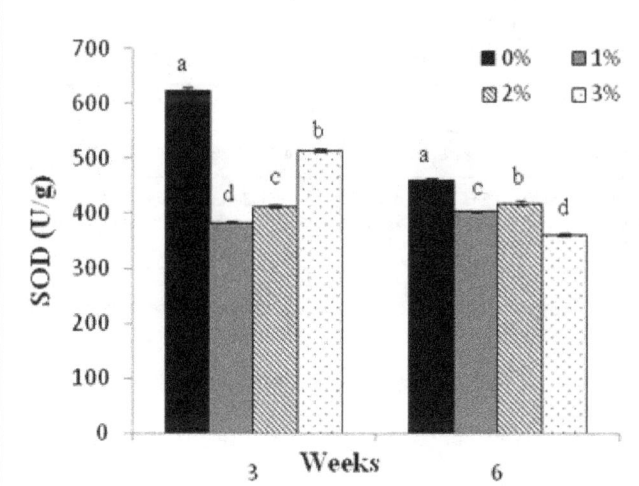

Figure 2: Superoxide dismutase (SOD) activity in liver of Nile tilapia fed different levels of dietary FOS. Data represent the mean ± S.E. Values with different superscript letters are significantly different among group at each sampling point.

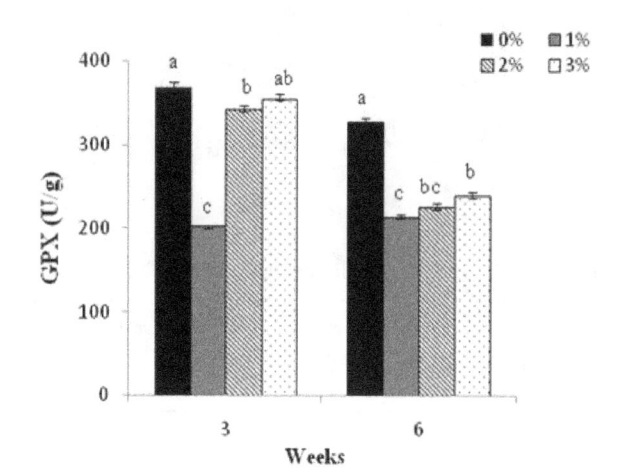

Figure 4: Glutathione peroxidase (GPx) activity in liver of Nile tilapia fed different levels of dietary FOS. Data represent the mean ± S.E. Values with different superscript letters are significantly different among group at each sampling point.

Serum lysozyme activity and IgM were significantly improved (P<0.05) with dietary FOS either after 3 or 6 weeks feeding compared to the control (Figures 5 and 6). It was observed that after 6 weeks feeding, fish received diet containing 3% FOS showed significant

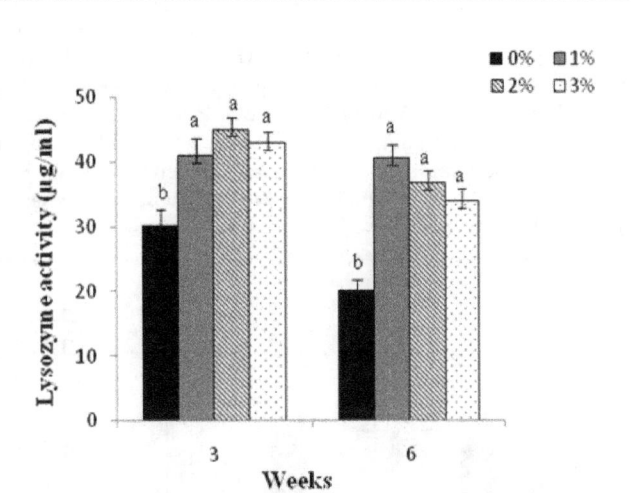

Figure 5: Serum lysozyme activity of Nile tilapia fed different levels of dietary FOS. Data represent the mean ± S.E. Values with different superscript letters are significantly different among group at each sampling point.

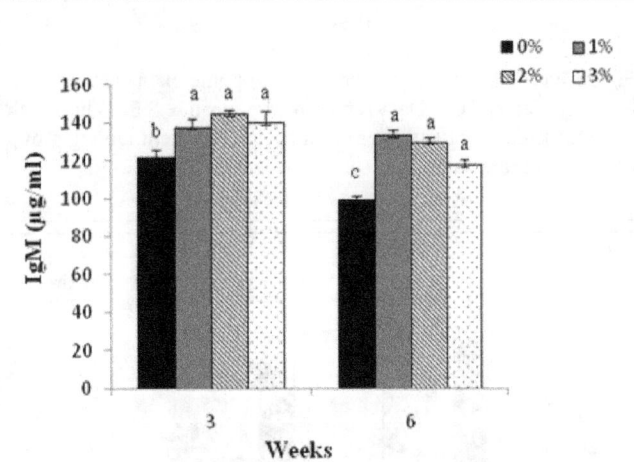

Figure 6: Serum immunoglobulin M (IgM) level of Nile tilapia fed different levels of dietary FOS. Data represent the mean ± S.E. Values with different superscript letters are significantly different among group at each sampling point.

The effect of dietary FOS levels on the growth performance of Nile tilapia after 6 weeks feeding was shown in Table 2. Dietary supplementation of FOS had no significant effect (P>0.05) on growth performance and health status of Nile tilapia. However, weight gain was significantly increased in group supplemented with 2% dietary FOS compared to the other treatment groups. There were no mortalities among all treated groups throughout the experiment.

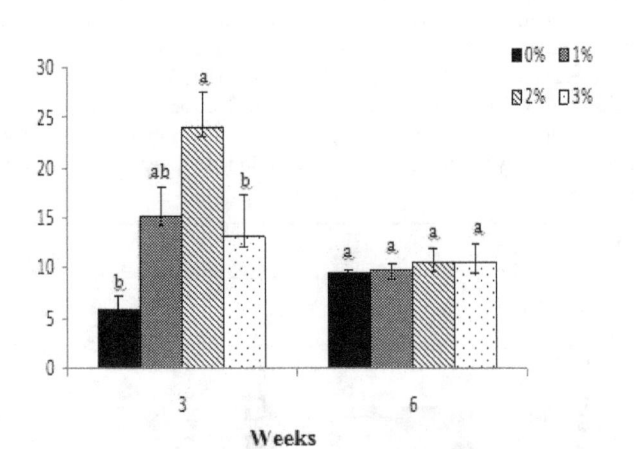

Figure 7: Nitric oxide (NO) activity (μ mol/l) in serum of Nile tilapia fed different levels of dietary FOS. Data represent the mean ± S.E. Values with different superscript letters are significantly different among group at each sampling point.

Parameters	0 %	1 %	2 %	3 %
Initial weight (g)	25.22 ± 2.19	25.22 ± 1.7	23.52 ± 2.15	24.4 ± 0.46
Final weight (g)	40.95 ± 3.35	45.2 ± 1.71	46.3 ± 1.21	43.51 ± 1.83
Weight gain (g)	15.74 ± 2.14^b	19.98 ± 0.99^{ab}	22.78 ± 2.68^a	19.11 ± 2.18^{ab}
FCR	2.18 ± 0.28	1.62 ± 0.15	1.49 ± 0.34	1.73 ± 0.21
CF (%)	1.43 ± 0.04	1.53 ± 0.05	1.5 ± 0.01	1.53 ± 0.03
HSI (%)	2.24 ± 0.25	2.31 ± 0.18	1.78 ± 0.39	2.32 ± 0.15
SSI (%)	0.39 ± 0.1	0.29 ± 0.04	0.25 ± 0.04	0.27 ± 0.03

Mean values (± S.E) in the same raw with different superscript letters are statistically different at (P<0.05) using one-way analysis of variance (ANOVA) and Duncan's multiple range test.

Table 2: Growth performance of *O. niloticus* fed different dietary levels of FOS for 6 weeks.

Discussion

Dietary supplement of prebiotics have shown great interest in aquaculture due to its growth performance enhancement and increasing resistance of fish to pathogens and environmental stressors through stimulating the host's immune response [8,36].

Malondialdehyde (MDA) is formed as an end product of lipid peroxidation, which is the initial step of cellular membrane damage caused by reactive oxygen species (ROS) [37]. In the present study, dietary supplementation of FOS significantly reduced liver MDA content after 3 and 6 weeks feeding. The decreased MDA indicated that FOS could inhibit the process of lipid peroxidation which is considered a biomarker of oxidative damage caused by free radicals [38]. A Similar result was observed in blunt snout bream, *Megalobrama amblycephala* fed 0.4% FOS for 8 weeks [12]. On the other hand, LPO level was not affected by dietary FOS in turbot, *Scophthalmus maximus* [13].

Antioxidant enzymes SOD, CAT and GPX are considered the first line of antioxidant defense and served as sensitive biomarkers of oxidative stress [39]. SOD is considered the first enzyme responsible for scavenging ROS and protecting cells from damage by free radicals process [40]. The present results showed that liver SOD activity significantly decreased in Nile tilapia fed with different dietary FOS levels along the experimental period (6 weeks). This indicated that FOS possesses antioxidant potentials that might effectively reduce ROS production and its adverse effects. This explanation was supported by the fact that the higher SOD activity, the more superoxide radicals need to be reacted [41]. Guerreiro et al. [13] also recorded significant decrease in SOD activity of turbot, *Scophthalmus maximus* fed 0.5 and 1% short chain fructooligosaccharides for 9 weeks. However, liver SOD activity was significantly increased by dietary FOS for 8 weeks in other studies [6,12,21]. This difference might be attributed to fish species, prebiotic concentrations, feeding duration and feeding modes. CAT and GPX activity in the current study revealed significant decrease in Nile tilapia fed FOS at a concentration of 1 and 2% for 3 weeks, indicating that FOS has capability to reduce peroxide radicals and converting it into oxygen and water [42]. There was no significance difference in CAT activity with 2 and 3 % dietary FOS for 6 weeks compared to the control. This result could be explained that the exogenous antioxidants supplementation could reduce the gene expression of CAT enzyme and consequently decreases its activity [43]. Similar no significant effect on CAT activity has also been reported in blunt snout bream [12] and triangular bream *Megalobrama terminalis* [6]. The significant decrease of antioxidant enzymes activity in *O. niloticus* indicated that dietary FOS could maintain the redox state in the cell and minimize the adverse effects of ROS [44].

Several parameters such as respiratory burst activity, nitric oxide synthase, lysozyme activity, bactericidal activity, immunoglobulin level, antibody response, etc. are served as a good immunological indicator of fish health status [45]. In the present study, dietary FOS showed beneficial effect on the non-specific immune response of Nile tilapia evidenced by the significant increase of serum lysozyme activity and IgM level. The higher lysozyme activity probable attributed to the high leukocyte production with dietary FOS [6] due to the fact that fish lysozyme is mainly produced by neutrophiles and macrophages [46]. The immunostimulatory effect of FOS could be ascribed to the growth stimulation of beneficial bacteria such as lactobacilli and befidobacteria, which possess lipopolysaccharides that have immunostimulatory properties [47]. Moreover, acetate, propionate and lactic acid as end products of FOS fermentation play a crucial role in modulating the immune system [48]. Furthermore, FOS could interact with toll like receptors (TLR2) expressed on macrophages [49] and up-regulated the expression of antimicrobial peptides (Leap) which have important role in innate immune defense and hence disease resistance of fish [12]. Parallel to this study in a previous investigations, enhanced lysozyme activity has been recorded in red drum [26], Caspian roach fry [16], turbot [13] and blunt snout bream [12,21]. However, dietary FOS showed no significant effect on lysozyme activity in other studies [10,22]. This contradictory may be attributable to the prebiotic dosage, life stage and/or fish species [50].

Nitric oxide had also been shown to be a very important molecule in regulating immune functions as well as a direct antimicrobial effect [51]. In the current study, nitric oxide was significantly increased in *O. niloticus* fed 2% dietary FOS for 3 weeks; probably due to the increase

of macrophage production by dietary FOS. Meanwhile, after 6 weeks feeding, serum nitric oxide showed no significant improvement in all treated groups compared to the control fed control diet. This result could be supported by Yoshida et al. [52] who reported that in African catfish fed 10 g kg^{-1} prebiotic mannanoligosaccharide, the number of activated neutrophils increased during the first 2 weeks and then decreased back to the control level after 45 days.

In the present study, growth performance of *O. niloticus* fed FOS showed improvement but not statistically differ. Weight gain in group fed 2% dietary FOS for 6 weeks revealed significant increase compared to the control fed basal diet. The improved weight gain could be attributed to the enhanced digestive enzymes activity and microvilli height with dietary FOS supplementation [12,19] or stimulation the growth of beneficial probiotic Bacillus spp. [53]. Growth improvement by dietary FOS has previously been reported in different fish species [12,16,17,54]. On the other hand, lack of growth response to dietary FOS was observed in other studies [7,15,22-24]. The discrepancy of growth response to dietary FOS seems to differ depending on species, fish size, different intestinal moropholgy, gut microbiota, FOS concentration and feeding duration [50]. Furthermore, negative effects of dietary prebiotics on fish performance have been recorded in beluga sturgeon [25] and Arctic charr [55]. The negative impacts of dietary prebiotics have been attributed to inability of intestinal microbiota to ferment excessive levels of prebiotics and subsequent accumulated in the intestine which might have deleterious effect to the enterocytes [25].

Condition factor (CF) and organ indices such as hepatosomatic (HSI) index, and spleenosomatic index (SSI) could be used as indicators of the health status of fish [56]. In the present work, there was no significant effect on CF, HSI and SSI of *O. niloticus* fed FOS indicating that FOS supplement had no detrimental effects on liver tissue or general fish health. This agreed with results of other studies [7,10,16,54]. These results supported by the assumption that increase of these indices such as hepatosomatic index could be resulted from increased production of endoplasmic reticulum for protein synthesis in the liver tissue or hypertrophy and hyperplasia of liver cell in fish exposed to stress [57].

The present study concluded that dietary FOS supplementation had an enhancement effects on non-specific immunity and growth performance of Nile tilapia as well as it protect cells against the adverse effects of ROS. Dietary supplementation with 2% FOS for 6 weeks is considered the most suitable dose for Nile tilapia to be involved in aquafeed to improve aquaculture industry.

Acknowledgment

The authors are grateful to Prof. Dr. Adel A. Shaheen, Head of Department of Fish Diseases and Management, Faculty of Vet. Med. Benha University, Egypt for his helpful scientific support, critical reading and revision of this manuscript. Also grateful thanks extend to Mr. Ayman Hashim, Director of a private fish farm, Kafer El Sheikh, Egypt, for providing fish.

References

1. Barcellos LJG, Nicolaiewsky S, Souza SMG, Lulhier F (1999) The effects of stocking density and social interaction on acute stress response in Nile tilapia Oreochromis niloticus (L.) fingerlings. Aquaculture Research 30: 887- 892.

2. Chandran MR, Aruna BV, Logambal SM, Michael RD (2002) Immunisation of Indian major carps against Aeromonas hydrophila by intraperitoneal injection. Fish Shellfish Immunol 13: 1-9.

3. Magnadottir B (2010) Immunological control of fish diseases. Mar Biotechnol 12: 361-379.

4. Sanderson H, Brain RA, Johnson DJ, Wilson CJ, Solomon KR (2004) Toxicity classification and evaluation of four pharmaceuticals classes: antibiotics, antineoplastics, cardiovascular, and sex hormones. Toxicology 203: 27-40.

5. Lim SJ, Jang E, Lee SH, Yoo BH, Kim SK, et al. (2013) Antibiotic resistance in bacteria isolated from freshwater aquacultures and prediction of the persistence and toxicity of antimicrobials in the aquatic environment. J Environ Sci Health B 48: 495-504.

6. Zhang CN, Li XF, Xu WN, Jiang GZ, Lu KL, et al. (2013) Combined effects of dietary fructooligosaccharide and Bacillus licheniformis on innate immunity, antioxidant capability and disease resistance of triangular bream (Megalobrama terminalis). Fish & Shellfish Immunology 35: 1380-1386.

7. Hoseinifar SH, Soleimani N, Ringo E (2014) Effects of dietary fructo-oligosaccharide supplementation on the growth performance, haemato-immunological parameters, gut microbiota and stress resistance of common carp (Cyprinus carpio) fry. Br J Nutr 112:1296-1302.

8. Song SK, Beck BR, Kim D, Park J, Kim J, et al. (2014) Prebiotics as immunostimulants in aquaculture: a review. Fish & Shellfish Immunology 40: 40-48.

9. Gibson GR, Roberfroid MB (1995) Dietary modulation of the human colonie microbiota: introduction the concept of the prebiotics. J Nutr 125: 1401-1412.

10. Ye JD, Wang K, Li FD, Sun YZ (2011) Single or combined effects of fructo-and mannan oligosaccharide supplements and Bacillus clausii on the growth, feed utilization, body composition, digestive enzyme activity, innate immune response and lipid metabolism of the Japanese flounder Paralichthys olivaceus. Aquaculture nutrition 17: 902-911.

11. Akrami R, Iri Y, Rostami HK, Mansour MR (2013) Effect of dietary supplementation of fructooligosaccharide (FOS) on growth performance, survival, lactobacillus bacterial population and hemato-immunological parameters of stellate sturgeon (Acipenser stellatus) juvenile. Fish Shellfish Immunol 35: 1235-1239.

12. Zhang CN, Li XF, Jiang GZ, Zhang DD, Tian HY, et al. (2014a) Effects of dietary fructooligosaccharide levels and feeding modes on growth, immune responses, antioxidant capability and disease resistance of blunt snout bream (Megalobrama amblycephala). Fish & Shellfish Immunol 41: 560-569.

13. Guerreiro I, Perez-Jimenez A, Costas B, Oliva-Teles A (2014) Effect of temperature and short chain fructooligosaccharides supplementation on the hepatic oxidative status and immune response of turbot (Scophthalmus maximus). Fish & Shellfish Immunol 40: 570-576.

14. Zhou Z, Ding Z, Huiyuan LV (2007) Effects of Dietary Short-chain Fructooligosaccharides on Intestinal Microflora, Survival, and Growth Performance of Juvenile White Shrimp, Litopenaeus vannamei. Journal of the World Aquaculture Society 38: 296-301.

15. Ai Q, Xu H, Mai K, Xu W, Wang J, et al. (2011) Effects of dietary supplementation of Bacillus subtilis and fructooligosaccharide on growth performance, survival, non-specific immune response and disease resistance of juvenile large yellow croaker, Larimichthys crocea. Aquaculture 317: 155-161.

16. Soleimani N, Hoseinifar SH, Merrifield DL, Barati M, Abadi ZH (2012) Dietary supplementation of fructooligosaccharide (FOS) improves the innate immune response, stress resistance, digestive enzyme activities and growth performance of Caspian roach (Rutilus rutilus) fry. Fish & Shellfish Immunol 32: 316-321.

17. Wu Y, liu WB, Li HY, Xu WN, He JX, et al. (2013) Effects of dietary supplementation of fructooligosaccharide on growth performance, body composition, intestinal enzymes activities and histology of blunt snout bream (Megalobrama amblycephala) fingerlings. Aquaculture nutrition 19: 886-894.

18. Abd El-Gawad AE, Abd El-latif MA, Amin AA, Abd-El-Azem MA (2015) Effect of dietary fructooligosaccharide on bacterial Infection, oxidative stress and histopathological alterations in Nile tilapia (Oreochromis niloticus). Global Veterinaria 15: 339-350.

19. Abd El-latif AM, Abd El-Gawad EA, Emam MA (2015) Effect of dietary fructooligosaccharide supplementation on feed utilization and growth performance of Nile tilapia (Oreochromis niloticus) fingerlings. Egyptian Journal of Aquaculture 5: 1-16.

20. Zhang CN, Tian HY, Li XF, Zhu J, Cai DS, et al. (2014b) The effects of fructooligosaccharide on the immune reponse, antioxidant capability and HSP70 and HSP90 expressions of blunt snout bream (Megalobrama amblycephala) under high heat stress. Aquaculture 433:458-466.

21. Zhang CN, Li XF, Tian HY, Zhang DD, Jiang GZ, et al. (2015) Effects of fructooligosaccharide on immune response antioxidant capability and HSP70 and HSP90 expressions of blunt snout bream (Megalobrama amblycephala) under high ammonia stress. Fish Physiology and Biochemistry 41: 203-217.

22. Grisdale-Helland B, Helland SJ, Gatlin III DM (2008) The effects of dietary supplementation with mannanoligosaccharide, fructooligosaccharide or galactooligosaccharide on the growth and feed utilization of Atlantic salmon (Salmo salar). Aquaculture 283: 163-167.

23. Guerreiro I, Enes P, Oliva-Teles A (2015a) Effects of short-chain fructooligosaccharides (scFOS) and rearing temperature on growth performance and hepatic intermediary metabolism in gilthead sea bream (Sparus aurata) juveniles. Fish Physiology and Biochemistry 41:1333-1344.

24. Guerreiro I, Enes P, Merrifield D, Davies S, Oliva-Teles A (2015b) Effects of short-chain fructooligosaccharides on growth performance and hepatic intermediary metabolism in turbot (Scophthalmus maximus) reared at winter and summer temperatures. Aquaculture Nutrition 21: 433-443.

25. Hoseinifar SH, Mirvaghefi A, Mojazi Amiri B, Rostami HK, Merrifield D (2011) The effects of oligofructose on growth performance, survival and autochthonous intestinal microbiota of beluga (Huso huso) juveniles. Aquaculture Nutrition 17: 498-504.

26. Zhou QC, Buentello JA, Gatlin III DM (2010) Effects of dietary prebiotics on growth performance, immune response and intestinal morphology of red drum (Sciaenops ocellatus). Aquaculture 309: 253-257.

27. AOAC (2006) Official Methods of Analysis. (18thedn), Washington, DC.

28. Brett JR (1973) Energy expenditure of Sockeye salmon Oncorhynchus nerka during sustained performance. Journal of the Fisheries Research Board of Canada 30: 1799-1809.

29. APHA (1998) Standard method for the examination of water and wastewater. American Public Health Association, Washington.

30. Ohkawa H, Ohishi N, Yagi K (1979) Assay for lipid peroxides in animal tissues by thiobarbituric acid reaction. Analytical Biochemistry 95: 351-358.

31. Nishikimi M, Rao NA, Yagi K (1972) The occurrence of superoxide anion in the reaction of reduced phenazine methosulphate and molecular oxygen. Biochemical Biophysical Research Communications 46: 849-854.

32. Sinha AK (1972) Colorimetric assay of catalase. Analytical Biochemistry 47: 389-394.

33. Paglia DE, Valentine WN (1967) Studies on the quantitative and qualitative characterization of erythrocyte glutathione peroxidase. Journal of Laboratory and Clinical Medicine 70: 158-169.

34. Osserman EF, Lawlor DP (1966) Serum and urinary lysozyme (muramidase) in monocytic and monomyelocytic leukemia. Journal of Experimental Medicine 124: 921-951.

35. Granger DL, Taintor RR, Boockvar KS, Hibbs JB (1996) Measurement of nitrate and nitrite in biological samples using nitrate reductase and Griess reaction. Methods in Enzymology 268: 142-151.

36. Ringo E, Olsen RE, Gifstad TØ, Dalmo RA, Amlund H, et al. (2010) Prebiotics in aquaculture: a review. Aquaculture Nutrition 16: 117-136.

37. Pascual P, Pedrajas JR, Toribio F, Lopez-Barea J, Peinado J (2003) Effect of food deprivation on oxidative stress biomarkers in fish (Sparus aurata). Chemico- Biological Interactions 145:191-199.

38. Livingstone D (2003) Oxidative stress in aquatic organisms in relation to pollution and aquaculture. Revue de Médecine Vétérinaire 154: 427-430.

39. Jiang WD, Feng L, Liu Y, Jiang J, Zhou XQ (2009) Myo-inositol prevents oxidative damage, inhibits oxygen radical generation and increases antioxidant enzyme activities of juvenile Jian carp (Cyprinus carpio var.Jian). Aquaculture research 40: 1770-1776.

40. Chien YH, Pan CH, Hunter B (2003) The resistance to physical stresses by penaeus monodon juveniles fed diets supplemented with astaxanthin. Aquaculture 216: 177-191.

41. Wang YJ, Chien YH, Pan CH (2006) Effects of dietary supplementation of carotenoids on survival, growth, pigmentation, and antioxidant capacity of characins, Hyphessobrycon callistus. Aquaculture 261: 641-648.

42. Atli G, Canli M (2007) Enzymatic responses to metal exposures in a freshwater fish Oreochromis niloticus. Comparative Biochemistry and Physiology 145C: 282-287.

43. Girao PM, Pereira da Silva EM, Melo MP (2012) Dietary lycopene supplementation on Nile Tilapia (Oreochromis niloticus) juveniles submitted to confinement: effects on cortisol level and antioxidant response. Aquaculture Research 43: 789-798.

44. Halliwell B, Gutteridge JMC (2007) Free Radicals in Biology and Medicine. Oxford University Press, UK.

45. Chakrabarti R, Srivastava PK, Verma N, Sharma J (2014) Effect of seeds of Achyranthes aspera on the immune responses and expression of some immune-related genes in carp Catla catla. Fish & Shellfish Immunology 41: 64-69.

46. Fischer U, Utke K, Somamoto T, Kollner B, Ototake M, et al. (2006) Cytotoxic activities of fish leucocytes. Fish & Shellfish Immunology 20: 209-226.

47. Manning TS, Gibson GR (2004) Microbial-gut interactions in health and disease. Prebiotics. Best Pract Res Clin Gastroenterol 18: 287-298.

48. Passos LML, Park YK (2003) Fructooligosaccharides: implications in human health being and use in foods. Ciencia Rural 33: 385-390.

49. Vogt L, Ramasamy U, Meyer D, Pullens G, Venema K, et al. (2013) Immune modulation by different types of b2 / 1-fructans is toll-like receptor dependent. PLOS One 8: e68367.

50. Ibrahem MD, Fathi M, Mesalhy S, Abd EA (2010) Effect of dietary supplementation of inulin and vitamin C on the growth, hematology, innate immunity, and resistance of Nile tilapia (Oreochromis niloticus). Fish & Shellfish Immunology 29: 241-246.

51. Villamil L, Tafalla C, Figueras A, Novoa B (2002) Evaluation of immunomodulatory effects of lactic acid bacteria in Turbot (Scophthalmus maximus). Clinical and Diagnostic Laboratory Immunology 9: 1318-1323.

52. Yoshida T, Kruger R, Inglis V (1995) Augmentation of nonspecific protection in African catfish, Clarias gariepinus (Burchell), by the long-term oral-administration of immunostimulants. Journal of Fish Diseases 18: 195-198.

53. Mahious AS, Gatesoupe FJ, Hervi M, Metailler R, Ollevier F (2006) Effect of dietary inulin and oligosaccharides as prebiotics for weaning turbot, Psetta maxima (Linnaeus, C. 1758). Aquaculture International. 14: 219-229.

54. Hui-yuan LV, Zhi-gang Z, Rudeaux F, Respondek F (2007) Effects of dietary short chain fruct-oligosaccharides on intestinal microflora, mortality and growth performance of Oreochromis aureus ♂× O. niloticus ♀. Chinese Journal of animal Nutrition 19: 1-5.

55. Olsen RE, Myklebust R, Kryvi H, Mayhew TM, Ringo E (2001) Damaging effect of dietary inulin on intestinal enterocytes in Arctic charr (Salvelinus alpines L.). Aquaculture Research 32: 931-934.

56. Adams SM, Brown AM, Goede RW (1993) A quantitative health assessment index for rapid evaluation of fish condition in the field. Transactions of the American Fisheries Society 122: 63-73.

57. Ayoola SO (2008) Histopathological effects of glyphosate on juvenile African catfish (Clarias gariepinus). American-Eurasian Journal of Agriculture and Environmental Sciences 4: 367-367.

25

On-board Breeding Trial of Hilsa (*Tenualosa ilisha*, Ham. 1822) and Testing of Larval Rearing in Bangladesh

Md. Anisur Rahman[1], Tayfa Ahmed[1], Md. Mehedi Hasan Pramanik[1], Flura[1], Md. Monjurul Hasan[1]*, M. G. S. Riar[2], Khandaker Rashidul Hasan[3], Masud Hossain Khan[1] and Yahia Mahmud[4]

[1]*Bangladesh Fisheries Research Institute, Riverine Station, Chandpur, Bangladesh*
[2]*Bangladesh Fisheries Research Institute, Freshwater Station, Mymensingh, Bangladesh*
[3]*Bangladesh Fisheries Research Institute, Freshwater Sub-station, Saidpur, Nilphamari, Bangladesh*
[4]*Bangladesh Fisheries Research Institute, Headquarter, Mymensingh, Bangladesh*

Abstract

The Hilsa shad, *Tenualosa ilisha* commonly known as Hilsa is one of the most commercially important fish species in South Asian countries. For the conservation of Hilsa it is necessary to establish a standard breeding and culture protocol along with the present Hilsa management activities. The on board breeding trial was conducted during 10 October 2016 to 02 November 2016 which was the peak breeding time of Hilsa for the year 2016. The male and female Hilsa broods were collected from the River Meghna using BFRI experimental net during afternoon and late evening of full moon and new moon time. A total of six breeding trials were conducted in which 13 pairs of selected Hilsa broods were used for the breeding trials. For the breeding trial, both eggs and milt were collected through stripping and then the eggs were mixed with the milt immediately. The fertilized eggs were transferred to a plastic hatching jar for incubation providing mild water circulation, aeration and shade to protect penetration of direct sunlight for controlling temperature i.e., to maintain congenial environment. During the incubation period, the eggs were observed for 24 hrs to study the embryonic and larval development stages of Hilsa. After 4.2-4.5 hrs of fertilization morula stage of embryonic development was identified and after 8-8.5 hrs. of fertilization 18-myotome stage of embryonic development was identified from the sixth breeding trial. After that, no embryonic development of egg was observed up to 12 hrs after fertilization and the fertilized eggs were found to be dead filled with fungus at the end. The water quality parameters were found in good range during the breeding trial though the temperature was found to be fluctuated from the optimum ranges needed for the incubation of fertilized eggs. Although it was not possible to be succeeded completely in artificial breeding of Hilsa, the experience of on board breeding trial in the major breeding ground of River Meghna will give the necessary insight for future works.

Keywords: *Tenualosa ilisha*; Breeding; Hilsa; Embryonic development; Meghna river; Bangladesh

Introduction

Hilsa (*Tenualosa ilisha*, Hamilton) is the most important single species fishery in Bangladesh. It is the most popular food fish to the people of Bay of Bengal region. It is the national fish of Bangladesh which contributes about 11% of total fish production and 1% to the GDP [1]. Due to over exploitation and indiscriminate killing of Hilsa from inland open waters specially from the nursery, feeding and breeding ground of Hilsa (Meghna estuary), the fishery was declining very sharply since the year 2000-2001. Number of factors is responsible for this declines including: (i) barrier to natural migration for breeding due to the sedimentation in rivers (ii) excessive fishing pressure at different life stages (iii) river water pollution altering the physico-chemical parameters of rivers (iv) habitat destruction [2]. As a result, to conserve Hilsa fishery it was then needed to implement the Hilsa Fishery Management Action Plan (HFMAP), which has ceased the decreasing trends of Hilsa production since 2004 to till date. Although, production of Hilsa has increased in recent years but yet unable to meet the high demand to the first growing population of the country. Farm production of market size Hilsa through Hilsa aquaculture using hatchery seed will be a management measure to increase Hilsa production. Furthermore, growing production rate of fishes (e.g. carps, pangas, shrimps etc.) through pond culture and getting the 4th position throughout the world [3], yet Hilsa culture has not been commercially started in this country due to its typical life pattern. To overcome these limitations, on board breeding trial of Hilsa was introduced as one of the most important management and conservation measures to improve its fishery resources in this region.

Why on board breeding?

The life cycle of Hilsa is very typical and for that reason artificial breeding outside its natural environment is almost impossible. Hilsa spawn in freshwater and deposit eggs diversely [4]. After spawning, when the larvae can swim, they try to find suitable nursery grounds, normally in the lower region of the rivers in coastal waters and become *Jatka* i.e., juveniles of Hilsa [5]. The *jatka* remain around the nursery grounds for about 5-6 months and attain a minimum size of 15-16 cm [6,7]. Generally, the *jatka* acquire the ability to tolerate saline water and move downstream to the estuary. There, they spend their young life stages in brackish water. Later, the young move offshore for feeding and grow to adult size. After maturation, the adult again migrates upstream for spawning following the same life pattern. As a result, for successful aquaculture, the life cycle of Hilsa will be necessary to take into account. For on-board breeding trial, it is very important to create semi natural environment. If finally, it is possible to breed *Tenualosa*

*Corresponding author: Md. Monjurul Hasan, Bangladesh Fisheries Research Institute, Riverine Station, Chandpur, Bangladesh
E-mail: mhshihab.hasan@gmail.com

Figure 1: Map showing sampling spots in the Meghna river for collecting matured gravid Hilsa.

ilisha like *Tenualosa toli* in Malaysia, then that would open a new window for Hilsa aquaculture in confined water.

Previous year's achievements

The first on-board breeding trial was conducted by the Riverine Station (RS) Hilsa Research Team of Bangladesh Fisheries Research Institute (BFRI) in the major breeding grounds of Hilsa at the River Meghna of Bangladesh during the period of 25 September 2015 to 09 October 2015. About 11 pairs of brood Hilsa were collected from the River Meghna using gill net in the afternoon and late evening. From the collected broods 08 pairs were tried for breeding by direct stripping and 03 pairs were tried by dopamine injection for artificial breeding. The collected eggs were fertilized by the milt immediately. The fertilized eggs were observed for 24 hrs to study the embryonic and larval development stages of Hilsa. After 6 hrs some fertilized eggs

were found to be denatured and 24 hrs later all the fertilized eggs were found to be dead. The water quality parameters were found to be in good range during the breeding trial of Hilsa [8].

Material and Methods

Timeframe selection

Breeding trial activities were carried out between 10 October to 2 November 2016 in the River Meghna of Bangladesh. A total of six breeding trials were completed by the Riverine Station (RS) Hilsa Research Team of Bangladesh Fisheries Research Institute (BFRI) with speed boat & research vessel *M. V. Rupali Ilish*.

The matured/adult Hilsa (1+ age group) congregates in the major spawning grounds of Hilsa in lower Meghna estuaries for spawning. However, Hilsa spawn year-round but considerably two times in a year,

Figure 2: Pictorial view of reserve tank, bottle hatchery and circular tank set up on research vessel for on-board breeding trial of Hilsa.

Figure 3: Aeration in hatching bottle.

Figure 4: Aeration in hatching jar.

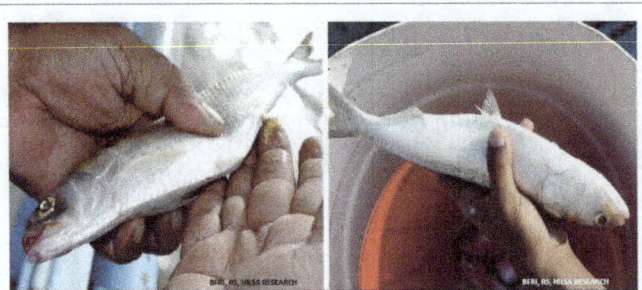

Figure 5: Sorting of matured female and male Hilsa suitable for breeding trial.

which are the 1st peak-spawning season (September-October) and in the 2nd peakspawning season (January-March). This was the major reason of selecting this time period for the conduction of on-board breeding trials.

Spot selection for artificial breeding

On the basis of availability of matured Hilsa, three sampling spots were identified by the Riverine Station (RS) Hilsa Research Team of Bangladesh Fisheries Research Institute (BFRI) (Figure 1):

a) Chairman ghat and Hatia (Noakhali)

b) Ramgoti (Laxmipur)

c) Katakhal and Charvoirobi (Chandpur)

Breeding protocol fix up

Breeding protocol has been made and standardized from the previous breeding trial experiences done by Riverine Station (RS) Hilsa Research Team of Bangladesh Fisheries Research Institute (BFRI) and from literature as well.

Preparation of breeding equipment's

Breeding equipment's such as bottle hatchery, circular tank, overhead tank, hatching jar and all other breeding apparatus were used for breeding operation. A portable model hatchery was established on the research vessel by M.V. Rupali Ilish to provide the breeders to breed in semi-natural condition. The portable hatchery, which was set up on the research vessel for on board breeding trial consists of four units (Figure 2) – a) Overhead tank b) Circular tank c) Bottle hatchery and d) Hatching jar.

Overhead tank: A plastic overhead tank of 1000 litres capacity was placed at 3.5-meter-high above the other operational hatchery unit's setup on the ground floor of the research vessel. River water of spawning grounds was stored in the overhead tank for the hatchery operation.

Circular tank: An iron sheet made circular tank of 500-liter capacity was setup on the ground floor of the research vessel. The collected broods were stocked in this tank for artificial breeding providing water supply, oxygenation, and disinfectant into the water.

Bottle hatchery: A 100 litre capacity incubation bottle was made (covered and surrounded by zero mesh sized net) for incubation of fertilized eggs. A fry-collecting outlet was also connected with the jar. The incubation jar was setup on the ground floor of the research vessel providing water supply with mild circulation and mild oxygenation by flow (Figure 3).

Hatching jar: A 3 litre capacity plastic jar with air stone was also used for the hatching of eggs (Figure 4). A 15-litre capacity sized plastic

Trial No.	Location	Brood Hilsa pair No.		Length Range (cm)		Weight Range (g)		Stripping Time (hr)	Date	Moon Phase
		Male	Female	Male	Female	Male	Female			
1st	Katakhali, Chandpur	2	4	26.5-34	26-35	250-510	220-370	12:30	13-10-2016	1st Quarter+4 days
2nd	Ramgoti, Laxmipur and chairman ghat, Noakhali	2	2	31-33	36-37	230-364	550-568	15:15	15-10-2016	Previous day of Full Moon
3rd	Moulavir Char, Hatia, Noakhali	2	1	26-26.5	31	174-196	266	19:15	20-10-2016	Full Moon+4 days
4th	Katakhali, Chandpur	3	1	28-34.5	25.5	190-452	162	16:30	29-10-2016	Previous day of New Moon
5th	Charvoirobi, Chandpur	4	2	27-28	26-32	192-298	142-363	17:00	1-11-2016	New Moon+1 day
6th	Katakhali, Chandpur	2	1	24-26	33	136-148	398	17:05	1-11-2016	New Moon+1 day

Table 1: Location wise breeding trials of Hilsa broods with different length and weight ranges.

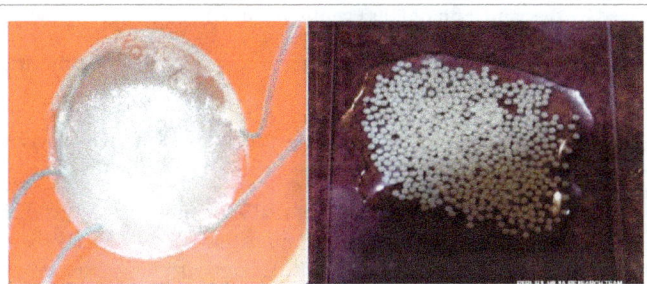

Figure 6: Pictorial view of fertilized Hilsa eggs and aeration of fertilized Hilsa eggs.

Fertilized eggs (non-adhesive and almost spherical in shape)

First cleavage takes in 1 hr. after fertilization and attains 2-cell stage

Further cleavage takes in 2 hr. 15 min after fertilization and attains 8-cell stage

Next cleavage takes in 3 hr. 20 min after fertilization and attains 16-cell stage

Morula stage attains in about 4 hr. 30 min followed by Blastula stage in 5 hr. 20 min

Gastrula stage attains in about 6 hr. 10 min followed by Yolk plug stage in 7 hr. 15 min

18 Myotome stage attains in about 8 hr. 30 min followed by 20 Myotome stage in 9 hr.15 min

Head and tail regions differentiate in about 10 hr. 15 min. of development

Optic vesicles appear after 11hr. of fertilization followed by 28 Myotome stage in 12 hr.

Auditory capsules develop after 13 hr. 10 min of fertilization

Faint twitching movements exhibit after 15 hr. of development

Hatching commences in 15 to 18 hr. and completes in 20 to 21 hr. at temperature 24-29^0c, Dissolved Oxygen 6.5 to 8.5 and pH 7.6 to 7.8

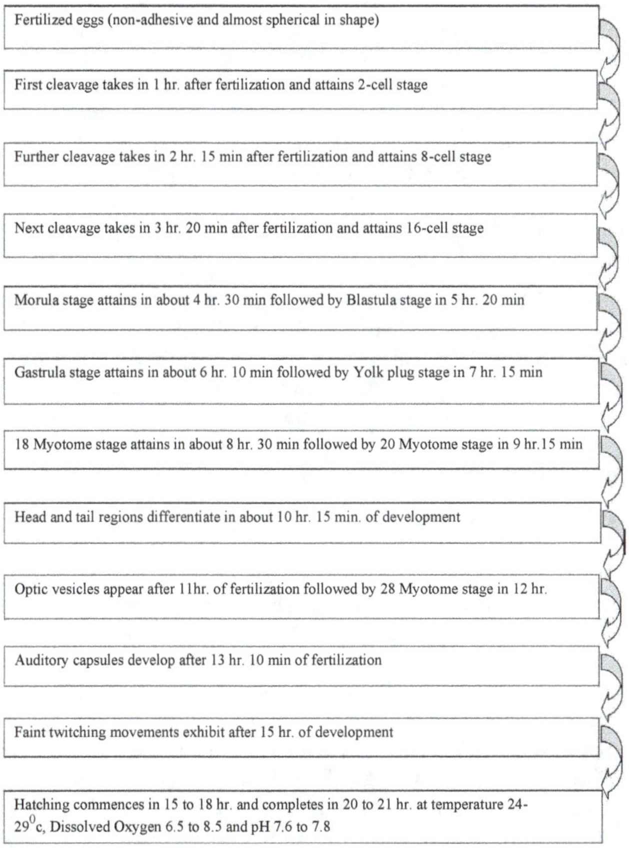

Figure 7: Flow diagram of Hilsa breeding protocol (Modified by BFRI, RS, Hilsa Research Team).

bowl was used for the collection of spawn/fry provided with aeration when needed.

BFRI Experimental Net operation for collecting suitable gravid Hilsa

For breeding experiment, BFRI Experimental Net was successfully operated and suitable gravid Hilsa were collected. BFRI Experimental Net is a 6.5 cm to 7.5 cm mesh sized gill net (multidimensional net).

Conditioning

After collection, selected matured broods were sorted (Figure 5) and kept in a bowl full of river water of spawning grounds for conditioning. To ready the fish for egg release, natural environmental condition was maintained by providing water circulation and oxygenation.

Breeding Process/framework

The whole breeding process was as follows

Live fish stripping: Altogether a total of 13 pairs of brood Hilsa (Table 1) with different length and weight ranges were tried for breeding collected from the major spawning grounds of Hilsa during the ban period. These breeding trials were done depending upon the collection of fully matured Hilsa broods.

Egg and Milt collection and mixing: Eggs from the matured female and milt from the matured male were collected through stripping and then mixed with the caudal fin of Hilsa.

Releasing of mixed solution (Egg and Milt) in plastic hatching jar: After that fertilized eggs were transferred to a plastic hatching jar for incubation providing mild water circulation, aeration and shade to protect the penetration of direct sunlight (which increase the water temperature also) as this species is light sensitive.

Temperature controlling: Temperature was controlled by restricting the open sunlight from the incubation jar. The incubation jar was also covered with black wet cloth.

Observation of fertilized eggs and embryonic development stages: Embryonic development stages of fertilized eggs were observed under compound microscope continuously until the dead cell was found (Figure 6).

Crosscheck for ensuring the embryonic developmental stages

Embryonic development stages of fertilized eggs were identified and crosschecked followed by observations on the embryonic and early larval development of Indian shad, *Tenualosa ilisha* [9]

Water quality observation

Important physico-chemical parameters *viz.* air temperature, water temperature, pH, Dissolved Oxygen, free Carbon dioxide and alkalinity and hardness of breeding grounds and reserved water of portable model hatchery were also observed during the period of breeding trial.

Results

Finalized breeding protocol

The breeding protocol was modified and finalized following [9,10] (Figure 7)

Collection and sorting of matured gravid Hilsa

A total of 259 gravid Hilsa were collected during the whole breeding operation. Most of the female broods were found to be matured and

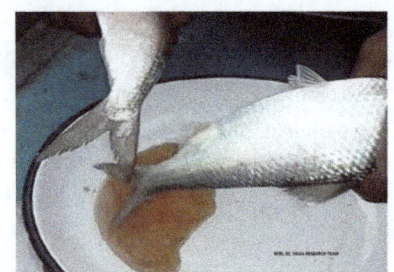

Figure 8: Mixing of eggs and milt collected from the female and male Hilsa broods through stripping.

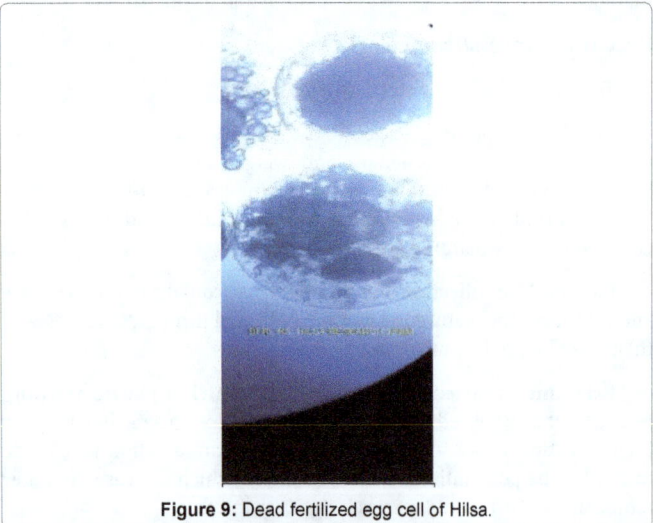

Figure 9: Dead fertilized egg cell of Hilsa.

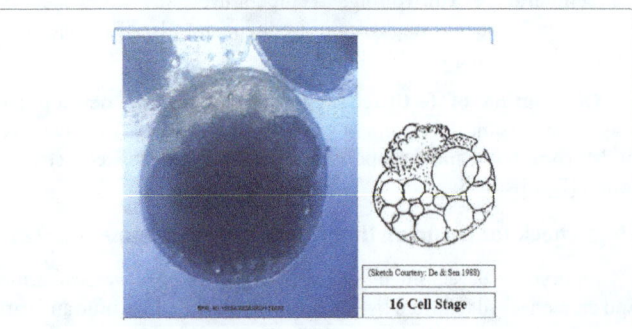

Figure 10: Sixteen (16)-cell stage of embryonic development of Hilsa egg.

their maturity stages were found between V to VI⁺. Large transparent ova that ooze out at the slightest pressure were marked stage VI⁺ as maturity stage in female Hilsa. Vas deferens full of milt which oozes out freely through little pressure were marked stage VI⁺ as maturity stage in male Hilsa.

Breeding trial

Breeding trials were completed at the time of ban period of Hilsa catching in Bangladesh at different locations following breeding protocols.

Mixing of eggs and milt

Immediately after getting the oozing female, the eggs were stripped into a bowl by wet method and the milt from the males were also stripped over the eggs for fertilization. Then the eggs and milt were mixed by the caudal fin of Hilsa gently (Figure 8).

Observation of fertilized eggs and embryonic developmental stages

A total of 6 breeding trials were conducted at the time of ban period of Hilsa catching at different locations.

First breeding trial: First breeding trial was conducted at Katakhal, Chandpur on 13 October 2016 where 2 males and 4 female Hilsa broods were utilized. Collected eggs and milt through stripping were mixed and released in the incubation jar provided with good water circulation and oxygenation. The eggs were fertilized; vitelline membrane and cytoplasmic membrane formed around the egg yolk. Immediately after the fertilization of eggs, these started swelling. The eggs were found very soft, smooth, non-adhesive and almost spherical in shape. After 1-1.2 hrs of fertilization vitelline membrane were found to be separated from the egg yolk and cleavage occurred which were observed under the electron (compound) microscope. This stage was identified as 2-cell stage of embryonic development. Then further cleavage was observed after 2.3-3 hrs of fertilization. This stage was identified as 8-cell stage of embryonic development. After that, no embryonic development was observed up to 6-7 hrs after fertilization and after that eggs were found dead (Figure 9).

Second breeding trial: Another breeding trial was operated at Ramgoti, Laxmipur and Chairman ghat, Noakhali on 15 October 2016 where 2 males and 2 female Hilsa broods were utilized. Collected eggs and milt through stripping were mixed and released in the incubation jar provided with good water circulation and oxygenation. No fertilized eggs were found under electron (compound) microscope which could be due to the less maturity of Hilsa broods.

Third breeding trial: The next breeding trial was conducted at Moulavir char, Hatia on 20 October 2016 where 2 males and 1 female Hilsa broods were utilized. Collected eggs and milt through stripping were mixed and released in the incubation jar provided with good water circulation and oxygenation. The eggs were fertilized; vitelline membrane and cytoplasmic membrane formed around the egg yolk. Immediately after the fertilization of eggs, these started swelling. The eggs were found very soft, smooth, non-adhesive and almost spherical in shape. After 1-1.3 hrs of fertilization vitelline membrane were found to be separated from the egg-yolk and cleavage occurred which was observed under the electron (compound) microscope. This stage was identified as 2-cell stage of embryonic development. Then further cleavage was observed after 2.3-2.6 hrs of fertilization. This stage was identified as 8-cell stage of embryonic development. After that, no embryonic development was observed up to 6-7 hrs after fertilization.

Fourth breeding trial: Another breeding trial was operated at Katakhal, Chandpur on 29 October 2016 where 3 male and 1 female Hilsa broods were utilized. Collected eggs and milt through stripping

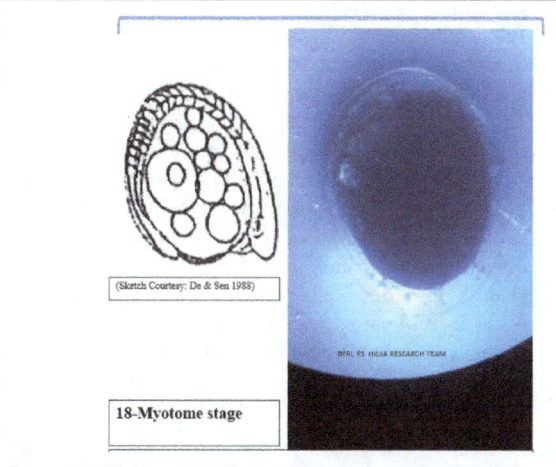

Figure 11: Morula stage of embryonic development of Hilsa egg.

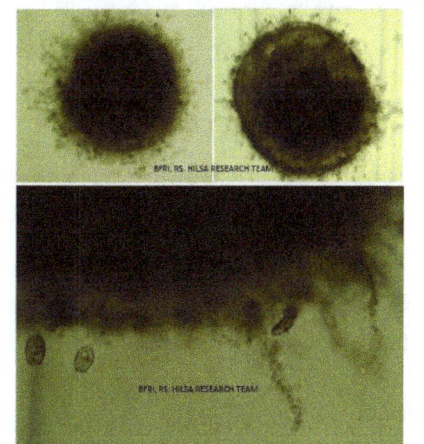

Figure 12: 'Eighteen (18)-Myotome' stage of embryonic development of Hilsa egg.

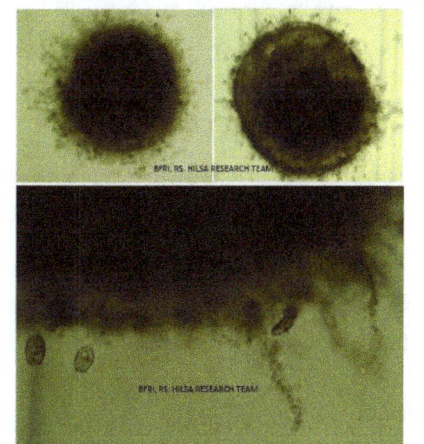

were mixed and released in the incubation jar provided with good water circulation and oxygenation. No fertilized eggs were found under electron (compound) microscope, which could be due to the less maturity of Hilsa broods.

Fifth breeding trial: The 5th breeding trial was conducted at Charvoirobi, Chandpur on 01 November 2016 where 2 males and 4 female Hilsa broods were utilized. Collected eggs and milt through stripping were mixed and released in the incubation jar provided with good water circulation and oxygenation. The eggs were fertilized; vitelline membrane and cytoplasmic membrane formed around the egg yolk. Immediately after the fertilization of eggs, these started swelling. The eggs were found very soft, smooth, non-adhesive and almost spherical in shape. After 3 hrs of fertilization no further embryonic development was observed and the eggs were found to be ruptured which could be due to the collection of less matured Hilsa broods and also may be the water temperature could not be maintained at desired level.

Sixth breeding trial: The final breeding trial was conducted at Katakhal, Chandpur on 01 November 2016 where 2 males and 1 female Hilsa broods were utilized. Collected eggs and milt through stripping were mixed and released in the incubation jar provided with good water circulation and oxygenation. The eggs were fertilized; vitelline membrane and cytoplasmic membrane formed around the egg yolk. Immediately after the fertilization of eggs, these started swelling. The eggs were found very soft, smooth, non-adhesive and almost spherical in shape. After 1.1-1.4 hrs of fertilization vitelline membrane were found to be separated from the egg yolk and cleavage occurred which was observed under the electron (compound) microscope. This stage was identified as 2-cell stage of embryonic development. Then further cleavage was observed after 2.2-2.5 hrs of fertilization. This stage was identified as 8-cell stage of embryonic development. After 3-3.5 hrs of fertilization further cleavage was observed and this stage was identified as 16-cell stage of embryonic development (Figure 10). Then 4.2-4.5 hrs. of fertilization morula stage of embryonic development was identified (Figure 11). 18-myotome stage of embryonic development was identified after 8-8.5 hrs. of fertilization (Figure 12). After that, no embryonic development of egg was observed up to 12 hrs after fertilization and the eggs were found to be dead filled with fungus at the end (Figure 13).

Water quality monitoring

Important physico-chemical parameters *viz.* air temperature; water temperature, pH, Dissolved Oxygen, free Carbon dioxide, alkalinity and hardness of breeding grounds and reserved water of portable model hatchery were also monitored during the breeding trials. Water quality parameters were found very congenial in range (Table 2). It was not possible to maintain the water temperature at desired level.

Discussion

Hilsa ascends the fresh water stretches of the rivers from coastal areas of sea mainly for breeding [4,11,12]. Presently, in almost all river systems of many countries the abundance of the species has drastically dwindled [13]. Many workers [12,14] have suggested the possibilities of Hilsa culture in confined water bodies. There is an urgent need for development of hatchery and grow out technology of Hilsa species due to the great national and international interest as well as enormous commercial interest [13]. Study from [13] revealed that the size of sexual maturity attained by male and female Hilsa vary from 160 mm to 400 mm and 190 mm to 430 mm respectively. Stages of maturity of female gonads have been studied by few workers [15-17] revealing that female below 300 mm size group hardly take part in spawning activity.

Figure 13: Dead filled 8 to 16-cell of embryonic development stage of Hilsa egg with fungus.

Location and trial	Katakhal		Ramgoti		Moulavir Char		Katakhal		Charvoirobi		Katakhal	
No.	1st trial		2nd trial		3rd trial		4th trial		5th trial		6th trial	
Breeding Units	BG	HJ	BG	HJ	BG	HJ	BG	HJ	BG	HJ	BG	HJ
Air Temp. (°C)	31	30	30	30	29.5	29	28	28	29	28.5	29	29
Water Temp. (°C)	29	29	29	28	28	28	29	27	27	26	27	27
pH	8.0	7.4	7.25	7.0	7.5	7.0	7.75	7.5	7.5	7.5	7.5	7.5
DO (mg/l)	5.6	4.8	5.0	5.0	5.3	5.1	5.3	5.2	4.8	5.0	5.0	5.1
CO_2 (mg/l)	24	20	20	16	18	12	14.2	13	15	16	12	10
Alkalinity (mg/l)	80	76	126	112	115	110	65	68	102	93	85	88
Hardness (mg/l)	92	85	111	108	98	92	81	72	69	61	71	64

BG: Breeding Ground; HJ: Hatching Jar

Table 2: Water quality parameters of two breeding units.

The spawning season of Hilsa is noticed during July-August to October-November in all river systems and lagoons i.e., Hoogly [10,12], Ganga [18], Chilka [4], Godavari [15], Padma and Meghna [19,20] but in Hooghly estuary the period of spawning is found to be prolonged and extended up to February and March. While in the river Brahmaputra, the peak spawning of Hilsa was observed from May to July [21]. Few workers [22,23] reported that Hilsa spawn throughout the year in Hooghly while [12] opined that the species spawn intermittently and the fish has two distinct breeding season but the same do not spawn twice during the year [24] reported that Hilsa spawn in rivers and estuaries.

Since 1908, several attempts towards artificial spawning of Hilsa have been made [25-28] but the major breakthrough came during late 70s [29].

The breeding trials were mostly conducted either in the afternoon or in the late evening supported by [17,30] who reported that the most suitable time for fertilization is during afternoon and evening. Immediately after fertilization of the eggs, this starts swelling and the color of the yolk turns to light greenish yellow from its original light yellow reported by Hora [14] which was also observed in the present study. About 15-20 min, after fertilization the eggs become almost colorless and attained a size of 1.95 mm to 2.10 mm diameter, the average being 2.02 mm. The eggs were very soft, smooth, non-adhesive and almost spherical in shape. The eggs were almost demersal in nature in still water and easily buoyed up and drifted by slight currents [14]. The yolk is roughly spherical with an average diameter of 0.88 mm [17].

Hilsa spawn in fresh water and deposited eggs demersally. The eggs hatch after 23-26 hrs at an average temperature of 23°C [4]. Detailed embryonic and larval development of the species were studied by many workers [17,28]. The rate of cleavage and embryonic development varied with the temperature of the water has been reported by the workers. The incubation period was found to be ranged between 18-20 hrs at 28.5°C [28], 18-20 hrs. under 23.5°C to 30.0°C [31] and 18-21 hrs under 24.0°C to 29.0°C [17]. The embryonic and early larval development of Hilsa were studied by [9,17] who stated that morula stage attains in about 3 hrs 30 min followed by blastula and then gastrula in 5 hrs 30 min. They also reported that myotomes appear after 10 hrs and hatching completes 20-21 hrs. of fertilization.

Conclusion

Hilsa shad (*Tenualosa ilisha*, Hamilton) is the national fish of Bangladesh having great socioeconomic and nutritional value. Since 2004, Hilsa production from wild fishery has increased significantly in Bangladesh, yet unable to meet the increased demand of Hilsa among common people of the country. Though the implementation of Hilsa Fisheries Management Action plan (HFMAP) [32], awareness building

among Hilsa fishers and mass people increased Hilsa production in the wild in Bangladesh but it is an urgent need to find out the alternate sources of Hilsa production like culture in captive or semi captive condition by artificial breeding. For the conservation of Hilsa fisheries both in river and sea, several management activities have already been taken both in country and neighboring countries. For example, Hilsha Fisheries Management Action plan (HFMAP) which was being implemented in Bangladesh territory resulting in huge Hilsa production; trans-boundary management plan which was also discussed among Bangladesh, India, Myanmar under the Bay of Bengal Large Marine Ecosystem (BoBLME) project. In captive condition, it was found that Hilsa has grown up slowly but not attained sexual maturity for induced breeding. Efforts on culture of Hilsa have been attempted in India for many years and in Bangladesh recently. Recent efforts by Bangladesh Fisheries Research Institute (BFRI) in Hilsa larval rearing are encouraging but need fine tuning of the existing protocols. As Hilsa has a typical life cycle and high sensitivity to oxygen depletion in water and sunlight, it is very much challenging to get the full success on artificial breeding of this species. To overcome this problem on-board breeding trial can be an important management measures for artificial breeding to produce mass seed production. Hilsa research team of Bangladesh Fisheries Research Institute (BFRI) has been partially successful in on-board breeding trial of Hilsa and reached 18-myotome stage of embryonic development stage of Hilsa eggs.

Challenges and Learning

The Hilsa Research Team faced many challenges to overcome during the experimental operations, at the same time many things has been learnt through these works which have been stated below:

1. Collection of alive broods (male and female Hilsa).

2. Availability of fully matured male and female Hilsa broods and also to make available at a time is a big challenge.

3. To keep them alive in captive condition longer.

4. Desired Hilsa fish is available in very remote and wild areas.

5. Milt preservation (like cryo-preservation etc.).

6. Supply of appropriate Dissolved Oxygen (DO) and maintaining optimum temperature throughout the incubation period.

7. Multi-dimensional models/methods need to apply for breeding of Hilsa.

Acknowledgement

This research work was supported by The United States Agency for International Development (USAID)-funded Enhanced Coastal Fisheries (ECOFISHBD) project of World Fish and the authors are thankful to them.

References

1. DoF (2016) National Fish week 2016 Compendium (In Bengali). Department of Fisheries, Ministry of Fisheries and Livestock, Bangladesh.

2. BOBLME (2014) Report of the hilsa fisheries assessment working group meeting, Kolkata, India, BOBLME.

3. FAO (2014) The state of world fisheries and aquaculture. Rome.

4. Jones S, Menon PMG (1951) Observations on the life history of the India Shad, Hilsa ilisha (Hamilton). Proc Indian Acad Sci 31: 101-125.

5. Haldar GC, Rahman MA (1998) Ecology of Hilsa, *Tenualosa ilisha* (Hamilton). Proceedings of BFRI/ACIAR/CSIRO Workshop on Hilsa Fisheries Research in Bangladesh Held on 3-4 March 1998 at Bangladesh Agricultural Research Council, Dhaka, Bangladesh. Fisheries Research Institute, BFRI Proceedings Series No. 6:11-19.

6. Raja BTA (1985) A review of the biology and fisheries of *Hilsa ilisha* in the Bay of Bengal. Bay of Bengal Programme, Marine Fishery Resource Management in the Bay of Bengal.

7. Mazid MA, Islam S (1991) Hilsa fishery development and management. A Report Published by Fisheries Research Institute, Mymensingh, Bangladesh.

8. Rahman MA (2015) Status of conservation and migration of Hilsa in the Meghna River Estuary and its potential of breeding for stock enhancement and aquaculture. ECOFISHBD Project. 1st Year Annual Research Progress Report, BFRI, RS, Chandpur.

9. De DK, Sen PR (1988) Observations on the embryonic and early larval development of Indian shad, *Tenualosa ilisha* (Ham.). J Inland Fish Society of India 18: 1-12.

10. De DK (1980) Maturity, fecundity and spawning of post-monsoon run of Hilsa. Hilsa ilisha in the upper stretches of the Hoogly estuarine system. J Inland Fish Soc 12: 54-63.

11. Hora SL (1938) A preliminary note on the spawning grounds and bionomics of the so called Indian shad, Hilsa ilisha (Ham.) in the river Ganges. Rec Indian Mus 40: 147-148.

12. Pillay TVR (1958) Biology of the hilsa, Hilsa ilisha (Ham.) of the river Hoogly. Indian Journal of Fisheries 5: 201-257.

13. Sahoo AK, Puvanendran V (2012) Status of Hilsa, Tenualosa Ilisha (Hma.) Aquaculture In India: A review. Regional Workshop on Hilsa: Potential for Aquaculture. Dhaka, Bangladesh.

14. Hora SL (1940) Dams and the problems of migratory fishes. Current Science 9: 406-407.

15. Pillay SR, Rao KY (1962) Observations on the biology and fishery of the Hilsa ilisha (Ham.) of the river Godavari. Proc Indo Pacific Fish Coun 10: 37-61.

16. Mathur PK (1964) Studies on the maturity and fecundity of the Hilsa, Hilsa ilisha (Ham.) in the upper stretches of the Ganga. Indian Journal of Fisheries 11: 426-448.

17. De DK (1986) Studies on the food and feeding habit of hilsa. Hilsa ilisha (Ham.) of the Hoogly estuarine system and some aspects of its biology. University of Calcutta, India.

18. Nair PV (1958) Seasonal changes in the gonads of Hilsa ilisha (Ham.). Philipp J Sci 255-276.

19. Quddus MMA (1982) Two types of Hilsa ilisha and population biology from Bangladesh water. Doctrol Thesis. The University of Tokyo.

20. Quddus MMA, Shimizu M, Nose Y (1984) Spawning and fecundity of two types of Hilsa ilisha in Bangladesh waters. Ibid 50: 177-181.

21. Rao KY, Pathak SC (1972) A note on the occurrence of spawning of Hilsa ilisha (Ham.) in the river Brahmaputra (Assam). Proc Nat Acad Sci India B 42: 231-233.

22. Hora SL, Nair KK (1940) Observations on the bionomics and fishery of the India shad. Hilsa ilisha (Ham.) in Bengal waters. Rec Indian Mus 42: 35-50.

23. Bhanot KK (1973) Observations on the spawning of Hilsa ilisha (Hamilton) in the Hoogly Estuary. J Inland Fish Soc, India 5: 50-54.

24. Blaber SJM, Milton DA, Brewer, DT, Salini JP (2001) The shads (Genus Tenualosa) of tropical Asia: An overview of their biology, status and fisheries, Proceeding of the international Terubok Conference, Malaysia.

25. Wilson HC (1909) Artificial propagation of Hilsa in the Coleroon. Government of Madras, India.

26. Raj BS (1917) On the habits of the hilsa and their artificial propagation in the Coleroon. J Proc Asiat Scc Bengal 13: 184.

27. Southwell T, Prashad B (1918) On Hilsa investigation in Bengal, Bihar and Orisssa. Bull Dept. Fisheries, Bengal, Bihar and Orissa, India.

28. Kulkarani CV (1950) Breeding habits, eggs and early life history of the India shad Hilsa ilisha (Ham.) in the Narbada River. Proc Nar Insl Sci India 15: I69-J 76.

29. Malhotra JC, Mathur PK, Kamal MY, Chandra R, Desai VR (1969) Successful artificial propagation of Hilsa ilisha (Ham.) near Alahabad. Current Sci 38: 429-430.

30. Sen PR, De DK, Nath D (1990) Experiments on artificial propagation of hilsa, Tenualosa ilisha (Ham.). Indian J Fish 37: 159-162.

31. Malhotra JC, Mathur PK, Kamal MY, Mehrotra SN (1970) Observations on the hatching of fertilized eggs of Hilsa ilisha (Ham.) in confined freshwater. Current Science 39: 538-539.

32. Haldar GC, Islam, MR, Akanda MSI (2004) 'Implementation Strategies of Hilsa Fisheries Conservation and Development Management (in Bengali) Fourth Fisheries Project, Department of Fisheries, Ministry of Fisheries and Livestock, Dhaka.

26

Effects of Different Carotenoids on Pigmentation of Blood Parrot (*Cichlasoma synspilum* × *Cichlasoma citrinellum*)

Tieliang Li[1], Chuan He[2], Zhihong Ma[1], Wei Xing[1], Na Jiang[1], Wentong Li[1], Xiangjun Sun[1] and Lin Luo[1]*

[1]*Beijing Fisheries Research Institute, Beijing Key Laboratory of Fishery Biotechnology, Beijing, PR China*
[2]*Extension Stations of Aquiculture Technology, Beijing Municipal Bureau of Agriculture, Beijing, PR China*

Abstract

A feeding experiment was carried out to determine the effects of dietary carotenoid source on body color, skin and scale pigmentation, antioxidant responses of blood parrot (*Cichlasoma synspilum* ♀ × *Cichlasoma citrinellum* ♂). Seven experimental diets were formulated as following: the control diet without carotenoids; PO diet with 4.0 g/kg paprika oleoresin, HP diet with 2.0 g/kg *Haematococcus pluvialis*, PR diet with 2.0 g/kg *Phaffia rhodozyma*, AS diet with 0.4 g/kg synthetic astaxanthin, CA diet with 1.0 g/kg β-carotene and POL diet with 2.0 g/kg paprika oleoresin + 3.0 g/kg lutein. Each experimental diet was fed to triplicate groups of fish to visual satiation twice per day for 8 weeks.

The results showed that AS diet turn fish body red higher and sooner, followed by PR diet and HP diet, successively (P < 0.05). PO and CA had no significant effect on improving body color, but POL diet can turn fish color yellow effectively (P < 0.05). Astaxanthin concentration in diet is positively correlated with redness *a**, but negatively with lightness *L**. Dietary lutein is positive linear correlation with yellowness *b**. All carotenoids can significantly reduce plasma superoxide dismutase (SOD), glutathione peroxidase (GSH-Px), and lipid peroxide (LPO) level, but increase total antioxidant capacity (T-AOC). The present results suggest that dietary inclusion of 0.4 g/kg synthetic astaxanthin can improve body color effectively and quickly, and all test carotenoid source can increase antioxidant activities of blood parrot.

Keywords: Blood parrot (*Cichlasoma synspilum* ♀ × *Cichlasoma citrinellum* ♂); Carotenoids; Pigmentation; Antioxidant capacity

Introduction

Blood parrot (*Cichlasoma synspilum* ♀ × *Cichlasoma citrinellum* ♂) is one of the most popular ornamental fish in China. It has bright red appearance and plump body. It is well known that in addition to body shape and size, body color is one of the most important quality criteria setting the market value of blood parrot [1]. Like other animals, fish are unable to perform *de novo* synthesis of carotenoids [2]. Fish use oxygenated carotenoids, one of the most important groups of natural pigments, for pigmentation of skin and flesh. Under intensive farming conditions and aquarium rearing, blood parrot will be color fading when they are fed exclusively on artificial diets without pigments. It is therefore necessary to study the suitable pigment additives for blood parrot. Commonly used carotenoids in fish feed include synthetic β-carotene, astaxanthin, cantaxanthin, zeaxanthin and lutein and natural sources of yeast, bacteria, algae, higher plants and crustacean meal [3,4]. For example, synthetic astaxanthin has been used to increase the skin coloration of blood parrot [1]. In order to use carotenoids additive suitable for blood parrot commercially on a large scale, different source of carotenoids should be studied. Natural astaxanthin extracted from the algae *Chlorella vulgaris* has been used to increase the skin coloration of goldfish, *Carassius auratus* [5]. Lutein, found naturally in marigold meal, is also an effective pigment for inducing the orange coloration of goldfish [6] and is now used commercially in ornamental fish food. An esterified source of astaxanthin from *Haematococcus pluvialis* increased the natural reddish hue of the skin of red porgy (*Pagrus pagrus*) better than Carophyll Pink˙ [7]. These carotenoids might have potential to be used at color improvement of blood parrot.

On the other hand, it had been reported that carotenoids can protect the body against oxidative damage [8]. Dietary astaxanthin has shown to enhance antioxidant status in Atlantic salmon *Salmo salar* and improve their health [9]. Dietary astaxanthin has shown

to enhance antioxidant status in prawn and increase its resistance to various environmental stresses [10]. Therefore, besides the function of pigmentation improvement, carotenoids may increase antioxidants capabilities. A relationship between different carotenoids sources and antioxidant status in blood parrot should be evaluated.

Taking all above into account, one main purpose of the present study was to evaluate the effectiveness of different carotenoids sources such as the algae *H. pluvialis*, artificial synthesis astaxanthin, P*phaffia rhodozyma*, lutein from *Tagetes erecta*, paprika oleoresin, and β-carotene in dietary with recommended concentrations, to appropriately develop a pigment for blood parrot body color. We aimed to determine the value of skin color chroma and the total carotenoid contents in skin and scales, to elucidate what kinds of dietary carotenoids are metabolically transformed in the skin of blood parrot. Given the important role of carotenoids in fish antioxidation, another purpose of the present work was to check whether dietary carotenoids would affect antioxidant capacity of blood parrot.

Materials and Methods

Experimental diets

A basal diet, containing 324 g/kg crude protein and 57 g/kg crude lipid, was composed of American white fishmeal 120 g/kg, soybean

*Corresponding author: Lin Luo, Beijing Fisheries Research Institute, Beijing 100068, PR China, E-mail: luo_lin666@sina.com

meal 300 g/kg, degossypolled cottonseed meal 120 g/kg, soybean oil 30 g/kg, wheat flour 200 g/kg, wheat middling 200 g/kg, vitamin premix 10 g/kg, and mineral premix 200 g/kg. Using this basal mixture, six pigmented diets were formulated by added pigments in accordance with recommended doses by the supplier. Diet PO contains 4 g paprika oleoresin/kg feed; diet HP contains 2 g algae *H. pluvialis*/kg feed; diet PR contains 2 g *P. rhodozyma* /kg feed; diet AS contains 0. 4 g astaxanthin/kg feed; diet CA contains 1 g β-carotene /kg feed; and diet POL contains 4 g paprika oleoresin /kg feed and 3 g lutein /kg feed. The ingredient and proximate analyses of the diets were shown in Table 1. The pigment materials were mixed with the basal diets completely and the experimental diets were made into floating extruded pellets (Ø2.0 mm) by TSE65S model twin screw steam extruder (Beijing Modern Yanggong Machinery S&T Development Co., Ltd, China). All diets were stored at -20°C to avoid oxidation of the pigmentations throughout the experiment.

Fish and rearing conditions

The homogeneous blood parrots were obtained from a commercial fish farm at Tianjin, China. Prior to the beginning of the experiment, fish were fed the control diet 2 weeks for acclimatization in laboratory culturing system. Hatchery reared fish, at roughly the same total body length of 10 cm, were randomly distributed into twenty-one groups of 30 fish each (seven treatments, with triplicate each) at the beginning of the test. The fish were cultivated in an aquarium measuring 90 cm × 40 cm × 50 cm and supplied with continuous aeration. The oxygen level was kept at 7.3 ± 0.1 mg/L. The water was changed three times weekly. One-third of the water was changed each time. The water temperature was 26 ± 1°C. The fish were satiation hand-fed twice daily at 10:00 and 16:00. The fish were subjected to a natural photoperiod during 56 days experimental period.

Biochemical and colour analysis

Feed analysis: Biochemical analysis of feed was conducted in triplicate according to AOAC (2005). Briefly, crude protein (N × 6.25) was determined by the Kjeldahl method after acid digestion using an Auto Kjeldahl System (2100-Auto-analyzer, Foss, Hillerød, Denmark). Crude lipid was determined by the ether extraction method using a Soxtec System HT (Soxtec System HT6, Foss, Hillerød, Denmark). Moisture was determined by oven-drying at 105°C for 24 h. Ash was determined by combustion at 550°C for 12 h.

Colour parameters analysis: From the beginning to the end of the trial, three fish of each treatment were taken randomly to determined skin color every week. Photos were taken every two weeks to record the change of body color. As an ornamental fish, blood parrots need simple, rapid and accurate color analysis alive. Fishes were individually measured for skin color using a Minolta CM-600 Chroma Meter (Minolta Camera Co. Ltd., Asaka, Japan) before commencement

of the feeding trial to establish baseline measurements (week 0) and then every week for the 8-week period. The Chroma Meter was set to take absolute measurements in the L^*, a^*, and b^* measuring mode (CIE1976) using D65 illuminant. Color measurements were taken in the centre of the lateral body. L^* is the lightness variable where L^* = 100 is white and L^* = 0 is black; a^* is the red chromaticity coordinates where + a^* stands for red and -a^* stands for green, and b^* is the yellow chromaticity coordinates where + b^* stands for yellow and -b^* stands for blue. To avoid the influence of day time on day body color, L^*, a^*, and b^* measurements were always done at 14:00 every time.

Analysis of carotenoids in feed, fish flesh, skin and scales: β-carotene, zeaxanthin, lutein and astaxanthin content from feed, skin, scale and flesh samples were obtained according to the following high performance liquid chromatography (HPLC) method: 1. Feed, flesh, skin or scale samples (1 g) was taken into a 250 mL Erlenmeyer flask individually and mixed with 30 mL ethanol, 5 mL 10% (mass/volume) ascorbic acid, 10 mL 50% (mass/volume) KOH. The mixture solution was heated and recycled in Hotplate Stirrer for Round-Bottom-Flasks for 30 mins, then cooled by putting the flasks into the cold water. 2. All above saponated mixture solution were taken into separating funnel and washed by 100 mL petroleum ether; then the separating funnel was shaken 3 min twice; water in the separating funnel was removed after the mixture solution layering. 3. Anhydrous sodium sulfate was added into the separating funnel, and then the filtrate was collected and vacuum drying. Dried sample was added an appropriate amount of mobile phase solution and then filtered through a 0.2 μm Millipore filter and stored in 4 mL brown vials. 4. Identification of carotenoid was accomplished by HPLC with standards. The volume of the solution was recorded at 448 nm by HPLC to calculate the β-carotene, zeaxanthin and lutein concentration, 470 nm to astaxanthin concentration.

The carotenoid concentration was calculated according to the formula:

$$X \ (\mu g/100 \ g) = (C \times V) / (m \times 100)$$

Total carotenoids concentration (mg / kg) = X (β-carotene) + X (zeaxanthin) + X (lutein) + X (astaxanthin)

Where X is carotenoid monomer concentration (mg / kg); C is carotenoid monomer concentration in sample solution (μg / mL); V is final constant volume of concentrate solution (mL); m is the sample weight (g).

Analysis of antioxidant parameters

At the end of feeding trial, blood samples were withdrawn from the caudal vein of three fish randomly taken from each aquarium using sterile syringes. Blood samples were prepared by mixing 14 μL heparinized solution containing 0.14 mg heparin sodium with 700 μL blood immediately after withdrawing the blood. The

	PO	HP	PR	AS	CA	POL	The control
Moisture (%)	89.1	89.3	89.3	89.5	89.4	88.9	89.5
Curde protein (%)	32.4	32.5	32.5	32.5	32.5	32.3	32.5
Crude fat (%)	5.7	5.7	5.7	5.7	5.7	5.7	5.7
Ash (%)	8.2	8.2	8.2	8.3	8.2	8.2	8.3
β-carotene(mg/kg)	60.7	15.4	186.3	2.6	5449	47.7	2.4
Lutein (μg/100 g)	362	99.2	61.4	51.9	22.5	776	21.7
Zeaxanthin (μg/100 g)	260	27.5	26.8	22.5	7.47	152.2	8.8
Astaxanthin (μg/100 g)	79.8	1065.2	1493.7	2076.4	59.8	88.4	5.2
Total carotenoid (μg/100 g)	762.5	1207.3	1768.2	2153.4	5539	1064.3	38.1

Table 1: Proximate analysis of the experimental diets.

samples were chilled at -80°C until being used for determination of Superoxide Dismutase (SOD), glutathione peroxidase (GSH-Px), total antioxidant capacity (T-AOC), and lipid peroxide (LPO). The different antioxidant parameters of SOD, GSH-Px, T-AOC and LPO were analyzed using Randox Laboratories kits (Crumlin, Co. Atrim, UK) by spectrophotometry (U-2000; Hitachi Ltd., Japan).

Statistical analysis

All data were subjected to a one-way analysis of variance (ANOVA) using the Statistica 7.0 software environment to test the effects of the experimental diets. All of the results are expressed as the means±S.E.M. Duncan's test was used to test differences among the individual means. The differences were regarded as significant when $P < 0.05$. The slopes of color parameters and scale and skin pigment responses to the diets were compared after fitting a linear regression model. Correlations were regarded as significant when the correlation coefficient $R > 0.5$.

Results

Effects of experimental diets on the body color

No mortality was observed during the feeding trial. The changing of lightness L^* of blood parrot during 56 days trial was shown in (Figure 1). At the end of the first week, fish fed AS diet had significantly lower lightness than that of fish fed the control diet ($P < 005$). Compared with the rest of fish groups, AS fish group kept the lowest lightness from day 21 until the end of trial ($P < 005$). Meanwhile, as time went on, the lightness of AS fish group was significantly lower since day 21 compared to the two first weeks ($P < 0.05$). And the lightness of PO fish group was significantly higher since day 42 compared to all the previous weeks ($P < 0.05$). Redness a^* of AS, PR and HP fish group was significantly higher than that of PO, CA, POL and the control fish groups from day 7, 14 and 21 to the end of the trial, respectively ($P < 0.05$) (Figure 2), and redness a^* of AS, PR, and HP fish groups order from high to low is AS > PR > HP ($P < 0.05$). On the other hand, redness of AS fish group increased with the days going on and got the highest value at day 56 ($P < 0.05$). Although redness of PR and HP fish group increased with time going on, they kept the same level from day 35 to 56 ($P > 0.05$). Redness of PO, CA and POL fish group had no significantly difference with the control group during the whole trial ($P > 0.05$).

Yellowness b^* of POL fish group was significantly higher than that of the rest fish group since 35 days until the end of the trial ($P < 0.05$) (Figure 3).

Visual differences in coloration between diet fish were shown in Figures 3 and 4. AS fish group was obviously redder than other fish groups, followed by PR and HP fish groups, successively. While POL fish group was obviously yellower. CA and PO fish groups were yellower than the control fish group, but the difference was not much clear.

Table 2 showed regression models of coloration efficacy. Regression analysis revealed that lightness L^* of fish body had negative linear relation with diet ($R = -0.53$) and skin ($R = -0.54$) astaxanthin levels, while redness a^* was significantly positively correlated to diet, skin and scale astaxanthin levels, with r values of 0.91, 0.91 and 0.85, respectively. Yellowness b^* of fish body positively and linearly related with diet lutein levels ($R = 0.52$).

Carotenoids in fish scale and skin

Carotenoids concentrations in the scale and skin of blood parrot after being fed for day 56 was shown in Table 3. β-carotene

concentration in scale of fish fed on HP had not been detected. β-carotene concentration in scale and skin of fish fed on CA diet were significantly higher than that of the rest fish groups ($P < 0.05$). Lutein and zeaxanthin concentrations in scale and skin of fish fed on POL diet were significantly higher than those of the rest fish groups ($P < 0.05$). AS fish group had the significantly highest astaxanthin concentration in scale and skin, followed by PR fish group, HP fish group, PO, CA and POL fish groups, the control fish group, orderly ($P < 0.05$). POL fish group had the highest total carotenoids in both scale and skin because of their especially higher lutein concentration ($P < 0.05$). PR and AS fish groups had significantly higher total scale carotenoids than PO and the control fish groups ($P < 0.05$). Total skin carotenoids of all fish groups fed pigment diets were significantly higher than that of the control fish group ($P < 0.05$); total carotenoids in AS and PR fish skin were significantly higher than that in PO and HP fish skin ($P < 0.05$);

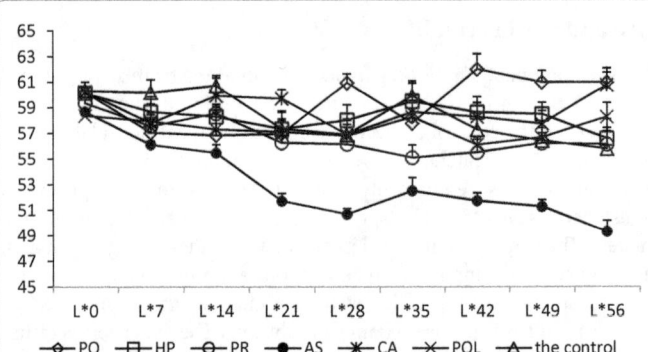

Figure 1: The changing of lightness (L*) of blood parrot during 56 days trial. Values are mean + SE.

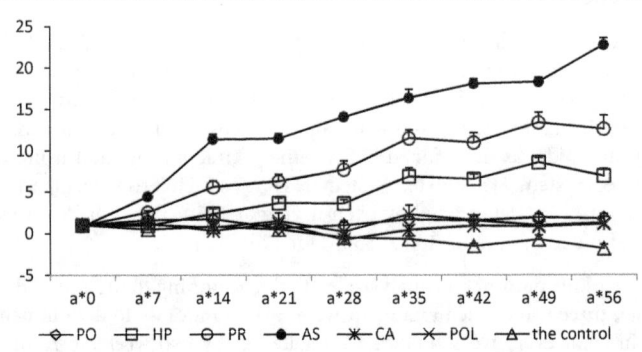

Figure 2: The changing of redness (a*) of blood parrot during 56 days trial. Values are mean + SE.

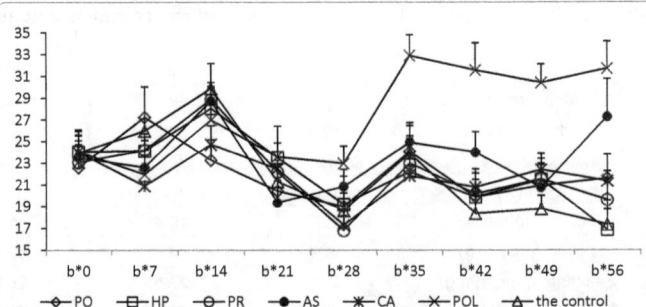

Figure 3: The changing of yellowness (b*) of blood parrot during 56 days trial. Values are mean + SE.

(P < 0.05). T-AOC value of fish fed POL diet was significantly lower than that of fish fed PO, HP, PR and AS diets (P < 0.05). The T-AOC value of CA fish was significantly lower than that of PR and AS fish (P < 0.05). POL fish had significantly lower LPO value than all other fish groups (P < 0.05), followed by AS fish group (P < 0.05). LPO values of PO, HP, PR and CA fish were significantly higher than those of POL and AS fish (P < 0.05).

Discussion

Paprika oleoresin (also known as paprika extract) is an oil-soluble extract from the fruits of *Capsicum annum Linn* or *Capsicum frutescens* (Indian red chillies), and is primarily used as a coloring and/or flavoring in food products. It is composed of capsanthin and capsorubin, the main coloring compounds (among other carotenoids) [11]. Based on the information from paprika oleoresin supplier, capsanthin and capsorubin in paprika oleoresin is about 8%. Although capsanthin and capsorubin contents had not been detected in the trial, total concentration of capsanthin and capsorubin in diet PO and POL would be 32000 μg/100 g and 16000 μg/100 g, theoretically. PO diet did not significantly improve body color in this trial. It is suggesting that capsanthin and capsorubin have no effective contribution on improving body color of blood parrot at least under their current contents in diets.

On the other hand, it could be seen that *a** and *b** of PO fish group had no difference with time, they kept the same level from the beginning to the end of trial (Figures 2 and 3). While POL diet made fish yellower than all other diets since day 35 to the end of trial. The difference between PO and POL diet was with or without lutein adding in diets. It is suggested that it is lutein but not zeaxanthin that turn blood parrot yellow effectively Meyers & Chen [12]. Classified the aquatic animals into 3 categories by their ability of converting carotenoids into astaxanthin. **Type** I: salmonoids or sea bream type, which cannot oxidize β-ionone of the carotenoid and can only use the oxidized carotenoid. Type II: carp type, which can use and convert

Figure 4: Digital images of blood parrot from each of the seven diet treatments after 56 days feeding: (Before feeding) original color of fish before feeding pigments (also showing sampling areas for color measurements (1) where all measurements were taken on the same side for all of the fish used in the experiment (side shown in image); (PO) *Paprika oleoresin*; (HP) *H. pluvialis*; (PR) *Phaffia rhodozyma;* (AS) astaxanthin; (CA) β-carotene; (POL) paprika oleoresin + lutein; (The control) the control without any pigments.

and total carotenoids in AS fish skin was also significantly higher than that in CA fish skin (P < 0.05).

The correlation between pigments in diet and pigments in scale and skin of fish had also been shown in Table 2. Regression analysis showed that skin β-carotene levels were significantly correlated with diet β-carotene (R = 0.91) and total carotenoids dose (R = 0.84). Scale β-carotene levels were also correlated with diet β-carotene (R = 0.59) and total carotenoids dose (R = 0.53). Both lutein in scale and skin were linearly related to diet lutein level, with R values of 0.69 and 0.68, respectively. Zeaxanthin in skin was correlated with the dietary dose of zeaxanthin (R = 0.56), while that in scale was correlated with the dietary dose of lutein (R = 0.53). Bothastaxanthin in scale and skin were significantly linearly related to the astaxanthin in the diets, with R values of 0.98 and 0.95, respectively.

Effects of diet pigments on fish antioxidant parameters

SOD value of the control fish was significantly higher than that of other fish groups (P < 0.05) (Table 4). The control fish had the highest GSH-P x and LPO activities, but the lowest T-AOC activity in all fish fed on pigmented diets (P < 0.05). GSH-Px of fish fed on POL diet was the significantly lower than that of fish fed on other experiment diets

Items	Correlation	Regression models
L* of fish body by diet astaxanthin	R = -0.53, P < 0.01	Y = 59.36-0.004X, R² = 0.29
a* of fish body by diet astaxanthin	R = 0.91, P < 0.01	Y = -0.44 + 0.01X, R² = 0.82
b* of fish body by diet lutein	R = 0.52, P < 0.01	Y = 20.82 + 0.01X, R² = 0.27
L* of fish body by astaxanthin in skin	R = -0.54, P < 0.01	Y = 59.13 - 0.01X, R² = 0.29
a* of fish body by astaxanthin in skin	R = 0.91, P < 0.01	Y = 0.24 + 0.03X, R² = 0.83
a* of fish body by astaxanthin in scale	R = 0.85, P < 0.01	Y = -2.31 + 0.25X, R² = 0.72
β-carotene in skin by diet β-carotene	R = 0.91, P < 0.051	Y = 1.96 + 0.02X, R² = 0.83
β-carotene in scale by diet β-carotene	R = 0.59, P < 0.01	Y = 3.073 + 0.004X, R² = 0.35
β-carotene in skin by diet total carotenoids	R = 0.84, P < 0.01	Y = -24.44 + 0.03X, R² = 0.71
β-carotene in scale by diet total carotenoids	R = 0.53, P < 0.01	Y = -1.114 + 0.004X, R² = 0.28
Lutein in skin by diet lutein	R = 0.69, P < 0.01	Y = 397.13 + 1.31X, R² = 0.48
Lutein in scale by diet lutein	R = 0.68, P < 0.01	Y = 30.19 + 0.20X, R² = 0.46
Zeaxanthin in skin by diet zeaxanthin	R = 0.56, P < 0.01	Y = 171.76 + 0.21X, R² = 0.31
Zeaxanthin in scale by diet lutein	R = 0.53, P < 0.01	Y = 15.51 + 0.03X, R² = 0.28

Table 2: Regression models of coloration efficacy and regression models of deposition efficacy for each pigment in skin and scale of fish.

Carotenoids (µg/100 g)		Diet notation						
		HP	PR	AS	CA	POL	The control	PO
scale	β-carotene	/	2.67 ± 1.28ᵃ	0.76 ± 0.23ᵃ	25.88 ± 6.32ᵇ	15.74 ± 5.96ᵃᵇ	1.98 ± 0.51ᵃ	1.64 ± 0.74ᵃ
	Lutein	35.90 ± 2.50ᵃ	42.07 ± 5.87ᵃ	34.44 ± 3.64ᵃ	45.03 ± 6.98ᵃ	220.83 ± 30.15ᵇ	50.18 ± 5.46ᵃ	42.99 ± 2.97ᵃ
	Zeaxanthin	13.51 ± 1.14ᵃ	20.62 ± 4.51ᵃ	14.54 ± 1.85ᵃ	23.42 ± 4.68ᵃ	43.40 ± 6.30ᵇ	17.91 ± 4.94ᵃ	16.88 ± 1.37ᵃ
	Astaxanthin	54.03 ± 2.97ᶜ	62.61 ± 3.02ᵈ	74.71 ± 4.28ᵉ	11.03 ± 1.29ᵇ	15.22 ± 1.37ᵇ	2.54 ± 0.07ᵃ	14.32 ± 1.12ᵇ
	Total carotenoid	103.44 ± 5.22ᵃᵇ	126.78 ± 9.⁴1b	123.86 ± 7.42ᵇ	105.37 ± 11.46ᵃᵇ	295.20 ± 37.97ᶜ	72.60 ± 6.38ᵃ	69.78 ± 6.46ᵃ
skin	β-carotene	2.09 ± 0.19ᵃ	3.99 ± 0.51ᵃ	2.40 ± 0.56ᵃ	136.54 ± 19.74ᵇ	7.00 ± 2.46ᵃ	2.24 ± 0.25ᵃ	1.74 ± 0.17ᵃ
	Lutein	437.56 ± 56.43ᵃᵇ	467.89 ± 35.97ᵃᵇ	470.33 ± 54.65ᵃᵇ	629.89 ± 34.45ᵇ	1697.33 ± 170.57ᶜ	266.78 ± 30.94ᵃ	639.89 ± 77.48ᵇ
	Zeaxanthin	156.86 ± 21.54ᵃᵇ	187.11 ± 13.17ᵃᵇ	152.84 ± 20.31ᵃ	238.89 ± 8.97ᵇ	378.89 ± 41.28ᶜ	217.78 ± 21.95ᵃᵇ	168.13 ± 23.62ᵃᵇ
	Astaxanthin	239.62 ± 17.23ᵇ	496.88 ± 18.06ᶜ	749.93 ± 24.44ᵈ	12.73 ± 0.44ᵃ	19.47 ± 0.78ᵃ	3.72 ± 0.22ᵃ	15.77 ± 1.5⁴a

Note: Data are mean ± SD. Means not bearing the same superscript letters in the same row are significantly different (P < 0.05).

Table 3: Carotenoids content in the scale and skin of blood parrot after being fed for 56 days.

Antioxidant parameter	Diet notation						
	PO	HP	PR	AS	CA	POL	The control
SOD (U/ml)	59.04 ± 1.23ᵃ	58.13 ± 1.13ᵃ	59.72 ± 0.74ᵃ	61.37 ± 1.18ᵃ	59.52 ± 1.25ᵃ	58.18 ± 0.57ᵃ	66.02 ± 1.18ᵇ
GSH-Px (U/ml)	1033.6 ± 14.2ᵇ	1025.0 ± 21.8ᵇ	1055.1 ± 19.3ᵇ	1065.9 ± 13.7ᵇ	1059.2 ± 7.6ᵇ	931.1 ± 46.6ᵃ	1293.4 ± 10.4ᶜ
T-AOC (U/ml)	12.40 ± 0.43ᶜᵈ	12.24 ± 0.36ᶜᵈ	12.69 ± 0.24ᵈ	12.72 ± 0.29ᵈ	11.39 ± 0.69ᵇᶜ	10.44 ± 0.37ᵇ	9.14 ± 0.32ᵃ
LPO (nmol/ml)	37.06 ± 2⁹0c	39.43 ± 1.67ᶜ	36.83 ± 1.47ᶜ	23.21 ± 1.74ᵇ	36.71 ± 2.58ᶜ	15.67 ± 1.04ᵃ	50.57 ± 3.98ᵈ

Note: Data are mean ± SD. Means not bearing the same superscript letters in the same row are significantly different (P < 0.05).

Table 4: Main effects of dietary carotenoid concentration on activity of antioxidants in blood parrot after 56 days feeding trial.

zeaxanthin into astaxanthin and store astaxanthin. Type III: crustacean type, which can convert β-carotene, zeaxanthin, canthaxanthin, and echinenone into astaxanthin. As one of carp types, common goldfish had been proved the ability to convert lutein to astaxanthin from the result of feeding [¹⁴C] lutein to the yellow variety [13]. Salmonids preferentially absorb and deposit more polar carotenoids, particularly astaxanthin, rather than canthaxanthin, zeaxanthin, or carotenes [14,15] indicated that rainbow trout (*Oncorhynchus mykiss*) could deposit more astaxanthin than β-carotene, a condition that could result from differences in absorption, tissue distribution, and retention. Our results coincide with these findings. In this trial, although lutein, zeaxanthin and β-carotene had been transport from diets and deposit in skin and scale because their content in skin and scale were lineal correlated with them in diet, it seemed that lutein, β-carotene and zeaxanthin in diet had no significant contribution to body red color or astaxanthin concentration in diet, skin and scale. Regression analysis of a^* revealed that a^* is just linearly related to astaxanthin in diet, skin and scale, respectively, whilst astaxanthin in skin and scale are directly come from diet astaxanthin with significant linear correlation. Above results suggest that astaxanthin is the most efficacious pigment in red pigmentation of blood parrot. The results from this study are suggesting that the type of blood parrot using carotenoids more likely similar with salmonoids or sea bream type.

In this trial, carotenoids in skin and scale are highly coincides with them in diets (Tables 2 and 3). But all flesh samples of blood parrot are colorless and translucent, and none of carotenoids has been tested in all flesh samples. Fish absorb dietary carotenoids through the intestinal mucosa transport them through the blood *via* serum lipoproteins [16] and deposit them into specialized skin cells called chromatophores [17]. Apparently, blood parrot deposits carotenoids into skin and scale instead of flesh to keep its body color as effective as possible [18].

As body color is the major factor influencing the commercial value of blood parrot, it is therefore becoming important to enhance body red color as soon as possible. In general, it is the sooner the better. So the time of getting red is very crucial, too. In the present trial, body color parameters were measured every week in order to find out which

pigment will turn fish red more quickly. This was most apparent after 7 days of feeding, when red chromaticity a^* values of AS fish group were significantly higher than other fish groups and red coloration continued to increase until 56 days of feeding on the AS diet. After being fed 14 days, PR fish group got significantly higher a^* values than the control fish group and kept this advantage until the end of the trial, too. The third one was HP fish group from 21 days to the end of the trial. The time of getting red also fits in with astaxanthin concentration in diet.

According above results, the intensity and time of getting red are strongly according with diet astaxanthin concentration. The more astaxanthin in diet, the more and sooner fish body will get red. It seems that PR and HP fish group would get the same a^* level if more *P. rhodozyma* or *H. pluvialis* were added into diet to reach the same astaxanthin concentration with AS diet. However, although commercial price of synthetic astaxanthin is about 3 times over of *P. rhodozyma* and *H. pluvialis*, its dosage is just one fifth of the latter pair. On turning blood parrot to red, synthetic astaxanthin will be more commercially competitive than natural astaxanthin sources from *P. rhodozyma* and *H. pluvialis*.

Lightness L^* of AS fish group was the lowest from 21 day feeding to the end of trial. It might be associated with their dark red appearance. The lightness of fish body decreased with the increasing of redness. This is in accord to Baron [19] who found that the skin of flame-red dwarf gourami (*Colisa lalia*) became darker when fed synthetic astaxanthin Lucantin˙ Pink. It can be seen that L^*, a^* and b^* of the control fish decreased with the time increasing, and their values are significantly lower at the end of trial compared with those at the beginning of the trial. There is not any pigments supply in the control diet will be the reason of fish fading. This result confirmed that pigments need add in blood parrot feed to make fish keeping good body color in actual production.

Owing to their structure, carotenoids possess free radical scavenging properties and therefore act as antioxidants. This has been demonstrated in both *in vivo* and *in vitro* systems, as well as in studies using membrane models [20-22]. SOD, a cytosolic enzyme that is specific for scavenging superoxide radicals, is involved in protective mechanisms within tissue injury following oxidative process and phagocytosis. The higher the SOD activity, the more superoxide radicals need to be reacted [23]. It was found that characins (*Hyphessobrycon callistus*) [23]. And olive flounder (*Paralichthys olivaceus*) [24] fed on carotenoid diets had lower SOD activities than fish fed the control diet. In this study, lower SOD activity of fish fed dietary carotenoids indicated that dietary carotenoid effectively reduced SOD activity in blood parrot.

GSH-Px exists in blood, liver, mitochondria, and cytoplasm and involves in the reaction of removal of H_2O_2 and is recognized as one of the most important antioxidant defenses against oxygen toxicity in organisms. The lower the GSH-Px value, the higher protection that cell has already been provided and reported [25,26] that much lower GSH-Px in characins fed on pigmented diets than that in control fish and the decreasing trend of GSH-Px with increasing dietary astaxanthin and β-carotene concentration. Our results are in accordance with previous findings. Carotenoid sources significantly decreased plasma GSH-Px activities of blood parrot. And the lowest GSH-Px activities of fish fed POL diet might be due to its extremely high total carotenoid concentration (Table 3). It is suggesting that carotenoids could reduce peroxide in cells and concomitantly the GSH-Px effectively, and their efficiency might be positively related with their concentration. The antioxidant capacity that T-AOC expresses includes enzymatic and non-enzymatic antioxidant activities. The higher T-AOC value, the higher antioxidant capacity it has [23]. In this study, all carotenoid diets were significantly higher than the control diet. It means that these carotenoids can improve antioxidant capacity of blood parrot effectively.

LPO is decomposed into a variety of oxygen-containing redicals. The resulting radicals attack almost all cell components, such as proteins, lipids, nucleic acids and membranes, initiate free radical chain reactions, and induce oxidative stress [27]. The serum LPO level is considered to be one of the sensitive indicators of tissue damage derived from oxidative stress [28]. The higher LPO level in serum, the more tissue damage derived from oxidative stress. LPO production in serum and muscle were observed to be significantly decreased after the red yeast administration [29]. Our research confirmed their results. In the present study, the plasma LPO of fish fed the control diet was observed to be significantly higher than that of all carotenoids diets (P < 0.05). POL and AS diets were shown more effectively inhibit the accumulation of LPO in plasma than the rest of carotenoids diets. These results suggest that all carotenoids in this study reduce the LPO level in blood parrot dramatically, and their efficiency maybe positively related with their concentration, too.

In conclusion, blood parrot is able to efficiently utilize synthetic and natural astaxanthin supplied in the diets. Deposit is mainly done in its skin and scale, which causes an acceptable red-colored natural color. All test carotenoid source could increase antioxidant activities of blood parrot effectively.

Acknowledgements

This work was financially supported by Beijing Innovation Consortium of The Ornamental Fish Research System, project No. GSY20160204.

References

1. Yang H, Mu X, Luo D, Hu Y, Song H, et al. (2012) Sodium taurocholate, a novel effective feed-additive for promoting absorption and pigmentation of astaxanthin in blood parrot (Cichlasoma synspilum ♀ × Cichlasoma citrinellum ♂). Aquaculture 350-353: 42-45.

2. Goodwin T (1984) The Biochemistry of Carotenoids, vol II, Chapman and Hall, London, UK.

3. Shahidi F, Metusalach A, Brown J (1998) Carotenoid pigments in seafoods and aquaculture. Crit Rev Food Sci Nutr 38: 1-67.

4. Kalinowski C, Robaina L, Fernandez-Palacios H, Schuchardt D, Izquierdo M, et al (2005) Effect of different carotenoid sources and their dietary levels on red porgy (Pagrus pagrus) growth and skin colour. Aquaculture 244: 223-231.

5. Gouveia, Rema P (2005) Effect of micro algal biomass concentration and temperature on ornamental goldfish (Carassius auratus) skin pigmentation. Aquaculture Nutrition 11: 19-23.

6. Mela M, Smullen R, Obra R (2002) Non-invasive methods for measuring the accumulation of carotenoids in common goldfish (Carassius auratus) fed by astaxanthin, canthaxanthin and lutein supplemented diet. In: Proceedings of the 10th international symposium on Nutrition and Feeding in Fish.

7. Tejera N, Cejas J, Rodríguez C, Bjerkeng B, Jerez S, et al. (2007) Pigmentation, carotenoids, lipid peroxides and lipid composition of skin of red porgy (pagrus pagrus) fed diets supplemented with different astaxanthin sources. Aquaculture, 270: 218-230.

8. Halliwell B, gutteridge J (1989) Free Radicals in Biology and Medicine (2ndedn) Clarendon Press, Oxford, USA.

9. Christinasen R, Torrissen O (1996) Growth and survival of Atlantic salmon, Salmo salar L. fed different dietary levels of astaxanthin Juveniles. Aquaculture Nutrition 2: 55-62.

10. Chien Y, Shian W (2005) The effects of dietary supplementation of algae and synthetic astaxanthin on body astaxanthin, survival, growth and low dissolved oxygen stress resistance of kuruma prawn, Marsupenaeus japonicas Bate. Journal of Experimental Marine Biology and Ecology 318: 201-211.

11. Pérez-Gálvez, A, Martin H, Sies H, Stahl W (2003) Incorporation of carotenoids from paprika oleoresin into human chylomicrons. Br J Nutr 89: 787-793.

12. Meyers S, Chen H (1982) Astaxanthin and its role in fish culture. Proceedings of the Warm water Fish Culture Workshop. Spec. Publ 3: 153-165.

13. Hsu W, Rodriguez D, Chichester C (1972) The biosynthesis of astaxanthin. VI. The conversion of [14C] lutein and [14C] β-carotene in goldfish. Int J Biochem 3: 333-338.

14. Schiedt K, Leuenberger M, Vecchi M, Glinz E (1985) Absorption, retention and metabolic transformations of carotenoids in rainbow trout, salmon, and chicken. Pure Appl Chem 57: 685-692.

15. Amar E, Kiron V, Satoh S, Watanabe T (2004) Enhancement of innate immunity in rainbow trout (Oncorhynchus mykiss Walbaum) associated with dietary intake of carotenoids from natural products. Fish and Shellfish Immunol 16: 527- 537.

16. Bowen J, Soutar C, Serwata R, Lagocki S, White D, et al. (2002) Utilization of (3S, 3'S)-astaxanthin acyl esters in pigmentation of rainbow trout (Oncorhynchus mykiss). Aqua Nutr 8: 59-68.

17. Chatzifotis S, Pavlidis M, Donate J, Vardanis G, Sterioti A, et al. (2005) The effect of different carotenoid sources on skin coloration of cultured red porgy (Pagrus pagrus). Aqua research 36: 1517-1525.

18. Furr H, Clark R (1997) Inestinal absorption and tissue distribution of carotenoids. Journal of Nutritional Biochemistry 8: 364-377.

19. Baron M, Davies S, Alexander L, Snellgrove D, Sloman K, et al. (2008) The effect of dietary pigments on the coloration and behavior of flame-red dwarf goruami, Colosa lalia. Anl behavior 75: 1041- 1051.

20. Kennedy T, Liebler D (1992) Peroxy adical scavenging by β-carotene in lipid bi-layers: effect of oxygen partial pressure. J Biol Chem 267: 4658-4663.

21. Nakagawa K, Fujimoto K, Miyazawa T (1996) β-Carotene as a high potency antioxidant to prevent the formation of phospholipid hydroperoxides in red blood cells of mice. Biochimica et Biophysica Acta 1299:110-116.

22. Someya K, Totsuka Y, Murakoshi M, Kitano H, Miyazawa T, et al. (1994) The antioxidant effect of palm fruit carotene on skin lipid peroxidation in guinea pigs as estimated by chemiluminescence-HPLC method. J Nutr Sci Vitaminol 40: 315-324.

23. Wang Y, Chien Y, Pan C (2006) Effects of dietary supplementation of carotenoids on survival, growth, pigmentation, and antioxidant capacity of characins, Hyphessobrycon callistus. Aquaculture 261: 641-648.

24. Pham K, Byun H, Kim K, Lee S (2014) Effectsof dietary carotenoid source and level on growth, skin pigmentation, antioxidant activity and chemical composition of juvenile olive flounder Paralichthys olivaceus. Aquaculture 431: 65-72.

25. Kappus H, Sies H (1981) Toxic drug effects associated with oxygen metabolism, redox cycling and lipid peroxidation. Experientia 37: 1233-1241.

26. Cohen G, Doherty M (1987) Free radical mediated cell toxicity by redox cycling chemicals. British J Cancer 55: 46-52.

27. Aoshima H, Satoh T, Sakai N, Yamada M, Enokido Y, et al. (1997) Generation of free radicals during lipid hydroperoxide-triggered apoptosis in PC12h cells. Biochim Biophys Acta 1345: 35-42.

28. Hata Y, Kaneda T, Fukuda H, Mino M (1984) Lipid hydroperoxide and nutrition, Kouseikan, Tokyo, Japan.

29. AOAC (2005) Official Methods of Analysis. (18thedn.) Association of Official Analytical Chemists, Gaithersburg, USA.

27

Effect of Chilling on Microbiological, Biochemical and Sensory Attributes of Whole Aquacultured Rainbow Trout (*Oncorhynchus mykiss* Walbaum, 1792)

George Ninan*, Lalitha K.V, Zynudheen A.A and Jose Joseph

Central Institute of Fisheries Technology, Kochi 682 029, India

Abstract

The effect of chilling (0-2°C) on the quality deterioration of whole ungutted aquacultured rainbow trout (*Oncorhynchus mykiss*, Walbaum,1792) was studied by integrated evaluations of microbiological, biochemical, and sensory attributes. The counts of aerobic mesophilic, psychrotrophic bacteria and *Pseudomonas* increased exponentially. An initial lag phase was noticed for H2S producing bacteria, *Aeromonas* and Enterobacteriaceae. Presence of pathogens such as *Aeromonas hydrophila* and *A. sobria* are of concern in the case of delay in icing or temperature abuse during storage. The pH values increased from an initial value of 6.74 to 7.13. PV showed fluctuations. Of the chemical indicators of spoilage, Thiobarbituric acid (TBA) values increased very slowly reaching final value of 16.56 µg MA g^{-1}. Total Volatile Base Nitrogen (TVB-N) values exceeded 27. 87 mg N 100 g^{-1} on day 14 when the psychrotrophic counts exceeded 10^7 cfu g^{-1} indicating that this value may be useful as a measure of degree of freshness for whole ungutted rainbow trout. Based on the TVB-N and microbiological limits, the shelf life of trout at 0-2°C was 9-12 days.

Keywords: Rainbow trout; Chilled storage; Spoilage bacteria, H$_2$S producing bacteria; Volatile bases; Quality; Shelf life

Introduction

Several species of fresh water fish including common carp (*Cyprinus carpio*), rohu (*Labeo rohita*), mrigal (*Cirrhinus mrigala*) and rainbow trout (*Oncorhynchus mykiss*) are being farmed in India in order to meet the increasing demand for fresh fish. Among the fresh water fishes, rainbow trout is farmed in many parts of India. Worldwide, farmed trout production is increasing. The demand for trout in the Country is estimated to be 800 T per annum [1]. Rainbow trout (*Oncorhynchus mykiss* Walbaum), an exotic cold water species, is considered to be a highly priced fish in India. The fish is marketed mainly as fresh in chilled condition. Recently it was found its way to supermarkets in metro cities as an exotic food item which has demand among the urban population. Trout fillets in the form of smoked and vacuum packaged products have high demand in northern European countries. Trout is a preferred table fish in the U.S and Europe. The increasing demand for high quality fresh seafood has intensified the search for methods and technologies for better utilization of fresh fish.

The quality of fresh fish is a major concern to the consumers and industry. Fish is extremely perishable and the shelf life of such products is limited in the presence of normal air by the chemical effect of atmospheric oxygen and the growth of aerobic spoilage microorganisms. It is generally accepted that the environment can influence the microflora associated with the skin, gills and intestines of finfish [2]. Therefore, fish grown in different aquaculture systems may harbour different numbers and species of bacteria. The culture practices such as pond fertilization, supplementary feeding with slaughter house waste or agricultural byproducts imposes a high probability of contamination on the aquacultured fish. Types and levels of bacterial populations associated with farmed fish are useful indicators of quality and safety of fish. Interaction between microbial metabolism and physic-chemical reactions accelerate fish quality deterioration as amines formation, lipid, nucleotide and protein degradation contributes to off odours, off flavours and texture softening [3,4] *Pseudomonas* spp. and S.

putrefaciens are the specific spoilage bacteria of marine and freshwater tropical fish stored in ice [5].

Although quality attributes of farmed rainbow trout from temperate counties were evaluated by many workers [6–11], few carried out quality assessment of tropical freshwater fish species [12]. Dawood et al. [13] reported a rapid deterioration in quality of headed and gutted rainbow trout (*Salmo gairdneri*) over a 14 day period of storage when fish had been held at high ambient temperature (30°C) for 6 h. Studies indicated the shelf life of rainbow trout as two weeks for gutted fish in ice [8], 15–16 days for whole ungutted fish, 10–12 days for fillets stored in ice [12] and 6 days for gutted vacuum packed and refrigerated samples [14].

With the increase in aquaculture of this species in India, it is important to study the storage capacity of fish under refrigerated condition following harvest. Due to the perishable nature of fish, there is an obvious need for development of efficient preservation methods, which allow shelf life extension of these products. However, collective works on various quality aspects (chemical, microbiological, textural and sensory) of trout of tropical region stored in ice are scarce. The objective of this study was to assess the quality of farmed rainbow rout stored in ice and kept at chilled condition by integrated evaluations of sensory, microbiological and biochemical attributes and to understand

---

*Corresponding author: George Ninan, Senior Scientist, Fish Processing Division, Central Institute of Fisheries Technology (ICAR), Kochi 682 029, India
E-mail: george66jiji@rediffmail.com

the spoilage microflora to develop efficient preservation methods which allow extension of shelf life.

Materials and Methods

Material

Rainbow trout (*O.mykiss*) of average weight 250g and average length 278mm were obtained from aquaculture farm located at Rajamallay near Munnar in the High Ranges in Southern India. It was harvested by aggregating into a corner of the pond and scooped by two persons using a drag net. The fish were killed by immersing in ice-cold water (hypothermia), and transported to the laboratory within 6 h of harvesting, in insulated polystyrene boxes containing ice. On reaching the laboratory the whole fish samples were repacked with flake ice (ice/fish ratio 1:1) in polystyrene boxes, provided with outlets for water drainage and stored in a chilled room at a temperature of 0 to 2°C. The ice/fish ratio was maintained constant throughout the experiment. Ten randomly chosen fish were removed from ice after 0, 3, 6, 9, 12, 14 and 15 days for analysis

Microbiological analysis

Twenty five gram muscle with skin were aseptically weighed and homogenized with 225 ml sterile physiological saline for 60s. in a stomacher (Lab Blender 400; Seward Medical, Norfolk, IP24, IXB, UK). The homogenates were serially diluted and 0.5 ml of appropriate serial dilutions was plated on the surface of appropriate media in duplicate by spread plate method and then incubated. For mesophilic and psychrotrophic bacteria, tryptic soy agar plates (TSA, Oxoid, U.K.) were used and plates were incubated at 37 and 7°C for 2 and 10 days respectively [15,16]. *Pseudomonas* were counted on Cetrimide-Fusidin-Cephaloridine (CFC) agar (Oxoid code CM 559, supplemented with SR 103; Oxoid U.K.) after 3 days incubation at 20°C. [17,18]. *Aeromonas* spp. were counted on starch ampicillin (SA) agar (Hi Media, India) containing 10ug/ml of ampicillin incubated at 28°C fir 48 h. [19]. *Brochothrix thermosphacta* was determined on Streptomycin sulfate-Thallous acetate – Actidione Agar (STAA Hi Media, India) after incubation at 20°C for 4days [20].

Enterobacteriaceae and H$_2$S-producing bacteria (including *Shewanella putrefaciens*) were counted on violet red bile glucose agar (VRBGA, Oxoid code CM 485) and Iron Agar (IA, Oxoid code CM 867), respectively by pour plate method and plates were incubated respectively at 30°C for 24 h. [21] and 20°C for 5 days [22].

Faecal *Streptococci* and *Staphylococcus aureus* counts were determined respectively on KF Streptococci Agar (Oxoid code CM 701) after incubation at 37°C for 2 days and on Baird Parker Agar (Oxoid code CM 275) incubated at 37°C for 2 days and typical colonies were confirmed [23]. Total coliforms, faecal coliforms and *Escherichia coli* were estimated by the three tube MPN method [24].

The dominant aerobic microflora at the final sampling points was determined by isolating and identifying 20% of the colonies from PCA (30 and 7°C) plates. 20-25 colonies were randomly selected from PCA plates. A total of 112 bacterial cultures were isolated from fresh and ice stored trout. Fifty two colonies were picked from PCA plates (30°C) sampled from the fresh prawn and identified. To identify spoilage flora, 60 colonies were randomly selected from PCA (7°C) plates poured on the day fish were considered unacceptable (sensory score 6) and characterized morphologically and bio-chemically. A 5-8 black colonies each were isolated and characterized from IA, VRBGA and SA agar plates at the final sampling points. They were then grouped

according to the taxonomic schemes proposed by several authors for identification (Dainty et al. [25]; Molin and Ternstorm [17]; Krieg and Holt [26]; Sneath et al. [27]; Kirov [28]; Brenner et al. [29]). The isolated cultures were identified and confirmed using API 20NE and API 20NE system (Biomerieux, France).

Biochemical analysis

Moisture, ash and total nitrogen and total fat content were determined using AOAC methods N. 950.46B, 920.153, 928.08 and 960.39 of AOAC, [16] respectively. pH was measured in fish homogenates (10g of fish per 10ml of distilled water) with a Cyberscan 510 pH meter (Eutech Instruments, Singapore). Thiobarbituric acid (TBA) value was determined according to the method of Tarladgis et al. [30] by mixing 10g of fish meat with 100 ml. 0.2 N HCl. TBA value was calculated and expressed in µg malonaldehyde / g of fish sample. Total Volatile Base Nitrogen (TVBN) and Trimethyl amine (TMA) was determined in triplicate by the micro diffusion method [31] from the trichloro acetic acid extract of the muscle. TMA-N and TVB-N was calculated and expressed in mg/ 100g of the sample. Peroxide Value (PV) and Free Fatty Acid (FFA) were determined according to Jacobs [32] and AOCS [33] respectively.

Sensory analysis

Sensory analysis of whole trout stored in ice were performed during storage by a ten member trained sensory panel composed of the staff from the laboratory. While drawing the samples for sensory evaluation, special attention was given to check any change in colour or odour. The panel assessed different attributes like appearance, odour, flavour and texture. The overall acceptance was determined by evaluating the attributes like odor, taste and texture of whole cooked fish (cooking steaks in boiling water containing 2% salt for two minutes) Each sample of the lot was classified using a 10 point hedonic scale, 4 being the acceptability limit.

Sensory analysis of whole trout was performed during iced storage according to the European Community (EC) grading scheme by ten trained panelists [34]. The panel assessed different attributes like appearance, odour, flavour and texture. The appearance of the skin, eyes, gills and internal organs, surface slime, and the odor and texture of each fish (whole) was assessed into four quality grades - excellent quality (perfect condition, E), high quality (slight loss of excellent characteristics, A), good quality (some deterioration, but fit for sale, B) and unfit for sale (C). Color analysis was performed with a Hunter lab Miniscan ® XE plus spectrocolorimeter (Hunter Associates Laboratory, Inc. Reston, Virginia, USA). Measurements were recorded using the L* a* b* colour scale [35]. Chroma (C*) and Hue (h*) also were calculated from the L* a* b* values. Three repetitions of the different colour parameters were recorded.

Statistical analysis

Experiments were replicated twice on different occasions with fish samples from the same farm. Results are presented as mean ± standard deviation and significance of the differences between the mean values was determined by One-way Analysis of Variance (ANOVA), followed by Duncan's test using SPSS software (version 10.0) for Windows. *p*-value lower than 0.05 was considered statistically significant.

Results and Discussion

Aerobic mesophilic and psychrotrophic bacteria grew exponentially from an initial load of 3-5 log$_{10}$cfu g^{-1} reaching 7.6 log$_{10}$cfu g^{-1} on day 15 (Figure 1). The initial mesophilic bacterial load of 4.7 log$_{10}$cfu g^{-1}

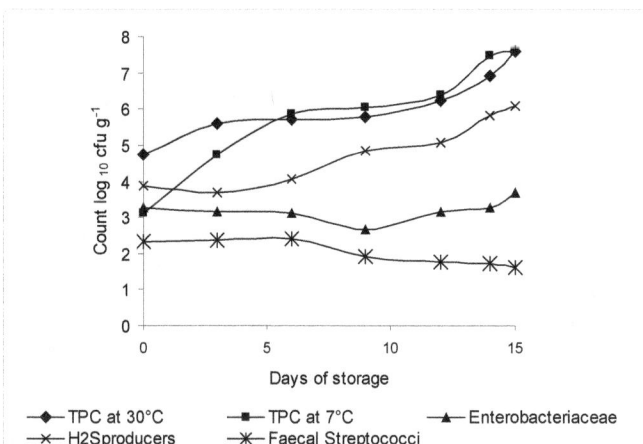

Figure 1: Changes in the population of aerobic mesophilic, bacteria, psychrophic bacteria, H2S producing bacteria, faecal streptococci and Enterobacteriaceae in trout stored at 0-2°C.

indicated good quality of trout as it ranged from 2-6 $\log_{10}$ cfu g^{-1}. [36]. These values obtained in this study are close to the values reported earlier for aquacultured fresh trout from Spain [37,41], Greece [38, 39] and U.K [40]. It is widely accepted that the initial microbial load of fresh water varies depending on water conditions and temperature. The initial and final aerobic mesophilic bacterial associations of trout stored in ice were found to be similar to those reported in the literature for trout stored aerobically.

In this study, mesophilic and psychrotrophic counts in fresh fish tissues were close to or lower than the m value (5×10^5 cfu/g) recommended by the International Commission of Microbiological Specification for Foods [42] for whole fresh water fish. Taking the 10^7 cfu g^{-1} psychrotrophic count as the spoilage level, the shelf life of chilled stored trout in this study was 9-12 days as reported earlier by Fik and Surówka [43] and Rezaei et al. [44]. For fresh fish, the microbiological limit (M) for human consumption proposed by ICMSF [42] is 10^7 cfu g^{-1}.

Of the bacterial groups examined in the present study, H2S-producing bacteria had the highest counts followed by *Pseudomonas* spp., Aeromonas, *B. thermosphacta* and Enterobacteriaceae (Figures 1 & 2). H2S producing bacterial counts constituted <5% of the total flora in fresh trout (3.8 $\log_{10}$ cfu g^{-1}), their levels increased significantly (P <0.01) during storage and its proportion in the total flora reached 10-15% at the end of storage indicating their role in the spoilage. H2S producing bacteria were identified as *Shewanella* and Aeromonas. *Pseudomonas* displayed the typical growth pattern of psychrotrophic bacteria without a lag phase (Figure 1) increasing from initial counts of 3.0 to 5.02 $\log_{10}$ cfu g^{-1} on day 15. These two bacterial groups were found to be the specific spoilage organism (SSO) in fish from temperature and tropical waters [5]. *Shewanella* spoilage is characterized by TMA and sulphides (H2S) whereas the *Pseudomonas* spoilage is characterized by absence of these compounds and occurrence of sweet, rotten sulphydryl odours. *Pseudomonas* and *Shewanella* isolates produced large amounts of TVBN, secreted huge amounts of proteolytic enzymes and intense off-odours and were identified as strong spoiling bacteria. Chytiri et al. [12] also reported dominance of *Pseudomonas*, H2S producing bacteria (including *Shewanella* putrefaciens) and *B. thermosphacta* in the spoilage microflora of whole ungutted and filleted trout over an 18-day storage period in ice.

Aeromonas was also found to be members of the microflora of farmed trout with an initial load of 3.18 $\log_{10}$ cfu g^{-1} as reported earlier by Nam and Joh [45]. A reduction in bacterial load was noticed during the first week of iced storage and growth was resumed after 12 days (Figure 2). Aeromonas isolates were identified as A. hydrophila and A. sobria. These bacterial species produced proteinases, reduced TMAO and produced off-odours indicating their spoilage potential. Although these bacteria are capable of growth at chill temperatures, they required a period of adaptation (i.e., the lag phase and slow growth phase) in this study. Lee et al.[46] isolated A. hydrophila from diseased trout from Korea and this bacterium is responsible for hemorrhagic septicemia, a disease affecting a wide variety of freshwater and marine fish [47]. Epizootic Ulcerative syndrome caused by A. sobria resulted in great damage to fish farms in Bangladesh and India [48]. Gonzalez et al.[49] noticed the strong potential spoilage activity of aeromonads in wild and aquacultured iced freshwater fish. A. hydrophila, A. veronii biovar veronii and A. veronii biovar sobria are the strains more often associated with gastroenteritis in humans and the enteropathogenic potential of these strains were comparatively high when grown at low temperatures than at 37oC[50,51]. The occurrence of A. hydrophila and A. sobria in trout farms and ice stored trout must be taken into consideration because it can cause gastroenteritis and wound infections.The ability of these organisms to grow at refrigeration temperatures indicates the potential food safety issues from such foods.

Brochothrix thermosphacta population in ice stored trout increased from an initial count of 2.32 $\log_{10}$ cfu g^{-1} to 3.9 $\log_{10}$ cfu g^{-1} (Figure 3). Similar counts for whole ungutted trout were reported by Chytiri et al. [12] on day 15 in iced storage.

Enterobacteriaceae were also part of the microflora of farmed rainbow trout which is in agreement with the findings of Arashisar et al. [10], Chytiri et al. [12] and Nerantzaki et al.[39] . Enterobacteriaceae counts decreased during the first week of storage from an initial value of 3.2 $\log_{10}$ cfu g^{-1}. At the end of storage, a count of 3.66 $\log_{10}$ cfu g^{-1} was noticed. The dominant species identified in this study were Citrobacter freundii, Hafnia alvei and Pantoea agglomerans. In this study, Enterobacteriaceae were found in high numbers in fresh trout and their

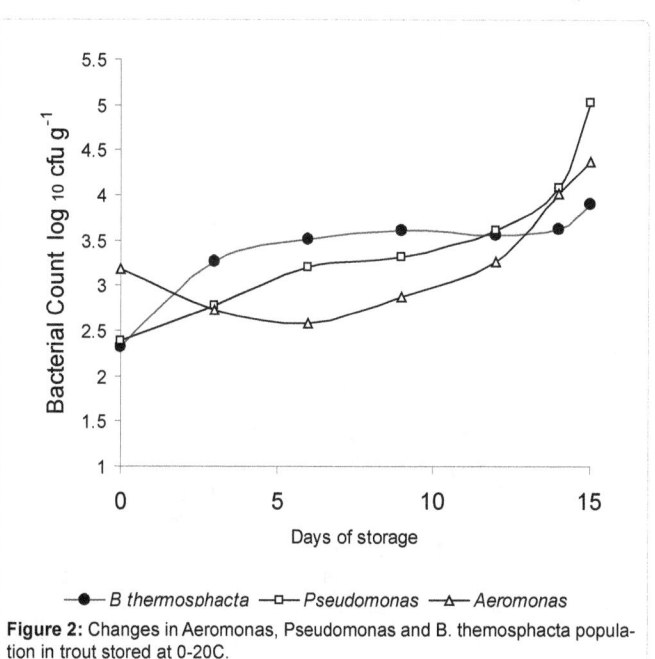

Figure 2: Changes in Aeromonas, Pseudomonas and B. themosphacta population in trout stored at 0-20C.

abundance decreased during ice storage, possibly because of their lower growth rate than that of other Gram-negative psychrotrophic spoilers. The contribution of Enterobacteriaceae to the microflora of trout and its potential to cause spoilage must be taken into consideration in case of delay in chilling after catch or temperature abuse during storage.

Among the indicator organisms, faecal streptococcal population was 2.3 log 10 cfu g⁻¹ initially and at the end of storage the count was ca. 1.6 log 10 cfu g⁻¹ (Figure 1). S. aureus numbers were low (1.3 log 10 cfu g⁻¹) in fresh trout and were within the acceptable limit. S. aureus were not detected in trout during iced storage. Icing affected populations of total coliforms, Faecal coliforms and E. coli levels (Figure 3) and 1-2 log reduction was noticed . E. coli counts in trout on day 3 (< 6 g⁻¹) was below the m limit (11 g -1) recommended by the ICMSF [42]) for good quality fish. High levels of faecal coliforms were previously reported for fish farms in India [52-54].

A total of 52 strains were isolated from 30^C PCA plates and identified. In fresh trout, majority of the isolates (70%) were gram-negative rods. The main bacterial groups identified among the 52 isolates randomly selected from TSA plates were i. Gram-negative aerobic coccobacilli and rods (Moraxella, Acinetobacter, Flavobacterium), ii. Gram-negative aerobic motile rods (*Pseudomonas*), iii.Enterobacteriaceae (Enterobacter, Citrobacter, Hafnia, Klebsiella), iv. Aeromonadaceae (Aeromonas), v. Micrococcaceae (Kocuria, Staphylococcus) andvi.Gram-positive spore forming bacteria (Bacillus). González et al. [41] reported predominance of Acinetobacter, *Pseudomonas*, Staphylococcus, Enterococcus and Bacillus in rainbow trout from Spain. On icing, the abundance of Enterobacteriaceae decreased. A total of 56 strains were characterized from 30^C PCA plates. The majority (>60%) of the trout isolates after 15 days in ice belonged to genera Moraxella, Acinetobacter, *Pseudomonas*, *Shewanella*, Aeromonas and Flavobacterium indicating that spoilage of fresh trout stored aerobically is due to the activity of more than one specific spoilage organism. Flavobacterium have been found in other

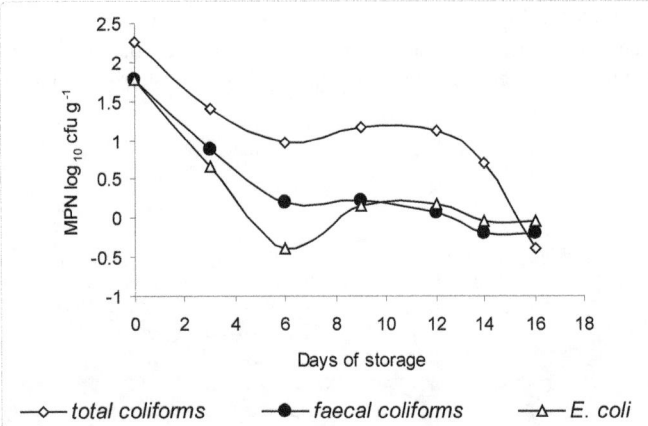

Figure 3: Changes in total coliform, faecal coliform and E coli population in trout stored at 0-2°C.

DAYS	Moisture (%)	Protein (%)	Fat (%)(dwb)	Ash (%)
0	78.04 ± 0.31	19.80± 0.65	1.60±0.12	0.61±0.23
3	79.06 ± 0.41	19.18±0.82	1.56 ±0.05	0.66 ±0.04
6	79.12± 0.23	19.06± 0.56	1.54 ± 0.06	0.64 ± 0.07
9	79.41± 0.32	18.48 ± 0.77	1.57± 0.14	0.66 ± 0.04
12	79.73± 0.45	18.09 ±0.54	1.47 ± 0.09	0.72 ± 0.06
14	80.24±0.42	18.06±0.56	1.45 ±0.04	0.65 ±0.04
15	80.64 ± 0.11	17.97 ±0.28	1.42 ± 0.04	0.61 ±0.03

Table 1: Proximate composition of trout stored at 0-20C(n= 3 ×2).

farmed fish species such as catfish and some are also the causative agent of bacterial cold water disease and rainbow trout fry syndrome [55,56]. Several investigations have concluded that Gram- negative rod-shaped bacteria (e.g., *Pseudomonas*, Moraxella and Acinetobacter) dominate on many fish caught in tropical waters [36,57].

Results of proximate analysis of whole ungutted trout stored at 0-2 0C during the 15-day storage are given in Table 1. Raw trout is a fish with a fat content of 1.60 ± 0.12 g/100 g edible meat, a protein content of 19.80 ± 0.65 g/100g edible meat and ash content of 0.61±0.23 g/100g edible meat. The levels of fat and ash were low compared to that reported for trout from other regions [58-60,41]. During storage in ice, no significant variations in the composition of major constituents viz. moisture, protein and fat were observed. A slight increase of 3.33% moisture content was observed after 15 days in chilled storage. The retention of good texture of fish muscle during chilled storage can be attributed to the minimum leaching of the major constituents and low water penetration into the flesh.

The changes in TVB-N and TMA levels in whole trout throughout the storage in ice are shown in Table 2. The results of this study confirmed the earlier studies of Rodriguez et al. [61] and Chytiri et al. [12] who reported TMA values of ≤ 1.0 mg/100g for whole fresh trout and values of ≤ 3.0 mg/100g on day 15 for ice stored fish indicating the low level of trimethylamine oxide (TMAO) in the flesh of this fish species. TMAO quantity in fish varies with the species and the environment. A wide range of TMA-N values have been reported by several investigators as acceptability limit i.e., 1-5mg N/100 g [62- 65]. Sikorski et al. [66] and Dalgaard et al. [67] reported that a population of 108-109 cfu/g of S. putrefaciens was considered crucial for TMA production. In this study, the count of H2S- producing bacteria (including S. putrefaciens) was low (6.07 log cfu/g) at the end of iced storage which could be the reason for low levels of TMA.

TVBN, including trimethylamine, dimethylamine, ammonia and other volatile basic nitrogen compounds, was produced mainly by bacterial decomposition of fish flesh. In this study, TVB-N levels increased to 31.25 mg N 100g⁻¹ on day 14. The values exceeded the limit of acceptability of 25 mg N 100g⁻¹ proposed by Stansby [68]. However, fish were acceptable based on sensory score and microbiological counts. Critical limits of 25, 30 and 35 mg N 100g⁻¹ of TVB-N were established for different groups of fishes [69]. However, no limit for acceptability has been established for rainbow trout. Hence, based on the TVB-N levels, microbiological counts and spoilage indicators, TVB-N limit of 27 mg N 100g⁻¹ may be proposed for rainbow trout as acceptable limit. As TMA production was low, ammonia probably accounted for the major portion of volatile bases. Giménez et al. [9] observed TVB-N value of 35 mg N 100g⁻¹ in trout fillet stored under air on day 8 and reported good correlation with bacterial counts (108 cfu g⁻¹). In contrast, Arashisar et al. [10] reported values of 40 mg N 100g⁻¹ in filleted trout at the end of 14 days storage. However, the values were < 20 mg N 100g⁻¹ when psychrotrophic bacterial levels exceeded 107cfu g⁻¹ on day 6 at 4 ± 1˚C. There is good correlation between TVB-N and microbiological parameters in the present study as reported earlier by Katikou et al. [70]. The results also suggest that since TVBN values exceeded limit of acceptability on day 14 when fish had a stale order, it may be useful as a measure of degree of freshness; although some reports differ in this context [71, 12].

While evaluating the spoilage potential of nine bacterial groups isolated from cold smoked salmon, Stohr et al. [72] have reported that Gram negative bacterial strains such as Aeromonas, *Shewanella* and Serratia produced TMA in concentrations ranging from 11.0 to 13.1 mg

Days	PV (meq O$_2$/Kg)	TBA (µg MA/g)	FFA (mg % oleic acid)	TMA (mg /100 g)	TVB-N (mg N/100 g)	pH
0	ND	4.97±0.27 [a]	3.28±0.26 [a]	1.23±0.11 [a]	14.6±0.26 [a]	6.77 ±0.03 [a]
3	9.11±0.08 [a]	4.5±0.1 [b]	2.92±0.04 [b]	1.5±0.1 [b]	19.67±0.29 [b]	6.75 ±0.02 [a]
6	5.34±0.42 [b]	9.55±0.11 [c]	2.33±0.09 [c]	1.6±0.2 [bc]	19.73±0.46 [b]	7.03±0.04 [b]
9	3.48±0.35 [c]	14.14±0.16 [d]	2.95±0.01 [b]	1.71±0.03 [bc]	21.08±0.16 [c]	7.12±0.01 [c]
12	4.62±0.17 [d]	15.47±0.06 [e]	3.4±0.36 [a]	1.68±0.03 [bc]	27.87 ±0.26 [d]	7.14 ±0.03 [c]
14	5.82±0.28 [b]	16.2±0.36 [f]	2.81±0.10 [b]	1.79±0.07 [c]	31.25±0.15 [e]	7.14 ±0.02 [c]
15	8.4 ±0.35 [e]	16.56±0.47 [f]	3.93±0.05 [d]	3.32 ± 0.13 [d]	31.20±0.44 [e]	7.13 ±0.03 [c]

*Within each column, means with the same superscript do not differ significantly (p > 0.05).

Table 2: Changes in Peroxide Value (PV), Thiobarbituric Acid value (TBA), Trimethyl amine (TMA), Free Fatty Acid value (FFA), Total Volatile Nitrogen (TVB-N) and pH of trout stored at 0-20C*(n= 3 ×2).

Days	Skin	Eyes	Gills	Flesh colour	Outer slime	EC Grade
0	Bright, shining ; firm	Translucent cornea; convex; absence mucus	Fresh odor; red color	Pinkish white	thin; transparent	E
3	Bright shining; firm	Translucent cornea; convex; absence mucus	Fresh odor; red color	Pinkish white	thin; transparent	E
6	Bright shining; firm	Translucent cornea; convex; absence mucus	Fresh odor; red color	Pinkish white	thin; transparent	E
9	Waxy; slight loss of shine; soft	Opalescent cornea; plane; moderate mucus	Fishy odor; red color	Whitish	Aqueous; transparent	A
12	Waxy; slight loss of shine; soft	Opalescent cornea; plane; moderate mucus	Stale odor; dark red color	Whitish	Opaque; thick; slight milky	B
14	Dull; some bleaching	Opalescent cornea; plane; excessive mucus	Stale odor; dark red color	Whitish with slight yellow stain	Opaque; thick; slight milky	B
15	Dull; some bleaching	Opalescent cornea; sunken; excessive mucus	Spoiled odor; dark red color	Whitish with slight yellow stain	Thick;Milky	C

Table 3: Sensory assessment of trout stored at 0-2°C.

N 100g^{-1} and high TVBN production was generally correlated with high TMA production. In this study, even with low TMA levels, high TVBN values of 31.25 mg N 100g^{-1} were obtained and this may be attributed to ammonia production.

pH values for whole ungutted trout samples increased with storage time from an initial value of 6.74 (Table 2) indicating bacterial growth and production of volatile basic compounds such as ammonia by fish spoilage bacteria. Increase in pH due to accumulation of alkaline compounds through autolytic activities and microbial metabolism has been reported in earlier studies [65,73,74]. Many microbes including *Pseudomonas* produce ammonia during amino acid metabolism.

Changes in PV, TBA and FFA for whole ungutted trout stored at 0-2° C during the 15-day storage are shown in Table 2. PV and TBA are indices to measure the first and second stages of oxidative rancidity respectively. PV measures peroxides and hydroperoxides and a value of above 10 -20 is an indication of rancidity [75]. In this study PV showed fluctuations during the chilled storage, but the values were very low to cause rancidity or off flavour at any point of time during the study.

TBA values for whole ungutted trout samples increased steadily from an initial value of 4.9 ± 0. 3 µg MA/g and reached a value of 14.1 µg MA/g on day 9 when fish retained high quality as per EC grade and the value was 16.6 µg MA/g when fish spoiled on day 15 (Table 2). The results of this study confirmed the earlier finding that oxidative rancidity remained relatively low in aquacultured whole ungutted trout throughout the entire period of storage in ice and oxidative rancidity

indices viz., PV and TBA are poor indicators of quality since lipids in trout are relatively stable during chilled storage[12]. The low level of lipid oxidation products suggest that whole ungutted rainbow trout stored at 0-2 0 C has some intrinsic factors to prevent oxidation. The lipids were found to better stable in whole than in gutted or filleted trout possibly due to the fact that it is harder for oxygen to penetrate into whole fish and there may be a higher accumulation of proteolysis products, acting as antioxidants, with time of storage [76]. Because the rainbow trout studied showed no increased lipid oxidation during the first week of storage (a decrease in PV relative to the initial level being observed), it may be suggested that the muscle tissue of rainbow trout, particularly in the whole fish, was predominantly a site of antioxidant over pro-oxidant activity at that time [77]. Fifteen days in chilled conditions has accumulated only 4% of FFA which indicates that rainbow trout has very low level of lipase activity during chilled storage.

Changes in sensory attributes of the whole ungutted trout in chilled storage were given in descriptive terms as recorded by given by the panelists (Table 3). Whole ungutted trout was in excellent condition up to six days in chilled conditions and the high quality was retained for nine days. The fish was still considered to be of good quality and fit for sale between 12–14 days of chilled storage. The fish samples were spoiled on 15 th day.

The overall acceptability scores for whole ungutted trout remained in the range of seven points up to nine days indicating that there was no significant loss of sensory attributes viz., odour, taste and texture.

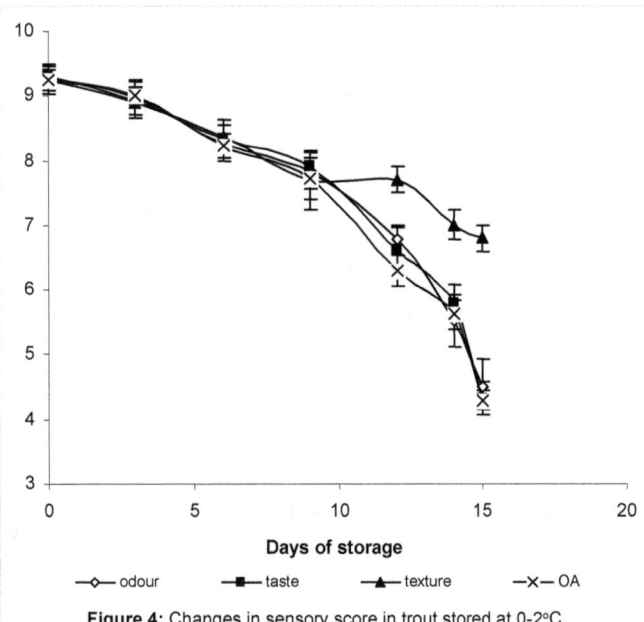

Figure 4: Changes in sensory score in trout stored at 0-2°C.

Days	L*	a*	b*	Hue	Chroma
0	47.26 ±.05	2.41 ± 0.15	9.32 ± 0.07	75.49 ± 0.97	9.63 ± 0.03
3	47.72 ± 0.14	2.23 ± 0.03	9.82 ± 0.02	77.19 ± 0.16	10.07 ± 0.03
6	48.52 ± 0.03	2.09 ± 0.02	10.23 ± 0.04	78.42 ± 0.09	10.44 ± 0.04
9	49.75 ± 0.23	1.95 ± 0.18	9.84 ± 0.64	78.79 ± 0.37	10.04 ± 0.66
12	50.89 ± 0.56	1.51 ± 0.07	10.42 ± 0.19	81.77 ± 0.37	10.53 ± 0.19
14	53.02 ± 0.19	1.25 ± 0.03	10.61 ± 0.09	83.29 ± 0.22	10.68 ± 0.09
15	55.65 ± 0.45	1.28 ± 0.07	10.79 ± 0.11	83.23 ± 0.32	10.86 ± 0.12

L* value (Lightness), a* (red) value, b*(yellowness)

Table 4: Changes in colour parameters of the mince of trout stored at 0-2°C (n= 3 ×2).

Acknowledgements

The authors are thankful to the Director, Central Institute of Fisheries Technology, for according permission to publish the paper. The assistance rendered by M/s Tata Tea Ltd, Munnar, Kerala and the High Range Angling Association in the collection of trout samples is gratefully acknowledged.

The limit of acceptability for odour and taste was reached by the 14 th day in chilled condition (Figure 4). However the chilled fish retained the texture with slight changes and the limit of acceptability was not reached for this attribute during the entire chilled storage period.

The sensory changes observed in rainbow trout during storage in ice were in agreement with the descriptions presented by other authors [11,66,78,79]. The quality deterioration in trout was correlated with reduced tastefulness of the cooked fish. The reduction in tastefulness was caused primarily probably by nucleotide decomposition [80,81]. Based on the sensory score, trout had a shelf life of 12-14 days in ice.

Change in colour parameters of mince prepared from chill stored rainbow trout is given in Table 4. During chill storage there is a gradual increase in L* value (Lightness) with a corresponding decrease in a* (red) value. The b*(yellowness) increased during this period. The instrumental colour values correlated well with the sensory observation of flesh given in Table1. Fresh sample (0 day) has pinkish white flesh which progressively turned paler during storage and has yellow stained white colour at the end of storage period. This finding indicated that freshness of trout, in relation to sensory analysis, was lost after total aerobic bacterial count reached limit count.

Conclusion

The results of the study indicate that the shelf life of whole gutted trout stored in ice as determined by the chemical and microbiological quality is 9-11 days. The whole ungutted rainbow trout stored in ice remained in excellent condition up to six days and retained high quality up to nine days. Based on sensory score, trout had a shelf life of 12-14 days in ice. The results obtained in the present study for trout tend to confirm the earlier observations that TMA, PV and TBA are of questionable use as quality indices. TVBN values exceeded the limit of acceptability when fish had a stale order and microbial count exceeded 3×10^6 cfu g^{-1} indicating that TVBN may be useful as a measure of degree of freshness. Detailed sensory evaluation is the effective and practical method to assess the freshness of chill stored whole ungutted rainbow trout.

References

1. Sehgal KL (1999) Fish and fisheries at higher altitudes: Asia FAO Fisheries Technical Paper. No.385. FAO, Rome. 304p.

2. Horsley RW (1973) The bacterial flora of the Atlantic salmon (Salmo salar L) in relation to its environment. J App Bacteriol 36: 377-386.

3. Ozogul Y, Ahmad JI, Hole M, Ozogul F, Deguara S (2006) The effects of partial replacement of fish meal by vegetable protein sources in the diet of rainbow trout (Onchorynchus mykiss) on post mortem spoilage of fillets. Food Chem 96: 549–561.

4. Hernandez MD, López MB, Alvarez A, Ferrandini E, Garcia B, et al. (2009) Sensory, physical, chemical and microbiological changes in aquacultured meagre (Argyrosomus regius) fillets during ice storage. Food Chem 114: 237-245.

5. Gram L, Huss H (1996) Microbiological Spoilage of Fish and Fish Products. Int J Food Microbiol 33: 121-137.

6. Randell K, Hattula T, Ahvenainen R (1997) Effect of packaging method on the quality of rainbow trout and Baltic herring fillets. Lebensm. Wiss Technol 30: 56–61.

7. Lyhs U, Hatakka M, Maki-Petays N, Korkeala H (1998) Prevalence of Listeria monocytogenes in Finnish vacuum-packaged fishery products In: Proc. 4th World Congress Foodborne Infections and Intoxications, 7 - 12 June 1998, Berlin. 2: 1051-1052.

8. Mills A (2001) Handling and Processing Rainbow Trout Torry Advisory Note no. 74 Ministry of Agriculture, Fisheries and Food, Torry Research Station, UK.8p.

9. Gimenez B, Roncales P, Beltran JA (2002) Modified atmosphere packaging of filleted rainbow trout. J Sci Food Agric 84: 1154–1159.

10. Arashisar S, Hisar O, Kaya M, Yanik T (2004) Effect of modified atmosphere and vacuum packaging on microbiological and chemical properties of rainbow trout (Oncorynchus mykiss) fillets. Int J Food Microbiol 97: 209-214.

11. Kolakowska A, Zienkowicz L, Domiszewski Z, Bienkiewicz G (2006) Lipid changes and sensory quality of whole- and gutted rainbow trout during storage in ice. Acta Ichthyol Piscat 36: 39-47.

12. Chytiri S, Chouliara I, Savvaidis IN, Kontominas MG (2004) Microbiological,chemical and sensory assessment of iced whole and filleted aquacultured rainbow trout. Food Microbiol 21 : 157-165.

13. Dawood AA, Roy RN, Williams CS (1986a) Effect of delayed icing on the storage of life of rainbow trout. J Food Tech 21: 159-166.

14. Rodríguez CJ, Besteiro I, Pascual C (1999) Biochemical changes in freshwater rainbow trout (Oncorynchus mykiss) during chilled storage. J Sc Food Agricult 79: 1473-1480.

15. Austin B, Al- Zahrani AMJ (1988) The effect of antimicrobial compounds on the gastrointestinal microflora of rainbow trout, Salmo gairdneri Richardson. J Fishery Biol 33: 1-14.

16. AOAC (2000) Official Methods of Analysis (17th ed.), Association of Official Analytical Chemists, Washiqton, DC, 1298p.

17. Molin G, Ternstrom A (1982) Numerical taxonomy of psychrotrophic pseudomonads. J Gen Microbiol 128: 1249–1264.

18. Mead GC (1985) Enumeration of pseudomonads using cephaloridine-fucidin-cetrimide agar (CFC). Int J Food Microbiol 2: 21-26.

19. Palumbo SE, Maxino SA, Williams F, Buchanan RL, Thayer DW (1985) Starch-ampicillin agar for the quantitative detection of Aeromonas hydrophila. Appl Environ Microbiol 50: 1027–1030

20. Gardner GA (1966) A selective medium for the enumeration of Microbacterium thermosphactum in meat and meat products. J Appl Bacteriol 29: 455–460

21. Mossel DAA (1987) Violet Red Bile Glucose (VRBG) agar. Int J Food Microbiol 5: 280- 281.

22. Gennari M, Campanini R (1991) Isolamento e caratterizzazione di Shewanella putrefaciens da pesce fresco ed alterato, carni fresche e alterate, prodotti lattiero- caseari, acqua e suolo. Industrial Alimenteria 30: 965-976, 988.

23. Food and Drug Administration (1998) FDA Bacteriological Analytical Manual. 8th Edition. AOAC International, Gaithersburg, MD, Chapter 12.

24. APHA (1998) Standard Methods for the examination of water and waste water , 20th edition, American Public Health Association, Inc, Washington D.C Part 9000-9221: 48-59

25. Dainty RH, Shaw BG, Hardinger CD, Michanie S (1979) The spoilage of vacuum packaged beef by cold tolerant bacteria, In Russell & R. Fuller, Cold tolerant bacteria in spoilage and the environment New york: Academic press, pp 83-110.

26. Krieg NR, Holt JG (1984) Bergey's Manual of Systematic Bacteriology.vol.1, Baltimore, USA: Williams and Wilkins. 964 p.

27. Sneath PHA, Mair NS, Sharpe ME, Holt JG (1986) Bergey's Manual of Systematic Bacteriology. vol.2, Baltimore, USA: Williams and Wilkins. 964 p.

28. Kirov SM (2001) Aeromonas and Plesiomonas species. In Doyle M, Beuchat L,Montiville T(Eds.). Food Microbiology: Fundamentals and Frontiers, ASM Press, Washington, D.C, pp 301-327.

29. Brenner DJ, Krieg NR, Staley JT (2004) Bergey's manual of systematic bacteriology, 2nd edn, Vol 2. Springer, USA.

30. Tarladgis BG, Watts BM, Younathan MT, Dugan L (1960) A Distillation Method for the quantitative determination of Malonaldehyde in Rancid Food. J Am Oil Chem Soc 37: 44-48.

31. Conway EJ (1962) Determination of Volatile Amines, Microdiffusion Analysis and Volumetric Error 5 th edn, Parch Croskey and Sockwood & Son Limited, London.

32. Jacobs MB (1958) In: The chemical analysis of foods and food products. Kreiger Pub.Co., New York. pp 393 – 394.

33. AOCS (1989) Official Methods of Recommended Practices of American Oil Chemists Society A.O.C.S., Chempaign, USA.

34. Howgate P, Johnston A, Whittle KJ (Eds.) (1992) Multilingual Guide to EC Freshness Grades for Fishery Products. Marine Laboratory, Scottish Office of Agriculture, Environment and Fisheries Department, Aberdeen.

35. Commision Internationale de L'Eclairage (CIE) (1986). Colorimetry. 2nd Ed. CIE 434 No. 15.2 CIE, Vienna.

36. Huss HH (1995) Quality and Quality changes in fresh fish. FAO Fisheries Technical Paper No.348, FAO, Rome.

37. Gonzalez-Rodriguez MN, Sanz JJ, Santos JA, Otero A, Garcia-Lopez ML (2001) Bacteriological quality of aquacultured freshwater fish portions in prepackaged trays stored at 3oC. J Food Prot 64: 1399-1404.

38. Savvaidis IN, Skandamis P, Riganakos KA, Panagiotakis N, Kontominas MG (2002) Control of natural microbial flora and Listeria monocytogenes in vacuum-packaged trout at 4 and 10 degrees C using irradiation. J Food Prot 65: 515-522.

39. Nerantzaki A, Tsiotsias A, Paleologos EK, Savvaidis IN, Bezirtzoglou E, et al. (2005) Effects of ozonation on microbiological, chemical and sensory attributes of vacuum-packaged rainbow trout stored at 4± 0.5 C. Eur Food Res Technol 221: 675–683.

40. Ozogul Y, Ozogul F (2002) Degradation Products of Adenine Nucleotide in Rainbow Trout (Oncorhynchus mykiss) Stored in Ice and in Modified Atmosphere Packaging. Turk J Zool 26: 127-130.

41. González CJ, López-Diaz TM, García-Lopez ML, Prieto M, Otero A (1999) Bacterial microflora of wild brown trout (Salmo trutta), wild pike (Esox lucius), and aquacultured rainbow trout (Onchorynchus mykiss). J Food Prot 62: 1270–1277.

42. ICMSF (International Commission on Microbiological Specifications for Foods) (1998) Microorganisms in Foods. 6. Microbial Ecology of Food Commodities. Blackie Academic and Professional, Baltimore.

43. Fik M, K Surowka (2004) Autoproteolysis rate of rainbow trout muscle proteins. Nahrung 48: 104-109.

44. Rezaei M, Hosseini SF (2008) Quality assessment of farmed rainbow trout (Onchorynchus mykiss) during chilled storage. J Food Sci 73: 93-96.

45. Nam IY, Joh K (2007) Rapid detection of virulence factors of Aeromonas isolated from a trout farm by hexaplex-PCR. J Microbiol 45: 297-304.

46. S, Kim S, Oh Y, Lee Y (2000) Characterization of Aeromonas hydrophila Isolated from Rainbow Trouts in Korea. J Microbiol 38: 1-7.

47. Paniagua C, Rivero O, Anguita J, Naharro G (1990) Pathogenicity factors and virulence for rainbow trout (Salmogairdneri) of motile Aeromonas spp. isolated from a river. J Clin Microbiol 28: 350-355.

48. Chacon MR, Figuras MJ, Castro-Escarpulli G, Soler I, Guarro J (2003) Distribution of Virulence genes in clinical and environmentl islates of Aeromonas spp. Antonie van Leeuwenhoek 84: 269-278.

49. González CJ, Santos J A, Garcia-Lòpez ML, González N, Otero A (2001) Mesophilic Aeromonads in Wild and Aquacultured Freshwater Fish. J Food Protect 64: 687-691.

50. Merino S, Camprubi S, Tomas J M (1992) Effect of growth temperature on outer membrane components and virulence of Aeromonas hydrophila strains of serotype O: 34. Infect Immun 60: 4343-4349.

51. Merino S, Rubires X, Knochel S, Tomas JM (1995) Emerging pathogens: Aeromonas spp. Int J Food Microbiol 28: 157–168.

52. PK, Thampuran N, Gopakumar K (1995) Microbial profile of cultured fishes and prawns viz a viz their spoilage and contamination. FAO Fisheries Report No.514 supplement, FAO, Rome, Italy, pp.1-12.

53. Lalitha KV, Surendran PK (2004) Bacterial microflora associated with farmed freshwater Prawn Macrobrachium rosenbergii (de Man) and the aquaculture environment. Aquac Res 35: 629-635.

54. Lalitha KV, Surendran PK (2006) Microbiological changes in farm reared freshwater prawn (Macrobrachium rosenbergii de Man) in ice. Food Control 17: 802-807.

55. Holt RA, Rohovec JS, Fryer JL (1993) Bacterial cold-water disease. In: Bacterial diseases of fish (V. Inglis, R. J. Roberts and N. R. Bromage, Ed), Blackwell Scientific Publications, London, pp 3–22.

56. Nematollahi A, Decostere A, Pasmans F, Haesebrouck F (2003) Flavobacterium psychrophilum infections in salmonid fish. J Fish Dis 26: 563-574.

57. Surendran PK, Joseph J, Shenoy AV, Perigreen PA, Mahadeva Iyer K, et al. (1989) Studies on spoilage of commercially important Tropical fishes under iced storage. Fish Res 7: 1-9.

58. Akhtar, N (1994) Fish composition and balance in population in Rawal Dam reservoir of Pakistan. Pakistan Journal of Zoology 61:111–118.

59. Unlusayin M, Kaleli S, Gulyavuz H (2001) The determination of flesh productivity and protein components of some fish species after hot smoking. J Sci Food Agri 81: 661–664.

60. Gokoglu N, YerlikayaP, Cengiz E (2004) Effects of cooking methods on the proximate composition and mineral contents of rainbow trout (Oncorhynchus mykiss). Food Chemistry 84: 19–22.

61. CJ, Besteiro I, Pascual C (1999) Biochemical changes in freshwater rainbow trout (Oncorhynchus mykiss) during chilled storage. J Sci Food Agri 79: 1473–1480.

62. Kyrana VR, Lougovois VP, Valsamis DS (1997) Assessment of shelf-life of maricultured gilthead sea bream (Sparus aurata)stored in ice. Int J Food Sci and Technol 32: 339–347.

63. Tejada M, Huidobro A (2002) Quality of farmed gilthead seabream (*Sparus aurata*) during ice storage related to the slaughter method and gutting. Eur Food Res Technol 215: 1–7.

64. Goulas AE, Kontominas MG (2007) Combined effect of light salting, modified atmosphere packaging and oregano essential oil on the shelf-life of sea bream (*Sparus aurata*): biochemical and sensory attributes. Food Chem 100: 287–296.

65. Lalitha KV, Sonaji ER, Manju S, Jose L, Gopal TK, et al. (2005) Microbiological and biochemical changes in pearl spot (*Etroplus suratensis* Bloch) stored under modified atmospheres. J Appl Microbiol 99: 1222–1228.

66. Sikorski ZE, Kolakowska A, Burt JR (1990) Postharvest biochemical and microbial changes. In: Sikorski Z.E. (ed.) Seafood: resources, nutritional composition and preservation. CRC Press, Boca Raton, pp 55-76.

67. Dalgaard P, Gram L, Huss HH (1993) Spoilage and shelf life of cod fillets packed in vacuum or modified atmospheres. Int J Food Microbiol 19: 283-294.

68. ME (1963) Analytical methods. In: Industrial Fishery Technology Krieger Publishing Co., Inc., New York, 367 p.

69. Dalgaard P (2000) Fresh and lightly preserved seafood. In: Man,C.M.D. and A.A. Jones (eds) Shelf-Life Evaluation of Foods. Aspen Publishers Inc., London, UK. pp. 110-139.

70. Katikou P, Ambrosiadis L, Georgantellis D, Koidis P, Georgakis AS (2007) Effect of lactobacillus cultures on microbiological, chemical and odour changes during storage of rainbow trout fillets. J Sci Food Agric 87: 477-484.

71. Dawood AA, Roy RN, Williams CS (1986b) Quality of rainbow trout chilled-stored after post-catch holding. J Sci Food Agric 37: 421–427.

72. Stohr V, Joffraud JJ, Cardinal M, Leroi F (2001) Spoilage potential and sensory profile associated with bacteria isolated from cold-smoked salmon. Food Res Int 34: 797 – 806.

73. Ruiz-Capillas C, Moral A (2001) Correlation between biochemical and sensory quality indices in hake stored in ice. Food Res Int 34: 441- 447

74. Pons-Sanchez-Cascado S, Veciana-Nogues MT, Bover-Cid S, Marine-Font A, Vidal-Carou MC (2006) Use of volatile and non-volatile amines to evaluate the freshness of anchovies stored in ice. J Sci Food Agric 86: 699-705.

75. Connel JJ (1995) Control of fish quality, 4th Ed. London: Fishing News Books Limited.

76. Je JY, Park PJ, Kim SK (2005) Antioxidant activity of a peptide isolated from Alaska pollack (*Theragra chalcogramma*) frame protein hydrolysate. Food Res Int 38: 45-50.

77. Han TJ, Liston J (1989) Lipid peroxidation protection factors in rainbow trout (*Salmo gairdnerii*) muscle cytosol. J Food Sci 54: 809-813.

78. Gould E, Peters JA (1971) On the testing the freshness of frozen fish. Fishing News (Books), London.

79. Olafsdottir G, Martinsdottir E, Oehlenschlager J, Dalgaard P, Jensen B, et al. (1999) Methods to evaluate fish freshness in research and industry. Tr Fd Sc & Tech 8: 258-265.

80. Huss HH (1988) Fresh fish quality and quality changes. FAO Fisheries Series, No. 29, FAO, Rome.

81. Regenstein JM (1996) Assuring the freshness and quality of aquacultured fish. In: Froid et al., Aquaculture Refrigeration and Aquaculture, 20-22 March 1996, Bordeaux, France, pp. 343-362.

28

Oxidative Stress Induction in Monosex Nile Tilapia (*Oreochromis niloticus, Linnaeus*, 1758): A Field Study on the Side Effects of Methyltestosterone

Alaa El-Din H. Sayed[1]*and Nasser S. Abou Khalil[2]

[1]*Zoology Department, Faculty of Science, Assiut University, 71516 Assiut, Egypt*
[2]*Medical Physiology Department, Faculty of Medicine, Assiut University, Assiut, Egypt*

Abstract

In this survey, fishes were obtained from four localities: Assiut as a control and Beheira, Alexandria and Kafr EL-Sheikh; three farms from each governate as farmed monosex produced using Methyltestosterone (MT). Serum MT, total antioxidant capacity (TAC), malondialdehyde (MDA), and total peroxides (TPX) were estimated, followed by calculation of oxidative stress index (OSI). MT concentration in the serum of fishes farmed at Assiut showed no detectable levels of hormonal residues, while the monosex farms showed high levels of MT concentration in the serum of the sampled fishes. In comparison with control fishes of Assiut farms, serum TAC levels of monosex fishes collected from farms of Beheira and Alexandria were significantly lower. Serum TPX content of the monosex fishes obtained from Alexandria farms were significantly higher than the wild fishes obtained from Assiut farms. Calculated from ratio of serum TPX content and TAC concentration, OSI illustrated a significant difference between control fishes collected from Assiut and monosex fishes collected from Beheira. Although few significant changes were found in the examined oxidative stress endpoints, the results of this work put our foot in the beginning of road to link hormonally sex reversal practice with oxidative stress induction, and provocative for other researchers to invade this field of research more deeply utilizing more specific relevant markers and cutting-edge techniques.

Keywords: Musculinizing inducer; Sex reversed tilapia; Malondialdhyde; Total antioxidant capacity; Total peroxide

Introduction

Recently, production of tilapia in Egypt exceeds that of common carp make it the most commonly cultured fish species owing to some favorable features appropriated for aquaculture industries such as high tolerance to adverse environmental conditions, resistance to disease, efficient food conversion, fast growth rate, good consumer acceptance, and ease of spawning [1,2].

Taking into account the wide spectrum of economic and rearing benefits, sex reversal techniques are paid greater attention over the time in aquacultured fish systems. Achievement of high growth rate, prevention of large energy dissipation into reproduction and courtship behaviour, reducing aggressiveness, uniformity in size, avoiding undesirable effects of sexual maturation on appearance and meat quality, and reducing undesirable environmental impacts are the main contributory factors in rising the interest about monosex fish cultures [3] One of the most practiced sex reversal techniques is hormonal induction especially using methyltestosterone that has been test in more than 25 fish species [4]. However, this technique faces several limitations and disadvantages like time and cost consumption, low survival of sex reversed male or female, delayed sexual maturity, sterility at high dose, paradoxical sex reversal, and carcinogenicity of hormonal residues and other health hazards [5,6].

Oxidative stress is a situation characterized by an imbalance between increased production of oxidant species and/or decreased efficacy of the antioxidant defense system [7] leading acromolecule to damage including lipid peroxidation, protein crosslinking, DNA damage, changes in growth and function of cells [8]. The investigation of this imbalance is difficult due to the limited availability of specific biomarkers of oxidative stress, and the fact that measurement of individual antioxidant may give misleading picture because antioxidants work in concert through chain breaking reactions. Therefore, analysis of total antioxidant capacity may be the most relevant investigation [9] and take into consideration the cumulative synergistic action of

all the antioxidants present in the sample providing an integrated parameter rather than the simple sum of measurable antioxidants [10]. Malondialdhyde (MDA) is the product of polyunsaturated fatty acid peroxidation and a key reflector of free radical-scavenging and production capability in many pathological conditions of fish species [11,12]. Oxidative stress index that is as a combined product of pro-oxidants/antioxidants ratio gives a bidirectional mirror image to the overall oxidative/reductive potency of the given specimen [13,14].

Oxidative stress in area of marine fish farming is important both for the health of farmed fish and for seafood quality. It adversely impacted fish welfare on several levels as growth [15] reproduction [16], and immunity [17].

Two face impacts of androgenic steroids on redox homeostatic status is emerged from literature. They range from pro-oxidant properties in pre-spawing brown trout (*Salmo trutta*) and zebra finch [18,19], or even oxidative stress inducers in some studies [20-22], to antioxidant influences in several *in vivo* and *in vitro* models [23-25]. It is worthy to note that pro-oxidant effects of androgen can be evoked indirectly by its metabolites, estradiol, whereas other metabolite, 5α–dihydrotestosterone, stimulates signal expression [26]. However, there is no available data until now about modulation of peroxidative and antioxidants diagnostic markers in monosex farms produced by methyltestosterone (MT) administration. Inclusion of such aspects

*Corresponding author: Alaa El-Din H. Sayed, Faculty of Science, Assiut University, 71516 Assiut, Egypt, E-mail: alaa_h254@yahoo.com

is of utmost importance to widen the scope in this area of research and resolve the controversies. Therefore, this paper tries to focus on a field relevant problem going a forward step in understanding the unfavorable impacts of hormonal sex reversal in fish populations with special emphasis on the potential induction of oxidative stress using methyltestosterone in monosex Nile tilapia. To accomplish this goal, water quality, serum and muscle MT, TAC as a cytoprotectant indicator, and both of MDA and TPX as pathogenetic effectors were measured, followed by calculation of oxidative stress index (OSI) as a reflector of the oxidant/antioxidant status in three aquafarms of four governments in Egypt using Assiut fish farms as a control to confirm our previous study results as genotoxic effects of MT [27,28].

Materials and Methods

Sample collection and characterization of water quality and morphometric outcome measures of the studied fishes

Healthy fishes of The Nile tilapia (*Oreochromis niloticus*) were caught from Assiut farms as control and three farms of Beheira, Alexandria and Kafr EL-Sheikh as monosex farms in Egypt. The data about water quality assessment and age, weight, and length of the collected fishes already published in previous study [28].

Serum and muscle methyltestosterone measurement

Their levels were determined by enzyme-linked immunosorbent assay reader (Stat Fax-200, Awareness Technology company, USA) using a commercially available kit (Art. Nr. R3601) according to the instructions of manufacturer (R-Biopharm GmbH, Darmstadt, Germany). Briefly, all reagents were brought to room temperature and reconstituted as recommended using the provided reagents. 50 μl of diluted enzyme conjugate was added to the bottom of each well of 96-well micotiter plate, followed by addition of 50 μl of standard and prepared sample to separate duplicate wells. After addition of 50 μl of diluted MT antibody to each well, the content of each well was mixed thoroughly, and incubated overnight (12-16 h) at 2-8°C. The liquid was poured out of the wells and the microwell holder was tapped upside down vigorously against absorbent paper to ensure complete removal of liquid from the wells. The wells were filled three times, each time with 250 μl of distilled water and the liquid was poured out as the previous step. 50 μl of each of substrate and chromogen was pipetted into each well followed by mixing the content thoroughly and incubation of the microtiter plate for 30 minutes at room temperature. After addition of 100 μl of stop solution to each well, the contents were mixed well and the absorbance was measured at 450 nm against air blank within 60 minutes. The mean values of the absorbance were obtained for the standards and the samples were divided by the absorbance value for the first standard (zero standard) and multiplied by 100. The zero standard was made equal to 100%, and the absorbance values were quoted in percentages. The values calculated for the standards were entered in a system of coordinates on semilogarithmic graph paper against MT concentration in (ng/l), after that MT concentration in ng/l corresponding to % absorbance of each sample was read from the calibration curve. The detection limit of this assay is 0.2 μ/l, and the coefficients of variation of the absorbance units obtained from 3 independent experiments entered against the corresponding MT concentrations are so low that good reproducibility (interassay variation) of the results is ensured. The specificity of the assay was determined by analyzing the cross-reactivity to corresponding substances.

Total antioxidant capacity measurement

Serum TAC was measured according to protocol given by Koracevic [29,30]. Briefly, 0.02 ml of distilled water was added in blank tube to 0.5 ml of R1 (H_2O_2 diluted 1000 times before use), whereas 0.02 ml of sample was added in sample tube. Then, the tubes were mixed and incubated 10 minutes at 37°C. Working reagent was prepared by mixing equal volumes of R2 (chromogen) and R3 (enzyme and buffer) immediately before use, and then 0.5 ml of the working reagent was added to both of blank and sample tubes. The contents were mixed and the tubes were shaken and then incubated 5 minutes at 37°C. The absorbance of blank and sample were read immediately against distilled water at 505 nm. Serum TAC, MDA, and TPX were measured colorimetrically using spectrophotometer (Spectronic 21, Moton Roy Company, USA).

Malondialdhyde measurement

According to Ohkawa procedure, 1 ml of chromogen was pipetted into tubes labeled as sample, standard, and blank. Then 0.2 ml of sample and standard were added to the corresponding tubes. The contents were mixed and the tubes were shaken, covered with screw cap, and then heated in boiling water bath for 30 minutes. After cooling the mixture, 0.2 ml of sample was added to blank tube and the content was mixed well. The absorbance of sample against blank, and standard against distilled water were read at 534 nm.

Total peroxide measurement

It was assessed following [31] R1 consists of 4.8 mg ammonium ferrous sulfate, 5 ml distilled water, 130 μl conc sulphuric acid, 4 mg xylenol orange, and 45 ml absolute ethanol containing 40 mg butylated hydroxyl toluene were added and dissolved by stirring. The blank solution was prepared by the same steps that of reagent without adding ammonium ferrous sulphate. 50 μl of serum was added to 1.25 ml of R1. The contents were mixed and the tubes shaken, and then incubated at room temperature for half hour, followed by centrifugation at 5000 rpm for 3 min in eppendrof and then measured against blank at 560 nm

Calculation of serum oxidative stress index

It is the percent ratio of TPX content to TAC concentration, and measured according to the following equation [31]:

$$OSI = (TPX, \mu M/L)/(TAC, \mu M/L) \times 100$$

Statistical analysis

The data were expressed as means ± standard error of the mean (SEM). Significant differences between groups were analyzed using one-way ANOVA followed by least significant difference (LSD) test for multiple comparisons using SPSS software, version 16. Differences were considered statistically significant at $P < 0.05$.

Ethical statement

The study was carried out in accordance with the Egyptian laws and University guidelines for the care of experimental animals. All procedures of the current work have been approved by the Committee of the Faculty of Science, Assiut University, Egypt.

Results

Sex reversed fishes characterized with a marked accumulation of MT in the serum and muscle

Figures 1a and 1b reveals MT concentrations in the serum and muscle of sampled fishes from the monosex farms in comparison with Assiut farms [27,28]. They showed no detectable levels of hormonal

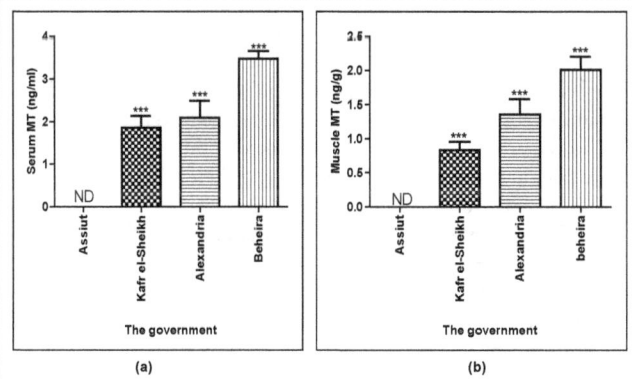

Figure 1: Graphic representation of methyltestosterone (MT) levels in blood serum **(a)** and muscle **(b)** of Nile tilapia (*Oreochromisniloticus*) collected from Assiut farms as control, and three farms of Beheira, Alexandria, and Kafr el-Sheikh as monosex farms.

ND: Non detectable.

residues in Assiut farms, while the monosex farms exhibited high levels of MT concentrations in the serum and muscle of the sampled fishes in a high significant (p < 0.001) manner versus control farms.

Oxidative stress induction under MT load in all male sex fishes

As shown in Table 1, Figures 2a and 2b serum TAC levels of monosex farms of Beheira (0.794 ± 0.102 µM/L) were significantly lower than those of Kafr EL-Sheikh (0.959 ± 0.029 µM/L) and control farms of Assiut (1.006 ± 0.030 µM/L) (P < 0.05 and P < 0.01, respectively). With respect to serum TPX content, the control fishes from Assiut farms were 1.520 ± 0.050 µM/L that were significantly (P < 0.01) lower than the monosex farms of Alexandria (1.759 ± 0.083 µM/L). Also, there was a significant difference (P < 0.01) between monosex farms of Beheira (1.531 ± 0.054 µM/L) and Alexandria. Supersingly, no significant difference could be monitored regarding serum MDA levels neither between monosex farms of the three governments (Kafr EL-Sheikh, Alexandria, and Beheira) and control farms of Assiut on one hand, nor between the monosex farms themselves on the other hand. Calculated from the ratio of serum TPX content and TAC concentration, OSI revealed an imbalance between reactive oxygen species (ROS) and antioxidant defense mechanisms in the given specimens. The only significant difference (P < 0.05) that could be detected was between control fishes from Assiut (151.715 ± 5.193%) and monosex fishes from Beheira (250.854 ± 59.894%) (Table 2).

Discussion

One expanding area of research not yet fully studied to date is the potential negative effects of androgens on aquacultures during production of monosex fish farms. Their detailed mechanistic pathways have not been completely elucidated and will be the subject of future experiments to expand the knowledge base and resolve discrepancies. Understanding their pathophysiological mechanisms pave the way for optimal therapeutical strategies for rising masculinized fishes without adverse impacts on fish health condition and welfare. With high levels of MT in their muscles, it may represent a major surprise to know that these currently studied masculinized fishes indeed are market-ready ones. Given that monosex fish is a vehicle for highly concentrated androgen, the problem limits may surpass fish burden to human food safety. Few significant changes in the parameters along with estimation

of general index of oxidative stress as MDA in this study make further confirming researches utilizing more specific and relevant oxidative stress markers requisite before come to the radical conclusion that application of MT in tilapia resulted in disruption of radicals-scavenging/production systems in all male production process. But this study provides a driving force for other researchers to uncover this key area of knowledge in more details.

Inclusion of MT in Nile tilapia feeding resulted in a slight but significant depletion of serum TAC levels of monosex fishes of Beheira farms giving insight into implication of this intervention on the overall enzymatic and non-enzymatic antioxidants and crosslinked with high MT residue in their fish serum and muscle. In accordance, brown trout (*Salmo trutta*, L.) with higher testosterone levels prior to spawning have higher levels of oxidative damage exemplified by decreased antioxidant capacity at the time of spawning. In addition, testosterone induced changes in the oxidative status of rat testis by reducing TAC [32]. Enhanced reactive oxygen species production, and depleted antioxidant enzyme levels and activities are linked to testosterone administration in several published articles [33-35]. It seemed that increased reactive oxygen species generation, as manifested by increased TPX and MDA levels in this study, consumed antioxidant reserves of sex reversed Nile tilapia. Genomic techniques, such as real-time quantitative polymerase chain reaction allow the simultaneous measurement of genes and their expression that involved in multiple biochemical pathways. Assessment of mRNA antioxidant enzyme genes expression in human vascular endothelial cells treated with supra-physiological doses of testosterone enanthate provides mechanistic information relating testosterone exposure to oxidative stress initiation [36]. In contrast, dehydroepiandrosterone as an adrenal androgen increased total glutathione content and reduced glutathione in young and old healthy rats [37], and glutathione-S-transferase in aged rats [38]. In addition, testosterone reversed the adverse effects of ovariectomy and 3-nitropropionic acid on reduced glutathione content and glutathione peroxidase activity in striatum of rats [39], increased activities of superoxide dismutase and glutathione peroxidase in cardiomyocytes of testicular feminized and castrated male mice [40], and total antioxidation capability in mice [41] *In vitro*, it enhanced catalase activity in undifferentiated mouse neuroblastoma cells [42], and glutathione reductase activity, and increased the content of thiol groups in human neutrophils [43]. Moreover, there are no marked changes in enzymatic antioxidant activities in male rat macrophages following testosterone supplementation [44].

Parameters/ Governments	Assiut	Kafr el- Sheikh	Alexandria	Beheira
Serum MT (ng/ml)	ND	2.03 ± 0.16***	2.32 ± 0.31***	3.47 ± 0.19***
Muscle MT (ng/g)	ND	0.93 ± 0.07***	1.54 ± 0.15***	2.01 ± 0.19***

Table 1: Methyltestosterone (MT) levels in blood serum and muscle of Nile tilapia (*Oreochromis niloticus*) collected from Assiut farms as control, and three farms of Beheira, Alexandria, and Kafr el-Sheikh as monosex farms.

Parameters/ Governments	Assiut	Kafr el- Sheikh	Alexandria	Beheira
TAC (µM/L)	1.006 ± 0.030	0.959 ± 0.029	0.846 ± 0.031	0.794 ± 0.102*
MDA (nmol/ml)	4.082 ± 1.201	4.393 ± 2.845	5.734 ± 1.422	5.630 ± 2.013
TPX (µM/L)	1.520 ± 0.050	1.625 ± 0.049	1.759 ± 0.083**	1.531 ± 0.054
OSI (%)	151.715 ± 5.193	171.160 ± 8.888	210.042 ± 11.983	250.854 ± 59.894*

Table 2: Impact of methyltestosterone supplementation on serum total antioxidant capacity (TAC), malondialdhyde (MDA), total peroxide (TPX), and oxidative stress index (OSI) of fish collected from Assiut farms as control, and three farms of Beheira, Alexandria, and Kafr el-Sheikh as monosex farms.

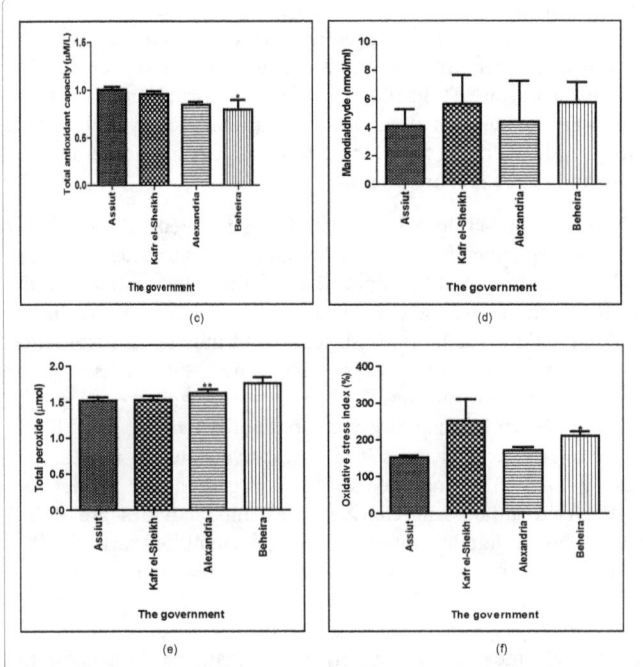

Figure 2: Graphic representation of changes in serum total antioxidant capacity (c), malondialdhyde (d), total peroxide (e), and oxidative stress index (f) of fishes collected from Assiut farms as control, and three farms of Beheira, Alexandria, and Kafr el-Sheikh as monosex farms.

Serum TPX concentrations of sex reversed fishes in Alexandria aquacultures were somewhat higher than untreated fishes in Assiut aquacultures. A clear vision to available publications supplies us with numerous examples for the close association between further oxidative stress inducing modulators and end products of peroxidative injury. For example, in Nile tilapia infected with *Aphanomyces laevis* and Phoma herbarum by water bath exposure and intramuscular injection, TPX and lipid peroxidation increased in both gills and mytomal muscles [45]. Similarly, mercuric chloride increased TPX in livers, kidneys, and gills of female catfish [46]. Cellular peroxide and lipid peroxidation end products increased markedly in gills of brown mussels (Perna perna) exposed for 7 and 21 days to zinc chloride [47].

Strikingly, there was no significant difference in serum MDA levels; however to some extent they were still higher in the monosex farms of the three governments (Kafr EL-Sheikh, Alexandria and Beheira) against control farms of Assiut. Indeed the data concerning androgen/ MDA relationship are inconclusive. Testosterone administration increased MDA in serum and testis of rabbits, liver, left ventricle and testicular homogenate of rats, and prostate of mice [48-51]. Conversely, it reduced levels of lipid peroxidation products in striatum of ovariectomized mice, liver of castrated and benzene treated rats, vanadium-induced testicular toxicity, intestinal ischemia/reperfusion, and brain tissues in rats [52-55]. We must keep in mind that MDA is not only the most studied indicator molecule of peroxidative damage, but also interacts with DNA and proteins to generate mutagenic and atherogenic by products [56] adding new dimension to elevation of this diagnostic marker.

OSI gives a summarized picture for the whole oxidant/antioxidant landmarks and goes behind the same successive order as that of serum and muscle MT offering an exactly copy/paste image. Among a wide range of oxidative stress indicators, it is a promising comparative index representing the interplay between the radical generating factors and

their counterplayers. Even though the only slight significant difference that could be traced in OSI was between control fishes from Assiut and monosex fishes from Beheira OSI of fishes of two other sex reversed culture medium was slightly higher, but not significantly, than that of non-manipulated fishes. Actually, this biodiagnostic reflector increased in additional several oxidative stress-linked models as hyperoxia in Atlantic cod, arsenite toxicity in goldfish, diabetes, hepatic ischemia/ reperfusion, and iron overload-induced cardiac dysfunction in rats [57-60]. Elevation of OSI of hormonally masculinized fishes could be due to increased TPX and decreased TAC. The present data give insight about oxidant/antioxidant balance shifting towards oxidant side but still need confirmation by other studies demonstrating the dark side of excess testosterone.

To sum up, undesirable impact of MT over dose on oxidants/ antioxidants systems in monosex Nile tilapia was manifested by slight but still significant overproduction of MDA and TPX with corresponding diminution of TAC resulting in increased OSI that correlated with serum and muscle MT concentrations. Estimation of more sensitive, specific, and reproducible bio-reflectors of oxidative stress is essential to prove finally the close relationship between MT application in fish aquacultures and peroxidative injury. Continuation in exploring the negative impacts of MT on fish welfare necessities further investigations on the potential oxidative stress-associated pathological conditions. Transportation of highly concentrated MT through aquatic food chain implicates the plausible human health affliction that needs epidemiological studies focus on relationship between MT and various aspects of health hazard in populations that depend on seafood as a main constituent of their diet. Also, these data open a new window for the researchers to look for alternative ideal sex reversal agents with minimal side effects on fish health conditions as a future recommendation.

References

1. Asad F, Ahmed I, Saleem M, Iqbal T (2010) Hormonal masculinization and growth performance in Nile Tilapia (Oreochromis niloticus) by androgen administration at different dietary protein levels. Int J Agric Biol 12: 939-943.

2. El-Saidy D, Gaber M (2005) Effect of dietary protein levels and feeding rates on growth performance, production traits and body composition of Nile tilapia, Oreochromis niloticus (L.) cultured in concrete tanks. Aquaculture Research 36: 163-171.

3. Beardmore J, Mair G, Lewis R (2001) Monosex male production in finfish as exemplified by tilapia: applications, problems, and prospects. Aquaculture 197: 283-301.

4. Pandian T, Sheela S (1995) Hormonal induction of sex reversal in fish. Aquaculture 138: 1-22.

5. Kavumpurath S, Pandian T (1993) Deterimination of labile period and critical dose for sex estrogens in Betta splendens. Indian J Exp Biol 31: 16-20.

6. Pandian T (1994) Endocrine sex reversal in fishes: Masculinization evokes greater stress and mortality. Curr Sci 66: 240-243.

7. Gosmaro F, Marco B, Silva B, Giorgio B, Enrico P, et al. (2013) Measurement of total antioxidant capacity of human plasma: setting and validation of the CUPRAC-BCS method on routine apparatus ADVIA 2400. Talanta 115: 526-532.

8. Ehsaei M, Mehdi K, Mohammad H, Daryoush H (2014) Prooxidant-antioxidant balance in patients with traumatic brain injury. Acta neurologica Belgica 115: 69-73.

9. Konuganti K, Hema S, Sameer Z, Wilma D, et al. (2012) A comparative evaluation of whole blood total antioxidant capacity using a novel nitroblue tetrazolium reduction test in patients with periodontitis and healthy subjects: A randomized, controlled trial. J Indian Soc Periodontol 16: 620-622.

10. Ghiselli A, Mauro S, Fausta N, Cristina S (2000) Total antioxidant capacity as

a tool to assess redox status: critical view and experimental data. Free Radic Biol Med 29: 1106-1114.

11. Feng M, Qu R, Ying L, Zhongbo W (2013) Biochemical biomarkers in liver and gill tissues of freshwater fish Carassius auratus following In Vivo exposure to hexabromobenzene. Environ Toxicol 29: 1460-1470.

12. Mozhdeganloo Z, Heidarpour M (2014) Oxidative stress in the gill tissues of goldfishes (Carassius auratus) parasitized by Dactylogyrus spp. J Parasit Dis 38: 269-72.

13. Karlsson A, Lene S, Rosseland B, Anders K (2011) Changes in arterial PO(2), physiological blood parameters and intracellular antioxidants in free-swimming Atlantic cod (Gadus morhua) exposed to varying levels of hyperoxia. Fish Physiol Biochem 37: 249-258.

14. Perez-Jimenez et al. (2012) The effect of dietary methionine and white tea on oxidative status of gilthead sea bream (Sparus aurata). Br J Nutr 108: 1202-1209.

15. Betancor M, Almaida P, Sprague M, Tocher R, Hemandez A (2015) Roles of selenoprotein antioxidant protection in zebrafish, Danio rerio, subjected to dietary oxidative stress. Fish physiology and biochemistry 41: 705-720.

16. Wilson S, Samantha M, Jessica J, Trisha A, David A, et al. (2014) Oxidative stress in Pacific salmon (Oncorhynchus spp.) during spawning migration. Physiological and Biochemical Zoology 87: 346-352.

17. Wei K, Yang J (2015) Oxidative damage of hepatopancreas induced by pollution depresses humoral immunity response in the freshwater crayfish Procambarus clarkii. Fish & shellfish immunology 43: 510-519.

18. Alonso-Alvarez C, Sophie B, Bruno F, Olivier C, Gabriele, et al. (2007) Testosterone and oxidative stress: the oxidation handicap hypothesis. Proceedings. Biological sciences/The Royal Society 274: 819-825.

19. Hoogenboom M, Metcalfe B, Ton G, Bonnie V, David C, et al. (2012) Relationship between oxidative stress and circulating testosterone and cortisol in pre-spawning female brown trout. Comp Biochem Physiol A Mol Integr Physiol 163: 379-387.

20. Goldfarb A, McIntosh K, Boyer T (1996) Vitamin E attenuates myocardial oxidative stress induced by DHEA in rested and exercised rats. Journal of applied physiology 80: 486-490.

21. Kumar D, Deepab P, Muthaiyan R, Periasamy M, Velliyur G, et al. (2012) Protective effect of Crataeva nurvala bark against MNU and testosterone induced oxidative stress in liver of male albino rats. Biomedicine & Aging Pathology 2: 94-98.

22. Prasad S, Neetu K, Shukla Y (2006) Modulatory effects of diallyl sulfide against testosterone- induced oxidative stress in Swiss albino mice. Asian J Androl 8: 719-723.

23. Chen J, Lin X, Congxin H (2014) DHEA inhibits vascular remodeling following arterial injury: a possible role in suppression of inflammation andoxidative stress derived from vascular smooth muscle cells. Mol Cell Biochem 388: 75-84.

24. Eleawa S, Saki H, Hussien M, Alkhateeb M (2013) Effect of testosterone replacement therapy on cardiac performance and oxidative stress in orchidectomized rats. Acta physiologica 209: 136-147.

25. Mostafa T, Rashed L, Kotb K, Taymour M (2012) Effect of testosterone and frequent low-dose sildenafil/tadalafil on cavernous tissue oxidative stress of aged diabetic rats. Andrologia 44: 411-415.

26. Casagrande S, Costantini D, Tagliavini J, Dell'omo G (2012) Differential effects of testosterone metabolites oestradiol and dihydrotestosterone on oxidative stress and carotenoid-dependent colour expression in a bird. Behav Ecol Sociobiol 66: 1319-1331.

27. Sayed A (2015) Erythrocytes alterations of monosex tilapia (Oreochromis niloticus, Linnaeus, 1758) produced using methyltestosterone. Egyptian Journal of Aquatic Research.

28. Sayed A, Moneeb R (2015) Hematological and biochemical characters of monosex tilapia (Oreochromis niloticus, Linnaeus, 1758) cultivated using methyl testosterone. The Journal of Basic and Applied Zoology 72: 36-42.

29. Koracevic D, Koracevic G, Djordjevic V, Andrejevic S, Cosic V, et al. (2001) Method for the measurement of antioxidant activity in human fluids. Journal of Clinical Pathology 54: 356-361.

30. Ohkawa H, Nobuko O, Kunio Y (1979) Assay for lipid peroxides in animal tissues by thiobarbituric acid reaction. Anal Biochem 95: 351-358.

31. Harma M, Muge H, Ozcan E (2005) Measurement of the total antioxidant response in preeclampsia with a novel automated method. European Journal of Obstetrics, Gynaecology, and Reproductive Biology 118, 47-51.

32. Tothova L, Celec P, Ostanikova D, Okuliarova M, Zeman M (2013) Effect of exogenous testosterone on oxidative status of the testes in adult male rats. Andrologia 45: 417-423.

33. Chainy G, Samantaray S, Samanta L (1997) Testosterone-induced changes in testicular antioxidant system. Andrologia 29: 343-349.

34. Klapcinska B, Jagsz S, Sadowska E, Jan G, Kempa K, Langfort J, et al. (2008) Effects of castration and testosterone replacement on the antioxidant defense system in rat left ventricle. The Journal of Physiological Sciences 58: 173-177.

35. Victor V, Milagros R, Celia B, Susana R, Marcelino G, et al. (2014) Mitochondrial impairment and oxidative stress in leukocytes after testosterone administration to female-to-male transsexuals. The Journal of Sexual Medicine 11: 454-461.

36. Skogastierna C, Maja H, Anders R, Lena E, et al. (2013) A supraphysiological dose of testosterone induces nitric oxide production and oxidative stress. European Journal of Preventive Cardiology 21: 1049-1054.

37. Jacob M, Daiane J, Alev S, Matheus P, Tarsila B, et al. (2011) Dehydroepiandrosterone improves hepatic antioxidant reserve and stimulates Akt signaling in young and old rats. The Journal of Steroid Biochemistry and Molecular Biology 127: 331-336

38. Jacob M, Daiane J, Alev S, Matheus P, Tarsila B, et al. (2010) Redox imbalance influence in the myocardial Akt activation in aged rats treated with DHEA. Experimental Gerontology 45: 957-963.

39. Tunez I, Montserat F, Juan A, Francisco J, Joise P, et al. (2007) Effect of testosterone on oxidative stress and cell damage induced by 3- nitropropionic acid in striatum of ovariectomized rats. Life Sci 80: 1221-1227.

40. Zhang L, Saizhu W, Yunjun R, Lei H, et al. (2011) Testosterone suppresses oxidative stress via androgen receptorindependent pathway in murine cardiomyocytes. Mol Med Rep 4: 1183-1188.

41. Jia J, Wu Y, Zhou X, Wang Y (2012) Effects of testosterone propionate on oxidative stress and the expression of spleen cytokine genes in endosulfan-treated mice. Journal of Environmental Pathology, Toxicology and Oncology 31: 17-26.

42. Chisu V, Manca P, Zedda M, Lepore G, Gadau S, et al. (2006) Effects of testosterone on differentiation and oxidative stress resistance in C1300 neuroblastoma cells. Neuro Endocrinology Letters 27: 807-812.

43. Marin D, Anaysa P, Rita D, Rui C, Roseman O, et al. (2010) Testosterone suppresses oxidative stress in human neutrophils. Cell Biochemistry and Function 28: 394-402.

44. Azevedo R, Lacava Z, Moyasaka C, Chaves S, Cun R, et al. (2001) Regulation of antioxidant enzyme activities in male and female rat macrophages by sex steroids. Brazilian Journal of Medical and Biological Research 34: 683-687.

45. Ali E, Hashem M, Salahy M (2011) Pathogenicity and oxidative stress in Nile tilapia caused by Aphanomyces laevis and Phoma herbarum isolated from farmed fish. Diseases of Aquatic Organisms. 94: 17-28.

46. Al-Salahy M (2011) Physiological studies on the effect of copper nicotinate (Cu-N complex) on the fish, Clarias gariepinus, exposed to mercuric chloride. Fish Physioly and Biochemistry 37: 373-385.

47. Trevisan R, Samira F, Jaco J, Marcio R, Afonso C, et al. (2014) Zinc causes acute impairment of glutathione metabolism followed by coordinated antioxidant defenses amplification in gills of brown mussels Perna perna. Comp Biochem Physiol C Toxicol Pharmacol 159: 22-30.

48. Aydilek N, Aksakal M (2005) Effects of testosterone on lipid peroxidation, lipid profiles and some coagulation parameters in rabbits. J Vet Med A Physiol Pathol Clin Med 52: 436-439.

49. Aydilek N, Aksakal M, Karakilcik A (2004) Effects of testosterone and vitamin E on the antioxidant system in rabbit testis. Andrologia 36: 277-281.

50. Sadowska-Krepa E, Slawomir J, Andrzej S, Stanislaw J, Pawel G, et al. (2011) High-dose testosterone propionate treatment reverses the effects of endurance training on myocardial antioxidant defenses in adolescent male rats. Cardiovasc Toxicol 11: 118-127.

51. Wachnik A, Biro G, Biro L, Korom M, Anna G, et al. (1993) Effect of sex

hormones on copper, zinc, iron nutritional status and hepatic lipid peroxidation in rats. Die Nahrung 37: 28-34.

52. Albayrak Y, Halici Z, Odabasoglu F, Unal D, Oral A, et al. (2011) The effects of testosterone on intestinal ischemia/reperfusion in rats. Journal of Investigative Surgery 24: 283-291.

53. Chandra A, Ghosh R, Chatterjee A, Sarkar M, et al. (2010) Protection against vanadium-induced testicular toxicity by testosterone propionate in rats. Toxicol Mech Methods 20: 306-315.

54. Guzman D, Gerardo B, Ivonne E, Emestina H, Daniel S, et al. (2005) Effect of testosterone and steroids homologues on indolamines and lipid peroxidation in rat brain. J Steroid Biochem Mol Biol 94: 369-373.

55. Verma Y, Rana S (2008) Modulation of CYP4502E1 and oxidative stress by testosterone in liver and kidney of benzene treated rats. Indian J Exp Biol. 46: 568-572.

56. Rio D, Amanda J, Nicoletta P (2005) A review of recent studies on malondialdehyde as toxic molecule and biological marker of oxidative stress. Nutrition, Metabolism & Cardiovascular Diseases. 15: 316-328.

57. Abd Allah E, Marwa A, Asmaa F (2014) Comparative study of the effect of verapamil and vitamin D on iron overload-induced oxidative stress and cardiacstructural changes in adult male rats. Pathophysiology 21: 293-300.

58. Bagnyukova T, Lida L, Igor P, Lushchak V (2007) Oxidative stress and antioxidant defenses in goldfish liver in response to short-term exposure to arsenite. Environ Mol Mutagen 48: 658-665.

59. Salum E, Jaak K, Priit K, Salum T, Zilmer K, et al. (2013) Vitamin D reduces deposition of advanced glycation end-products in the aortic wall and systemic oxidative stress in diabetic rats. Diabetes Res Clin Pract 100: 243-249.

60. Tufek A, Orhan T, Ibrahim A, Ulas A, Osman E, et al. (2013) The protective effects of dexmedetomidine on the liver and remote organs against hepatic ischemia reperfusion injury in rats. Int J Surg 11: 96-100.

29

Effects of Different Dietary Carbohydrate/Lipid Ratios on Growth, Feed Utilization and Body Composition of Early Giant Grouper *Epinephelus Lanceolatus* Juveniles

Weifeng Li[1,2], Xiaoyi Wu[1,2*], Senda Lu[1,2], Shuntian Jiang[1,2], Yuan Luo[1,2], Mingjuan Wu[1,2] and Jun Wang[1,2]

[1]*Key Laboratory of Tropical Biological Resources of Ministry of Education, Hainan University, Haikou, China*
[2]*Department of Aquaculture, Ocean College of Hainan University, Haikou, China*

Abstract

An 8-week growth trial was undertaken to determine effects of different dietary carbohydrate (CHO)/lipid (L) ratios on growth, feed utilization and body composition of early giant grouper *Epinephelus lanceolatus* juveniles. Five isoenergetic (4.1 kcal/g) and isonitrogenous (50% CP, dry-matter basis) experimental diets were formulated to contain different crude lipid (CL) levels (22%, 19.8%, 17.6%, 15.4% or 13.2%, dry-matter basis) and together different corn starch levels (0%, 4.95%, 9.9%, 14.85% or 19.8%), thereby forming different dietary CHO/L ratios. Groups of 41 early giant grouper juveniles (average initial weight of 0.397 g/fish) were stocked into small floating cages (L 120 cm × W 70 cm × H 50 cm). Triplicate groups of fish were fed each dietary treatment three times daily to apparent satiation.

Increasing dietary CHO/L ratio from 0.13 to 1.25 had no significant influences (P ≥ 0.05) on growth performance of early giant grouper juveniles, but when dietary CHO/L ratio was increased to 1.88, Weight gain (WG) of experimental fish was significantly decreased (P < 0.05). Feed conversion ratio (FCR) values among fish fed CHO/L ratios ranging from 0.13 to 1.25 were not significantly different but lower than that of FCR in fish fed the CHO/L ratio of 1.88. No significant differences in protein productive value (PPV) were observed among fish fed CHO/L ratios of 0.13, 0.4, 0.76 and 1.25. Fish fed a CHO/L ratio of 1.88 had significantly lower PPV than fish fed other CHO/L ratios. Hepatosomatic index (HSI) was increased with dietary CHO/L ratio increasing. Fish fed the CHO/L ratio of 0.13 or 0.4 had significantly higher intraperitoneal fat (IPF) ratios than fish fed the CHO/L ratio of 0.76, 1.25 or 1.88. Whole-body lipid content was decreased with the increasing dietary CHO/L ratio. Muscle moisture, protein as well as lipid contents were not influenced by dietary CHO/L ratios. Liver protein and lipid contents had a decreasing trend with dietary CHO/L ratio increasing. Liver glycogen content in fish fed the CHO/L ratio of 1.25 or 1.88 was significantly higher than fish fed the CHO/L ratio of 0.13, 0.4 or 0.76. Based on quadratic broken line model to WG values, 1.30 of CHO/L ratio, corresponding to 14.79% dietary crude lipid and 19.23% dietary CHO, was proved to be optimal for early giant grouper juveniles.

Keywords: *Epinephelus lanceolatus;* Carbohydrate; Lipid; Growth

Introduction

In aquatic feeds, soluble carbohydrate (CHO) and lipid are usually used as energy sources for aquatic animals. Carbohydrate after absorbed can provide fish with equal amounts of energy as protein. Capacities of fish to utilizing CHO depend on the species as reviewed [1,2]. For instance, only 10% of CHO was acceptable for yellowtail kingfish [3], but juvenile humpback grouper were able to efficiently utilize about 20% of readily digestible CHO [4,5], and for juvenile sunshine bass, up to 42% dextrin in the diet could be efficiently utilized [6]. However, excessive supplementations of digestible CHO to diets could compromise fish growth performance [3] and disrupt liver function and increase susceptibility to infectious diseases [2].

Compared to CHO, lipid can provide fish with more energy value per unit and be better utilized by most fish species, and moreover, lipid is the source of essential fatty acids required by fish for normal growth, development and maintaining health [7]. As one of the macronutrients in aquatic feed, lipid has many advantages for fish growth, but in comparison to CHO, it is more expensive and less available, especially so for fish oil. Excess lipids in diets usually increase lipid deposition in fish carcass [8,9], lead to a substantial decline in performance, affect gut health [10] and increase susceptibility to autoxidation and tissue lipid peroxidation, which may also adversely affect the immune response and disease resistance of fish [11].

Optimizing of dietary CHO and lipid levels are beneficial not only to improving fish quality but also to sparing feed cost [12] reported

that dietary lipid could be partially replaced by CHO without reducing productivity or carcass quality of sunshine bass. Based on the best growth performance or health status, suitable dietary CHO/Lipid ratios have been established in some fish species such as walking catfish (3.38) [8] blunt snout bream (3.58) [13], yellow catfish (2.45-5.58) [14] and yellowfin seabream (0.62) [15].

Giant grouper *Epinephelus lanceolatus* has been widely cultured in China in recent years due to its faster growth compared to other grouper species [16]. To date, available information on nutrition of *Epinephelus lanceolatu* is quite limited, with only one published study [16] evaluating effects of choline on lipid metabolism and stress tolerance of juvenile giant grouper. The aim of this study was to determine effects of different dietary CHO/L ratios on growth, feed utilization and body composition of early giant grouper *Epinephelus lanceolatus* juveniles.

*Corresponding author: Xiaoyi Wu, Department of Aquaculture, Ocean College of Haiman University, Haikou, P.R. China, E-mail: wjurk@163.com

Materials and Methods

Experimental diets and designs

In this trial, five isoenergetic (4.1 kcal/g) and isonitrogenous (50% CP, dry-matter basis) experimental diets were formulated to contain different crude lipid (CL) levels (22%, 19.8%, 17.6%, 15.4% or 13.2%, dry-matter basis) and together different corn starch levels (0%, 4.95%, 9.9%, 14.85% or 19.8%), thereby forming different dietary CHO/L ratios of 0.13, 0.40, 0.76, 1.25 and 1.88, respectively (Table 1). Corn starch was used as the main carbohydrate source. The 50% dietary protein level designed in this study was according to the study of dietary digestible energy was calculated or estimated using physiological fuel values of 4.0, 4.0 and 9.0 kcal/g (16.7, 16.7 and 37.7 kJ/g) for carbohydrate, protein [16] and lipid, respectively [17,18]. It was reported that fish larvae utilize dietary phospholipids more efficiently than neutral lipids [19] and have high requirements of phospholipids such as at least 9.5% for pikeperch larvae [20] and 6.95-8.51% for large yellow croaker larvae [21] so in this study, 10% of soy lecithin was supplemented to all experimental diets. The 4.1 kcal/g diet of digestible energy level was closed to those of values in diets for pikeperch larvae [20] and European sea bass larvae [22].

Fishmeal was well ground, and all dry ingredients were weighed and mixed in a Hobart mixer (A-200T Mixer Bench Model unit, Resell Food Equipment Ltd., Ottawa, Canada) for 30 min. Thereafter, oil was gradually added, while mixing constantly. Then, 30-50 mL of water was slowly blended into the mixture for each 100 g of dry matter, resulting in suitably textured dough. The diets were pelletized into a noodle-like shape of 1.0-mm diameter using a twin-screw extruder (Institute of Chemical Engineering, South China University of Technology, Guangzhou, PR China) and then all diets were air-dried for 24 hrs, sieved and stored at -20°C until fed.

Experimental procedures

30-day post hatching (dph) giant grouper juveniles purchased from a commercial marine fish hatchery (Yangpu, Hainan) were put into small floating cages (L 120 cm × W 70 cm × H 50 cm) at a density of 50 fish per cage and acclimated using a commercial grouper micro-diet (Crude protein: 50%, crude lipid: 12%) together with ground muscle from trash fish for 4 days. During the acclimation, fish were hand fed to satiation three times daily (08:00, 12:00 and 17:00) and ground muscle was gradually reduced. After the micro-diet was completely accepted by experimental fish, groups of 41 early giant grouper juveniles (average initial weight of 0.397 g/fish) were randomly distributed into 15 small cages which were labelled and located in five connective 6-m³ indoor concrete ponds (L 3 m × W 2 m × H 1 m) with 3 cages occurring in each pond. All ponds received flowing sea water (salinity: 33.1 g/L) from the same reservoir at a rate of 3 g/L.

During the experimental period, each dietary treatment had three replicates, and each replicate cage was in different ponds. Water temperature (27-28°C), total ammonia (0-0.20 mg/L) and dissolved oxygen (5.9-6.2 mg/L) were monitored daily. Fish were exposed to a 12 hrs. light: 12 hrs. dark cycle and hand-fed each dietary treatment three times daily (08:00, 12:00 and 17:00) to apparent satiation until pellets were first seen to sink to bottom of the pond. Feed intake was recorded daily, and experimental ponds and cages were cleaned once a week. The growth trial was continued for 8 weeks.

Sampling and analysis

At the end of the trial, two fish from each cage were collected for whole-body composition analysis. Three fish per cage were individually weighed and dissected to obtain liver, intestine and intraperitoneal fat (IPF) weights for computing body condition indices including hepatosomatic index (HSI) ((liver wt/live wt)*100) and IPF ratio ((IPF wt/live wt)*100), respectively. Intraperitoneal fat was obtained by removing and weighing the fat from the abdominal cavity as well as that adhering to the intestine of the fish. Condition factor (CF) also was computed as (bodyweight × 100)/(body length)³. Muscle and liver samples for compositional analysis also were taken from these three fish. Livers for glycogen analysis were quickly dissected from another two randomly selected fish which were removed from each replicate cage, and dissected livers were wrapped in aluminium foil, frozen in liquid nitrogen, and stored at 80°C until analyzed.

Crude protein was estimated by measuring total nitrogen by the Dumas method [23] and multiplying by 6.25. Dry matter was determined by heating at 125°C for 3 hrs, and ash was quantified after heating at 650°C for 3 hrs [24]. Crude lipid was determined by chloroform and methanol extraction [25]. Glycogen contents in liver were analyzed [26]. Carbohydrate content of diets was analyzed by the 3′5-dinitro salicylic acid method [27,28].

Statistical analysis

Experimental data obtained for response parameters were tested by subjecting the data to one-way analysis of variance and Tukey's test using SPSS (Version 16.0). Significance was set at $P < 0.05$.

	Dietary CHO/L ratios				
	0.13	0.40	0.76	1.25	1.88
Peruvian fishmeal (Anchovy)[a]	34.03	34.03	34.03	34.03	34.03
Casein	25	25	25	25	25
Yeast meal	2	2	2	2	2
Fish oil (Salmon)	9.42	7.22	5.02	2.82	0.62
Soy lecithin	10	10	10	10	10
Corn starch	0	4.95	9.9	14.85	19.8
Vitamin mixture[b]	4	4	4	4	4
Mineral mixture[c]	2	2	2	2	2
Carboxymethyl cellulose	2.5	2.5	2.5	2.5	2.5
Cellulose	11.05	8.30	5.55	2.80	0.05
Analyzed composition[d]					
Moisture %	9.26	9.17	9.27	9.78	10.32
Crude protein %	49.2	49.8	49.7	49.2	48.9
Crude lipid %	22.6	20.0	17.6	15.1	13.0
Carbohydrate %	3.05	7.96	13.41	18.87	24.43
Ash %	5.66	5.53	5.13	5.28	5.72
Estimated digestible energy[e] (kcal/g)	4.12	4.11	4.11	4.08	4.10
CHO/L (g/g)	0.13	0.40	0.76	1.25	1.88

[a]Yongsheng Feed Corporation, Binzhou, China; proximate composition (% dry matter): moisture: 8.9; crude protein: 73.6; crude lipid: 10.7.

[b]Vitamin mixture contained (as g kg⁻¹ vitamin mixture): Choline concentrate 50%, 200 g; vitamin E(500 IU/g) 10 g; vitamin D3 (500,000 IU/g) 0.50 g; vitamin B3 1 g; vitamin B5 2 g; vitamin B1 100mg; vitamin B2 0.4 g; vitamin B6 300 mg; vitamin C 20 g; vitamin B9 100 mg; vitamin concentrate B12 (1 g/kg), 1 g; vitamin K3 1 g; meso-inositol 30 g; cellulose, 732.1 g. (from Mazurais et al. 2009).

[c]Mineral mixture contained (as g kg⁻¹ mineral mixture): 90 g KCl, 40 mg KIO₃, 500 g CaHPO₄·2H₂O, 40 g NaCl, 3 g CuSO₄·5H₂O, 4 g ZnSO₄·7H₂O, 20 mg CoSO₄·7H₂O, 20 g FeSO₄·7H₂O, 3 g MnSO₄·H₂O, 215 g CaCO₃, 124 g MgSO₄·7H₂O, and 1 g NaF (from Mazurais et al. 2009).

[d]Values represent means of duplicate samples.

[e]By calculation.

Table 1: Formulations and analyzed composition of experimental diets.

Results

Growth performance and feed utilization

Dietary CHO/L ratios ranging from 0.13 to 1.25 did not significantly affect WG of experimental fish (Table 2), but fish growth performance was significantly reduced as dietary CHO/L ratio was increased to 1.88. The analysis of quadratic broken line model to WG values showed that 1.30 of dietary CHO/L was optimal for early giant grouper juveniles (Figure 1). Fish fed the diet with a CHO/L ratio of 1.88 had a significantly higher FCR than fish fed diets with a CHO/L ratio of 0.13, 0.4, 0.76 or 1.25. No significant differences in FCRs of experimental fish were observed when dietary CHO/L ratio was increased from 0.13 to 1.25. Fish fed diets with a CHO/L ratio of 1.88 had significantly lower PPVs than fish fed diets with a CHO/L ratio of 0.13, 0.4, 0.76 or 1.25.

Body condition indices

Fish fed the diet with a CHO/L ratio of 0.13 had a significantly lower condition factor (CF) than fish fed a CHO/L ratio of 1.88 (Table 3). Fish fed a CHO/L ratio of 1.88 had significantly higher hepatosomatic index (HSI) than fish fed other CHO/L ratios. Intraperitoneal fat (IPF) ratio of fish fed CHO/L ratios of 0.13 and 0.4 were significantly higher than those of fish fed CHO/L ratios of 0.76, 1.25 and 1.88, while this parameter showed no significant differences between fish fed CHO/L ratios of 0.13 and 0.4 as well as between fish fed CHO/L ratios of 0.76, 1.25 and 1.88.

	Dietary CHO/L ratios				
	0.13	0.40	0.76	1.25	1.88
FBW[2]	36.94 ± 0.65[a]	36.93 ± 1.29[a]	36.59 ± 1.36[a]	37.48 ± 1.83[a]	31.45 ± 0.15[b]
Daily FI[3]	1.42 ± 0.01	1.42 ± 0.17	1.48 ± 0.25	1.47 ± 0.12	1.77 ± 0.10
WG%[4]	9204 ± 163[a]	9202 ± 325[a]	9115 ± 342[a]	9340 ± 460[a]	7822 ± 38[b]
FCR[5]	0.73 ± 0.00[b]	0.73 ± 0.05[b]	0.69 ± 0.03[b]	0.75 ± 0.04[b]	0.90 ± 0.03[a]
PPV[6]	47.05 ± 0.37[a]	46.21 ± 3.26[a]	48.67 ± 2.04[a]	45.89 ± 1.95[a]	36.90 ± 1.44[b]
Survival %	64 ± 1	64 ± 4	72 ± 2	62 ± 6	73 ± 5

[1]Treatment means (± SEM) represent the average values of three tanks per treatment, and values within the same row with different letters are significantly (P<0.05) different.

[2]FBW: Final mean body weight (g).

[3]Daily Feed Intake: 100 × feed offered/average total weight/days.

[4]Weight Gain: 100 × (final mean weight - initial mean weight)/initial mean weight.

[5]Feed Conversion Ratio: g dry feed/g weight gain (included the dead fish).

[6]Protein Productive Value: 100 × retained protein (g)/protein fed (g).

Table 2: Growth performance and feed utilization of early giant grouper *Epinephelus lanceolatus* juveniles fed diets contained different CHO/L ratios for 8 weeks[1].

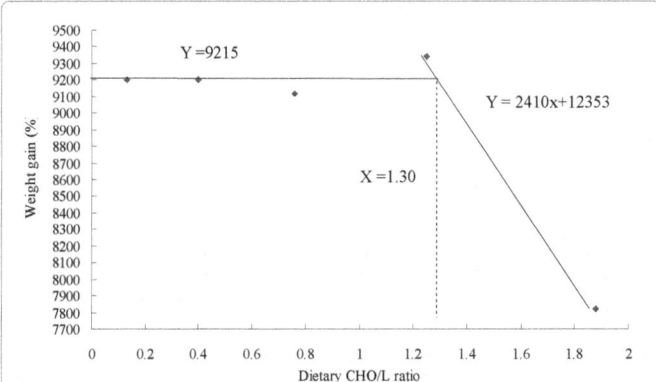

Figure 1: The analysis of quadratic broken line model based on WG values showed that 1.30 of dietary CHO/L was optimal for early giant grouper juveniles.

	Dietary CHO/L ratios				
	0.13	0.40	0.76	1.25	1.88
CF[2]	1.83 ± 0.08[b]	2.07 ± 0.04[ab]	2.01 ± 0.08[ab]	1.98 ± 0.00[ab]	2.13 ± 0.06[a]
HSI[3]	2.96 ± 0.29[a]	3.82 ± 0.29[a]	3.95 ± 0.40[b]	3.78 ± 0.21[b]	6.33 ± 0.29[a]
IPF Ratio[4]	4.61 ± 0.12[a]	4.89 ± 0.14[a]	3.61 ± 0.29[b]	3.73 ± 0.30[b]	3.63 ± 0.02[b]

[1]Treatment means (± SEM) represent the average values of three tanks per treatment, and values within the same row with different letters are significantly (P < 0.05) different.

[2]CF: Body weight × 100/(Body length)[3].

[3]HSI: Liver weight × 100/fish weight.

[4]IPF ratio: Intraperitoneal fat weight × 100/fish weight.

Table 3: Body condition indices of early giant grouper *Epinephelus lanceolatus* juveniles fed diets contained different CHO/L ratios for 8 weeks[1].

Dietary CHO/L ratios					
	0.13	0.40	0.76	1.25	1.88
Whole-body composition					
Moisture %	69.7 ± 0.4[b]	69.5 ± 0.1[b]	70.1 ± 0.2[b]	70.2 ± 0.4[b]	71.9 ± 0.3[a]
Protein %	16.89 ± 0.11[a]	16.68 ± 0.08[a]	16.65 ± 0.01[a]	16.90 ± 0.11[a]	16.22 ± 0.15[b]
Lipid %	9.16 ± 0.42[a]	9.16 ± 0.33[a]	8.80 ± 0.37[a]	7.95 ± 0.19[b]	6.35 ± 0.24[c]
Muscle composition					
Moisture %	73.97 ± 0.49	75.11 ± 0.94	74.97 ± 0.18	73.93 ± 0.74	75.94 ± 0.32
Protein %	20.80 ± 0.22	20.84 ± 0.71	21.23 ± 0.20	21.60 ± 0.68	20.97 ± 0.20
Lipid %	10.33 ± 1.05	9.06 ± 1.06	8.74 ± 0.52	9.27 ± 0.09	8.67 ± 0.75
Liver composition					
Moisture %	59.18 ± 1.27	62.03 ± 0.63	62.46 ± 1.11	62.84 ± 0.87	62.56 ± 0.46
Protein %	9.48 ± 0.09[a]	8.56 ± 0.26[b]	8.14 ± 0.18[b]	8.29 ± 0.42[b]	6.42 ± 0.10[c]
Lipid %	21.04 ± 1.28[a]	16.41 ± 1.00[b]	14.80 ± 1.49[bc]	14.88 ± 0.47[bc]	13.41 ± 0.35[c]
Glycogen%	11.61 ± 0.19[b]	11.95 ± 0.45[b]	15.76 ± 0.23[a]	16.76 ± 0.63[a]	17.29 ± 0.54[a]

[1]Treatment means (± SEM) represent the average values of three tanks per treatment, and values within the same row with different letters are significantly (P<0.05) different.

Table 4: Whole-body, muscle and liver composition (fresh-wt. basis) of early giant grouper *Epinephelus lanceolatus* juveniles fed diets contained different CHO/L ratios for 8 weeks[1].

Whole-body, muscle and liver compositions

Values of whole-body lipid content were not significantly different among fish fed CHO/L ratios of 0.13, 0.4 and 0.76, but whole-body lipid contents of fish fed CHO/L ratios of 0.13, 0.4 or 0.76 significantly higher than fish fed CHO/L ratios of 1.25 and 1.88 (Table 4). Fish fed a CHO/L ratio of 1.88 had significantly higher whole-body moisture but lower whole-body protein compared to fish fed a CHO/L ratio of 0.13, 0.4, 0.76 or 1.25. Muscle moisture, protein as well as lipid contents were less affected by dietary CHO/L ratios. Liver moisture showed no significant differences among different experimental treatments. Liver protein and lipid contents of fish fed the diet with a CHO/L ratio of 0.13 were significantly higher than fish fed a CHO/L ratio of 0.4, 0.76, 1.25 or 1.88. Fish fed a CHO/L ratio of 0.76, 1.25 or 1.88 had significantly higher glycogen in liver than fish fed a CHO/L ratio of 0.13 or 0.4.

Discussion

Results of the present study demonstrated that 1.30 of CHO/L ratio, corresponding to 14.79% dietary crude lipid and 19.23% dietary CHO, was proved to be optimal for early giant grouper juveniles, indicating that reduction in dietary lipid from 22.6% to 14.79%, with concomitant increase in CHO level from 3.05% to 19.23%, corresponding to CHO/L ratios of 0.13 to 1.30, did not negatively affect the growth of early giant grouper *Epinephelus lanceolatus* juveniles, but when fish were fed the

24.32% CHO and 13.0% lipid contained diet with a CHO/L ratio of 1.88, WG was significantly reduced. This indicated that early giant grouper juveniles could effectively utilize 19.23% CHO at 50% dietary CP and 14.79% dietary CL levels, and dietary CHO could replace about 30% lipid without sacrificing fish growth. Previous studies on juvenile humpback grouper initiating ~ 5 g [4] or 8 g [5] of body weight showed that juvenile humpback grouper can efficiently utilize about 20% of readily digestible dietary CHO, which was in line with the results obtained in this study. Successful replacements of partial lipid by digestible CHO have been reported in other fish species such as hybrid striped bass [12], rainbow trout [5], hybrid *Clarias* catfish [29] and yellowfin seabream [15] In this study, fish fed the low-CHO (3.05%) diet did not display poorer growth performance than fish fed high-CHO (7.96%, 13.41%, 18.87%) diets, meaning that dietary CHO was not essential for growth of early giant grouper juveniles. This disagreed with the results of some other studies [8,29,30].

Dietary lipids are the preferable source of metabolic energy for the development of fish and the main source of essential fatty acids for maintaining rapid growth of larval fish, especially marine fish [31]. The significantly lower WG of fish fed a CHO/L ratio of 1.88 (3.0% fish oil) compared to fish fed CHO/L ratios of 0.13, 0.4, 0.76 and 1.25 (12.6%, 10%, 7.6% and 5.1% fish oil) were perhaps due to the deficiency of essential fatty acids resulted from the decrease of fish oil. Reduced growth and poor feed conversion efficiency in fish fed diets with high CHO/L ratios have also been reported in chinook salmon [32], channel catfish [17] red drum [6,33] and hybrid *Clarias* catfish [29].

Fish fed dietary CHO/L ratios ranging at 0.13-1.25 (15.1-22.6% CL) had similar growth performance, demonstrating that 15.1% of dietary lipid is adequate for fast growth of early giant grouper juveniles. Optimal dietary lipid requirements of several other grouper species have been reported such as juveniles of *E. coioides* (10%) [34,35], *E. malabaricus* (7.9-12.5%) [36,37], *E. aeneus* (10.5-14.7%) [38] and *C. altivelis* (16-26%) [5].

The higher feed conversion ratios (FCRs) and lower PPV of fish fed a CHO/L ratio of 1.88 compared to fish fed a CHO/L ratio of 0.13, 0.4, 0.76 or 1.25 were attributed to their poorer growth compared to fish fed low CHO/L ratios. Similar reduced growth, feed efficiency, and protein retention have also been observed in juvenile yellowfin seabream [15] and red drum [6,33] fed a high carbohydrate and low lipid diet.

In the present study, all experimental diets had similar protein and gross energy contents, so the higher IPF ratios in fish fed diets with low CHO/L ratios (0.13 and 0.4) compared to fish fed diets with high CHO/L ratios (0.76, 1.25 and 1.88) maybe result from their limited ability of utilizing dietary CHO as energy and/ or de novo synthesis of lipid. This is in agreement with the results reported in yellowfin seabream [15] and hybrid striped bass [39]. In the present study, increasing dietary lipid to 17.6% or higher (20.0%, 22.6%) resulted in increased whole-body lipid. Similar results have been reported in rainbow trout [18], red drum [33,40], striped bass [41], channel catfish [42,43], hybrid *Clarias* catfish [29] walking catfish [8], *Tilupia zillii* [43], common carp [44,45] and brown-marbled grouper [46].

It is reported that liver enlargement may result from increased lipid or glycogen deposition [47]. The higher HSI values in fish fed high CHO/L ratios compared to those in fish fed low CHO/L ratios were mainly due to their higher hepatic glycogen contents which resulted from the higher CHO levels contained in the high CHO/L diets. Similar results were also reported in striped bass [48] and Asian seabass [49]. Fish fed the diet with low CHO content (3.05%) had

lower CF value than fish fed the high CHO contained diet, which was possibly due to the lower HSI observed in fish fed the CHO/L ratio of 0.13. No differences in muscle protein and lipid contents among all experimental treatments showed that muscle composition of fish was less influenced by dietary CHO/L ratios. Fish fed high-lipid diets had higher liver lipid content than fish fed low-lipid diets, agreeing with the reports in hybrid striped bass [50,51].

In conclusion, results of this study showed that at 50% dietary crude protein level, 1.30 of CHO/L ratio, corresponding to 14.79% dietary crude lipid and 19.23% dietary CHO, was optimal for growth performance of early giant grouper juveniles, when dietary CHO/L ratio was increased to 1.88 (at 13.0% dietary crude lipid level and 24.32% CHO level), fish growth significantly reduced; Dietary lipid in excess, resulted in increased lipid deposition in the body.

References

1. Wilson R (1994) Utilization of dietary carbohydrate by fish. Aquaculture 124: 67-80.

2. Hemre G, Mommsen T, Krogdahl A (2002) Carbohydrates in fish nutrition: effects on growth, glucose metabolism and hepatic enzymes. Aquaculture Nutrition 8: 175-194.

3. Booth M, Moses M, Allan G (2013) Utilisation of carbohydrate by yellowtail kingfish *Seriola lalandi*. Aquaculture 376-379: 151-161.

4. Shiau S, Lin Y (2002) Utilization of glucose and starch by the grouper *Epinephelus malabaricus* at 23°C. Fish. Sci 68: 991-995

5. Suwirya K, Giri A, Marzuqi M, Trijoko (2004) Utilisation of dietary dextrin by juvenile humpback grouper (*Cromileptes altivelis*). In: Rimmer MA, McBride S, Williams, KC (eds.), Advances in Grouper Aquaculture. ACIAR Monograph, vol. 110. Australian Centre for International Agriculture Research, Canberra, Australia, pp. 107-109

6. Serrano J, Nematipour G, Gatlin D (1992) Dietary protein requirement of the red drum (*Scianops ocellatus*) and relative use of dietary carbohydrate and lipid. Aquaculture 101: 283-291.

7. Lin X, Shiau S (2003) Dietary lipid requirement of grouper, *Epinephelus malabaricus* and effects on immune responses. Aquaculture 225: 243-250.

8. Erfanullah, Jafri A (1998) Effect of dietary carbohydrate-to-lipid ratio on growth and body composition of walking catfish (*Clarias batrachus*). Aquaculture 161: 159-168.

9. Hanley F (1991) Effects of feeding supplementary diets containing varying levels of lipid on growth, food conversion, and body composition of Nile tilapia *Oreochromis niloticus* L. Aquaculture 93: 323-334.

10. Bonvini E, Parma L, Mandrioli L, Sirri R, Brachelente C, et al. (2015) Feeding common sole (*Solea solea*) juveniles with increasing dietary lipid levels affects growth, feed utilization and gut health. Aquaculture 449: 87-93.

11. Dias J, Rueda-Jasso R, Panserat S, Conceiçao L, Gomes E, et al. (2004) Effect of dietary carbohydrate to lipid ratios on growth, lipid deposition and metabolic hepatic enzymes in juvenile Senegalese sole (*Solea senegalensis*, Kaup). Aquac Res 35: 1122–1130.

12. Nematipour G, Brown M, Gatlin D (1992) Effects of dietary carbohydrate:lipid ratio on growth and body composition of hybrid striped bass. J World Aquacult Soc 23: 128-132.

13. Li X, Wang Y, Liu W, Jiang G, et al. (2013) Effects of dietary carbohydrate / lipid ratios on growth performance, body composition and glucose metabolism of fingerling blunt snout bream *Megalobrama amblycephala*. Aquaculture Nutrition 19: 701-708.

14. Wang L, Liu W, Lu K, Xu W, Cai D, et al. (2014) Effects of dietary carbohydrate/ lipid ratios on non-specific immune responses, oxidative status and liver histology of juvenile yellow catfish *Pelteobagrus fulvidraco*. Aquaculture 426-427: 41-48.

15. Hu Y, Liu Y, Tian L, Yang H, Liang G, et al. (2007) Optimal dietary carbohydrate to lipid ratio for juvenile yellowfin seabream (*Sparus latus*). Aquaculture Nutrition 13: 291-297.

16. Yeh S, Shiu P, Guei W, Lin Y, Liu C (2013) Improvement in lipid metabolism and

stress tolerance of juvenile giant grouper, *Epinephelus lanceolatus* (Bloch), fed supplemental choline. Aquaculture Research 46: 1-12

17. Garling D, Wilson R (1977) Effect of dietary carbohydrate to lipid ratio on growth and body composition of fingerling channel catfish. Prog Fish-Cult 39: 43-47

18. Lee D, Putnam G (1973) The response of rainbow trout to varying protein/energy ratios in a test diet. J Nutr 103: 916-922.

19. Cahu C, Zambonino I, Barbosa V (2003) Effect of dietary phospholipid level and phospholipid: neutral lipid value on the development of sea bass (*Dicentrarchus labrax*) larvae fed a compound diet. Journal of Nutrition 90: 21-28.

20. Hamza N, Mhetli M, Khemis I, Cahu C, Kestemont P, et al. (2008) Effect of dietary phospholipid levels on performance, enzyme activities and fatty acid composition of pikeperch (*Sander lucioperca*) larvae. Aquaculture 275: 274-282.

21. Zhao J, Ai Q, Mai K, Zuo R, Luo Y, et al. (2013) Effects of dietary phospholipids on survival, growth, digestive enzymes and stress resistance of large yellow croaker, *Larmichthys crocea* larvae. Aquaculture 410-411: 122-128.

22. Mazurais D, Glynatsi N, Darias M, Christodoulopoulou S, Cahu C, et al. (2009) Optimal levels of dietary vitamin A for reduced deformity incidence during development of European sea bass larvae (*Dicentrarchus labrax*) depend on malformation type. Aquaculture 294: 262-270.

23. Ebeling M (1968) The Dumas method for nitrogen in feed. J Assoc Anal Chem 75: 401-413.

24. AOAC (1990) Official Methods of Analysis. AOAC (Association of Official Analytical Chemists) 1298.

25. Folch J, Lees M, Sloane-Stanley G (1957) A simple method for the isolation and purification of total lipides from animal tissues. J Biol Chem 97: 383-394.

26. Hassidh W, Abraham S (1957) Chemical procedures for analysis of polyssacarides. Methods of Enzymology 3: 34-50

27. Yu S, Olsen C, Marcussen J (1998) Methods for the assay of 1, 5-anhydro-D-fructose and A-1,4-glucanlyase. Carbohyd Res 305: 73-82.

28. Brauge C, Corraze G, Medale F (1993) Combined effects of dietary lipid carbohydrate ratio and environmental factors on growth and nutritional balance in rainbow trout Aquaculture Research 21: 220-221.

29. Jantrarotai W, Sitasit P, Rajchapakdee S (1994) The optimum carbohydrate to lipid ratio in hybrid Clarias catfish (*Clarias macrocephalus* × *C. glariepinus*) diets containing raw broken rice. Aquaculture 127: 61-68.

30. McGoogan B (1998) Effects of dietary protein and energy manipulations on growth and aspects of nitrogen metabolism of red drum, *Sciaenops ocellatus*. Ph.D. dissertation, Texas A&M University, College Station, TX 135.

31. Sargent J, Tocher D, Bell J (2002) The lipids. Fish Nutrition. Academic Press, Elsevier, San Diego 181-257.

32. Buhler D, Halver J (1961) Nutrition of salmonid fishes: IX. Carbohydrate requirements of chinook salmon. J Nutr 74: 307-318.

33. Ellis S, Reigh R (1991) Effects of dietary lipid and carbohydrate levels on growth and body composition of juvenile red drum, *Sciaenops ocellatus*. Aquaculture 97: 383-394.

34. Luo Z, Liu Y, Mai K, Tian L, Liu D, et al. (2004) Optimal dietary protein requirement of grouper *Epinephelus coioides* juveniles fed isoenergetic diets in floating net cages. Aquac Nutr 10: 247-252.

35. Luo Z, Liu Y, Mai K, Tian L, Liu D, et al. (2005) Effect of dietary lipid level on growth performance, feed utilization and body composition of grouper *Epinephelus coioides* juveniles fed isonitrogenous diets in floating net cages. Aquac Int 13: 257-269.

36. Tuan L, Williams K (2007) Optimum dietary protein and lipid specifications for juvenile malabar grouper (*Epinephelus malabaricus*). Aquaculture 267: 129-138.

37. Chen H, Tsai J (1994) Optimal dietary protein level for the growth of juvenile grouper, *Epinephelus malabaricus*, fed semipurified diets. Aquaculture 119: 265-271.

38. Lupatsch I, Kissil G (2005) Feed formulations based on energy and protein demands in white grouper (*Epinephelus aeneus*). Aquaculture 248: 83-95.

39. Gaylord T, Gatlin D (2000) Dietary lipid level but not L-carnitine affects growth performance of hybrid striped bass (*Morone chrysops* ♀ × *M. saxatilis* ♂). Aquaculture 190: 237-246.

40. Williams C, Ronbinson E (1998) Response of red drum to various dietary levels of menhaden oil. Aquaculture 70: 107-120.

41. Millikin M (1983) Interactive effects of dietary protein and lipid on growth and protein utilization of age-O-striped bass. Trans Am Fish Sot 122: 185-193.

42. Page J, Andrews J (1973) Interactions of dietary levels of protein and energy on channel catfish (*Ictaluruspunctatus*). J Nutr 103: 1339-1346.

43. El-Sayed A, Garling D (1988) Carbohydrate to lipid ratios in diets for *Tilapia zilli* fingerlings. Aquaculture 73: 157-163.

44. Dabrowski K (1977) Protein requirements of grass carp, *Ctenopharyngodon idella*. Aquaculture 12: 63-73.

45. Takeuchi T, Watanabe T, Ogino C (1979) Availabilty of carbohydrate and lipid as dietary energy sources for carp. Bull Jpn Sot Sci Fish 45: 977-982.

46. Shapawi R, Ebi I, Yong A, Ng WK (2014) Optimizing the growth performance of brown-marbledgrouper, *Epinephelus fuscoguttatus* (Forskal), by varying the proportion of dietary protein and lipid levels. Animal Feed Science and Technology 191: 98-105.

47. Berger A, Halver J (1987) Effect of dietary protein, lipid and carbohydrate content on the growth, feed efficiency and carcass composition of striped bass, *Morone saxatilis* (Walbaum), fingerlings. Aquacult Fish Manage 18: 345-356.

48. Rawles S, Gatlin D (1998) Carbohydrate utilization in striped bass (*Morone saxatilis*) and sunshine bass (*M. chrysops* ♀ × *M. saxatilis* ♂). Aquaculture 161, 201-212.

49. Catacutan M, Coloso R (1997) Growth of juvenile Asian seabass, *Lates calcarifer*, fed varying carbohydrate and lipid levels. Aquaculture 149: 137-144.

50. Wu X, Castillo S, Rosales M, Burns A, Mendoza M, et al. (2015) Relative use of dietary carbohydrate, non-essential amino acids, and lipids for energy by hybrid striped bass, *Morone chrysops* ♀ × *M. saxatilis* ♂. Aquaculture 435: 116-119.

51. Williams K, Irvin S, Barclay M (2004) Polka dot grouper *Cromileptes altivelis* fingerlings require high protein and moderate lipid diets for optimal growth and nutrient retention. Aquac Nutr 10: 125-134.

30

Effect of El-Sail Drain Wastewater on Nile Tilapia (*Oreochromis niloticus*) from River Nile at Aswan, Egypt

Ali SM[1], Yones EM[2], Kenawy AM[3], Ibrahim TB[3] and Abbas WT[3]*

[1]*Microbiology Department, National Institute of Oceanography and Fisheries, Aswan Research Station, Egypt*
[2]*Fish Diseases Lab, National Institute of Oceanography and Fisheries, Alexandria, Egypt*
[3]*Hydrobiology Department, Veterinary Research Division, National Research Center, Dokki, Giza, Egypt*

Abstract

This study demonstrates the impact of wastewater of El-Sail Drain on the health of *Oreochromis niloticus* collected from two sites of River Nile at Aswan Governorate. One of these sites is before (I) and the other is after (II) the disposal point of El-Sail drain. The physicochemical parameters of water (pH, electric conductivity, total dissolved solids, dissolved oxygen, biological and chemical oxygen demands, nitrite, nitrate and ammonia) were determined. Heavy metals (Cu, Pb, Cd and Ni) concentrations in water and fish tissues (gills, muscles, liver and gonads) were detected. The microbiological, parasitological and pathological conditions of fish were also investigated. Higher values of pH, EC, BOD and COD were detected in site II than from site I. In contrast to DO, nitrite, nitrate and ammonia which were lower in site II. Heavy metals concentrations in water of both sites, especially Ni, Pb and Cd exceeded the permissible limits and its abundance followed the order: Pb>Ni>Cd>Cu. Total bacterial count, total coliform, *Salmonella* sp., *Shigella* sp. and *E. coli* were detected in higher numbers in water samples from site II. Moreover, the fish caught from that site revealed higher bacterial and parasitic infection. The bioaccumulation of Ni and Pb exceeded the maximum permissible limit; however, Cu and Cd concentrations were below the permissible limit in different tissues. The bioaccumulation factor of Cu showed its highest value in liver. The histopathological lesions were more prominent in fish collected from site II. So, consuming fish caught from the studied sites around El-sail drain disposal point represents serious hazard on human health.

Keywords: El-Sail drain; Waste water; *Oreochromis niloticus*; River Nile; Aswan

Introduction

River Nile is the main source of fresh water in Egypt. It receives a lot of pollutants. Nevertheless, the river is still able to recover in virtually all the locations, with very little exceptions [1].

El Sail drain (Kima drain) is considered as one of the major sources of pollution of the River Nile at Aswan governorate. It is used for the disposal of either treated or not sewage wastewater, household solid waste, as well as industrial wastewater (Kima factory). These pollutants can elevate some water parameters, increase levels of BOD and COD in River Nile water and also increase the incidence of pathogenic bacteria, toxic organic compounds and heavy metals; El-Sail drain discharges high amount of organic matter estimated by 10.1 tons/day COD, 3.2 tons/day BOD, 0.03 tons/day heavy metals and 3.25 E+04 MPN/100 ml faecal Coliform bacteria [2]. Kima drain wastewaters exhibit high concentrations of dissolved salts, particularly close to where the waste of the Kima factory enters and decrease substantially near the end of the Kima drain [3].

Heavy metals are the most-active polluting substances; they affect the quality of the environment, with its long-term impact on living organisms. Determination of trace metals concentration in natural water system has received increasing attention for monitoring the environmental pollution, due to the fact that some metals are not biodegradable and accumulated in different organs of animals and human [4]. They can cause serious impairment to circulatory, metabolic, physiological and even structural systems when high concentrations are present in aquatic ecosystems [5].

Unfavorable environmental conditions are the main contributors to stress phenomenon that languish fish immunity and opens the pathway to pathogens and parasites [6]. Many previous studies have stated the presence of *Salmonella*, *Shigella* and *Escherchia coli* in fish harvested from water polluted with human and animal wastes. Also, representatives of family Enterobacteriaceae are usually found among the most prevalent bacteria on the fresh water fish. These microorganisms adsorbed on the surfaces of fish and may be found also in their intestinal contents [7].

Fish frequently serve as intermediate or transport host for larval parasites of many animals, including humans. Most of fish parasites are believed to cause little or no harm to their host under the natural environmental conditions. However, their mere presence often renders fish undesirable by consumers [8]. While; severe parasitic infection is becoming a threat for fish health management and production throughout the world. It cause decrease in growth rate, weight loss, spread human and animal diseases, postpone sexual maturity of fish and increase fish mortalities [9]. Moreover, the parasitic infection in fish may be detrimental to the fish industry because it lowers the quality, quantity and the economic value of fish [10].

Fish can be used as a monitoring tool for the quality of the aquatic environment and fish histopathology, with a broad range of causes, is increasingly being used as indicator of environmental stress since it provides a definitive biological end-point of historical exposure [11].

*****Corresponding author:** Wafaa T Abbas, Department of Hydrobiology, National Research Centre, El-Bohooth Street, Dokki 12622, Giza, Egypt
E-mail: wtabbas2005@yahoo.com

As well as histopathology can be used as indicators for the effects of various anthropogenic pollutants on organisms and are a reflection of the overall health of the entire population in the ecosystem [12].

Although, Aswan is fortunate in having good quality fresh water from River Nile comparing to the upward Cairo Province, but it receives domestic, agriculture and industrial wastewaters from different drains. Therefore, this work was planned to determine the effect of El-Sail drain discharge on River Nile water at Aswan and its impact on heavy metals concentrations, microbial and parasitological infection and also the histopathological alterations in different tissues of *Oreochromis niloticus* fish.

Material and Methods

The experimental area

The studied area is selected to be under the influence of El-Sail drain effluent at Aswan city during autumn 2013. Water and fish samples were collected from two sites of the River Nile which were 141 m before the disposal point of El-Sail drain (site I, N:24° 06` 54.94"; E: 32° 53` 54.74") and 245 m after the disposal point (site II, N:24° 07` 05.14"; E: 32° 53` 50.33") as shown in Figure 1. Water samples were collected from the surface water, (ca. <1 m ashore), in sterile brown bottles (200 ml capacity) for microbiological analysis, and in one liter polyethylene bottle for chemical analysis. Samples of Nile Tilapia (*Oreochromis niloticus*) were collected from each site (80 fish/site). Fish were transposed a live back after catching to the laboratory for subsequent analysis. Means of fish total lengths and total weights were 16.98 ± 1.94 cm and 80.15 ± 16.05 g respectively.

Physicochemical analysis of water

Water samples have been subjected to various analyses including pH value, electric conductivity (EC) and total dissolved solids (TDS) by using portable devices (pH meter model HI 8314 and digital conductivity meter HI2300 Hanna Ins. Romania). Dissolved oxygen (DO) was measured using the modified Winkler method and biological

Figure 1: Satellite image (A) and photos (B,C) locating El-Sail drainage effluent into River Nile at Aswan, Egypt. X: output of El-Sail drain into the River Nile.

oxygen demand (BOD) with the five-day incubation method [13]. Chemical oxygen demand (COD) was carried out using the potassium permanganate method [14]. Colorimetric methods were used to determine ammonia and nitrite [13] and nitrate [15].

Heavy metals analysis

Heavy metals (Cd, Pb, Cu and Ni) in water samples and Nile tilapia organs (gills, muscles, liver and gonads) were determined using atomic absorption spectrometry (Perkin-Elmer 3110, USA) with graphite atomizer HGA-600, after using the digestion technique by nitric acid according to the standard methods for examination of water and wastewater [16]. The bioaccumulation factor was estimated according to Authman and Abbas [17] as the following equation: Bioaccumulation factor (BAF)=(Pollutant concentration in fish organ (mg/kg)/Pollutant in water (mg/l)).

Bacteriological analysis

Ten ml of each water sample was subjected to serially dilutions (10^{-1} to 10^{-5}) with sterile physiological saline (0.85% wt/vol. NaCl) in deionized water. The pour plate technique and the nutrient agar were used for the enumeration of total bacterial counts at both 22°C and 37°C incubation temperatures [13]. For total spore-forming bacteria, water samples and its successive dilutions were pasteurized for 15 min at 80°C, prior to plating in nutrient medium and incubating at 30°C. MPN technique was used for enumerated total and faecal coliforms (using MacConky broth medium) and faecal streptococci (using Azide dextrose broth medium), then incubated at 37°C for 24-48 h [13]. Eosin Methylene Blue (EMB) was used for enumerated Escherichia coli and Violet Red Bile agar (VRB) for enumerated Enterobacterease. As well as Salmonella Shigella agar (SS agar) was used for enumerated *Salmonella* sp. and *Shigella* sp. [16].

Also the fish body surface was wiped with 70% ethanol, and parts of gills, skin, dorsal muscles, liver, gut and gonads were taken from each fish. 10 g of each organ was aseptically transferred in to 90 ml of sterilized 0.85% normal saline, homogenized and centrifuged for 2.5 min at 14000 rpm and then allowed to stand for about five min. Ten-fold serial dilutions up to 10^7 were done. Nutrient agar was used for total plate count at 22°C & 37°C, EMB for Escherichia coli, VRB for Enterobacterease spp., and SS agar for *Salmonella* sp., and *Shigella* sp. [16].

Parasitological examination

Examination of gills and branchial cavity: Branchial cavity was dissected and examined by naked eye for the presence of large cysts. The gill arches were isolated and put in a saline solution; monogenean worms and crustacean parasites were collected and permanently mounted un-stained in glycerol jelly [6].

Examination of muscles: Small snips of muscles (about one gram) were taken from fish samples, compressed between 2 large glass slides (compressorium) and examined under the binocular dissecting microscope for the presence of metacercariae. Isolated metacercariae were withdrawn by fine long tipped pipette into 0.5% saline solution, stained by acetic acid alum carmine stain, mounted and examined under the microscope [18].

Examination of gastrointestinal tract: The alimentary canal of each fish was separated, dissected and divided into small pieces, washed with physiological saline for several times to get rid of mucus and coarse particles that may be adherent to the parasites, then each part was opened and examined in a Petri dish under binocular dissecting microscope, the helminthes were collected by Pasteur pipette, stained with carmine stain, mounted and examined under the microscope [6].

Histopathological examination

After dissecting the fish, gills, skin, liver, spleen and gonads were carefully removed and small pieces were fixed in 10% formalin, dehydrated in ascending grades of alcohol and cleared in xylene. The fixed tissues were embedded in paraffin wax and sectioned at 5 microns. Sections were stained according to Harris Hematoxylin and Eosin method [19], examined microscopically and photographed by using a microscopic camera.

Statistical analysis

Data were statistically analyzed using analysis of variance [20], using the STATISTICA (6.0) computer programs.

Results

Determination of water quality

The physicochemical, heavy metal and microbiological analysis of the water samples collected from two sites in River Nile; after and before El-Sail drain were showed in Table 1.

The physicochemical analysis revealed that, pH value, EC, COD and BOD were significantly higher in site II (8.01, 259 μm/l 2.4 mg/l and 1.06 mg/l) than site I (7.82, 253 μm/l, 1.2 mg/l and 0.83 mg/l), respectively. On the other hand, the lower DO value was recorded at site II (1.92 mg/l) and also the soluble forms of nitrogen; nitrite, nitrate and ammonia showed significant reduction in site II (0.009, 0.097 and 0.043 mg/l) comparing to site I (0.012, 0.128 and 0.091 mg/l), respectively.

In respect to the heavy metal analysis, in general, the abundance of

Parameter	Site I	Site II
Microbiological analysis		
Total Bacterial Count at 22°C (cfu ml^{-1})	2.3×10^{4b}	13.9×10^{4a}
Total Bacterial Count at 37°C (cfu ml^{-1})	3.1×10^{4b}	21.9×10^{4a}
Total Spore-Forming bacteria (cfu ml^{-1})	9^a	7^a
Total Coliform (MPN/100 ml)	350^b	1600^a
Fecal Coliform (MPN/100 ml)	50^b	275^a
Fecal Streptococcus (MPN/100 ml)	110^a	110^a
Salmonella sp (cfu ml^{-1})	2^a	4^a
Shigella sp (cfu ml^{-1})	30^b	57^a
E coli (cfu ml^{-1})	16^b	35^a
Physico-chemical analysis		
pH	8.01^a	7.82^b
EC (μm^{-1})	253^b	259^a
Total dissolved solids (mg l^{-1})	162.2^b	165.7^a
DO (mg l^{-1})	2.04^a	1.92^b
BOD (mg l^{-1})	0.83^b	1.06^a
COD (mg l^{-1})	1.2^b	2.4^a
NO$_2$-N (mg l^{-1})	0.012^a	0.009^b
NO$_3$-N (mg l^{-1})	0.128^a	0.097^b
NH$_3$-N (mg l^{-1})	0.091^a	0.043^b
Heavy metals (ppm)		
Cu	0.26^b	0.41^a
Ni	1.81^a	1.9^a
Pb	2.63^a	2.59^a
Cd	1.04^a	0.13^b

Site I: before El-sail drain disposal point. Site II: after El-sail drain disposal point. Means followed by the same letter are not significantly different (p ≥ 0.05).

Table 1: Water quality of the both studied sites.

tested heavy metals in water followed the order: Pb>Ni>Cd>Cu. Site II had lower Cd concentration and higher Cu concentration comparing to site I, while, Ni and Pb had convergent concentrations in both studied sites. Ni, Pb and Cd concentrations in water of both sites were exceeded the allowable limits.

The microbiological analysis of water samples showed increase in total bacterial counts in site II than Site I, while there was no difference between the two sites in the spore-forming bacterial count. Coliform bacteria were detected in the two sites with four fold increases in site II than site I, similarly the faecal coliform increased to about five folds in site II than site I. On the other hand, there was no significant difference in fecal Streptococcus count in both sites. The FC: FS ratio was 0.5 in site I and 2.5 for site II. Pathogenic bacteria (Salmonella sp., Shigella sp., and E. coli) were detected in water samples of both sites with higher values in site II than site I.

Concentration of heavy metals in fish tissues

The bioaccumulation of different heavy metals in fish tissues were in the following order, Cu: liver>gonads>gills>muscles; Ni: gills>liver>gonads>muscles; Pb: gills>liver>muscles>gonads and Cd: gonads>muscles>gills>liver. Generally, the highest metal concentrations were recorded in liver and muscles of fish from site II whereas, site I recorded the highest values of metals in gills and gonads. Moreover, the highest bioaccumulation factor (BAF) was that of Cu in liver, followed by Ni and Pb in gills and then Cd in gonads. However, the lowest values of BAF were Cu and Ni in muscles, followed by Pb in gonads and Cd in liver (Table 2).

Microbial load for various organs of Nile tilapia

In general, the highest bacterial load of various fish organs was recorded at site II, followed the order: gut>gills>skin>muscles. The highest TBC recorded in site II (5-2570 cfu×10^4/g) compared with that of site I (1-124 cfu×10^4/g). As well as, the highest count recorded in skin and gut and the lowest recorded in gonads. Similarly, the total spore-forming bacteria (TSF) counts showed increase in site II compared to site I; Gonads and gills of fish from site II recorded the highest TSF count comparing with other organs. Enterobacteriaceae spp. ranged 1–2309 cfu×10^3/g and its number was the highest in gut organs at site II than other organs. E. coli detected in all studied organs in site II and detected only in 33% from the examined organs in site I. E. coli load in site II was >10^3, the highest count recorded in gills (3218 cfu/g) and skin (2356 cfu/g). Salmonella sp detected in 50% of tissue samples collected from site I while, it detected in all tissue samples collected from site II at a count of (17-215 cfu/g) in the following order: gills >gut>gonads>skin>liver>muscles. Shigella sp. ranged from 1-131×10^2 cfu/g, high count showed in skin of fish collected from site II after El-Sail drain disposal point (Table 3).

Parasitic infection

Different types of external and internal fish parasites were recorded in both studied sites with higher prevalence in site II than site I. The monogenean and crustaceans parasites were recorded in gills with percentages of 70% and 30% in site II and I respectively (Figures 2a and 2b). Regarding the internal parasites, Clinostomum sp. trematodes were detected in gills and branchial cavity of fish in lower infestation, 10% in both sites (Figure 2c). While, nematodes and Acanthocephala sp. were detected in the intestine of fishes with high percentage (40%) in site II and with low rate (10%) in site I (Figures 2d and 2e). Also, Diplostomum sp. encysted metacercareae were recorded in higher number in fish from

Organ	Site	Heavy metals concentrations (ppm)							
		Cu	BAF	Ni	BAF	Pb	BAF	Cd	BAF
Gills	SI	5.2	20	8.65	4.78	11.85	4.51	0.35	0.34
	SII	3.05	7.44	7.45	3.92	10.45	4.03	0.35	2.69
Muscles	SI	1.85	7.12	5.2	2.87	7.7	2.93	1	0.96
	SII	3.15	7.68	6.4	3.37	8	3.09	0.35	2.69
Liver	SI	46.1	177.31	8.4	4.64	7.2	2.74	0.45	0.43
	SII	68.05	165.98	7.7	4.05	11.3	4.36	0.05	0.38
Gonads	SI	17.4	66.92	7.7	4.25	7.25	2.76	0.8	0.77
	SII	0	0	5	2.63	6.35	2.45	0.95	7.31
Permissible level		30 mg/kg		0.4 mg/kg		2 mg/kg		2 mg/kg	

Site I: before El-sail drain disposal point. Site II: after El-sail drain disposal point. BAF: bioaccumulation factor.

Table 2: Heavy metals concentrations and bioaccumulation factor in different organs of *Oreochromis niloticus* from the both studied sites.

Organs	Sites	TBC at 37°C (cfu x10⁴/g)	SFB (cfu/g)	Enterobacteriaceae (cfu x 10³/g)	E. coli (cfu/g)	Salmonella sp. (cfu/g)	Shigella sp. (cfux10²/g)
Muscles	SI	17^d	83^c	3^b	35^f	0^e	1^e
	SII	80^{cd}	338^{bc}	4^b	419^{de}	17^{de}	5^e
Skin	SI	124^{cd}	256^c	127^b	0^f	12^{de}	46^{cd}
	SII	2570^a	758^{bc}	128^b	2356^b	43^c	131^a
Gills	SI	1^d	487^{bc}	2^b	0^f	36^{cd}	19^{de}
	SII	240^c	2159^{ab}	19^b	3218^a	215^a	91^b
Liver	SI	2^d	58^c	1^b	0^f	0^e	11^e
	SII	92^{cd}	399^{bc}	7^b	759^c	29^{de}	22^{de}
Gut	SI	32^d	139^c	162^b	0^f	115^b	61^{bc}
	SII	2258^b	263^c	2309^a	471^d	122^b	77^{bc}
Gonads	SI	1^d	190^c	3^b	193^{ef}	0^e	9^e
	SII	5^d	3401^a	4^b	794^c	96^b	18^{de}

Site I: before El-sail drain disposal point. Site II: after El-sail drain disposal point. TBC: Total bacteria counts. SFB: Spore-forming bacteria. Means followed by the same letter within the same columns are not significantly different (p ≥ 0.05).

Table 3: Microbial load in various organs of *Oreochromis niloticus* collected from both studied sites.

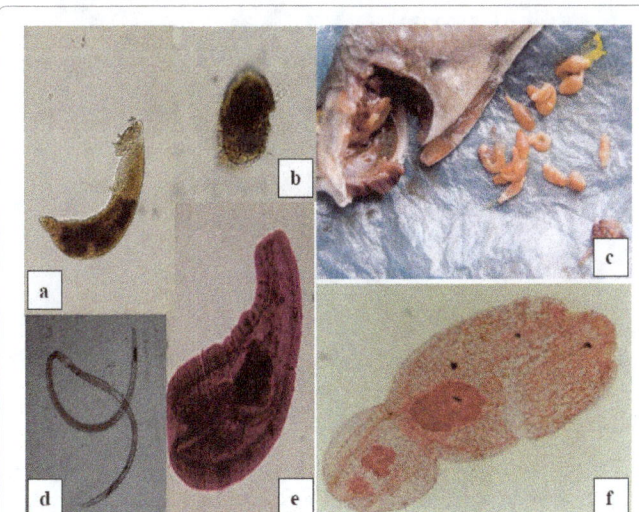

Figure 2: Parasitic infection in different fish organs. **a:** monogenea from gills, **b:** crustacean from gills, **c:** Clinostomatid excysted metacercariaefrom gills and branchial cavity, **d:** nematode larva from intestine, **e:** *Acanthothephala* sp. from intestine and **f:** Diplostomatidexcysted metacercarea from muscles.

site II (10/gm) than that in site I (3/gm) (Figure 2f).

Histological alterations

Fish collected from site II showed much greater damage in their histopathological examination compared to site I which revealed mild changes. The skin of fish collected from site I revealed slight accumulation of melanophores in the epidermal layer (Figure 3a), while severe hyperplasia, hypertrophy and excessive accumulation of melanophores were observed in the epidermal layer of fish collected from site II (Figure 4a). Gills exhibited slight congestion in the lamellar and branchial blood vessels in site I fish (Figure 3b), where gills of site II fish showed edema with epithelial lifting and telangiectasis (Figure 4b) and severe lamellar fusion and epithelial hyperplasia with external parasites were shown in between the gill tissues (Figures 4c and d). Liver in site I fish showed moderate vacuolar degeneration with mononuclear inflammatory cells infiltration in between hepatic parenchyma (Figure 3c) accompanied with slight congestion and widening of the hepatic blood (Figure 3d), on the other hand severe degenerative and necrotic changes in the hepatocytes and pancreatic tissues with aggregation of mononuclear inflammatory cells were observed in site II fish (Figure 4e). Spleen of site II fish had hyper activation of melanomacrophage cells (Figure 4f) accompanied with depletion of the lymphocytic tissues and blood vessels hyperplasia (Figure 4g). Ovary of site I fish revealed normal oocytes with different stages of maturation, stromal edema, (Figure 3e), however site II fish ovary seen in ripe stage with histological changes of oocytes as liquefaction of cytoplasm, nucleus loses and degeneration in oocyte wall with liquefaction of the yolk sphere (Figure 4h). Testis of site I fish exhibited degeneration and necrosis leading to decrease in the number of spermatogenic cells in the numerous siminiferous tubules (++) (Figure 3f), but site II fish testis showed severe degeneration in semniferous tubules, associated with decrease in the number of spermatocytes or spermatids (+++) (Figure 4i).

Discussion

River Nile is the donor of life to Egypt; it represents the principle freshwater source that meets nearly all the demands for drinking water

0.41 mg/l) were lower than the permissible levels (1 mg/l) permitted by the Egyptian Organization for Standardization [27]. On the other hand, a high Pb concentration (2.59–2.63 mg/l) which exceed the Egyptian Standards of the Environmental Laws no. 48/1982 [22] (the maximum Pb concentration in water is 0.05 mg/l), Pb can find its way to the water of the River Nile through the leaching of gasoline from the fishery boats and the tour ships travels from Aswan to Sudan. Ni concentrations of water in the two sites were also exceeded the permissible level of EOS (0.07 mg/l) [27]. The Cd concentrations in the water of tested site (0.13–1.04 mg/l) are higher than the permissible level (0.01 mg/l) recommended by the Egyptian Organization for Standardization. Generally, the increase in heavy metals concentrations at the two studied sites around the drain can be attributed to the huge quantities of sewage and industrial wastes via El-Sail drain. Since metals are regarded as serious pollutants of the aquatic environment because of their environmental persistence and tendency to be concentrated in

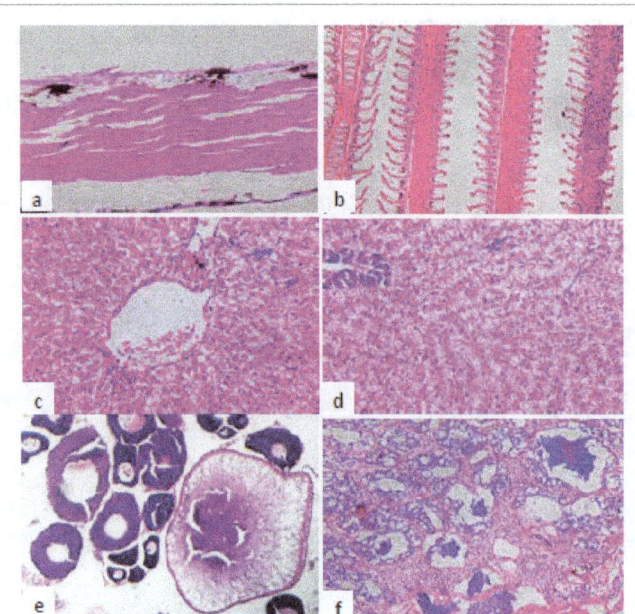

Figure 3: Histopathological changes in *Oreochromis niloticus* collected from site I showed: **a.** Skin with slight accumulation of melanophores in the epidermal layer. **b.** Gills showing congestion in the lamellar and branchial blood vessels associated with hyperplasia at the base of secondary lamellae. **c.** Liver showing moderate vacuolar degeneration with mononuclear inflammatory cells infiltration and slight congestion. **d.** Liver showing widening of hepatic blood vessels and slight congestion. **e.** Ovary showing stromal edema, degenerated and necrotic follicles and oocytes with different stages of maturation. **f.** Testis showing degeneration and necrosis leading to decrease in the number of spermatogenic cells in the numerous siminiferous tubules, and vacuolar degeneration in the germinal epithelium. (H&E, X400).

and irrigation. In recent years, the problems of pollution in water bodies have become a point of local concern. The water quality released from the High Dam in Aswan shows little degradation and it remains remarkably clean from chemical pollution until it reaches to the Delta [21]. Water quality of the River Nile in Aswan is generally meeting the water quality standards stipulated by the Egyptian governmental law 48/1982 [22]. While there are some polluted drains in Aswan which have very bad effect on the water quality; El Sail (KIMA), El Ganayen and El Berba drains. Many health problems were determined previously especially with the high bacterial counts in that water [2].

The present study indicated that, El Sail drain wastewater caused many changes in water quality of River Nile and consequently affects the fish health. Although physicochemical valuesare within the allowable values stated in the Egyptian law 48/1982 [22] but, there is more or less differences comparing the two studied sites after and before the drain disposal point. In general, El-Sail drain waste water causes oxygen content reduction in site II compared with site I, this may be due to the presence of high load of organic pollutants that consumes the dissolved oxygen during oxidation processes, this result was agree with that revealed by other researchers [23,24]. As well as, the BOD values of water from site II was lower than COD which may reflect the hardly biodegradable of discharge compounds [25]. Also, this study reported a decrease in the soluble nitrogen in water of site II comparing to site I, this may be attributed to the decrease in biological activities of aquatic organisms and nitrification in the water column due to the presence of pollution [26].

Regarding the heavy metal concentration, the present study indicated that the concentration of Cu in the water of two sites (0.26-

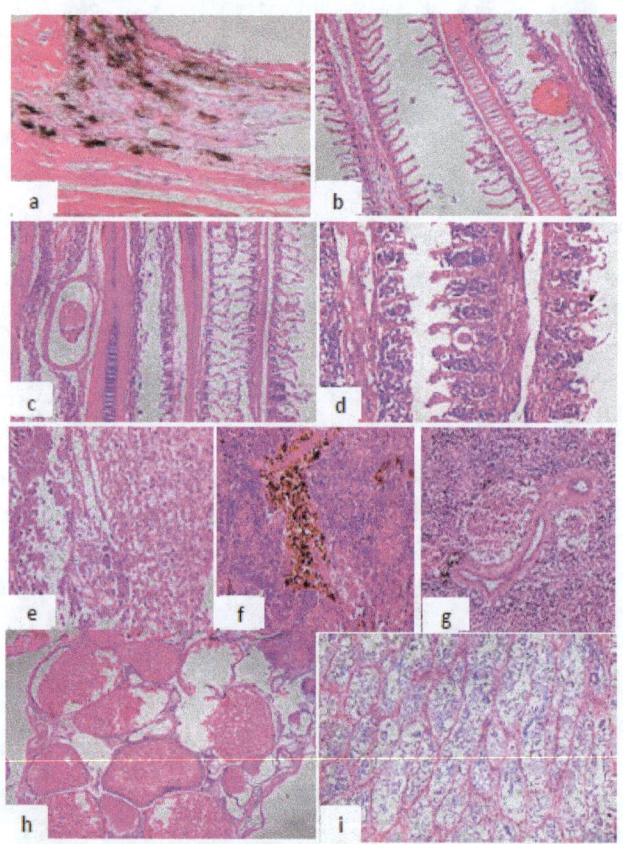

Figure 4: Histopathological changes in *Oreochromis niloticus* collected from site II showed: **a:** Skin with severe hyperplasia and hypertrophy in the epidermal layer and excessive accumulation of melanophores in the dermal layer. **b:** Gills showed edema with epithelial lifting, telangiectasia. **c&d:** Gills showed lamellar fusion, epithelial hyperplasia, proliferation of hypertrophic cells, external parasites in between the gill tissues surrounded with fibrous connective tissue sheet. **e:** Liver showed severe degenerative and necrotic changes in the hepatocytes and in the pancreatic tissues with aggregation of mononuclear inflammatory cells. **f:** Spleen showed congestion and hyperactivation of melanomacrophes cells. **g:** Spleen showed depletion of the lymphocytic tissues, congestion and hyperplasia in the wall of blood vessels. **h:** Ovary showed in ripe stage with liquefaction of cytoplasm of oocyte, nucleus loses and degeneration in wall of oocyte with liquefaction of the yolk sphere with large vacuoles of ripe stage and irregular wall of oocytes. **i:**Testis showing degeneration and necrosis in the primary spermatocytes in semniferous tubules and decreased number of spermatogenic cells. (H&E,X40).

aquatic organisms; heavy metals in fish tissues can reach concentrations up to 20000 fold higher than its concentrations in the surrounding water environment [28]. The difference in the tendency of different organs to accumulate different metals depends upon the target organ, fish species as well as the metal type [29]. The present results showed that the order of metal concentrations was Pb>Ni>Cu>Cd in the gills and muscles and Cu>Pb>Ni>Cd in the liver and gonads. The present data also showed that liver accumulated higher amounts of Cu and this may be due to its ability to retain and store Cu [30]. Similarly, some previous studies found that Cu exhibited its highest levels in the liver and the lowest values in the muscles [31,32]. The high accumulation of Cu in the liver could be attributed to the specific metabolic processes and enzyme catalyzed reaction involved Cu that taking place in the liver. The sulfur legends in the liver also have a great tendency to co-ordinate with Cu via oxygen carboxylate amino group nitrogen and/or sulfur of the mercapto group in the metalothionin protein which is in the highest concentration in the liver [33]. The concentrations of Cu in the muscles of the studied fish are still below the permissible level for Cu (30 mg/kg) recommended by the National Health and Medical Research Council [34]. On the other hand, Pb concentrations in all organs of fish are higher than US FDA maximum permissible level (2.0 mg/kg) and its accumulation in the tissues was in the following order: gills>liver>muscles>gonads. This high Pb accumulation was in agreement with its high concentration in water due to the gasoline pollution caused by boat traffic in this location. The concentrations of Cd in all fish organs are still below the WHO permissible level (2.0 mg/kg) [35].

Regarding the bacterial load in water, the present study indicated an increase in bacterial count in water from site II compared to site I. This may be attributed probably to the sewage disposal from El-Sail drain which spills its untreated waste water directly to river near this site. The ratio of TBC at 22°C to TBC at 37°C not exceeded 10 in both two sites, which explain the high pollution from El-Sail drain wastewater. The bacterial load in water increases by increasing the water temperature and the organic matter [36]. The total bacterial count of skin and gut of fish caught from site II was exceeded the permitted limit. Also, the total spore-forming bacteria, Enterobacteriacae, E. coli, Salmonella sp. and Shigella sp. was detected in all fish organs and it was the lowest in fish muscles from the two sites comparing with other organs. Microorganisms adsorbed on the surfaces of the fish and that found in their intestinal contents; They do not affect the fish during life but after death saprophytic and commensally residents invade the flesh and bring about its decomposition [37], as well as it can induce disease to humans when handling and consuming such flesh. The recommended total count of bacteria is106 per gram as a maximum permitted limit for fish, and the fish must be free from Salmonella sp. and Shigella sp. as documented by EGASQC [38]. Faecal coliform bacteria, a subgroup of the total coliform population, had a direct correlation with faecal contamination from warm-blooded animals. This results were emphasizes on a large amounts of sewage disposal directly in the River Nile without any treatments. Fish and fish products have long been considered a vehicle of food-borne bacterial and parasitic infections leading to human illness. Further research is needed to elucidate the behavior of bacterial contaminants in tilapia fish as well as in River Nile ecosystem.

Parasites are attracting increasing interest from parasite ecologists as potential bioindicators of environmental quality due to the variety of ways in which they respond to anthropogenic pollution [39]. Certain parasites can provide valuable information about the chemical state of their environment not only through their presence or absence but also

through their ability to concentrate environmental toxins within their tissues in much more concentrations than found in their host. So the present study revealed that the higher percentage of parasitic infestation in site II may be attributed to the water pollution with heavy metals. This is in accordance with other studies which indicated that heavy metal pollution affecting the prevalence of internal parasitic diseases in cultured fishes [40]. In addition to the higher bacterial infection in that site which open the way to the second parasitic infection through weaken the fish and lower its immunity [6].

Concerning the histopathological examinations, fish collected from site II revealed much higher incidence of pathological alterations. The effect of sewage in the present study can be detected using an analysis of pathologies, where the pathological alterations in fish are the net result of adverse biochemical and physiological changes within the organism. Histopathologies are clear symptoms of in situ exposure to pollution in the form of structural alterations [41]. Also such pathological changes in histological structure can substantially impair the function of tissues and organs in fish [42]. A practical advantage of using fish histopathology in environmental assessments is that multiple organs can be examined; this increases the sensitivity at which pollution impacts can be detected. Fish exposed to sewage have more frequent and severe epitheliocystis [43]. Many toxicants (e.g. hydrocarbons, organochlorines and ammonia) cause a wide range of gill pathologies that include telangiectasis and lamellar fusion, epithelial hyperplasia, hypertrophy of chloride and mucus cells, and hyperplasia, as well as higher infestation rates with ectoparasites [44]. The liver is particularly susceptible to damage from a variety of toxicants; it is the major storage site of lipids in fish, liver metabolism is a potential target for the toxic action of chemicals [45]. Fish exposed to contaminated sediments are frequently affected by liver and kidney damage [44,46]. Structural abnormalities can result in the suppression or inhibition of physiological function, irrespective of whether the pathologies are caused by chemical, physical or secondary parasitic irritation. In this study, fish exhibited multifocal pathologies in the gills, liver, kidney, spleen and skeletal muscle tissue, indicating sublethal changes and potential reduction in the functional efficiency of these organs and hence affect the fish health.

Conclusion

This study provides evidence that of water quality and fish health is poorer in the area after disposal point (site II), affected by El-sail drain wastewater which spill the untreated sewage directly to the River Nile at Aswan. Thus, great efforts and cooperation between different authorities are needed to protect the River Nile from pollution and reduce environmental risk at this area which may achieved by treatment of industrial and sewage discharge and regular evaluation of pollutants.

References

1. Wahaab RA, Badawy MI (2004) Water quality assessment of the River Nile system: an overview. Biomed Environ Sci 17: 87-100.

2. Ezzat MN, Shehab H, Hassan AA, El-Sharkawy M, El Diasty A, et al. (2002) A Survey of Nile system pollution sources.

3. Soltan ME (1995) Effect of Kima drain wastewaters on Nile River waters. Environ Int 21: 459-464.

4. Saad SMM, El-Deeb AE, Tayel SI, Al-Shehri E, Ahmed NAM (2012) Effect of heavy metals pollution on histopathological alterations in muscles of Clarias gariepinus inhabiting the Rosetta branch, River Nile, Egypt.

5. Yang JL, Chen HC (2003) Effects of gallium on common carp (Cyprinus carpio): acute test, serum biochemistry, and erythrocyte morphology. Chemosphere 53: 877-882.

6. Eissa IAM (2002) Parasitic fish diseases in Egypt. Dar El-Nahda El-Arabia Publishing, 23 Abd El-KhalakTharwat St. Cairo, Egypt.

7. Austin B, Austin DA (1987) Bacterial Fish Pathogens: disease in farmed and wild fish.

8. Ryan L, Joseph E (2000) Biology, Prevention, and Effects of Common Grubs (Digenetic Trematodes) in Freshwater fish.

9. Chandra KJ (2006) Fish parasitological studies in Bangladesh: A Review. J Agric Rural Dev 4: 9-18.

10. Kaddumukasa M, Kaddu JB, Maranga B (2006) Occurrence of nematodes in the Nile Tilapia Oreochromis niloticus (Linne) in Lake Wamala, Uganda. Uganda J Agri Sci 12: 1-6.

11. Begum G (2004) Cabofuran insecticide induced biochemical alterations in liver and muscle tissues of the fish Clarias batrachus (Linn.) and recovery response. Aquat Toxicol 66: 83-92.

12. Abd El-Aziz, Eman A, El- Habashi, Nagwan, Yones EM (2011) Histopathological alterations of Nile Tilapia (Oreochromis niloticus) induced by environmental Contamination in Lake Burullus, Egypt. Egypt. J Comp Path Clinic Path 24: 93-117.

13. APHA (American Public Health Association) (1995): Standard methods for the analysis of water and wastewater.

14. Golterman HL (1971) Methods for chemical analysis of freshwaters.

15. Mullin JB, Riley JP (1955) The spectrophotometric determination of nitrate in natural waters, with particular reference to sea-water. Ann Chim Acta 12: 464-480.

16. APHA (American Public Health Association) (1998): Standard methods for the examination of water and wastewater.

17. Authman MM, Abbas HH (2007) Accumulation and distribution of copper and zinc in both water and some vital tissues of two fish species (Tilapia zillii and Mugil cephalus) of Lake Qarun, Fayoum Province, Egypt. Pak J Biol Sci 10: 2106-2122.

18. Pritchard MH, Kruse GOW (1982) The collection and preservation of animal parasites. University of Nebraska Press, Lincoln, Nebraska, United States of America.

19. Bernet D, Schmidt H, Meier W, Burkhardt-Holm P, Wahli T (1999) Histopathology in fish: Proposal for a protocol to assess aquatic pollution. J Fish Dis 22: 25-34.

20. Freed RE, Eisensmith SP, Goetz S, Reicosky D, Smail VW, Wolberg P (1990) STAT, a microcomputer program for the design, management and analysis of agronomic research experiments, version 4.0. Michigan State University, USA.

21. Saad MAH, Goma RH (1994) Effects of the High Dam and Aswan Cataract on the Chemical Composition of the Nile Waters. I. Major Anions. Verh. Internat. Verein. Limnol. 25: 1812-1815.

22. Egyptian Governmental Law No. 48 (1982) The implementer regulations for law 48/1982 regarding the protection of the River Nile and water ways from pollution.

23. Kakuta I, Murachi S (1997) Physiological response of carp, Cyprinus carpio, exposed to raw sewage containing fish processing wastewater. Environ Toxicol Water Qual 12: 1-9.

24. Ahmed NAM (2012) Biochemical studies on pollution of the River Nile at different stations of Delta barrage (Egypt).

25. Clark BG, Micheal AU (1972) Waste water engineering in the water Resources and Environmental Engineering, McGrow-Hill New York.

26. Saad SMM, El-Deeb AE, Tayel SI, Ahmed NAM (2011) Haematological and histopathological studies on Clarias gariepinus in relation to water quality along Rossetta branch, River Nile, Egypt. Egypt J Exp Biol (Zool.) 7: 223-233.

27. EOS (Egyptian Organization for Standardization) (1993) Egyptian standard, maximum levels for heavy metal concentrations in food.

28. Altindağ A, Yiğit S (2005) Assessment of heavy metal concentrations in the food web of lake Beyşehir, Turkey. Chemosphere 60: 552-556.

29. Mohamed FA, Aboul-Ezz AS (2006) Distribution pattern of some heavy metals in tissues of some fish species from the Mediterranean Sea. Afr J Biol Sci 2: 105-119.

30. Salanki I, Katalin V, Berta E (1982) Heavy metals in animals of Lake Balaton. Water Res 16: 1147-1152.

31. Soltan M, Moalla S, Rashed M, Fawzy E (2005) Physicochemical characteristics and distribution of some metals in the ecosystem of Lake Nasser, Egypt. Toxicol Environ Chem 87: 167-197.

32. Fernandes C, Fontainhas-Fernandes A, Cabral D, Salgado M (2008) Heavy metals in water, sediment and tissues of Liza saliens from Esmoriz-Paramos lagoon, Portugal. Environ. Monit Assess 136: 267-275.

33. Abdel-Baky TE (2001) Heavy metals concentrations in the catfish Clarias gariepinus (Burchell, 1822) from River Nile, El-Salam Canal and Lake Manzala and their impacts on cortisol and thyroid hormones. Egypt J Aquat Biol Fish 5: 79-98.

34. Marks PJ, Plaskett D, Potter I, Bradly J (1980) Relationship between concentration of heavy metals in muscle tissues and body weight of fish from the Swan-Avon estuary, Western Australia. Aust J Marine Freshwater Res 31: 783-793.

35. FAO (1992) Committee for inland fisheries of Africa: Working Party on Pollution and Fisheries.

36. Al-Harbi AH (2003) Faecal coliforms in pond water, sediments and hybrid tilapia Oreochromis niloticus and Oreochromis aureus in Saudi Arabia. Aquacult Res 34: 517-524.

37. Ayres JC, Mundt JO, Sandine WE (1980) Microbiology of Foods. W.H. Freeman and Company, San Francisco.

38. EGASQC (Egyptian General Authority of Standardization and Quality Control) (2000): Cairo, Egypt

39. Ogut H, Palm HW (2005) Seasonal dynamics of Trichodina spp. on whiting (Merlangius merlangus) in relation to organic pollution on the eastern Black Sea coast of Turkey. Parasitology Res 96: 149-153.

40. Overstreet RM (1993) Parasitic diseases of fishes and their relationship with toxicants and other environmental factors.

41. Hinton DE, Lauren DL (1990) Liver ultrastructural alterations accompanying chronic toxicity in fishes: potential biomarkers of exposure.

42. Couch JA, Fournie JWE (1993) Advances in Fisheries Science.

43. Nowak BF, LaPatra SE (2006) Epitheliocystis in fish. J Fish Dis 29: 573-588.

44. Mondon JA, Duda S, Nowak BF (2001) Histological, growth and 7-ethoxyresorufin O-deethylase (EROD) activity responses of greenback flounder Rhombosolea tapirina to contaminated marine sediment and diet. Aquatic Toxicol 54: 231-247.

45. Hinton DE, Segner H, Braunbeck T (2001) Toxic responses of the liver.

46. Hansen JA, Lipton J, Welsh PG, Calcela D, MacConnell B (2004) Reduced growth of rainbow trout (Oncorhynchus mykiss) fed a live invertebrate diet pre-exposed to metal-contaminated sediments. Environ Toxicol Chem 23: 1902-1911.

31

Impact of Formulated Diets on the Growth and Survival of Ornamental Fish *Pterophyllum Scalare* (Angel Fish)

A. Hyder Ali, A. Jawahar Ali, M. Saiyad Musthafa[*], M.S. Arun Kumar, Mohamed Saquib Naveed, Mehrajuddin War and K. Altaff

Department of Zoology, The New College, Chennai, India

[*]**Corresponding author:** M. Saiyad Musthafa, P.G & Research Department of Zoology, The New College, Chennai 6000 14. India
E-mail: saiyad_musthafa@rediffmail.com

Abstract

A feeding trail was conducted on juvenile of angel fish *Pterophyllum scalare* to investigate the effect of three different diets such as animal based protein, plant based protein and mixed protein on growth and survival rate of the fish. Juvenile *Pterophyllum scalare* were divided into three groups, fed with three different protein based diets along with control group. Before the feeding trail, the initial length and weight were measured. During the 4 weeks of experiment, fish were fed 3% body weight at a daily rate. The findings of the present study indicated that the growth and survival rate of angel fish *Pterophyllum scalare* varied significantly ($P<0.05$) with different protein based diets compared with control fish group. Fish fed with animal based protein diet showed better growth performance in terms of length and weight (19.3 ± 0.72 mm and 0.13 ± 0.01 mg) respectively and better survival rate (92%) than those fed with other protein diets ($P<0.05$). The fish fed with mixed protein diet showed higher specific growth rate (0.43%) than other diets provided. This study clearly demonstrates that animal based protein diet can be used as formulated feed for angel fish without any adverse effect on fish growth.

Keywords: *Pterophyllum scalare*; Animal based protein; Plant based protein; Growth performance

Introduction

The fresh and marine water ornamental fish production and trade is a profitable alternative in the aquaculture sector [1]. Apart from the economic importance, the nutritional strategy for ornamental fish is scanty and often few or even no data of the nutritional requirements is available [1,2]. According to Lovell [3], fish can regulate and maintain their food intake in natural conditions and therefore their nutritional requirements, reducing the possibility of suffering nutritional deficiencies; however, this problem can be observed when the fish are subject to confinement conditions. Most of the information is not specific to ornamental fish because it has been based on results from farm fish kept under different farming conditions, nutritional requirements and feeding habits. Therefore, the limited information about nutrient digestibility in ornamental fish increases the maintenance costs and the water pollution (Sales and Janssens). Nutrition is one of most important factors influencing the ability of cultured fish to exhibit its genetic potential for growth and reproduction. They are also greatly influenced by factors such as behavior of fish, quality of feed, daily ratio size, feed intake or water temperature. In contrast to the culture of edible fish, information on the dietary requirements and feeding practices of ornamental fishes are limited [4-7]. Angel fish (*Pterophyllum scalare*) a cichlid is in great demand due to its elegance, reproductive capacity and adaptability to captivity with high economic value [8]. In the wild the angelfish is found in the central Amazon River of Brazil and tributes into Peru, Colombia, Guyana, French Guiana and eastern Ecuador. They inhibit swamps or flooded grounds where the aquatic vegetations are dense and the water is either clear or silty. The aim of this study was to evaluate the impact of formulated feeds (animal based protein, plant based protein and mixed protein) on growth and survival of angel fish.

Materials and Methods

Experimental fishes

Juvenile of *Pterophyllum scalare* (Angel fish) used in this experiment were procured from a commercial ornamental fish farm, Kulathoor, Chennai. The animals were acclimatized for 1 week to the experimental conditions and diets. Before the experiment, the fishes were starved for 24h. Thirty juvenile (for each experimental group) angel fish were randomly selected with an average weight 0.05 ± 0.01 mg and length 0.99 ± 0.08 mm and stocked into each aquarium fibre tanks ($80 \times 30 \times 40$ cm) with three replicates and gentle aeration was provided by air stones. During the experiment, the water quality parameters were maintained and mean values for temperature, dissolved oxygen, pH and salinity were $27 \pm 2°C$, pH 6.5–7.8, and 0.2 mg/l, respectively. The photoperiod used was 12h light/12h dark cycle. The length and weight of the each fish was taken after completion of the experiments.

Cleaning and siphoning

To maintain hygienic condition and prevent pollution caused by remaining food and faeces, the aquarium were cleaned every day prior to feeding time in morning by siphoning out the excreta and 80% of the water was exchanged to prevent sudden increase in water temperature as the experiment was conducted in summer months. The dead fish, if any, were removed and recorded for calculating the survival rate.

Experimental diet and feeding

Three types of formulated diets were prepared (animal based protein, plant based protein and mixed protein) in the feed mill and were evaluated against control diet (Kayal plus). The composition and proximate analysis of the diets are given in Tables 1 and 2. Proximate compositions of feeds were analyzed following the standard methods as described in AOAC [9]. The diets were offered twice a day (morning and evening) for 30 days. All groups of fish were fed daily at 3% BW in two instalments at 08:00 and 16:30 hours (30 days for all the experiments). Fishes were exposed to the diet for 3h during each ration thereafter the uneaten feed was siphoned out, stored and weighed for calculating feed conversion ratio (FCR). Fish were bulk weighed at an interval of 15 days with feeding ration adjusted accordingly. Uneaten feed was siphoned out and stored separately for calculating FCR.

S.No.	Ingredients	Animal Protein Based Feed	Plant Protein Based Feed	Mixed (plant + animal) protein Based Feed
1	Fish Meal	400	-	200
2	Soya Meal	-	400	200
3	Wheat Floor	400	400	400
4	Corn Floor	100	100	100
5	Starch	150	150	150
6	Fish Oil	40	40	40
7	Vitamins	5	5	5
8	Minerals	5	5	5

Table 1: Composition of Animal protein and plant protein based formulated feed (gm/kg)

Parameters	Animal Protein Based Feed	Plant Protein Based Feed	High Protein Based Feed	Low Protein Based Feed
Moisture	7.6	8.2	10.3	9.9
Protein	38.9	24.5	37.1	24.6
Lipid	15.7	11.9	6.1	4.2
Carbohydrate	11.1	8.2	12.5	14.9

Table 2: Proximate Composition of different formulated feed (%)

Growth performance

The growth performance, specific growth rate (SGR), feed conversion ratio (FCR) and protein efficiency ratio (PER) for each group was determined by Olmedosanchez et al. [10]. Specific growth rate (SGR)=(Ln Final weight - Ln Initial weight)÷No of days in trial×100 Feed conversion ratio (FCR)=Feed given (dry wt)÷Weight gain (wet weight) Protein efficiency ratio (PER) = wet weight gain by fish (g)÷Protein intake (g) Statistical Analysis Data are presented as mean ± SD. Statistical significance of data were analyzed following one-way ANOVA and Duncan's multiple range tests were used to compare differences among individual means [11]. All the results were treated significant at the 5% level. The statistical package used for the analysis of data was SPSS 16 ver.

Results and Discussion

Fish growth is a complex process governed by many parameters like fish species, nutrients present in the feed, feed additives and rearing environment, individually or in combination. The growth parameters like percentage of weight gain, food conversion ratio (FCR), feed efficiency ratio (FER) and specific growth rate (SGR) were recorded.

In the present study, selected protein based experimental diets were eagerly consumed by three fish groups. Overall growth performances of angel fish fed with different experimental diets are shown in Table 3. One way ANOVA showed significant differences in length and weight of the angel fishes when fed with different formulated diets at $P<0.05$ level. Juvenile total length (19.3 ± 0.72 mm) and weight gain (0.13 ± 0.01 mg) were significantly high in animal based protein diet followed by mixed protein diet (19.2 ± 0.77 mm and 0.13 ± 0.01 mg), Plant based protein diet (12.4 ± 1.45 mm and 0.02 ± 0.01 mg) and control (19.2 ± 0.68 mm and 0.13 ± 0.01 mg). The fish fed with mixed protein diet showed higher specific growth rate (0.43%) than other diets. Lower specific growth rate (0.1%) was observed in plant based diet (Table 3).

S. No	Experimental diets	Total Length (mm)	Weight gain (mg)	SGR
1	Control	19.2 ± 0.68[ab]	0.13 ± 0.01 mg[ab]	0.43
2	Animal based protein diet	19.3 ± 0.72[ab]	0.13 ± 0.01 mg[ab]	0.36
3	Plant based protein diet	12.4 ± 0.77[a]	0.02 ± 0.01[a]	0.1
4	Mixed protein diet	19.2 ± 0.77[ab]	0.13 ± 0.01[ab]	0.43

Table 3: Growth performance of angel fish (*Pterophyllum scalare*) fed in experimental diets. Each value is the mean ± SD of three observations.

Groups with different alphabetic superscripts differ significantly at $P<0.05$.

Angel fish fed on different formulated diets viz., animal based protein diet, plant based protein diet and mixed protein diet showed significantly high survival rate 92%, 80% and 91%, respectively, and low survival rate (76%) was noticed in control diet during 30 days of experimental period. The survival rate of angelfish significant differs with different experimental diets (Figure 1).

In the present study, the inadequate micro and macro nutrients of the plant based protein diet fed fish group exhibit minimum length and weigh; therefore, plant protein based ingredients is not desirable to the juveniles of angel fish. The level of protein content, lipids as well as essential amino acids and fatty acids might not be at sufficient level in the plant protein based diet. Sajad Hassan [12], reported that, some of the anti-nutritional factors are present in the plant protein based diet and also not easily digestible by the monogastric animals like fishes.

The protein efficiency ratio was moderately higher in 34% CP (Crude Protein) containing diets fed fish groups whereas gradual decreases of PER was noted in 30 and 26% CP comprised diets fed fish groups meanwhile there are no significant differences observed for the productive performance parameters among protein and energy levels Zuanon et al. [6]. Similarly, Ribeiro et al. [4] reported that protein efficiency ratio was not showing significant difference on freshwater angelfish fry fed diets with 26, 28, 30 and 32% CP, this may be associated with decreases in the amounts of nitrogen wasted by fish into the water [13], which are especially important for ornamental fish, which are frequently raised in small tanks.

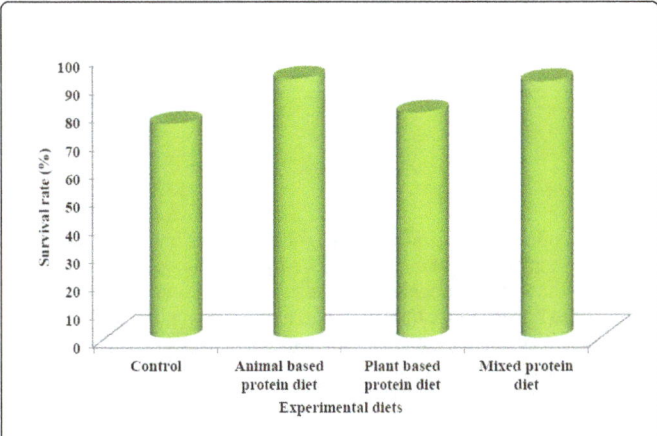

Figure 1: Survival rate (%) of angel fish fed on experimental diets

The weight gain values obtained in the present study showed variation compared to the observation of Rodrigues & Fernandes [5] and Luna-Figueroa [14] for the same species. This variation in weight gain for the same species may be related to factors including quality of the protein used (amino acid composition and digestibility), feeding frequency, amount of diet supplied and the animal development stage. The low values observed for the specific growth rate (SGR) of fish from the present study can be explained by the fish development phase, since the SGR decreases as fish increases in size [15]. Higher SGR values were reported by Luna-Figueroa [2], Rodrigues & Fernandes [5] and Zuanon et al. [6] (4.34, 2.04 and 2.47% per day, respectively), for fish of the same species, but smaller sizes than the fish from the present study.

Leger et al. [16] explained clearly in his review that, selection of fish diet ingredients and the ingestion of food by fish is probably affected by its size and palatability. Size is thereby considered as one of the most important aspects. As food preference is closely related to the match between food and mouth size, the selected food changes with the growth of fish [17]. Ulloa and Romero [8], while evaluating the different commercial diets on growth and survival of *P. scalare* juveniles under controlled conditions reported that the differences in size and presentation, all tested diets were consumed by fish at the different sampling times registered along the experiment, and consequently, at different growing stages (from 1.65 ± 0.21 to 3.12 ± 0.34 mm of mean initial and final standard lengths, respectively). In the case of the latest, the relationship between digestibility and biochemical composition of diet is considered as a crucial selecting criterion. Due to its high affinity to be metabolized and retained, protein is considered the most important energy component for fish growth by Halver [18]. In the case of cultured carnivorous fish, dietary protein requirement usually accounts for 40 to 50% of feed dry matter [19]. The findings of present feeding experiment of angelfish suggest that animal based protein can be used as formulated feed for effective rearing of angel fish. The present study also suggests that animal based protein along with plant based protein can be used without any adverse effect on fish growth and survival rate.

Acknowledgement

The authors are thankful to the Principal, Management and HOD, P.G & Research Department of Zoology, The New College, Chennai-14 for providing necessary facilities.

References

1. Chong A, Hashim R, Ali A (2003) Assessment of soybean meal in diets for discus (Symphysodon aequifasciata HECKEL) farming through a fishmeal replacement study. Aquacult Res 34: 913-922.

2. Blom JH, Dabrowski K (2000) Vitamin C requirements of the Angelfish Pterophylum scalare. J World Aquacult Soc 31: 115-118.

3. Lovell RT (2000) Nutrition of ornamental fish. Kirk's Current Veterinary Therapy XIII-Small Animal Practice. W.B. Saunders, Philadelphia, USA.

4. Ribeiro FAS, Rodrigues LA, Fernandes JBK (2007) Desempenho de juvenis de acarábandeira (Pterophyllum scalare) com diferentes níveis de proteína bruta na dieta. Boletim do Instituto de Pesca 195-203.

5. Rodrigues LA, Fernandes JBK (2006) Influence of the feed processing on production performance of Angelfish (Pterophyllum scalare). Acta Scientiarum Animal Science 113-119.

6. Zuanon JAS, Salaro AN, Moraes SSS, de Oliveira ALM, Balbino EM, et al. (2009) Dietary protein and energy requirements of juvenile freshwater angel fish. R Bras Zootec 38: 989-993.

7. Jaleel MA, Musthafa MS, Ali AJ, Mohamed MJ, Arun Kumar MS, et al. (2015) Studies on the growth performance and immune response of Koi carp fingerlings (Cyprinus carpio Koi) fed with Azomite supplemented diet. Journal of Biology and Nature 4: 160-169.

8. García-Ulloa M, Gómez-Romero HJ (2005) Growth of angel fish *Pterophyllum scalare* juveniles fed inert diets. Avances en Investigación Agropecuaria 9: 49-60.

9. AOAC (1995) Official methods of analysis of the association and official analytical chemists. (16thedn), AOAC International, Arlington, USA.

10. Olmedosanchez JA, Flores AC, Orozco JR (2009) The effect of an herbal growth promoter feed additive on shrimp performance. Res J Biol Sci 4: 1022-1024.

11. Zar JH (1984) Biostatistical analysis. Prentice Hall Englewood Chiffs, NJ.

12. Hassan S (2009) Ontogenic development of Asian Sea bass Lates calcarifer (Bloch) and feeding trials using microbial phytase in the diet of Milkfish Chanos chanos (Forskal). Thesis University of Madras 1-125.

13. Ruohonen K, Koskela J, Vielma J, Kettunen J (2003) Optimal diet composition for European whitefish (Coregonus lavaretus): analysis of growth and nutrient utilization in mixture model trials. Aquaculture 225: 27-39.

14. Luna-Figueroa J (2003) Pterophyllum scalare (Pisces: Cichlidae): Influencia de alimento vivo en la reproduccion y el crecimiento. 11 Congreso Iberoamericano Virtual de Acuicultura, CNA 2003.

15. Sunde LM, Imsland AK, Folkvord A (1998) Effects of size grading on growth and survival of juvenile turbot at two temperatures. Aquaculture International 19-32.

16. Leger PH, Bengston DA, Sorgeloos P, Simpson KL, Beck AD (1987) The nutritional value of Artemia: a review. Artemia research and its applications. Universal Press, Wetteren, Belgium.

17. Verreth J (1994) Nutrition and related ontogenetic aspects in larvae of the African catfish Clarias gariepinus. D.Sc thesis, Department of Fish Culture and Fisheries, Wageningen Agricultural University, Wageningen, The Netherlands.

18. Halver JE (1972) Fish Nutrition. Academic Press, Inc. London, UK.

19. NRC (1993) Nutrient Requirements of Fish. National Research Council, National Academy Press, Washington, D.C.

32

Effect of Feeding Frequency on Growth Performance and Survival of Nile Tilapia (*Oreochromis niloticus* L. 1758) in a Cage Culture System in Lake Hora-Arsedi, Ethiopia

Tewodros Abate Alemayehu[1]* and Ababe Getahun[2]

[1]*Debre Berhan University, Department of Biology, Debre Berhan, Ethiopia*
[2]*Department of Zoological science, Faculty of Life Science, Addis Ababa University, Addis Ababa, Ethiopia*

Abstract

In this study, the growth performance and survival rate of Nile tilapia (*Oreochromis niloticus*) subjected to different feeding frequencies were evaluated in cage culture. Juveniles with mean initial weight of 35.99 ± 0.23g were stocked in 1 m³ net cages and assigned to a duplicate of 50 fish in a completely randomized design in six treatments. T1 were fed 3% of their body weight divided into four equal meals per day for the first three months and then allowed to feed two times a day for the next three months; T2 and T3 were fed 3% of their body weight divided equally at frequency of four and two feedings/day, respectively, throughout the experiment. Feed was given once a day (without dividing) for T4 and once every other day (without dividing) for T5 throughout the experiment. All treatments were fed pelleted diet except the control groups in which fish were provided with only the natural food. The mean specific growth rates (SGR), Feed conversion ratio (FCR) and Feed conversion efficiency (FCE) were statistically similar for T1 and T2, but they were higher than T3, T4 and T5. However, mean weight gain, mean daily gain and Condition Factor (CF) showed a significant difference ($P<0.05$) among experimental groups. In conclusion, growth performance and net yield were increased with increased feeding frequency, so frequent feeding was recommended for optimum result of *O. niloticus* in cage culture. It was also revealed that cage culture at experimental level has no effect on the water quality and plankton abundance.

Keywords: Cage culture; Feeding frequency; Growth performance; *Oreochromis niloticus*; Survival

Introduction

Diet supplementation and selection of appropriate species for culturing environment are important criteria in aquaculture. Nile tilapia (*Oreochromis niloticus*) is a widely-cultured species all over the world [1-3], as it is easily spawned, tolerance to handling, and resistance to disease, efficient conversion of natural and prepared feeds, controllable reproduction, good marketability, tolerates poor water quality and grows rapidly at warm temperature [3-5]. Since the feed cost accounts approximately 50% of the operating costs in intensive culture systems [4,5], the economic viability of the culture operation depends on appropriate use of feed [6].

Nutrition is one of the most important factors influencing performance of cultured fish and is influenced by factors such as behaviour of fish, stocking density, quality of feed, daily ration size, feeding frequency and water temperature. Feeding frequency mainly depends on species cultured, age, size, feed quality and environmental factors [5,7-11]. These characters of species and environmental factors influence gastric evacuation time (return of appetite) of cultured organisms and gastric evacuation time of cultured organism on the other hand influences feeding frequency [12]. De Silva and Anderson [7], Tran et al. [13] and Malcolm et al. [14] reported that determining optimum ration size and feeding frequency is an important step in aquaculture operation since they are important to ensure maximal Feed Conversion Ratio (FCR) of cultured organism.

Several researches were carried out on effects of feeding frequency on growth of different fish species at different life stages, environmental conditions and culture conditions; but optimal feeding frequency is highly variable from species to species [15-22]. On the other hand, there is little information about optimum feeding frequency of farmed *O. niloticus*. Therefore, the aim of this study was to investigate effect of feeding frequency on growth performance and survival of *O. niloticus*.

Materials and Methods

Study site and experimental design

The study was conducted in Lake Hora-Arsedi, one of the Crater Lakes in Ethiopia and located at altitude of 1850 m asl and 8° 46" N and, 38° 59" E, is 45 km southeast of Addis Ababa (the capital of Ethiopia). The jetty was constructed from wood (eucalyptus) (at site 1) which is perpendicular to the water current [4,23-25]. It has an average depth between 6-7 m, 25 m length and 1m width, which is 7 meter away from the shore. The control site (site 2) was selected to sample plankton and for measuring physical parameters and compare that with the experimental site. Cages with the size of 1 m³ (1m × 1m × 1m) were constructed from frame (PVC type 50, tube of 10 cm with 1 mm polyethylene material) and the enclosure nylon netting material with mesh size of 4 mm as an enclosure material. The cages were placed side by side in rows under the jetty with equal interval (0.75 meter apart).

Mixed sex juveniles of *O. niloticus* were collected from Lake Hora-Arsedi using beach seine hauls 50 m × 2.5 m (with stretching mesh size of 20 mm). Immediately after screening, the fingerlings were transported to experimental cages by plastic barrel half-filled with lake water. The total length (TL. by measuring board) and total weight (TW. digital balance) were measured and fingerlings with length 115

---

*Corresponding author: Tewodros Abate Alemayehu, Department of Biology, Debre Berhan University, Debre Berhan, P. O. Box 445, Ethiopia
E-mail: ttabate@gmail.com

mm to 138 mm and weight of 30-40 gm were selected as experimental juveniles. Equal numbers of mixed sex juveniles (50 fingerlings) were stocked in duplicates of six treatments (T1, T2, T3, T4, T5 and T6) in a completely randomized manner.

Supplementary feed and feeding frequencies

Experimental diet was formulated to contain 30% crude protein which was optimum for *O. niloticus* as suggested by El-Sayed [5], and was prepared from locally available materials; Niger seed (*Guizotia abyssinica*) cake (20%), mill sweeping (16%), meat and bone meal (28%), wheat bran (32%) and wheat flour (4%). Water stability of the pellet was tested in fishery lab of Addis Ababa University. The average analyzed proximate nutritional compositions of food types are listed in Table 1.

The first four treatments (T1, T2, T3 and T4) were fed extruded feed (sinking pellet) 3% of their body weight daily, but T5 were fed 3% of their body weight every other day. Control groups (T6) were fed directly from the natural environment only. The feeding frequency and timing in different treatments over the experimental period is shown in Table 2. All treatments had free access to natural foods. Feed ration was placed in feeding trays, which was suspended at the midpoint in each cage. The amount of feed was adjusted every two weeks according to the new mean fish weight in each treatment.

Data collection

Water temperature was measured with thermometer *in-situ* monthly at 25 cm below the surface of water at experimental and control sites. Concentration of dissolved oxygen (DO) and pH of

the water were measured *in-situ* using oxygen meter and pH meter respectively monthly from January 16 to July 14, 2012. Euphotic depth of the lake was also estimated using Secchi disc at two sites to evaluate the effect of cage culture on water quality.

Weight and length of stocks were recorded starting from January 16, 2012 for each cage and dead fish were removed and recorded. During sampling, 30% of the stocked fish in each cage were scooped from each treatment randomly by use of scoop net every two weeks till July 14, 2012. The fish length and weight were measured using measuring board and digital balance respectively, and recorded. At the end of the experiment, the fish were harvested, counted; the weight and length of all the fish were measured.

Data analysis

Growth performances and feed utilizations were calculated in terms of Weight Gain, Daily Growth Rate, Specific Growth Rate, Feed Conversion Ratio, Food Conversion Efficiency, Survival Rate and Net yield based on the following relationships:

$$Weight\ Gain\ = Final\ weight\ (W_2)\ - Initial\ Weight\ (W_1)$$

$$Daily\ Growth\ Rate\ (DGR,\ g\ /\ day)\ = \frac{Final\ Weight - Initial\ Weight}{Cultured\ days}$$

$$Specific\ Growth\ Rate\ (SGR,\ g\ \%/day) = \frac{(In(W2) - In(W1))}{-No\ of\ Cultured\ days} \times 100$$

where W_1 and W_2 are initial and final weight (g) respectively

$$Survival\ rate\ (SR,\ \%) = \frac{N2}{N1} \times 100$$

where N_2 = No. of fish harvested and N_1 = No. of fish stocked

$$Food\ Conversion\ Ratio\ (FCR) = \frac{Total\ weight\ of\ dry\ feed\ given}{Total\ weight\ gain\ by\ fish}$$

$$Food\ Conversion\ Ratio\ (FCR) = \frac{Total\ weight\ of\ dry\ feed\ given}{Total\ weight\ gain\ by\ fish}$$

$$Food\ Conversion\ Efficiency\ (FCE, \%) = \frac{Gain\ in\ wet\ weight\ in\ fish}{Feed\ fed} \times 100$$

The well-being of fish was studied by calculating the Fulton Condition Factor (FCF);

$$FCF\ (\%\ in\ gm\ /\ cm^3) = \frac{TW}{TL^3} \times 100$$

where TW is total weight (gm) and TL = total length (cm)

The significance of relationships of growth performance data were statistically tested using one-way ANOVA by SPSS statistics software version 20. Moreover, one-way ANOVA was used to check the variation in physical parameters, zooplankton and phytoplankton abundance between site 1 and site 2. All statistical tests were considered significant at $p < 0.05$.

Results

Physical features of lake hora-arsedi

During the study period maximum water temperature (25.8°C) occurred in April in site two, while the minimum (22.6°C) was measured in February in sites 1. The DO measured at the two sites varied from 5.16 to 8.05 at site1 to 5.59 to 8.07 at site 2. The pH of the water ranged from 8.4 to 8.8 at site 1 and 8.3 to 8.8 at site 2. Secchi depth ranged from 91.5 to 110 cm at Site 1 and 91 to 111.5 cm at Site

Feed types	Nutrient Compositions by %					References
	Moisture	Crude Protein	Crude fat	Crude fiber	Ash	
Wheat flour	8.60	19.29	2.10	6.25	0.80	Tekeba Eshetu in 2005 [10]
Wheat bran	11.00	18.00	4.80	11.00	7.00	Stanton and Levallecy in 2010 [35]
Meat and bone meal	10.83	52.48	11.36	4.18	20.86	Asfaw Alemayehu in 2010 [21]
Niger seed cake	13.95	33.80	9.10	19.00	11.00	Tadelle Dessie and Ogle in 1997 [9]

Table 1: Feed types used and their nutrient proximate composition.

Treatments	Feeding frequency	Timing
T₁	A restricted daily ration divided into four equal meals and given four times a day for the first three months	8:00 a.m., 11:00 a.m., 2:00 p.m., 5:00 p.m.
	A restricted daily ration divided into two equal meals and given two times a day for the second three months	8:00 a.m., 5:00 p.m.
T₂	A restricted daily ration divided into four equal meals and given four times a day throughout the experimental time	8:00 a.m., 11:00 a.m., 2:00 p.m., 5:00 p.m.
T₃	A restricted daily ration divided into two equal meals and given two times a day throughout the experimental time	8:00 a.m., 5:00 p.m.
T₄	A restricted daily ration given once a day (without dividing) throughout the experimental time	8:00 a.m.
T₅	A restricted daily ration given once every other day (without dividing) throughout the experimental time	8:00 a.m.
T₆	Controls (Fed directly from the natural environment only)	-

Table 2: Feeding frequency and timing in different treatments over the experimental period.

2 (Table 3). All physical parameters were not significantly affected by culture conditions (p>0.05) but were affected by experimental dates alone (p<0.05).

Growth performances

The highest mean weight (205.17 g) was recorded at T1 and 201.07 g at T2 followed by 172.39 and 168.79 g at T3 and T4, respectively. The lowest mean weight was recorded at feeding frequency of T5 but all were better than T6 (non-feeding group) (Table 4). The maximum mean (0.947 gday^{-1}) and minimum mean (0.557 gday^{-1}) daily growth rates (DGR) were observed in T1 and T5 among feeding treatments, respectively. The least DGR of (0.302 gday^{-1}) were recorded for control group (non-feeding group). Daily growth rate decreased with decreasing feeding frequency and without supplementary feeding (Figure 1). Maximum mean specific growth rate (SGR) of 0.979% day^{-1}fish^{-1} and 0.953% day^{-1}fish^{-1} were recorded for treatment T1 and T2, respectively; whereas the minimum mean SGR was recorded as 0.743% day^{-1}fish^{-1} in treatment T5. The mean specific growth rate also decreased with decreasing of feeding frequency (Table 4).

The best mean FCR, FCE, net yield and annual total net production was recorded at T1 and T2 whereas the lowest was recorded at T5. There was no significant difference in food conversion ratio in the T1 and T2 before twelve weeks, while the least FCR was observed in T5 until week 12. Based on this, FCR for T1 and T2 differed significantly (p<0.05) from T5 and was better but not significant than T3 and T4 until week 12. However, mean FCR was not significantly different (p >0.05) among treatments at the end of the experiment (Table 4). The Fulton condition factor decreased as feeding frequency decreased and the lowest Fulton condition factor was observed in T5 from feeding treatments and in control group T6 (Figure 2). There was no significant difference observed in T1 and T2 during the experimental period (p >0.05), but they were superior and varied significantly from T3, T4, T5 and T6 (p<0.05). There was also no significant difference observed between T3 and T4, however they differed significantly from T5 and T6 (p<0.05). Survival rate was not significantly affected by feeding frequency and experimental period among feeding treatments (p> 0.05) but it showed significant variation from control group (p<0.05).

Discussion

It was revealed from data during the study period that water quality parameters were within the range that provides good growth for *O. niloticus* in cage culture and which were recommended by Stickney [3] and El-Sayed [5]. There was no significant difference observed in water temperature, pH, dissolved oxygen and Secchi depth during the study period at the two sites (p>0.05). These showed that cage culture has no effect on water quality at experimental level. It was also revealed from monthly data that there was insignificant variation (p>0.05) in the abundance of planktons between two sites. This is due to small amount

Sampling dates	Water temperature (°C)	pH	Dissolved oxygen (mg/l)	Secchi depth (cm)	Euphotic depth (cm)	Stations
16, Jan 2012	22.7	8.6	7.60	101.5	304.5	Site 1
	22.7	8.5	7.20	102.0	306.0	Site 2
15, Feb 2012	22.6	8.8	8.05	94.0	282.0	Site 1
	23.0	8.5	8.07	94.5	283.5	Site 2
16, Mar 2012	23.0	8.8	5.38	91.5	274.5	Site 1
	23.4	8.8	5.65	91.0	273.0	Site 2
15, Apr 2012	25.5	8.4	6.50	92.5	277.5	Site 1
	25.8	8.3	6.72	94.5	283.5	Site 2
15, May 2012	25.1	8.6	5.16	99.0	297.0	Site 1
	25.2	8.6	5.59	98.5	295.5	Site 2
14, Jun 2012	24.1	8.6	6.30	104.5	313.5	Site 1
	24.5	8.8	6.35	106.0	318.0	Site 2
14, Jul 2012	24.0	8.5	6.76	110.0	330.0	Site 1
	24.1	8.4	6.79	111.5	334.5	Site 2

Table 3: Physical parameters in Lake Hora-Arsedi during the experimental period.

	Treatments				
Parameters	T$_1$	T$_2$	T$_3$	T$_4$	T$_5$
Mean weight of initial stock (gm)	35.06 ± 0.539^a	36.16 ± 0.562^a	36.30 ± 0.539^a	36.11 ± 0.515^a	35.80 ± 0.613^a
Mean weight of final stock (gm)	205.17 ± 5.124^d	201.07 ± 3.521^d	172.39 ± 4.092^c	168.79 ± 4.799^b	135.56 ± 5.150^a
Mean length of initial stock (mm)	132.27 ± 0.911^a	133.63 ± 1.406^a	131.63 ± 0.835^a	133.77±1.097^a	132.2 ± 0.999^a
Mean length of final stock (mm)	201.03 ± 5.242^d	198.10 ± 5.898^c	196.33 ± 5.782^b	196.43 ± 5.197^b	194.07 ± 4.961^b
Weight of feed (kg)	71.58	71.37	61.45	60.39	45.97
Food conversion Ratio (FCR)	3.73^a	4.21^a	4.66^a	4.79^a	4.69^a
Specific growth rate (SGR) %day^{-1}fish^{-1}	0.98^c	0.95^c	0.87^b	0.86^b	0.74^a
Mean daily growth (gmday^{-1})	0.95^c	0.92^c	0.78^b	0.75^b	0.56^a
Fulton condition factor (FCF) (% in gm/cm³)	2.55^a	2.54^a	2.12^b	2.11^b	1.82^c
Survival rate (SR, %)	100	100	100	100	98
Stocking density (fish/m³)	50	50	50	50	49
Total net yield (kgyear^{-1})	17.25	16.73	13.79	13.44	9.94
Total weight gain (kgcage^{-1})	8.51	8.25	6.80	6.80	4.90

Values in the same rows with different superscripts are significantly different (p<0.05).

Table 4: Growth parameters, total amount of feed supplied, food conversion ratio and total net yield of feeding treatments during experimental period.

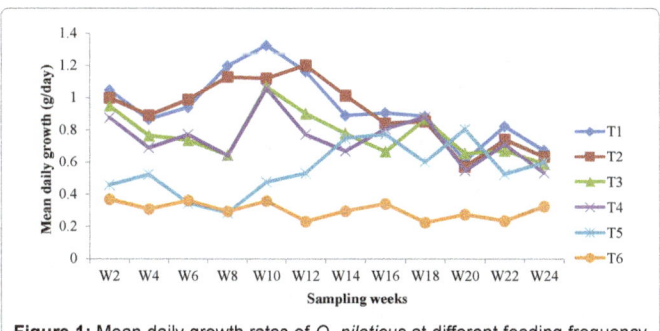

Figure 1: Mean daily growth rates of *O. niloticus* at different feeding frequency and control group.

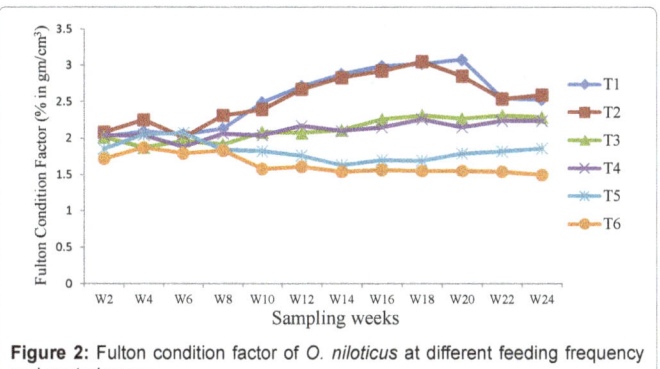

Figure 2: Fulton condition factor of *O. niloticus* at different feeding frequency and control group.

of nutrients entering in to the lake at experimental level compared to huge cage culture industries.

In this study feeding frequency had a significant effect ($p < 0.05$) on growth of *O. niloticus* and there is a positive relation between growth and increasing feeding frequency of this species. This agrees with the justification of Riche and Garling [26], Pillay and Katty [4], El-Sayed [5], Lim et al. [27] and Suresh and Bhujel [28]. They were explained that due to continuous feeding behaviour and smaller stomach capacity, tilapia respond better to more frequent feeding than other finfish and benefited from multiple daily feeding. Parker [29] also reported that tilapia cannot consume their daily requirement of feed for maximum growth in a single meal of a short duration, thus more than one feeding is needed each day. The work of Gaber and Hanafy [30] also confirmed this result. *O. niloticus* showed better growth at a feeding frequency of four times a day than two times a day without depending on the protein sources (fish meal protein or soybean meal protein) when fed restricted daily ration. Lim et al. [27] and El-Sayed [5] further explained that dividing the daily ration of tilapia reared in cages would probably reduce the exposure time of feed in the water and reduce the leaching rate of pellet and loss of nutrients. Thus, they recommend that dividing daily ration help to assure rapid and complete ingestion of the offered amount of feed and result in maximum FCR and good growth performance in this species.

Craig and Helfrich [31], Nandlal and Pickering [32] and El-Sayed [5] noted that feeding frequency should be lowered as fish grow. El-Sayed [5] noted that during larval stages, tilapia requires a daily ration of about 20% to 30% of their body weight divided into six to eight feedings. It must feed 3% to 4% of their body weight divided three to four times a day at the age of fingerlings. Board on Agriculture [33] in USA and Suresh and Bhujel [28] noted the nutrient requirements of fishes and identified the daily feeding allowance and frequencies for various species of fishes at different sizes. They stated that feeding

allowance and feeding frequency varied with the size of fish and species under the culture condition. For tilapia species for instance, for fish size ≥ 20 g and ≤ 100 g, feeding allowance must be 4% to 3% of their body weight at frequency of three to four times a day. However, for fish size >100 g, 3% to 2% of their body weight must be provided at frequency of two to three times a day for good growth. In this study however, T1 (changed frequency from four times a day to two times a day) has shown insignificant difference in weight gain than T2, which was fed four times a day throughout the experiment.

Since we don't have data on the amount of waste feed and the amount of feed used by fish for growth, it is difficult to calculate FCR in cage culture. However, in this research we compared the FCR of treatments based on the amount of feed given daily for different treatments. Based on the above assumption the FCR of T2 and T1 varied significantly ($p < 0.05$) from other treatments until the twelve week. This might be due to the somatic growth stage of the fish. Since the stomach at this stage was small, they benefited from frequent feeding of small amount and used extracted nutrients to build new tissue. The values obtained for Feed Conversion Efficiency (FCE) were also reflection of FCR values. FCE was >50% in early stage for T1 and T2 indicating that increasing feeding frequency maximizes efficient conversion of feed and good growth at early stage of the fish. The mean highest FCE was also observed in feeding frequency of T1 and T2. These results were in agreement with the finding of Siraj et al. [34,35] on red tilapia, (*Oreochromis mossambicus*) hybrid. They found that feeding frequency has significant effect on FCR of this species when feed restricted daily ration frequently when they are small sized.

However, the mean FCR of all the fish was not significantly affected by feeding frequency in the present study ($p > 0.05$). This agreed with the study of Gaber and Hanafy [30] on *O. niloticus* in concrete tanks. They reported that feeding frequency has no significant effect on FCR of the fish if the species was fed restricted daily ration. Gokcek, et al. [18] study on *Barbus luteus*, Abid and Ahmed [36] on juvenile *labeo rohita*, Ayo-Olalusi and Ugwumba [37] on Juvenile *Clarias gariepinus* and Tiril and Alagil [19] on rainbow trout, showed similar result to this experiment. They noted that feeding frequency has no significant effect on the FCR of those species.

Mean maximum SGR was attained for T1 and T2 and the minimum was observed in T5 from feeding treatments. This showed that supplementary feeding and increasing feeding frequency needed to ensure maximum percentage body weight increase per day. Similar result was observed in research done by Gaber and Hanafy [30] on *O. niloticus*. They reported that increasing feeding frequency has positive effect on SGR on this species. Nekoubin and Sudagar [22] on juveniles of Grass Carp and Priestley et al. [17] on Common Gold fish also confirmed that increasing feeding frequency to certain extent is important to attain better percentage body weight increase per day.

It was also revealed from this study that increased feeding frequency maximized the well-being of this species. The possible justification for better FCF for T1 and T2 was that they were fed frequently small amount of feed and this made them attain maximum FCE and FCR which resulted in good growth performance than other treatments. However, all treatments were in better condition than T5 and T6 (a non-feeding group). T5 were deprived of feed for three days per week and didn't feed 72 days from a total of 180 stocking days. This might have resulted in growth retardation in this group however; the lowest FCF in control group was due to absence of supplementary feed in this treatment. Contrary to this result, Gaber and Hanafy [30] findings on *O. niloticus* cultured on concrete tanks show that condition factor is not affected by feeding frequency if the fish were fed four times or twice a day.

It was found out that *O. niloticus* benefited from frequent feeding of small amount and used extracted nutrients for growth with better FCR, SGR, MDG, and FCF than less frequent feeding. In addition, net yield and total weight gain (kgcage⁻¹) were directly related to feeding frequency. However, feeding frequency has no effect on survival of this species at this experimental condition.

Acknowledgments

The authors would like to thank Development Partnership in Higher Education (DelPHE) project under the British Council for financial support.

References

1. Xie S, Cui Y, Yang Y, Liu J (1998) Effect of protein level in supplemental diet on the growth of cage-cultured Nile tilapia in East Lakes, P. R. China. Asian Fish Sci 10: 233-240.

2. McAndrew BJ (2000) Evolution, phylogenetic relationship and biogeography. In: Tilapias: Biology and Exploitation, Kluwer Academic publishing, The Netherlands.

3. Stickney RR (2000) Encyclopaedia of Aquaculture. John Wiley & Sons, New York, USA.

4. Pillay RTV, Katty MN (2005) Aquaculture principles and practices. Blackwell, Publishing, UK.

5. El-Sayed AFM (2006) Tilapia Culture. CABI Publishing, London, UK.

6. Tucker CS, Hargreaves JA, Boyd CE (2008) Better management practice for freshwater pond aquaculture. In: Environmental best management practice for aquaculture. Blackwell Publishing, USA.

7. De Silva SS, Anderson TA (1995) Fish Nutrition in Aquaculture. Chapman & Hall, London, UK.

8. Storebakken T (2002) Atlantic salmon, *Salmo Salar*. In: Nutritional requirements and feeding of finfish for aquaculture. CABI Publishing, UK.

9. Tadelle D, Ogle B (1997) Effect of maize (zea maize) and Noug (GuizoticaAbyssinica) cake supplementation on egg production performance of local birds under – condition in the central highlands of Ethiopia. Proceedings INFPO, Workshop, Senegal Dakar.

10. Tekeba E (2005) Compatibility of quality protein maize and seasame as substitute for synthetic amino acid broiler ratios. Thesis, Haramaya University, Haramaya.

11. De Silva S, Giovanni T, Francis D (2012) Nutrition. In: Aquaculture: Farming aquatic animals and plants, Wiley-Blackwell, UK.

12. Riche M, Haley DI, Oetker M, Garbrecht S, Garling DL (2004) Effect of feeding frequency on gastric evacuation and the return of appetite in tilapia *Oreochromis niloticus* (L.). Aquaculture 234: 657-673.

13. Tran LD, Dink TV, Ngo TP, Fotedar R (2011) Tilapia. In: Recent Advances and New Species in Aquaculture. Wiley- Blackwell, UK.

14. Malcolm J, Alanara A, Kadri S, Huntingford F (2012) Feeding biology and foraging. In: Aquaculture and behaviour, Wiley- Black well, UK.

15. Cho SH, Lim YS, Lee JH, Lee JK, Park S (2003) Effects of Feeding Rate and Feeding Frequency on Survival, Growth, and Body Composition of Ayu Post-Larvae, Plecoglossusaltivelis. J World Aquacult Soc 34: 85-91.

16. Davies OA, Inko-Tariah MB, Amachree D (2006) Growth response and survival of Heterobranchus longifilis fingerlings fed at different feeding frequencies. Afr J Biotechnol 5: 778-780.

17. Priestley MS, Stevenson A, Alexander GL (2006) The Influence of Feeding Frequency on Growth and Body Condition of the Common Goldfish (Carassiusauratus). J Nutr 136: 1979-1981.

18. Gokcek CK, Mazium Y, Akyurt T (2008) Effect of feeding frequency on the growth and survival of Himri Babel, Barbus luteus (Heckel, 1843), Fry under Laboratory Conditions. Pakistan J Nutr 7: 56-69.

19. Tiril US, Alagil F (2009) Effects of feeding frequency on nutrient digestibility and growth performance of rainbow trout (Oncorhynchus mykiss) fed a high lipid diet. Turkish Journal of Veterinary and Animal sciences 33: 317-322.

20. Ajani F, Dawodu MO, Bello-Olusoji OA (2011) Effects of feed forms and feeding frequency on growth performance and nutrient utilization of *Clarias gariepinus* fingerlings. African Afr J Agric Res 6: 318-322.

21. Asfaw A (2010) Effect of feed quality on growth performance and water quality in cage culture system for production of Nile tilapia [(Oreochromis niloticus), (Linnaeus, 1758)] in Lake Hora- Arsedi, Ethiopia. Thesis, School of Graduate Studies, AAU. Addis Ababa, Ethiopia.

22. Nekoubin H, Sudagar M (2012) Effects of Feeding Frequency on Growth Performance and Survival Rate of Grass Carp (Ctenopharyngodon idella). World Appl Sci J 8: 1001-1004.

23. Barnabe G (1994) The production of aquaculture organisms. In: Aquaculture: biology and ecology of cultured species, Ellis Horwood Litd, New York.

24. Masser MP (1997) Cage Culture: Site Selection and Water Quality. Southern Regional Agricultural Center Extension. Service. SRAC Publication No. 161. USA.

25. Beveridge M (2004) Cage Aquaculture (3ʳᵈ edn) Blackwell Publishing, UK.

26. Riche M, Garling D (2003) Feeding Tilapia in Intensive Recirculating Systems. North Central Regional Aquaculture Center.

27. Lim CE, Webster CD, Li MH (2006) Feeding practices. In: Tilapia: Biology, Culture and Nutrition, Haworth Press Inc, New York.

28. Suresh V, Bhujel R (2012) Tilapias. In: Aquaculture: farming aquatic animals and plants (2ⁿᵈ edn), Blackwell publishing, UK.

29. Parker R (2002) Aquacultural Science (2ⁿᵈ edn). Delmar, New York, USA.

30. Gaber MAM, Hanafy MA (2008) Relationship between dietary protein source and feeding frequency during feeding Nile Tilapia, Oreochromisniloticus (L.) cultured in concrete tanks. J App Aquac 20: 200-212.

31. Craig S, Helfrich AL (2002) Understanding Fish Nutrition, Feeds, and Feeding. Virginia Cooperative Extension.

32. Nandlal S, Pickering T (2004) Tilapia fish farming in Pacific Island countries,Tilapia Growout in ponds. Pacific Community and Marine Studies Program, the University of the South Pacific.

33. Board on Agriculture (BOA) (1993) Nutrient requirements of fish. National Academic Press, Washington, USA.

34. Siraj SS, Kamaruddin Z, Satar AKM, Kamarudin SM (1988) Effects on feeding frequency on growth, food conversion and survival of red tilapia (Oreochromis mossambicus, O. niloticus) hybrid fry. In: The Second International Symposium on Tilapia in Aquaculture, Manila, Philippines.

35. Stanton LT, Levallecy BS (2010) Feed composition for cattle and sheep. Colorado State University, Colorado.

36. Abid M, Ahmed SM (2009) Efficacy of feeding frequency on growth and survival of Labeo rohita (ham.) fingerlings under intensive rearing. The J Animal Plant Sci 2: 111-113.

37. Ayo-Olalusi IC, Ugwumba AAA (2009) Influence of Feeding Frequency on Feed Intake and Nutrient Utilization of Juvenile *Clarias gariepinus*. J Aquacult Feed Sci Nutr 2: 39-43.

33

Red Tilapia Broodstocks and Larval Production Under Different Water Salinities Without Acclimation

Ghada R Sallam[1], Walied A Fayed[2]*, Mohamed A El-Absawy[3], Hadir A Aly[1] and Zeinab A El-Greisy[3]

[1]Aquaculture Division, Fish Rearing laboratory, National Institute of Oceanography and Fisheries, Alexandria Branch, Alexandria, Egypt
[2]Department of Animal and Fish Production, Faculty of Agriculture (Saba Basha), Alexandria University, Alexandria, Egypt
[3]Aquaculture Division, Fish Reproduction laboratory, National Institute of Oceanography and Fisheries, Alexandria Branch, Alexandria, Egypt

Abstract

The direct transfer of fish to marine water without acclimation is considered critical on survival rate of fish. Florida red Tilapia, *Oreochromis* sp., was introduced to four salinity levels (9‰, 18‰, 24‰ and 36‰) and a control freshwater treatment in pre-spawning period to investigate the tolerance of the offspring to direct transfer to marine water without acclimation. Fish were stocked at a rate of 25/m^3 with initial body weight of 29.4 ± 0.12 g for six weeks as acclimation period. Broodstocks after acclimation were then stocked at 5/m^3 and fed to satiation with 25% crude protein commercial diet for 24 weeks as spawning period. Offsprings survival and growth were compared for their tolerance to three salinity levels (9‰, 18‰ and 36‰) in indoor concrete tanks with stocking density of 1000/ m^3, and fed on 30% crude protein (470 kcal ME/100 g) diet for 8 weeks. The results implied that best growth for broodstocks was observed with (36‰) and no significant difference in survival. However, the least number of fry/ kg were produced from broodstocks reared in 36‰ salinity, and the highest was observed with 18‰. Consequently, fry delivered from broodstocks reared in high salinity level (36‰) tolerated high salinity levels (18‰ and 36‰) with high survival rate (90% and 92%) respectively, and with highest growth rate. This study highlights the importance of rearing Florida red tilapia broodstocks in saltwater in order to have offspring tolerable to marine environment.

Keywords: Florida red tilapia; Salinity levels; Tolerance; Survival; Growth performance; Production

Introduction

The shortage in freshwater in many countries, and the competition for it with agriculture and other urban activities had increased the pressure to develop aquaculture in brackish water and seawater. Though, using salt water instead of freshwater in fish farming is a worldwide priority [1]. Freshwater encompasses much lower salts and ions concentration than brackish water and seawater. Control of salt and water balance within a narrow limit is critical to life in all multicellular organisms, including teleost fishes [2]. Salt tolerance is a term describing the overall fitness, or productivity, of the fish in a saline environment [3]. It is a combination of different quantitative traits, such as metabolism, growth [4], osmoregulation [5], immune-competence and fecundity [6]. Besides, Cnaani and Hulata [6] targeted from various culture management practices and nutrition as well as physiology and genetics to propose the best approaches for improvement of salinity tolerance in tilapias. Though tilapias normally live in freshwater, few species show high salt tolerance and could be raised in brackish water or seawater [7,8].

Due to the increasing lack of freshwater in the world, it would be beneficial to culture tilapia stocks in brackish or saline rearing environments to ensure a source of cheap and high-quality animal protein into the future [9]. In general, it is well established that salinity conditions during incubation and rearing are highly relevant for embryonic development, affecting variables such as hatching rate, and even later causing a lower survival rate and deformities in larvae [10], or affecting larval size, particularly when salinity is above the species tolerance, producing smaller fish when reared at higher salt concentrations [11]. Also, there is an increasing commercial interest in tilapia species or hybrids that can tolerate salinity and still exhibit acceptable growth [12].

To date, the consumption of saltwater tilapia fish has been increased because of their tasty flesh and not too strong fishy taste rather than freshwater tilapia [13]. Tilapias are popular cultured species because of their high environmental tolerant characteristics. The rapid growth of tilapia, poor quality resistance, ability to grow under sub-optimal nutritional conditions, and high fecundity, make them well suited for aquaculture [8]. Tilapia is one of the important fish species which has several good qualities and can face wide range of salinity and other environmental conditions and can grow well in water salinities ranging from 11 ppm to 29000 ppm, tolerate temperatures between 8°C to 42°C and can survive in low dissolved oxygen (DO) levels (0.1 ppm) [14]. *Oreochromis mossambicus* and its hybrids, including red tilapia are the major representatives of these euryhaline cichlids in aquaculture [15].

Based on the growth performance in saltwater, *O. mossumbicus* and red tilapia are competent strains for breeding tilapia in saltwater [15]. While the suitability of the Florida red tilapia strain for seawater grow-out has been demonstrated by high growth rates and feed conversion efficiencies, the hatchery phase of production remains restricted to water of lower salinities. The need for low-salinity water for maintaining broodstock and fry, thus affecting the ability of farmers to obtain fingerlings, restricts the establishment of future hatcheries in low-salinity water areas. Methods for seawater adaptation have been developed that minimize reliance on low-salinity water during the hatchery phase of production and that maximize survival and growth following transfer to seawater [16]. Other tilapias are generally less euryhaline and can tolerate water salinities ranging from about 20 to 35‰. Most of these tilapias grow, survive and reproduce at 0-29‰, depending on the species and acclimation period.

*Corresponding author: Walied A Fayed, Department of Animal and Fish Production, Faculty of Agriculture (Saba Basha), Alexandria University, Alexandria, Egypt, E-mail: fayedwal@gmail.com

Salinity tolerance of tilapia is also affected by fish sex and size. Perschbacher and McGeachin [17] evaluated the salinity tolerance of red tilapia (*O. mossambicus* × *O. urolepis hornorum*) fry, juveniles and adults. Adult fish were more salt-tolerant than fry and juveniles. Fry and juveniles tolerated direct transfer to 19%, without apparent stress and mortality, but 100% mortality occurred at 27‰. On the other hand, adult fish tolerated a direct transfer to 27‰, with 100% mortality at 37‰. Similarly, Watanabe et al. [18] studied the ontogeny of salinity tolerance in Nile tilapia, blue tilapia and hybrids tilapia (*O. mossambicus* female × *O. niloticus*) was referred to body size than to chronological age. The median lethal salinity-96 h (MLS-96) for Nile tilapia and blue tilapia over an age of 7-120 days post-hatching (dph) was 18.9 and 19.2‰. In contrast, MLS-96 of tilapia hybrids changed with age and increased from 17.2‰ at 30 dph to 26.7‰ at 60 dph [18]. Also, Watanabe et al. [18] reported also that male tilapia tend to be more salt tolerant than females. It has also been reported that tilapia hybrids descended from salt-tolerant parents (such as *O. mossambicus* and *O. aureus*) are highly salt-tolerant [19]. This may explain why Taiwanese red tilapia [20] and Florida red tilapia [21] grow faster in seawater and brackish water than in freshwater.

The sudden salinity changes may impact the physiological condition of the fish and the tolerance limits of the fish will cause stress and lead to decrease the immune system level. Rearing Tilapia in saltwater could have no different between freshwater. The Tilapia appeared normal and healthy through the external observation in saltwater but the level of stress with the sudden introduced to different salinity level unknown [13]. However, previous research by Sharif et al. [22] demonstrated that survival of red tilapia fry decreased with increasing salinity from 16‰ to 40‰ with acclimation after 12 days. Moreover, final body weight of Florida red tilapia fry was found to decrease significantly with increasing salinity from 16‰ to 32‰ after gradual acclimation period [23], and increases non-significantly with increased salinity from 0‰ to 25‰ [24]. Also, feed conversion rate of Florida red tilapia fry elevated significantly with increasing salinity from 16‰ to 32‰ [23].

Therefore, the objectives of the present study have two main approaches: first, determine the growth performance of Florida red tilapia broodstocks in different salinity levels; and second, is to produce Florida red tilapia fry that could tolerate and survive the direct transfer to marine water without being acclimated.

Materials and methods

Experimental location and fish species

This study was conducted on Florida red tilapia, *Oreochromis* sp., fingerlings for 32 weeks in El-Max Research Station, National Institute of Oceanography and Fisheries (NIOF), Alexandria, Egypt.

Experimental fish and acclimatization condition: Four hundred fingerlings with initial body weight (IBW) of 29.4 ± 0.12g were acclimated in fiberglass tanks for 6 weeks with a density of 10/m³ on four salinity (part per thousand, ppt) levels (9‰, 18‰, 24‰, and 36‰) and a freshwater as a control treatment. Females and males were placed separately and fed to satiation on 30% CP 470 Kcal ME/100g diet during the acclimation period. Florida red tilapia, *Oreochromis* sp. were gradually acclimated to the respective treatment of salinities by raising the salinity at the rate of 4‰ daily [18].

Experimental design: This study was designed to embrace two phases in which the first reveal the effect of different salinity levels (9‰, 18‰, 24‰, and 36‰) on broodstocks growth performance and fry production. Whereas the second phase involve the effect of

broodstocks acclimation salinity levels on the tolerance of produced fry to survive and grow in different salinity levels (9‰, 18‰, and 36‰) without acclimation.

Following the acclimatization period, the fish were transferred to ten concrete tanks with dimensions of 4 m × 2 m × 1 m (length × width × depth) representing 5 treatments each with two replicates. The tanks were continuously aerated to maintain dissolved oxygen level to 6.5 mg/l, and pH level of 7.5-9, and temperature 28 ± 2°C. Each replicate was stocked with 5 fish/m³ (total of 40 fish) both females and males with ratio of (3:1) allocated for spawning. Fish were fed 1% of their body weight on commercial diet encompassing 25% crude protein (CP) twice daily 7 days a week for 16 weeks. Every 10 days' fish were weighed and female brooders were checked for fry.

Twelve weeks after the onset of spawning, fry was collected from broodstocks subjected to the five salinity levels for the onset of the second phase of the experiment. Each batch of fry from a broodstock of different level of salinity was exposed to three different salinity levels, low, medium and high (9‰, 18‰, and 36‰, respectively) for 8 weeks without initial acclimatization period. The offspring were stocked with rate of 1000 fry/m³ in 30 (15 treatments × 2 replicates) concrete tanks (1 m³ each tank). Feed offered to fry to satiation on an isocaloric diet (30% CP and 470 Kcal ME/100 g diet). Fry were observed for survival and mortality rates were recorded.

Fish growth and feed utilization parameters

Growth performance and feed utilization of Florida red tilapia was determined, and was calculated as follows:

$$Specific\ Growth\ Rate\ (SGR)\ (in\ percent\ per\ day) = \frac{100[LnW_2 - LnW_1]}{T}$$

Where W1 and W2 are the initial and the final weights, respectively, and T is the number of days of the experiment;

$$Feed\ Conversion\ Ratio\ (FCR) = \frac{feed\ intake\,[in\ grams]}{weight\ gain\,[in\ grams]}$$

$$Protien\ Efficiency\ Ratio\ (PER) = \frac{weight\ gain\,[in\ grams]}{protien\ intake\,[in\ grams]}$$

Statistical analysis

Results of growth parameters, feed utilization parameters, and survival rate (%) of broodstocks and fry of the experimental treatments were treated using the ANOVA test (analysis of variance) and Tuckey test to a minimum significance (P<0.05). The results are expressed as means ± SEM. Statistical analysis was performed using one-way ANOVA according to Assaad et al. [25] and EXCEL (windows 10, 2015).

Results

Broodstocks growth performance and feed utilization

Acclimatization period of 6 weeks, fingerlings with initial body weight of 29.4g reached maturation and started to spawn after 7 weeks of acclimation. Fish were subjected to four salinity levels (9‰, 18‰, 24‰, and 36‰) and a freshwater as control treatment. The final body weight (FBW) and weight gain (WG) of fish after 24 weeks was significantly different (P<0.05) between treatments, showing increase with increasing salinity level (Table 1). Consequently, average daily gain (ADG), and specific growth rate (SGR%) revealed similar patterns as FBW and WG. The highest ADG, and SGR% was observed

Variables	C	9‰	18‰	24‰	36‰
FBW	163 ± 1.75ᵈ	187 ± 9.13ᶜᵈ	232 ± 0.83ᵇᶜ	257 ± 16.6ᵇ	364 ± 16.6ᵃ
WG	94.8 ± 0ᵈ	119 ± 8.58ᶜᵈ	161 ± 0.78ᵇᶜ	185 ± 15.6ᵇ	285 ± 15.6ᵃ
ADG	0.53 ± 0ᵈ	0.66 ± 0.05ᶜᵈ	0.9 ± 0.004ᵇᶜ	1.03 ± 0.09ᵇ	1.58 ± 0.09ᵃ
SGR%	0.49 ± 0ᵈ	0.56 ± 0.023ᶜᵈ	0.66 ± 0.002ᵇᶜ	0.71 ± 0.03ᵇ	0.85 ± 0.02ᵃ
SVR%	92.5 ± 0.5	96 ± 1	81 ± 12	100 ± 0	100 ± 0
FCR	1.35 ± 0.02ᵇ	1.4 ± 0.01ᵇ	1.6 ± 0.015ᵃ	1.64 ± 0.025ᵃ	1.7 ± 0.015ᵃ
PER	2.96 ± 0.044ᵃ	2.86 ± 0.02ᵃ	2.49 ± 0.02ᵇ	2.45 ± 0.04ᵇ	2.35 ± 0.021ᵇ
Total no. Fry	896 ± 0ᶜ	1140 ± 58.9ᵇᶜ	1680 ± 6.29ᵃ	1270 ± 85.3ᵇ	1300 ± 61.2ᵇ
Fry/kg	6.88 ± 0ᶜ	7.68 ± 0.07ᵇ	9.37 ± 0ᵃ	6.53 ± 0ᵈ	4.88 ± 0.03ᵉ

Values are means ± SEM, n=2 per treatment group.
Means in a row without a common superscript letter differ (*P*<0.05) as analyzed by one-way ANOVA and the TUKEY test.

Table 1: Mean values of final body weight (FBW, g fish⁻¹), weight gain (WG, g fish⁻¹), average daily gain (ADG, g fish⁻¹ day⁻¹), specific growth rate (SGR, %), survival rate (SUR, %), protein efficiency ratio (PER), feed conversion rate (FCR), total number of fry (TF, fry female⁻¹), and number of fry kg⁻¹ of Florida red tilapia *Oreochromis* sp. broodstocks reared in different salinity levels and control (C) freshwater treatment.

Variables	C	9‰	18‰	24‰	36‰
Low	97.5 ± 0.5ᵃ	87.9 ± 0.48ᵇ	90.7 ± 1.44ᵇ	87.2 ± 1.96ᵇ	70.8 ± 0.25ᶜ
Medium	79.5 ± 0.5ᵇ	86 ± 5.23ᵃᵇ	91.2 ± 0.96ᵃᵇ	94.4 ± 2.19ᵃ	96.9 ± 0.45ᵃ
High	--	18.8 ± 4.19ᶜ	73 ± 3.84ᵇ	95.7 ± 0.67ᵃ	96.8 ± 0.23ᵃ

Values are means ± SER, n=2 per treatment group.
Means in a row without a common superscript letter differ (*P*<0.05) as analyzed by one-way ANOVA and the TUKEY test.

Table 2: Mean values (± SER) of survival percentage (%) of Florida red tilapia fry reared in three salinity levels (low, medium, and high) delivered from broodstocks subjected to the 4 salinity levels and control (C) freshwater treatment.

with (36‰) level with values of 1.58 ± 0.09 g/fish/day, and 0.85 ± 0.02, respectively. Whereas for broodstock survival rate, revealed no significant difference between treatments, however, the (24‰, and 36‰) treatments demonstrated high survival rate followed by (9‰) then the control treatment (freshwater), and the lowest was found to be for the broodstocks subjected to (18‰) salinity level (Table 1).

In spite of that the growth performance of Florida red tilapia showed a linear relationship with the salinity levels, but the FCR of broodstocks demonstrated a significant (P<0.05) inverse correlation with salinity levels. The broodstocks in freshwater treatment showed an improved significant FCR than with broodstocks in (9‰, 18‰, 24‰, and 36‰) respectively (Table 1). On the same manner, the PER showed similar trend as the FCR and implying that feed utilization parameters was significant lower with (18‰, 24‰, and 36‰) than the control and (9‰) salinity level (Table 1).

Fry production

However, broodstocks showed different trend than FBW and WG in significance between treatments for average total number of fry produced and mean fry produced per kilogram fish (Table 1). The results displayed that (18‰) salinity level had the significant highest total number of fry and fry/kg with values of 1680 ± 6.29 fry/fish and 9.37 ± 0 fry/kg (Table 1). Although, the lowest significant (P<0.05) number of fry produced (896 ± 0 fry/fish) was observed for the control treatment (freshwater), but the lowest significant (P<0.05) 4.88 ± 0.03 fry/kg was obtained from 36‰ salinity level treatment (Table 1).

Fry survival rate

The fry survival is one of the main objectives for the present study, where the results obtained revealed that fry delivered from freshwater broodstock when exposed to high (36‰) salinity level had mortality of 100% after the first week of exposure (Table 2). However, same treatment fry (control) had survival rate of 97.5% and 79.5% by the end of the experimental study, when subjected to 9‰ and 18‰ salinity levels, respectively (Table 2). However, when comparing survival rate of fry delivered from broodstock of different salinity levels (9‰,

18‰, 24‰, and 36‰) with fry from control treatment broodstock (freshwater) at low (9‰) salinity level, the results demonstrates that fry from freshwater had the highest significant (P<0.05) survival rate than all others.

Moreover, when increasing salinity level that fry are subjected to medium, the survival rate is enhanced for fry from broodstock exposed to (9‰, 18‰, 24‰, and 36‰), showing that the highest significant (P<0.05) survival rate (96.9 ± 0.45%) was for fry delivered from broodstocks (36‰) and the lowest (79.5 ± 0.5) for fry from control treatment broodstock. Similarly, the survival rate of fry delivered from broodstock (36‰) showed the same trend when subjected to high salinity level compared to all other treatments (Table 2). This could be explained that when fry of Florida red tilapia broodstocks are reared in marine water their offspring could possibly tolerate high salinity (marine environment) levels without being acclimatized. And when those fry are exposed to lower salinity than that of their broodstocks their survival might be affected and could alleviate their mortality rate. Therefore, fry of Florida red tilapia should be acclimatized if reared in salinity levels different than that of their broodstocks, thus, tolerating salinity for offspring may be considered a maternal effect.

Fry growth performance and feed utilization

Correspondingly, the FBW and ADG of fry showed a parallel significant results as survival rate trend, which at low salinity level fry delivered from broodstocks exposed to freshwater and (9‰) salinity had significant (P<0.05) higher FBW and ADG than those delivered from (18‰, 24‰, and 36‰) respectively (Tables 3 and 4). However, the FBW and ADG trends are inversed when medium and high salinity levels are imposed.

Meanwhile, the FCR and PER of the broodstock showed an inverse significant relation with the salinity level (Tables 5 and 6), but the offspring delivered from broodstock subjected to elevated salinity levels had a better significant (P<0.05) FCR and PER when exposed to medium and high salinity levels than those from control and (9‰) broodstocks. Therefore, the present study determines the importance of culturing Florida red tilapia broodstocks in marine water that induces

Variables	C	9‰	18‰	24‰	36‰
Low	11.3 ± 0^a	9.77 ± 0.22^{bc}	10 ± 0.18^b	9.11 ± 0.005^c	9.88 ± 0.11^b
Medium	7.38 ± 0.39^d	9 ± 0.12^c	12.3 ± 0.28^b	13.5 ± 0.05^{ab}	14.7 ± 0.33^a
High	--	7.5 ± 0.38^c	12.5 ± 0.105^b	14.1 ± 0.12^a	14.9 ± 0.39^a

Values are means ± SEM, n=2 per treatment group.
Means in a row without a common superscript letter differ ($P<0.05$) as analyzed by one-way ANOVA and the TUKEY test.

Table 3: Mean values (± SER) of final body weight (g fish^{-1}) of Florida red tilapia fry reared in three salinity levels (low, medium and high) delivered from broodstocks subjected to the 4 salinity levels and control (C) freshwater treatment.

Variables	C	9‰	18‰	24‰	36‰
Low	0.2 ± 0^a	0.17 ± 0.004^{bc}	0.18 ± 0.003^b	0.16 ± 0.00001^c	0.18 ± 0.002^b
Medium	0.13 ± 0.01^d	0.16 ± 0.002^c	0.22 ± 0.01^b	0.24 ± 0.001^{ab}	0.26 ± 0.01^a
High	--	0.13 ± 0.01^c	0.22 ± 0.002^b	0.25 ± 0.002^a	0.27 ± 0.01^a

Values are means ± SEM, n=2 per treatment group.
Means in a row without a common superscript letter differ ($P<0.05$) as analyzed by one-way ANOVA and the TUKEY test.

Table 4: Mean values (± SER) of average daily gain (g fish^{-1} day^{-1}) of Florida red tilapia fry reared in three salinity levels (low, medium and high) delivered from broodstocks subjected to the 4 salinity levels and control (C) freshwater treatment.

Variables	C	9‰	18‰	24‰	36‰
Low	1.61 ± 0.05	1.6 ± 0.05	1.59 ± 0.01	1.52 ± 0.04	1.56 ± 0.04
Medium	1.6 ± 0.03^a	1.58 ± 0.03^{ab}	1.53 ± 0^{ab}	1.42 ± 0.03^{bc}	1.28 ± 0.05^c
High	--	1.49 ± 0.05^a	1.39 ± 0.05^a	1.38 ± 0.03^a	1.12 ± 0.01^b

Values are means ± SEM, n=2 per treatment group.
Means in a row without a common superscript letter differ ($P<0.05$) as analyzed by one-way ANOVA and the TUKEY test.

Table 5: Mean values (± SER) of feed conversion rate (FCR) of Florida red tilapia fry reared in three salinity levels (low, medium and high) delivered from broodstocks subjected to the 4 salinity levels and control (C) freshwater treatment.

Variables	C	9‰	18‰	24‰	36‰
Low	2.49 ± 0.08	2.5 ± 0.08	2.52 ± 0.02	2.62 ± 0.06	2.57 ± 0.07
Medium	2.5 ± 0.05^b	2.54 ± 0.04^b	2.61 ± 0.00^b	2.83 ± 0.05^{ab}	3.13 ± 0.12^a
High	--	2.69 ± 0.09^b	2.88 ± 0.104^b	2.91 ± 0.053^b	3.57 ± 0.032^a

Values are means ± SEM, n=2 per treatment group.
Means in a row without a common superscript letter differ ($P<0.05$) as analyzed by one-way ANOVA and the TUKEY test.

Table 6: Mean values (± SER) of protein efficiency ratio (PER) of Florida red tilapia fry reared in three salinity levels (low, medium, and high) delivered from broodstocks subjected to the 4 salinity levels and control (C) freshwater treatment.

and enhances the offspring growth performance and feed utilization in marine environment without being acclimatized.

Discussion

Pre-acclimation to salt water and gradual transfer to high salinity have a significant effect on tilapia growth and survival, as has been reported by Al-Amoudi [26]. Acclimation period varies between tilapia species where *O. aureus*, *O. mossambicus* and *O. spilurus* required shorter acclimation time (4 days) for a transfer to full-strength seawater than *O. niloticus* and *O. aureus* × *O. niloticus* hybrids (8 days) [26]. The author suggested that the physiological changes associated with seawater acclimation in tilapia are short-term, energy demanding and may account for as much as 20% of total body metabolism after 4 days in seawater. However, Florida red tilapia has been acclimated for 6 weeks in the present study on gradual salinity increase to reach full-strength seawater salinity in order not to pause any stress on fish.

Many researchers handled different aspects of acclimation of tilapia to elevated salinity levels through water salinity or diets containing various salt levels. Turingan and Kubaryk [27] supplemented Taiwanese red tilapia (*O. mossambicus* × *O. niloticus*) broodstock

with diets containing higher salt levels may produce seeds with better adaptability to water salinity. They found that egg hatchability was higher in seawater than in freshwater. The hatchability and larval growth were highest in seawater than in freshwater when fish received 12% salt in their diets.

Rengmark and Lingaas [28] investigated the role of transferrin, an iron-binding glycoprotein known to have an important role in the immune system, on salinity tolerance. They cloned and sequenced entire transferrin gene of tilapia and compared the expression levels in saltwater and freshwater reared tilapia using real-time PCR. Rengmark and Lingaas [28] observed that transferrin showed an 85% upregulation in tilapia kept in saltwater compared to freshwater, suggesting that transferrin or closely-linked genes may be involved in saltwater tolerance. In the present study, the survival rate of Florida red tilapia broodstocks was high and did not differ significantly for all treatments which implied that acclimation process was successfully applied with no stress or osmoregulatory failure in fish. Moreover, growth performance of Florida red tilapia herein increased significantly with the increase of water salinity that might be due to minimal metabolic energy diversion into osmoregulation. In support, Watanabe et al. [29] found that the daily feed consumption of Florida red tilapia fed a 32% CP diet increased with increasing salinity from 0 to 32‰.

The increase in the metabolic energy diverted into osmoregulation, with increasing water salinity has also been reported in *O. mossambicus* and *O. spilurus* [30], *O. niloticus* × *O. aureus* and common carp [31]. Previous studies indicated that growth rates of Florida red tilapia reared at different salinities increased with increasing temperature within the range 22°C to 32°C [32]. In accordance to the present study, experimental period was conducted in similar range of temperatures which favored the growth performance of Florida red tilapia.

Moreover, the better growth performance in saline water might be attributed to higher osmoregulation energy costs in freshwater than in brackish water or seawater for *O. mossambicus* × *O. hornorum* hybrid [33], suppressed territorial aggression by salinity [21], and inhibitory effects of aggressive behavior which varied among different salinities [20]. These results support the common assumption that growth of euryhaline teleosts is increased at salinities near iso-osmotic, since osmoregulation costs are minimal under these conditions [33]. On the contrary, Watanabe et al. [29] denoted that the growth of Florida red tilapia at salinities near iso-osmotic was found to be lower than that at higher salinities which was attributed to increased food consumption and lowered food conversion ratio in high salinity. Although, the results of the broodstocks FCR obtained in the present study were found to match that of Watanabe et al. [29] in which broodstocks from lower salinity treatments showed an improved significant FCR than that of higher salinities. But the broodstocks growth performance in the present study contradicts with the findings of Watanabe et al. [29] and corresponds with Febry and Lutz [33].

In addition, many researchers demonstrated that at equivalent salinity early exposure of tilapia broodstock to high salinity produced progeny with high salinity tolerance than those spawned in freshwater and hatched at high salinity. Respectively, produced fry in the present study delivered from freshwater broodstock when exposed to high salinity level had mortality of 100% after the first week of exposure. However, when comparing survival rate of fry delivered from broodstock of different salinity levels (9‰, 18‰, 24‰, and 36‰) with fry from control treatment broodstock (freshwater) at low (9‰) salinity level, the results demonstrates that fry from freshwater had the highest significant ($P<0.05$) survival rate than all others. Moreover, when increasing salinity level for fry the survival rate is enhanced showing that the highest significant ($P<0.05$) survival rate was for fry delivered from broodstock (36‰) and the lowest for fry from control treatment broodstock. Therefore, early exposure of broodstocks to high salinity benefits produced fry to tolerate direct exposure to marine water without acclimation. Thus, the Florida red tilapia might be cultured successfully in marine water with demanding freshwater.

On the other hand, Hassan et al. [13] observed that the mortality rate of red tilapia fingerlings increased with increasing salinity level from 0‰ to 35‰, and reached 100% mortality after 4 days of exposure. In 1989, Reference [16] assessed the influence of spawning salinity on survival and growth in brackish or seawater, growth of juveniles, suggesting that progeny spawned under elevated salinities are better adapted for growth in brackish and seawater. In the same manner, Sharif et al. [22] revealed that, survival of red tilapia decrease from 98% to 79.2% with increasing salinity from 16‰ to 40‰ after 12 days. Moreover, Hassan et al. [13] indicated that the direct transfer of freshwater adapted red tilapia to marine water causes 100% mortality due to respiratory distress and osmoregulatory exhaustion, which leads to increase osmotic concentration of blood serum and change in ionic contents. And since the chloride cells are the ionic regulator and extrusion in gill epithelia of sea adapted fish [34,35], Sharif et al. [22] noticed the formation of the chloride cells in gills of marine water adapted red tilapia and its absence in freshwater red tilapia. Moreover, Sharif et al. [22] stated that the chloride cells were observed in fish after 2 weeks in marine water. The author indicates that the acclimation of freshwater red tilapia needed about 2 weeks to be able to live in marine water. Correspondingly, this could explain the increased mortality of red tilapia fry from broodstock reared in freshwater when exposed to high salinity levels without acclimation in the present study, and the ability of fry descended from broodstocks reared in high levels of salinity to tolerate and survive without acclimatization.

Concurrently, Hibiya [36] disclosed that activation of chloride cells when fresh eel adapted to sea water is noted within 2 to 4 days and gills take one month to be completed of the sea water type. Low-salinity water requirements during the hatchery phase of production may be reduced by acclimating stocks to seawater at early stages of development [18]. Despite of survival rate, broodstock from brackish water (18‰) displayed the highest total number of fry and fry/kg, but the lowest number of fry produced was observed for the control treatment, and the lowest fry/kg was from seawater treatment.

In support to obtained results in the present study, Watanabe et al. [16] observed that fry production per unit female weight declined at salinities above 18‰ suggesting that Florida red tilapia broodstock may be maintained under salinities as high as 18‰ without impairing fry production. Although, FBW and FCR of fry showed a linear trend as survival rate, which at low salinity level fry delivered from broodstocks exposed to freshwater and low salinity had higher FBW and enhanced FCR than those from brackish and seawater. However, the FBW and FCR trends are inversed when medium and high salinity levels are imposed. However, Vũ [24] implied that final body weight of Florida red tilapia fry increased non-significantly with increased salinity from freshwater to 25‰.

On the contrary, El-Zaeem et al. [23] stated that final body weight of Florida red tilapia fry decreased significantly from 33.38 g to 18.68 g with increasing salinity from 16‰ to 32‰, respectively, after gradual acclimation period. Moreover, El-Zaeem et al. [23] found that FCR was elevated significantly Florida red tilapia fry with increasing salinity from 16‰ to 32‰.

Conclusion

It could be concluded from the present study that growth performance and survival rate of Florida red tilapia, *Oreochromis* sp., broodstocks increases with increasing salinity after gradual acclimation. Also, when fry of Florida red tilapia from acclimated broodstocks are reared in marine water they could possibly tolerate high salinity (marine environment) levels without acclimation. Therefore, the study highlights the possibility of acclimation of Florida red tilapia broodstocks on high salinity levels produces seeds that could tolerate the direct transfer to marine water without requiring acclimation.

Ethical Issues

We certify that all data collected during this study is presented in this manuscript and no data from the present study has been or will be published separately or elsewhere.

References

1. El-Sayed MA (2006) Tilapia culture in salt water: environmental requirements, nutritional implication and economic potentials. Avances en Nutrició Acuícola VIII. VIII Simposium internacional de Nutrición. Acuícola. Universidad Autónoma de Nuevo León, Monterrey, Nuevo León, Mexico.

2. Jeanette CF, Amy K, Liza M, Larry GR, Paul HY, et al. (2007) Effects of environmental salinity and temperature on osmoregulatory ability, organic osmolytes, and plasma hormone profiles in the Mozambique tilapia (*Oreochromis mossambicus*). Comp Biochem Physiol A Mol Integr Physiol 146: 252-264.

3. Stickney RR (1986) Tilapia resistance of saline waters: A review. Pro Fish-Cult 48: 161-167.

4. Sakamoto T, McCormick SD (2006) Prolactin and growth hormone in fish osmoregulation. Gen and Comp Endocri 147: 24-30.

5. Mancera JM, McCormick SD (2007) Role of prolactin, growth hormone, insulin-like growth factor I and cortisol in teleost osmoregulation. Fish Osmoregulation, Science Publishers, Enfield, NH.

6. Cnaani A, Hulata G (2011) Improving salinity tolerance in tilapias: Past experience and future prospects. The Isra J Aquac-Bamidg 63: 533-554.

7. Kamal AM, Mair GC (2005) Salinity tolerance in superior genotypes of tilapia, *Oreochromis niloticus*, *Oreochromis mossambicus* and their hybrids. Aquaculture 247: 189-201.

8. Lawson EO, Anetekhai MA (2011) Salinity tolerance and preference of hatchery reared Nile Tilapia, *Oreochromis niloticus* (Linnaeus 1758). Asian J Agri Sci 3: 104-110.

9. Mateen A (2007) Effect of androgen on the sex reversal, growth and meat quality of tilapia, *Oreochromis niloticus*. Department of Zoology and Fisheries, University of Agriculture, Faisalabad, Pakistan.

10. Takuma O, Tadahide K, Koichiro G, Koji M, Kazuharu N (2009) Influence of salinity on morphological deformities in cultured larvae of Japanese eel, Anguilla japonica, at completion of yolk resorption. Aquaculture 293: 113-118.

11. Fielder DS, Bardsley WJ, Allan DL, Pankhurst PM (2005) The effects of salinity and temperature on growth and survival of Australian snapper, *Pagrus auratus* larvae. Aquaculture 250: 201-214.

12. Armas-Rosales, MA (2006) Genetic effects influencing salinity and cold tolerance in tilapia. Louisiana State University, Agricultural and Mechanical College, USA.

13. Hassan M, Zakariah MI, Wahab W, Muhammad SD, Idris N (2013) Histopathological and behavioral changes in *Oreochromis* sp. after exposure to different salinities. J Fish Livest Prod 1: 103.

14. Pullin, RSV, Lowe-McComell RH (1982) The biology and culture of tilapias. International Center for Living Aquatic Resources Management, Manila, Philippines.

15. Tayamen MM, Reyes RA, Danting MJ, Mendoza AM, Marquez EB, et al, (2002) Tilapia broodstock development for saline waters in the Philippines. NAGA-ICLARM Quarterly 25: 32-36.

16. Watanabe WO, Ernst DH, Olla BL, Wicklund RI (1989) Aquaculture of red Tilapia (*Oreochromis* sp.) in marine environments state of the art. Adv in Trop Aquac 487-498.

17. Perschbacher PW, McGeachin RB (1988) Salinity tolerance of red hybrid tilapia fry, juvenile and adults. Proceedings of the Second International Symposium on Tilapia in Aquaculture ICLARM Conference Philippines.

18. Watanabe WO, Kuo CM, Huang MC (1985) The ontogeny of salinity tolerance in the tilapias *Oreochromis aureus, O. niloticus,* and an *O. mossambicus* × *O. niloticus* hybrid, spawned and reared in freshwater. Aquaculture 47: 353-367.

19. Romana-Eguia MR, Eguia RV (1999) Growth of five Asian red tilapia strains in saline environments. Aquaculture 173: 161-170.

20. Liao IC, Chang SL (1983) Studies on the feasibility of red tilapia culture in saline water. Proceeding of the International Symposium on Tilapia in Aquaculture. Tel Aviv University, Tel Aviv, Israel.

21. Watanabe WO, French KE, Ellingson LJ, Wicklund RI, Olla BL (1988b) Further investigations on the effects of salinity on growth of Florida red tilapia: evidence for the influence of behavior. The second International Symposium on Tilapia in Aquaculture, Philippines.

22. Sharaf MM, Sharaf SM, El-Marakby HI (2004) The effect of acclimatization of freshwater red hybrid tilapia in marine water. Pak J Biol Sci 7: 628-632.

23. El-Zaeem SY, Ahmed MM, Salama ME, Darwesh DMF (2012) Production of salinity tolerant tilapia through interspecific hybridization between Nile tilapia (*Oreochromis niloticus*) and red tilapia (*Oreochromis* sp.). Afr J Agri Res 7: 2955-2961.

24. Vũ AP (2013) Effect of salinity on growth and feed utilization of Nile tilapia (*Oreochromis niloticus*) and red tilapia (*Oreochromis* sp.). Vietnam National University, Vietnam.

25. Assaad H, Zhou L, Carroll RJ, Wu G (2014) Rapid publication-ready MS-Word tables for one-way ANOVA. Springer Plus 3: 474.

26. Al-Amoudi MM (1987) Acclimation of commercially cultured *Oreochromis* species to sea water-an experimental study. Aquaculture 65: 333-342.

27. Turingan JE, Kubaryk JM (1992) The effect of high salt diet on survival and hatchability of Taiwanese red tilapia (*O. mossambicus* × *O. niloticus*) eggs upon direct transfer to seawater. Aquac 92- Book of Abstract. W Aquac Soc Baton Rouge, Louisiana, USA.

28. Rengmark AH, Lingaas F (2007) Genomic structure of the Nile tilapia (*Oreochromis niloticus*) transferrin gene and a haplotype associated with saltwater tolerance. Aquaculture 272: 146-155.

29. Watanabe WO, Ellingson LJ, Wicklund RI, Olla BL (1988a) The effects of salinity on growth, food consumption and conversion in juvenile monosex male Florida red tilapia. The Second International Symposium on Tilapia in Aquaculture, Philippines.

30. Payne AL, Ridgway J, Hamer JL (1988) The influence of salt (NaCl) concentration and temperature on the growth of *Oreochromis spilurus, O. mossambicus* and a red tilapia hybrid". The Second International Symposium on Tilapia in Aquaculture. Philippines.

31. Payne AL (1983) Estuarine and salt tolerant tilapia. In: Proceedings, International Symposium on Tilapia in Aquaculture. Tel Aviv University, Israel.

32. Watanabe WO, Ernst DH, Chasar MP, Wicklund RI, Olla BL (1993) The effects of temperature and salinity on growth and feed utilization of juvenile, sex reversed male Florida red tilapia cultured in a recirculating system. Aquaculture 112: 309-320.

33. Febry R, Lutz P (1987) Energy partitioning in fish: the activity-related cost of osmoregulation in a euryhaline cichlid. J Exp Biol 128: 63-85.

34. Foskett JK, Schaffer C (1982) The chloride cells: A definitive identification as the salt-secretory cell in teleost. Science 215: 164-166.

35. McCormick SD (1995) Hormonal control of gill Na+, K+-ATPase and chloride cell function. In: Cellular and Molecular Approaches to fish ionic regulation. Fish. Academic Press, London.

36. Hibiya T (1983) An atlas of fish histology. Normal and pathology features. Koddasha LTD. Tokyo.

34

Survival and Immunity of Marron *Cherax cainii* (Austin, 2002) Fed *Bacillus mycoides* Supplemented Diet under Simulated Transport

Ambas I[1,3], Fotedar R[1] and Buller N[2]*

[1]*Sustainable Aquatic Resources and Biotechnology, Department of Environment and Agriculture, Curtin University, Bentley 6102, Australia*
[2]*Department of Agriculture and Food Western Australia, Animal Health Laboratories, South Perth, WA, Australia*
[3]*Faculty of Marine Science and Fishery, Department of Fishery, Hasanuddin University, Makassar, 90225, Indonesia*

Abstract

The present study examined the effect of simulated transport on marron, *Cherax cainii*, (Austin, 2002) after a 10 week feeding trial using basal diet or customised probiotic, *Bacillus mycoides* supplemented diet by measuring intestinal bacterial population, total haemocyte count (THC), bacteraemia, morbidity, dehydration and mortality.

Packing steps followed the established standard packing method for live transportation of marron. Each treatment group consisted of six polystyrene boxes ($65 \times 30 \times 40$ cm³), and each box contained 30 marron from each feeding group. The sealed boxes were placed on a trolley at room temperature to give the effect of transportation. Boxes were opened at 24th and 48th hour post simulated transport and marron from each treatment group were returned to the culture tank. After temperature acclimation, the marron were observed for mortality and samples were collected to assess marron health and immunity.

The results demonstrated that no mortality was observed at 24 h of transport both in basal diet and probiotic diet fed marron, however at 48h of transport the survival of probiotic fed marron was significantly higher (100 ± 0.0%) compared to survival (93.3 ± 2.8%) of basal diet fed marron. The higher survival rate of probiotic fed marron was also sustained by superior health and immune status indicated by higher intestinal bacterial population, higher total haemocyte count and lower haemolymph bacteria (bacteraemia) level. In brief, supplementation with host origin customized probiotic *B. mycoides* significantly improved marron tolerance to a live transport stress test, which resulted in no mortality up to 48h of transport.

Keywords: Simulate transport; Intestinal bacteria population; THC; Bacteraemia; Mortality

Introduction

Practices and methods used for crustacean handling and live trade may lead to serious physiological stress responses in aquatic animals [1-3] including marron [4-6]. Marron, *Cherax cainii* (Austin, 2002), is endemic to Western Australia and recently was introduced into South Africa, Zimbabwe, Japan, USA, China and the Caribbean as a commercial aquaculture species [6-8].

The reasons for live trade of crustaceans include for consumption, grow-out, restocking, and the aquarium trade; hence survival and quality of the animals is extremely important for welfare and economic reasons. The duration of the stressors encountered in the live trade process leads to short and long term changes in immune parameters as stress response shifts from adaptive to maladaptive [2,9]. Beyond this point the physiological stress response may reduce disease resistance and growth reduced quality and eventually death [2,10]. Therefore, improving immunity, stress tolerance and optimising health conditions of crustaceans during storage and live transport is of fundamental importance [3].

The successful culture and stocking of marron relies on better understanding the factors affecting their well-being during transport and their recovery afterwards [5]. Marron are an economically important aquaculture species in Western Australia and they show significant environmental stress tolerance post handling and live transport. Jussila observed no mortality of marron post 24 h handling and simulated transport, whereas detected no mortality of marron up to 36 h under simulated transport [5,6]. Moreover, marron may undergo live transport up to 72 h without mortality, however longer periods of transportation resulted in an average dehydration of 4.5% of body weight [4,5].

Feed additives such as probiotics and prebiotics can improve stress tolerance and immunity of aquatic animals including marron [6,11-16]. Our previous studies indicated that *Bacillus mycoides*, a marron origin customised probiotic, possessed a number of favourable probiotic properties [17], significantly improved gastrointestinal (GIT) health of marron [18] and improved immunity against the pathogenic bacterium *Vibrio mimicus* [19], a dominant bacterial pathogen of freshwater crayfish in aquaculture [20-22].

Cruz et al. suggested that aquatic animals should be treated with probiotics before exposure to transport and environmental stressors [23]. To date, improved stress tolerance and immunity by feeding probiotics have been documented in fish [11,24,25], shrimps [26-29] and molluscs [30,31] however information on probiotic-fed marron under practical transport conditions is not available. The present study evaluated the effect of simulated transport conditions on marron fed the probiotic *B. mycoides* by measuring intestinal bacterial population, total haemocyte count (THC), bacteraemia, morbidity, dehydration and mortality.

*Corresponding author: Buller N, Department of Agriculture and Food Western Australia, Animal Health Laboratories, South Perth, WA, Australia
E-mail: irfanambas@yahoo.com

Material and Methods

Acclimation and feeding trial

Marron juveniles were supplied by Blue Ridge Western Australia Pty Ltd and then acclimated to the culture tanks, fed using basal diet for 2 weeks, and distributed into six experimental culture tanks at a density of 12 marron per tank.

The experimental system consisted of three standing units of steel racks with three shelves in each unit. The experimental units were cylindrical plastic tanks (80 cm diameter and 50 cm high and 250 L in capacity) filled with freshwater running continuously at a rate of approximately 3 L/min. using a recirculating biological filtration system (Fluval 205, Askoll, Italy). Each tank was supplied with constant aeration and equipped with a submersible thermostat set to 24°C. PVC pipes of appropriate diameter were added to the tanks as shelters for the marron.

Prior to the simulated transport test, a feeding trial using basal and probiotic supplemented diet was conducted for 10 weeks. Each tank was stocked with 12 marron where each treatment consisted of five replicate tanks. The test diets were given to marron every day in the late afternoon at a rate of 1% of the total biomass and adjusted weekly after determination of biomass at the end of each week.

Experimental diet and set up

The experimental diets used in this feeding trial were (1) basal diet of a marron commercial feed supplied by Specialty Feed Pty Ltd, WA and (2) the basal diet supplemented with customised probiotic *B. mycoides*. Before use, the pelleted diet was homogenised with a blender to obtain a desirable pellet size before supplementation with the *B. mycoides* at 10^8 cfu/g of feed. The density of *B. mycoides* was based on the density used in other *Bacillus* species studies [32-36] and from results of our previous studies [18,19].

Supplementation of the probiotic followed established methods [37]. In brief, the basal diet was placed on tray covered with sterilised aluminium foil and sprayed with 20 mL of fish oil per kg of feed to improve attachment of probiotic bacteria. The feed was wrapped in individual sterilised aluminium foil packs containing the amount adjusted to marron biomass for each tank, and stored at 4°C until used. The diet was prepared each week and the feeding rate adjusted according to the marron weight.

Simulated live transport

After feeding with the test diet for 10 weeks, the animals were subjected to a simulated live transport following the "Code of Practice for the Harvest and the Post-Harvest Handling of Live Marron for Food" established by Department of Fishery Western Australia [38] and the standard packing of marron commonly used by marron growers. Feeding ceased one day before the commencement of simulated transport trial.

In brief, healthy marron of equal size (12.3 ± 0.5 g) from probiotic fed and a control basal diet were selected and placed in a polystyrene box (60 × 40 × 30 cm) for 24 h and 48 h simulated transport. Each box contained sufficient ice-gel bags covered by a moist foam layer and a temperature data logger (Onset HOBO). Marron from each treatment group were placed in a ventilated container prior to placing in the polystyrene boxes. Placing the marron in the ventilated container not only avoided the marron from mixing with different treatment groups, but also protected the marron from the moist foam layer and ice-gel bags, and was based on the method used in a previous study [6]. Subsequently, another moist foam layer and ice bag were placed over the marron ventilated containers, before the outer polystyrene box was sealed with a lid. The sealed boxes were placed on a trolley at room temperature and being moved intermittently to give simulated transport effects.

Twenty four and forty eight hours post simulated live transport, the animals were returned to the culture tanks once the parameters for intestinal bacteria population, total haemocyte count (THC), bacteraemia, morbidity, dehydration and mortality were measured and recorded.

Measurements of the parameters

Intestinal bacteria population: Bacterial population of the marron was measured before and during simulated transport at 24h and 48h. In brief, marron from each treatment group were sacrificed by placing them at -20°C for 5 minutes before aseptic removal of the GIT. The marron dorsal shell was cut-off horizontally from tail to head until the intestines were exposed. The intestine from individual animals was collected aseptically and placed in a sterilised pestle, weighed and then homogenised. The homogenates of intestines were serially (10^{-1}, 10^{-2}, 10^{-3}, 10^{-4}, 10^{-5} and 10^{-6}) diluted using sterile normal saline. Fifty microliters of each serial dilution was inoculated onto a blood agar (BA) plate and incubated overnight in a CO_2 incubator at 25°C. A colony count was performed for each dilution to determine the total number of aerobic bacteria [39].

Total haemocyte count (THC): The total haemocyte count was measured following the established methods used in western rock lobsters *Panulirus cygnus* George [40] and marron [6]. In brief, 0.5 mL of haemolymph withdrawn from the second last ventral segment of marron was inserted into a haemocytometer (The Neubauer Enhanced Line, Munich, Germany) counting chamber and immediately viewed under 100-fold magnification on a camera-equipped microscope and images were taken for THC counts. Cells were counted in both grids, and the mean was used as the haemocyte count. For each treatment group, the procedure was repeated using ten different animals. The total haemocyte count was calculated as THC = (cells counted x dilution factor ×1000)/volume of grid (0.1 mm³).

Haemolymph bacteria (Bacteraemia): Bacteraemia of marron was determined following the established method described by sang et al. [6] with a minor modification. Briefly, the haemolymph was withdrawn into a sterile syringe and placed onto a sterile glass slide to avoid bubbles before a 0.05 mL aliquot was lawn inoculated onto a BA plate. The plates were placed in a sterilised container and incubated overnight at 25°C. The total colony forming units (CFU) for each plate and CFU/mL were calculated on the basis of a total volume of 0.05 mL/ plate.

Dehydration: Dehydration of marron was measured using the established method [5]. Ten marron from each treatment group of equal size were weighed prior to the commencement of the simulated transport, then weighed at 24 h and 48 h post transport and the percentage of weight loss was recorded.

Morbidity and survival rate (%)

Morbidity (vigour index) of marron was measured following the established method proposed by Jussila et al. [5]. In this study, morbidity of marron was identified based on the response to stimuli at a time after simulated transport, and the time of recovery was recorded after being returned to the culture tanks.

Mortality of marron from each treatment group was measured at 24 h and 48 h post-simulated transport up to one day they were returned to the culture tanks. Determination of survival rate following the established equation;

$$SR = (Nt/No) \times 100$$

where SR is the survival rate (%); Nt is the number of marron at time t and No is the number of marron at the commencement (0), respectively.

Data analysis

The data were analysed using SPSS statistical package version 22.0 for Windows and Microsoft Excel. The difference between means was determined using one way analysis of variance (ANOVA) and a t-test. All significant tests were performed at P < 0.05 level. All data were presented as mean ± SE, unless otherwise indicated.

Results

Intestinal bacteria population

Intestinal bacterial population of marron declined at 24 and 48 h simulated transport both for basal diet and probiotic fed marron. Reduction of the intestinal bacterial population occurred at 24 h and 48 h of transport; however a significant reduction of more than half the initial population levels were observed at 48 h, both in basal diet and probiotic fed marron (Table 1). This result suggests that the longer the stress disturbance, the greater the reduction of intestinal bacterial population. Nevertheless, at 48 h post-transport, the bacterial population (646 ± 16.4) of probiotic fed marron was comparable to the initial bacterial population (626.7 ±19.7) of basal diet fed marron.

Total haemocyte count (THC)

Prior to the simulated transport test, the THC of marron fed probiotic and basal diets for 10 weeks was measured. The THC of marron fed probiotic supplemented diet was significantly higher compared to THC of basal diet fed marron at the commencement of simulated transport test (Figure 1). This result indicates that the health status of marron fed probiotic was higher at the initiation of the simulated transport test.

After 24 h and 48 h post transport, the THC in both treatment groups declined indicating that transport affects the THC in marron.

Haemolymph bacteria (Bacteraemia)

Haemolymph bacteria were observed in basal diet and probiotic fed marron after feeding with the test diets for ten weeks prior to the simulated transport test. Total bacterial count in the haemolymph of probiotic fed marron was significantly (P<0.05) lower compared to basal fed marron, indicating the customised probiotic of host origin may induce protection from bacteria and other foreign particles in the haemolymph. This result was strongly related to THC of marron in each treatment group, as haemocytes play an important role in removal of bacteria and foreign particles from the haemolymph of crayfish [41,42].

Dehydration

Dehydration occurred in marron fed either the basal diet or the probiotic supplemented diet. During the first 24 h, dehydration in the basal diet fed marron was 3.8 percent, whereas in probiotic fed marron it ranged between 3.0 and 3.7%. After 48 h of transport, the dehydration still occurred in both treatment groups but in the basal diet it was 0.45% while in probiotic fed marron the extra dehydration was 0.55%. There

was no significant different (P>0.05) in the mean dehydration between basal diet and probiotic fed marron after 24h and 48h of transport.

Morbidity and survival rate (%)

Marron showed a very weak response to stimuli after 24 h of transport and this condition was more noticeable after 48 h of transport. Marron started to show response to stimuli after 30 minutes during the temperature acclimation in the boxes. As most marron in the boxes were actively crawling and swimming, they were returned to the tanks for mortality observation. No mortality was observed in marron fed probiotic supplemented diet; while in basal diet fed marron mortality (6.7%) occurred at 48 h of transport. No clinical signs were observed on dead marron shells (Figure 2).

Discussion

A number of immune and physiological parameters are involved in the stress response following handling and live transport of crustaceans and fish such as behaviour, morbidity and vigour, THC, blood glucose, dehydration , oxygen uptake, blood composition, pH, hormones and ion [1,3]. In marron, the common selected parameters for testing following handling and simulated transport include dehydration [4] THC, haemolymph/plasma glucose, serum protein, dehydration [5], proportion of granule cells, clotting time and bacteraemia [6].

The circulating haemocytes of crustaceans are an essential part of the immune system, and perform functions such as phagocytosis, encapsulation, and lysis of foreign cells and much research in the defensive role of haemocyte in crustacean is being conducted [26,40,43]. The results suggest total haemocyte count (THC) is a reliable indicator of stress in crustaceans [2,44] including in marron [5,6].

In the present study, THC was investigated as an indicator for stress tolerance. THC of marron fed probiotic was significantly higher compared to basal diet fed marron both at 24 h and 48 h post simulated transport, indicating that the customised probiotic B mycoides was able to improve marron immunity. Enhancement of THC in probiotic fed marron leads to increased stress tolerance and diseases resistance, which results in a significantly higher survival rate (100%) over 24 h and 48 h post live transport. Higher THC of probiotic fed marron also provides better protection to the gill from parasites and bacteria pathogens which may cover and reduce respiration efficiency. Tinh et al. [45] suggested that probiotic bacteria can also be active on the gills and skin of the host. Inefficient respiration during handling and live transport may contribute to marron mortality in basal diet fed marron. Thus prior to transport, purging is generally essential in freshwater crayfish [3] to evacuate the GIT content and clean the gills and skin.

Handling and live transport creates physiological stresses which reduce THC in many crustacean species such as American lobster Homarus americanus) [1], crab Cancer pagurus) [2], western rock lobster Panulirus cygnus [2,46] marron[5,6], and a mollusc abalone Haliotis tuberculata [47]. Therefore, increasing the THC by feeding probiotic supplemented diets may improve stress tolerance and protect the animals from pathogens. This has been demonstrated in crayfish, Pacifastacus leniusculus, where a higher THC corresponded

Treatment	Intestinal bacteria population (million CFU/g of gut)		
	0h	24h	48h
Basal diet	626.7 ±19.7[a,1]	531.1 ± 15.7[a,1]	260.0 ±67.1[b,1]
B. mycoides diet	1656 ± 167.7[a,2]	1318 ± 131.5[a,2]	646 ± 16.4[b,2]

Table 1: Intestinal bacteria population (million CFU ± SE) of marron fed different diets at 24 h and 48 h post transport.

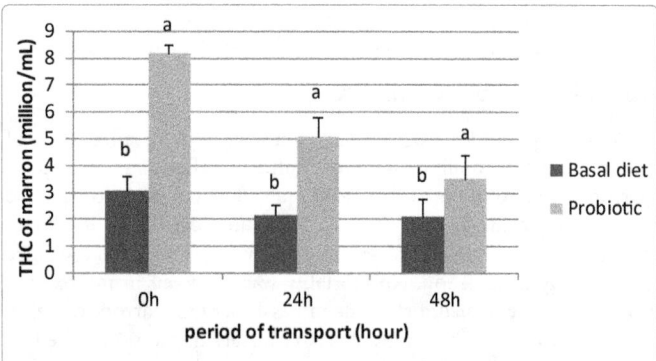

Figure 1: THC of marron fed basal and probiotic diets at 24h and 48h of transport.

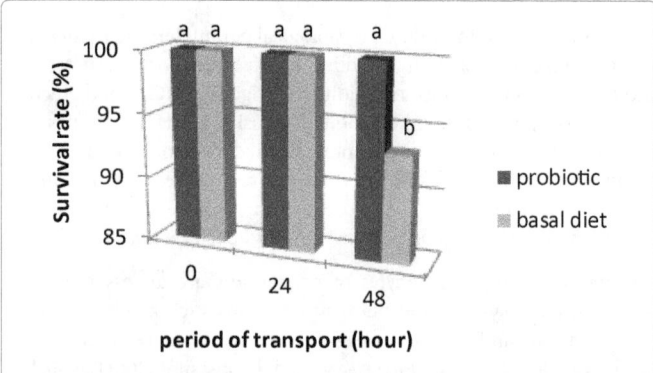

Figure 2: Survival rate (%) of marron fed basal and probiotics diets at 24 h and 48 h of transport.

to improved defence reactions of the animal when infected with the Oomycete fungus *Aphanomyces astaci* [48] and in marron where a higher THC corresponded to reduced bacteraemia and improved immunity against the pathogen *Vibrio mimicus*[6,19].

Bacterial population and diversity in the GIT is an important health component for aquatic animals [49,50] significantly affected by acute stress [51]. Stress due to high stocking density could also affect the performance of probiotics [24]. In the present study, the bacterial population level decreased both in probiotic and basal diet fed marron at 24h and a significant reduction was observed at 48h of transport indicating that prolonged stress significantly reduces intestinal bacteria of the animal. However, the higher bacterial population level of probiotic fed marron prior to transport resulted in the bacterial population remaining higher at 48h of transport compared to the population level of basal diet fed marron at the initiation of the transport test. The reduction of intestinal bacterial population due to handling and transport stress was also determined in other species. In Atlantic salmon, *Salmo salar* and rainbow trout, *Oncorhynchus mykiss* (Walbaum), adherent bacterial population in the mid-gut and hindgut were significantly reduced following acute handling stress, whereas the level increased in faeces, which suggests that considerable amounts of mucus was lost following stress[51,52]. This significant reduction of the intestinal bacterial population implies decline in the health status of the animal. The microbiota within the GIT can be considered a metabolically active organ, is an essential health component, provides protection against infection and instructs mucosal immunity [50]. Beneficial bacteria such as lactobacilli and bifidobacteria decrease following a stress response [50] and this may provide opportunistic pathogens to become established [2,44,53].

Haemolymph bacteria (bacteraemia) of marron fed probiotic diet was significantly lower compared to basal diet fed marron suggesting that greater THC plays an important role in reducing bacteria and foreign particles in marron haemolymph. Once pathogens or foreign particles enter the haemocoel, the haemocytes initiate phagocytosis [54]. In crayfish, hyaline cells are chiefly involved in phagocytosis, whereas semi-granular cells are active in encapsulation [41]. Crustacean haemocytes contain antibacterial activity [55,56] which can reduce the viable count of bacteria within 4 hours, however the antibacterial potency (per unit protein) varies from species to species [42].

Morbidity and mortality post live transport often occurs as a result of stress [3]. In the present study, morbidity and weakness indicated by no response to stimuli were observed both in basal diet and probiotic fed marron at 24 h and 48 h post simulated transport. However, after returning to the culture tanks the probiotic fed marron fully recovered and swam normally in less than 30 minutes, whereas basal diet fed marron took more time to recover and had several mortalities.

Other than mortality, marron also could be losing weight through dehydration from tissue and gill chambers while out of water during handling and live transport [3-5]. The present study indicated that dehydration of marron was observed in both test diets at 24 h and 48 h post transportation, however the dehydration was not significantly different between the two treatments. The results revealed that the dehydration of marron was comparable to the previous marron handling and live transport studies indicating that dehydration of 4 to 5% of the body weight over 24 to 72 h transportation is a common phenomenon. Jussila et al. observed a minor dehydration of marron during the first 4 hours that remained at 4.0 to 4.5% up to 24 h post handling and transportation, whereas Morrissy et al. observed wet dehydration of 3.9% during the first 24 h, with a further additional loss of 0.4% over the next 24 h [4,5]. Acute stress requires high energy which could reduce the hepatosomatic indices and contribute to the dehydration of the animal [23]. In marron, hepatopancreas significantly reduced after 24 h of transport [5]. Therefore, dehydration should be considered when crayfish are going to be transported for a long distance [5].

Probiotics have improved stress tolerance and immunity of various aquatic animal groups such as fish species including sea bream, *Sparus aurata* [11], grouper *Epinephelus coioides* [57], nile tilapia *Oreochromis niloticus* [58]; shrimp such as western king prawns, *Penaeus latisulcatus* [59] shrimp, *Litopenaeus vannamei* [60-62], and tiger shrimp, *Penaeus monodon* [63]. Among the bacterial genera evaluated for probiotic use, *Bacillus* species have been successful in improving the stress tolerance in these aquatic animal hosts and more recently in black swordtail, *Xiphophorus helleri* [16]. Our previous work using *B. mycoides* also showed improved marron immunity when experimentally challenged with the pathogen *Vibrio mimicus* [19].

Overall, the present study demonstrated that supplementation with host origin customized probiotic *B. mycoides* significantly improved the health status of marron by increasing their tolerance to a live transport stress test, which resulted in no mortality up to 48 h of simulated live transport.

Acknowledgement

The authors wish to The Directorate General of Higher Education (DIKTI) of Indonesia and The Department of Fishery, Hasanuddin University of Makassar Indonesia which financially supported this study. Thanks also are due to Mr. Peter McGinty of Aquatic Resource Management Pty Ltd Manjimup, Western Australia for providing the marron for this research. My gratitude are due to Mr. John Fielder of the Learning Centre Curtin University for the proof reading of the manuscript.

References

1. Lorenzon S, Giulianini PG, Martinis M, Ferrero EA (2007) Stress effect of different temperatures and air exposure during transport on physiological profiles in the American lobster Homarus americanus. Comp Biochem Physiol 147: 94-102.

2. Lorenzon S, Giulianini PG, Libralato S, Martinis M, Ferrero EA (2008) Stress effect of two different transport systems on the physiological profiles of the crab Cancer pagurus. Aquaculture 278: 156-163.

3. Fotedar S, Evans L (2011) Health management during handling and live transport of crustaceans: A review. Journal of Invertebrate Pathology 106: 143-152.

4. Morrissy NM, Walker PCF, Moore W (1992) An investigation of dehydration of marron (Cherax tenuimanus) in air during live transportation to market. Bernard Bowen Fisheries Research Institute, Western Australia Marine Research Laboratories. Fisheries Department of Western Australia. Fisheries Research Report 99: 1-21.

5. Jussila J, Paganini M, Mansfield S, Evans LH (1999) On physiological responses, plasma glucose, total hemocyte count and dehydration, of marron Cherax tenuimanus (Smith) to handling and transportation under simulated conditions. Freshwater Crayfish 12: 154-167.

6. Sang HM, Le KYT, Fotedar R (2009) Dietary supplementation of mannan oligosaccharide improves the immune responses and survival of marron, Cherax tenuimanus (Smith, 1912) when challenged with different stressors. Fish shellfish immunology 27: 341-8.

7. Morrissy NM, Evans L, Huner JV (1990) Australian freshwater crayfish: Aquaculture species. Journal of the World Aquaculture Society 21: 113-122.

8. Rouse DB, Kartamulia I (1992) Influence of salinity and temperature on molting and survival of the Australian freshwater crayfish (Cherax tenuimanus). Aquaculture 105: 47-52.

9. Barton BA, Iwama GK (1991) Physiological changes in fish from stress in aquaculture with emphasis on the response and effects of corticosteroids. Annu Rev Fish Dis 1: 3-26.

10. Ige BA (2013) Probiotics use in intensive fish farming. African Journal of Microbiology Research 7: 2701-2711.

11. Rollo A, Sulpizio R, Nardy M, Silvi S, Orpianesi C, et al.(2006) Live microbial feed supplement in aquaculture for improvement of stress tolerance. Fish Physiol Biochem 32: 167-177.

12. Zhang J, Liu Y, Tian L, Yang H, Liang G, et al. (2012) Effects of dietary mannan oligosaccharide on growth performance, gut morphology and stress tolerance of juvenile Pacific white shrimp, Litopenaeus vannamei. Fish & Shellfish Immunology 33: 1027-1032.

13. Soleimani N, Hoseinifar SH, Merrifield DL, Barati M, Abadi ZH (2012) Dietary supplementation of fructooligosaccharide (FOS) improves the innate immune response, stress resistance, digestive enzyme activities and growth performance of Caspian roach (Rutilus rutilus) fry. Fish & Shellfish Immunology 32: 316-321.

14. Lund I, Skov PV, Hansen BW (2012) Dietary supplementation of essential fatty acids in larval pikeperch (Sander lucioperca); short and long term effects on stress tolerance and metabolic physiology. Comp Biochem Physiol A Mol Integr Physiol 162: 340-348.

15. Wongsasak U, Chaijamrus S, Kumkhong S, Boonanuntanasarn S (2015) Effects of dietary supplementation with β-glucan and synbiotics on immune gene expression and immune parameters under ammonia stress in the Pacific white shrimp. Aquaculture 436: 179-187.

16. Hoseinifar SH, Roosta Z, Hajimoradloo A, Vakili F(2015) The effects of Lactobacillus acidophilus as feed supplement on skin mucosal immune parameters, intestinal microbiota, stress resistance and growth performance of black swordtail (Xiphophorus helleri). Fish & Shellfish Immunology 42: 533-538.

17. Ambas I, Buller N, Fotedar R (2015a) Isolation and screening of probiotic candidates from marron, Cherax cainii (Austin, 2002) gastrointestinal tract (GIT) and commercial probiotic products for the use in marron culture. Journal of Fish Diseases 38: 467-476.

18. Ambas I, Fotedar R, Buller N (2015b) Bacillus mycoides Improves Health of Gastrointestinal Tract in Marron (Cherax cainii, Austin 2002). J Aquac Mar Biol 2: 1-7.

19. Ambas I, Suriawan A, Fotedar R (2013) Immunological responses of customised probiotics-fed marron, Cherax tenuimanus, (Smith 1912) when challenged with Vibrio mimicus. Fish & shellfish immunology 35: 262-270.

20. Eaves LE, Ketterer PJ (1994) Mortalities in red claw crayfish Cherax quadricarinatus associated with systemic Vibrio mimicus infection Diseases of Aquatic Organism 19: 233-237.

21. Wong FYK, Fowler K, Desmarchelier PM (1995) Vibriosis due to Vibrio mimicus in Australian freshwater crayfish. Journal of Aquatic Animal Health 7: 284-291.

22. Evans L, Edgerton BF (2002) Pathogens, Parasites and Commensals, in Biology of Freshwater Crayfish, Blackwell Science Ltd, Osney OX2 0EL, UK.

23. Cruz PM, Ibanez AL, Hermosillo OAM, Saad HCR (2012) Use of Probiotics in Aquaculture. ISRN Microbiology.

24. Nayak SK (2010) Probiotics and immunity: A fish perspective. Fish and shellfish immunology 29: 2-14.

25. Tapia-Paniagua ST, Vidal S, Lobo C, Prieto-Alamo MJ, Jurado J, et al (2014) The treatment with the probiotic Shewanella putrefaciens Pdp11 of specimens of Solea senegalensis exposed to high stocking densities toenhance their resistance to disease. Fish shellfish immunology 41: 209-221.

26. Bachere E (2000) Shrimp immunity and disease control - Introduction. Aquaculture 191: 3-11.

27. Farzanfar A (2006) The use of probiotics in shrimp aquaculture. FEMS Immunology and Medical Microbiology 48: 149-158.

28. Ninawe AS, Selvin J (2009) Probiotics in shrimp aquaculture: Avenues and challenges. Critical Reviews in Microbiology 35: 43-66.

29. Lakshmi B, Viswanath B, Gopal DVRS (2013) Probiotics as Antiviral Agents in Shrimp Aquaculture. Journal of Pathogens.

30. Kesarcodi-Watson A (2009) Screening for probiotics of Green shell mussel larvae, Perna canaliculus, using a larval challenge bioassay. Aquaculture 296: 159-164.

31. Kesarcodi-Watson A, Kaspar H, Lategan MJ, Gibson L (2012) Performance of single and multi-strain probiotics during hatchery production of Greenshell™ mussel larvae, Perna canaliculus. Aquaculture 354-355: 56-63.

32. Keysami MA, Saad CR, Daud HM, Sijam K, Ar A (2007) Effect of Bacillus subtilis on growth development and survival of larvae Macrobrachium rosenbergii (de Man). Aquaculture nutrition 13: 131-136.

33. Keysami M, Mohammadpour M, Saad CR (2012) Probiotic activity of Bacillus subtilis in juvenile freshwater prawn, Macrobrachium rosenbergii (de Man) at different methods of administration to the feed. Aquaculture international 20: 499-511.

34. Li J, Tan B, Mai K (2009) Dietary probiotic Bacillus OJ and isomaltooligosaccharides influence the intestine microbial populations, immune responses and resistance to white spot syndrome virus in shrimp (Litopenaeus vannamei). Aquaculture 291: 35-40.

35. Zhang Q, Tan B, Mai K, Zhang W, Ma H, et al. (2011) Dietary administration of Bacillus (B. licheniformis and B. subtilis) and isomaltooligosaccharide influences the intestinal microflora, immunological parameters and resistance against Vibrio alginolyticus in shrimp, Penaeus japonicus (Decapoda: Penaeidae). Aquaculture research 42: 943-952.

36. Liu CH, Chiu CH, Wang SW, Cheng W (2012) Dietary administration of the probiotic, Bacillus subtilis E20, enhances the growth, innate immune responses, and disease resistance of the grouper, Epinephelus coioides. Fish shellfish immunology 33: 699-706.

37. Hai NV, Buller N, Fotedar R (2010) Effect of Customized Probiotics on the Physiological and Immunological Responses of Juvenile Western King Prawns (Penaeus latisulcatus Kishinouye, 1896) Challenged with Vibrio harveyi. Journal of Applied Aquaculture 22: 321-36.

38. Fisheries WA (2002) Code of Practice for the Harvest and the Post Harvest Handling of Live Marron for Food.

39. Buller NB (2004) Bacteria from Fish and Other Aquatic Animals: A Practical Identification Manual. CABI Publishing, Oxford shire, UK.

40. Fotedar S, Tsvetnenko E, Evans L (2001) Effect of air exposure on the immune system of the rock lobster Panulirus cygnus. Marine and Freshwater Research 52: 1351-1355.

41. Johansson MW, Keyser P, Sritunyalucksana K, Soderhall K (2000) Crustacean haemocytes and haematopoiesis. Aquaculture 191: 45-52.

42. Chisholm JRS, Smith VJ (1995) Comparison of antibacterial activity in the hemocytes of different crustacean species. Camp Biochem Physiol A Physiol 110: 39-45.

43. Soderhall K, Cerenius L (1992) Crustacean immunity. Annu Rev Fish Dis 2: 3-23.

44. Lorenz S, Francese M, Smith VJ, Ferrero EA (2001) Heavy metals affect the circulating haemocyte number in the shrimp Palaemon elegans. Fish shellfish immunology 11: 459-472.

45. Tinh NT, Dierckens K, Sorgeloos P, Bossier P (2008) A review of the functionality of probiotics in the larviculture food chain. Mar. Biotechnol 10: 1-12.

46. Jussila J, Mcbride S, Jago J, Evans LH (2001) Hemolymph clotting time as an indicator of stress in western rock lobster (Panulirus cygnus George). Aquaculture 199: 185-193.

47. Cardinaud M, Offret C, Huchette S, Moraga D, Paillard C (2014) The impacts of handling and air exposure on immune parameters, gene expression, and susceptibility to vibriosis of European abalone Haliotis tuberculata. Fish shellfish immunology 36: 1-8.

48. Persson M, Cerenius L, Soderhall K (1987) The Influence of Haemocyte Number on The Resistance of The Freshwater Crayfish, Pasifastacus leniusculus Dana, to The Parasitic Fungus, Aphanomyces astacii. Journal of fish diseases 10: 471-477.

49. Gomez GD, Balcazar JL (2008) A review on the interactions between gut microbiota and innate immunity of fish. FEMS Immunol Med Microbiol 52:145-154.

50. Gaggìa F, Mattarelli P, Biavati B (2010) Review Probiotics and prebiotics in animal feeding for safe food production. International Journal of Food Microbiology 141: S15-28.

51. Olsen RE, Sundell K, Mayhew TM, Myklebust R, Ringø E (2005) Acute stress alters intestinal function of rainbow trout, Oncorhynchus mykiss (Walbaum). Aquacult 250: 480-495.

52. Olsen RE, Sundell K, Hansen T, Hemre GI, Myklebust R, et al. (2002) Acute stress alters the intestinal lining of Atlantic salmon, Salmo salar L.: An electron microscopical study Fish Physiology and Biochemistry 26: 211-221.

53. Ringo E, Lovmo L, Kristiansen M, Bakken Y, Salinas I, et al. (2010) Lactic acid bacteria vs. pathogens in the gastrointestinal tract of fish: a review. Aquaculture Research 41: 451-467.

54. Li CC, Yeh ST, Chen JC (2010) Innate immunity of the white shrimp Litopenaeus vannamei weakened by the combination of a Vibrio alginolyticus injection and low-salinity stress. Fish & Shellfish Immunology 28: 121-127.

55. Van de Braak CB, Botterblom MH, Liu W, Taverne N, Van der Knaap WP, et al. (2002) The role of the haematopoietic tissue in haemocyte production and maturation in the black tiger shrimp (Penaeus monodon). Fish & shellfish immunology 12: 253-72.

56. Haug T, Kjuul A, Stensvåg K, Sandsdalen E, Styrvold O (2002) Antibacterial activity in four marine crustacean decapods. Fish and Shellfish Immunology 12: 371-85.

57. Sun YZ, Yang HL, Ma RL, Lin WY (2010) Probiotic applications of two dominant gut Bacillus strains with antagonistic activity improved the growth performance and immune responses of grouper Epinephelus coioides. Fish and Shellfish Immunology 29: 803-809.

58. Pirarat N, Pinpimai K, Endo M, Katagiri T, Ponpornpisit A, et al. (2011) Modulation of intestinal morphology and immunity in nile tilapia (Oreochromis niloticus) by Lactobacillus rhamnosus GG. Research in Veterinary Science 91: e92-e97.

59. Hai NV, Buller N, Fotedar R (2009b) The use of customised probiotics in the cultivation of western king prawns (Penaeus latisulcatus Kishinouye, 1896). Fish Shellfish Immunology 27: 100-104.

60. Tseng DY, Ho PL, Huang SY, Cheng SC, Shiu YL, et al. (2009) Enhancement of immunity and disease resistance in the white shrimp, Litopenaeus vannamei, by the probiotic, Bacillus subtilis E20. Fish Shellfish Immunology 26: 339-344.

61. Li K, Zheng T, Tian Y, Xi F, Yuan J, et al. (2007) Beneficial effects of Bacillus licheniformis on the intestinal microflora and immunity of the white shrimp, Litopenaeus vannamei. Biotechnology Letters 29: 525-30.

62. Liu KF, Chiu CH, Shiu YL, Cheng W, Liu CH (2010) Effects of the probiotic, Bacillus subtilis E20, on the survival, development, stress tolerance, and immune status of white shrimp, Litopenaeus vannamei larvae. Fish Shellfish Immunology 28: 837-844.

63. Rengpipat S, Rukpratanporn, S, Piyatiratitivorakul S, Menasaveta P (2000) Immunity enhancement in black tiger shrimp (Penaeus monodon) by a probiont bacterium (Bacillus S11). Aquaculture 191: 271-288.

35

Shell Growth Performance of Hatchery Produced *Pinctada margaritifera*: Family Effect and Relation with Cultured Pearl Weight

Chin-Long KY* and Gilles LE MOULLAC

Ifremer, UMR 241 Ecosystèmes Insulaires Oceaniens (EIO), Labex Corail, Centre du Pacifique, BP 49, 98719 Taravao, Tahiti, French Polynesia

Abstract

Size is the most important and valuable quality trait of cultured pearls produced by the black-lipped pearl oyster, *Pinctada margaritifera*. In French Polynesia, several breeding programmes have been started that aim to improve this size trait, which is highly related to shell growth rate in both recipient and donor oysters. Shell growth rate dictates the time of grafting, size of implanted nuclei and bio-mineralisation potential of the mantle and pearl sac. We assessed shell growth rate through routine digital shell biometric analysis on 22 hatchery families produced between 2005 and 2008. These included full-sib families and half-sib families derived from polyandry (one dam crossed with two or more sires). Results showed that: 1) a significant family effect was recorded for growth performance, analysed according to the Von Bertalanffy model, 2) a significant male effect was observed for some of the half-sib families and 3) a relationship was found between the shell growth performances of five families randomly selected and used as graft donors in a grafting experiment and the final weight of the cultured pearls produced. These results have important implications for the breeding of pearl oysters with high growth capacities: it may be possible to select oyster lines for the potential to produce large pearls using shell equivalent diameter estimated by the digital method as a selection criterion.

Keywords: *Pinctada margaritifera*; Pearl oyster; Hatchery-produced families; Growth performance; Shell diameter; Cultured pearl size

Introduction

The black-lip pearl oyster *Pinctada margaritifera* is a widespread species, occurring across the Indian and Pacific Oceans and in the eastern Mediterranean Sea. It also has a wide latitudinal range, from 30°N to 28°S, reaching its greatest abundance in the lagoons of French Polynesia [1]. It typically inhabits oligotrophic waters of low turbidity, attaching itself by byssal threads to coral reef substrates [2,3]. Today, the black pearl industry remains the second most important economic activity in French Polynesia, after tourism, and is the largest export industry (7.8 billion CFP Francs in 2013). By 2013, this industry had developed on 25 islands and atolls, with 517 pearl farms located across three archipelagos: Tuamotu (398 farms), Gambier (79 farms), and Society (40 farms) [4]. However, this industry has been declining since 2001 and is now facing a critical situation due to a combination of several economic factors: slowdown of the world economy, overproduction of pearls, and poor average pearl quality. It is estimated that only 5% of harvested pearls can be classed as grade A quality according to local regulatory standards [5]. In this context, there is a need to take measures that will enhance pearl quality, thus increasing the proportion of high value pearls.

One of the most important traits determining the commercial value of pearls is their weight (for equivalent grade quality); with the heaviest pearls generally commanding the highest prices. Weight of cultured *P. margaritifera* pearls typically ranges from 0.75 g to 14.48 g. Heavier specimens are usually issued from a *surgreffe* operation: insertion of a new nucleus following pearl harvest [6,7]. Shell and cultured pearl formation result from the bio-mineralization activities of two distinct tissues in a grafted pearl oyster: the mantle of the recipient oyster and the pearl sac built from graft tissues from the donor [5]. The pearl weight finally attained depends on several factors, including whether or not the donor oyster originated from the wild [8] or from a hatchery-produced family [9,10]. Some studies have shown phenotypic correlations between pearl weight and recipient oyster shell traits, e.g., for *Pinctada fucata* [11-13]. Wada and Komaru [14] studied the phenotypic correlation

between pearl size (and thus weight) and recipient shell valve weight in *P. fucata martensii* and found a positive relationship whereby oysters with heavier valves produced larger pearls. Another study, on *Hyriopsis cumingii*, showed that improving body length or weight in culture of this species indirectly led to improved pearl weight and size [15]. Jerry et al. [16], revealed genotype by environment (G*E) interaction for pearl weight in *Pinctada maxima* reared at two commercial grow-out locations. Indeed, environmental influences have been shown to be important factors to consider in *P. margaritifera* aquaculture [17,18].

The most important environmental parameters influencing bivalve growth are temperature and food availability [19-21], which is typical of many aquatic invertebrates. The influence of food availability on *P. margaritifera* in French Polynesia was demonstrated by Pouvreau et al. [22,23]. Growth studies are of interest for pearl farming, since growth represents the integrated response of the organism's entire physiological activity. Shell growth rates can provide precious information on pearl growth in *P. margaritifera* since shell increment and deposition of nacreous matter on the implanted nucleus are strongly correlated [24]. Pouvreau & Prasil [23] also demonstrated geographic variability in *P. margaritifera* growth among different sites in French Polynesia. This bivalve grows rapidly up to a phase of its development where size increase levels off and then falls to zero. On the basis of growth modelling (Von Bertalanffy model [25]), the time required for a pearl oyster to reach the size necessary to be grafted as a recipient (height=100

*Corresponding author: Chin-Long KY, Ifremer, UMR 241 Ecosystèmes Oceaniens (EIO), Labex Corail, Centre du Pacifique, BP 49, 98719 Taravao, Tahiti, French Polynesia, E-mail: chinky@ifremer.fr

mm) depends on pearl farming site as, in French Polynesia, the annual growth rate in shell size varies from 19.7 mm to 31.8 mm year[-1] [23].

Selective breeding programmes have been shown to be effective for improving performance of many aquaculture species. A selective breeding programme on *P. margaritifera* is currently underway at Ifremer (French Research Institute for the Exploitation of the Sea) in French Polynesia, with the main goal of establishing pearl oyster lines selected for growth. The main objective of the present study was to analyse the growth performance of 22 G1 families by routine digital growth measurements on their shells and then assess the impact of growth and family effects on pearl weight through a grafting experiment using donor oysters from some of these families. This study will help the development of breeding programmes to produce oyster lines that will enhance cultured pearl weight.

Materials and Methods

Breeding and rearing of *P. margaritifera*

Twenty-two families (full and half-sib) were produced in the Ifremer hatchery in Vairao (Tahiti, French Polynesia) between 2005 and 2008 by spawning non-selected females (N=14) and males (N=22) from wild broodstock (Table 1). Spawning was triggered by thermal shock [25] whereby the pearl oysters were placed in cooled seawater at 20°C for one night before being plunged into seawater at 31°C to 32°C. Immediately spawning started, the male and female oysters were placed in separate containers for gamete collection. To minimize the risk of contamination, oocytes were thoroughly rinsed out of the mantle cavities of the spawning females. The oocytes were fertilized with the spermatozoa of males selected on the basis of sperm motility. Rearing was conducted in a static system without antibiotics. The larval

G1 Families		Crosses		Growth parameters		
Name	Age (year)	Female	Male	k	Eq. D.	R²
O58	3.80	x	y	0.61	11.46	0.62
F610	3.12	K43	O24	0.64	10.65	0.97
F612	3.12	I21	O40	0.90	8.96	0.89
F613	3.02	FA7	FA1	0.35	14.62	0.94
F615	3.02	EQ1	EZ8	0.43	12.43	0.94
F618	2.53	ED2	FF8	0.33	15.41	0.88
F617	2.53	ED2	GC1	0.34	12.55	0.80
F616	2.53	ED2	BL3	0.43	9.52	0.98
F620	2.47	C8	ED4	0.25	18.92	0.82
F622	2.47	C8	B99	0.40	12.57	0.94
F621	2.47	C8	GV8	0.77	11.42	0.94
F619	2.47	C8	B92	0.36	11.02	0.91
F701	2.30	279	367	0.82	8.61	0.85
F702	2.28	O48	W2	0.47	12.22	0.91
F703	2.28	O48	268	0.51	11.16	0.90
F704	2.28	237	251	1.05	7.34	0.90
F732	1.51	BS7	AY8	0.27	17.59	0.97
F733	1.43	AV6	GE9	0.23	26.07	1.00
F801	1.32	BX2	CD9	0.24	21.39	1.00
F806	1.32	AS8	350	0.21	22.04	0.96
F805	1.32	AS8	270	0.21	21.41	0.99
F804	1.32	AS8	CC2	0.30	18.76	0.98

Table 1: Stocklist of the 22 first generation (G1) families of *P. margaritifera* produced in the hatchery from wild broodstocks. Crosses shown in boxes used the same female crossed with several males (4 polyandry matings that produced half-sib families). Growth parameters are: 1) the k constants and 2) equivalent shell diameters eq. D, following the Von Bertalanffy bivalve growth model. R² value corresponds to the relationship between shell height measured by calliper and eq. D.

and juvenile rearing procedures used were as described in [26]. After 3 months, the young spat was placed in plastic trays (**Aquapurse**) and reared in Vairao lagoon, where they were suspended on long lines. The trays were maintained at a depth ranging between 6 and 10 metres and cleaned every 3 months.

Digital shell growth analysis of the *P. margaritifera* families

A sample of pearl oysters (N=478) in the size range size 0.7-10.2 cm shell height was used to validate the digital shell growth analysis method described above. Size of all pearl oysters was measured every four months. The oysters were spread out over a cream-coloured PVC plate, and pictures were taken using a digital camera (Olympus μ790 SW, 7.1 megapixels) fixed on a base at 1 metre above the oysters. This method made it possible to simultaneously measure a hundred young oysters or ten large ones. The digital biometry was first compared with the classic measurement method for shell height, using a calliper on a large range of oyster sizes. The calibration of pictures was made using black circles of 10 cm diameter. Pictures were first treated with Photoshop (©Adobe Systems CS3, version 10.0). ImageJ software (©Broken Symmetry, version 1.4.3.67) was then used to obtain the area of the oysters, which allowed the calculation of the equivalent diameter and the major and minor axes of the ellipse of adjustment. Correlations were first tested between the height data collected with callipers and the digitally measured parameters: area and major and minor axes. The growth of pearl oysters was described using the equivalent diameter (eqD). For each of the 22 hatchery-produced families, 50 individuals were measured during cleaning process, every 3 months.

Experimental grafting procedure

In order to evaluate the link between growth of donor families and the corresponding harvested pearls, a graft experiment was designed. As the grafting operation, itself may introduce factors that influence cultured pearl quality [27], all grafts were undertaken by a single professional technician to minimise variability in technique. The donor oysters used were from five families (O58, F612, F613, F615 and F622), randomly selected from the 22 hatchery-produced families. A total of 50 donors (10 per family) were used to perform 500 grafts (10 grafts per donor) over a 4-day period (28-31 May 2008) in Mangareva (Gambier Archipelago, French Polynesia) under the same conditions as used for a commercial graft (see [28] for a description of this method). All recipient pearl oysters came hatchery-produced oysters from a same family (other than the 22 studied families for growth) with mean (± SE) antero-posterior measurement of 75.38 (± 5.64) mm and dorso-ventral measurement of 80.08 (± 7.72) mm. They were selected based on good visible health status (colour of the visceral mass and gills), shell size appearance, and muscle resistance when the shells were pried open for grafting. Each recipient was grafted using a 2.0 BU nucleus (6.06 mm diameter; Nucleus Bio, Hyakusyo Co. Japan). The grafted pearl oysters were reared in kangaroo pocket baskets in Mangareva lagoon, following standard commercial practices until harvest of the pearls 14 months later (15-17 July 2009).

Cultured pearl weight measurement

After 14 months of culture, the cultured pearls were harvested and placed in a compartmented box that allowed traceability between the pearls and the corresponding donor oysters. Some *keshi* (small irregular shaped nacreous but non-nucleated pearls that form during the culture period after nuclei have been rejected) were also harvested, but not graded. Cultured pearls were then cleaned by ultrasonication in soapy water (hand washing) with a LEO 801 laboratory cleaner (2 L capacity, 80 W, 46 kHz) according to Ky et al. [9]. Each pearl was then individually weighed with an Ohaus Explorer EP214D analytical balance (0.1 mg sensitivity).

Statistical analysis

Data from shell growth were modelled according to the nonlinear regression of Von Bertalanffy [29], which has been widely used for bivalve growth [25], using XLStat 2008.6.08 software. The Von Bertalanffy equation is: $H = H_\infty\left[1 - e^{-k\,(t1-t0)}\right]$, where H is the height in millimetres at time t, H_∞ is the asymptotic height in millimetres, t is the age in years and k is the rate at which the asymptotic value is approached in year^{-1}. H and H_∞ were replaced by eqD and eqD_∞ as described in the above "digital shell growth analysis" section. The shell heights are presented on the figures with vertical bars representing the 95% confidence intervals. One-way analysis of variance (ANOVA) was used to test for differences for pearl weights among the families. If the overall F-test was significant, Tukey multiple comparisons were performed among all pairs of families. Correlations between the shell growth donor k value and pearl size were tested using the critical value table for Pearson's correlation coefficient at the 5% alpha level.

Results and Discussion

Family effect on shell growth performance

The equivalent diameter (eqD value) of oyster shells was significantly correlated with the height measured among the 478 pearl oysters in our study (r=0.994). This made it possible to monitor shell growth performance of the 22 hatchery families using the digital method. The constant values (k and eqD_∞) of the non-linear regressions of the Von Bertalanffy model are presented in Table 1 and the corresponding modelled growth in Figure 1. Great variability was observed for growth potential among the 22 families studied. Comparison between pairs of families of the same age reared simultaneously in the same location and conditions (F616-F621 and F732-F733) show significant differences in shell growth capacity (Figure 1). The growth difference between families F616 and F621, measured over one year (from 1.5 to 2.5 years old), was 35 to 44% in favour of family F621. This corresponds to growth constants of k=0.43 and 0.77 for F616 and F621, respectively (Table 1). In contrast, between the younger families F732 and F733, in which growth was monitored from the age of 0.84 to nearly 1.5 years old, the growth differences were 2% to 20%, respectively at these measurement

times. These growth differences correspond to eqD_∞ values of 17.6 and 26.07 for F732 and F733 families, respectively (Table 1).

In general, shell growth rate is directly correlated with the biological age of an individual P. margaritifera pearl oyster. Growth rate (in terms of shell height) is rapid until the third year, and then decreases [23]. The specificity of this growth pattern has already been reported in other species in the Pinctada genus [1,30,31]. The decrease in growth rate in relation to age is highly correlated with the metabolic cost associated with sexual maturation, as the energy previously allocated to growth performance is instead diverted mainly to sexual maturation and reproductive effort [32]. The bivalves nevertheless attain their genetically-determined species-specific maximal size. Our results suggest there is a family effect for growth performance, which is particularly visible in the F616-F621 and F732-F733 pairs that were produced in the same period and therefore reared in identical environmental conditions. The growth parameters proposed by the Von Bertalanffy non-linear regression model may be used as performance descriptors. The parameters in our study should be considered as preliminary ones, however, because the growth of the families did not cover the entire life of the pearl oysters, these individuals were relatively young, and the growth asymptote was not reached. The observed differences between the growth of the families could result from three types of growth dynamic: 1) the same Eq D∞=11.79 x $(1-e^{(-0.47\,x\,age)})$ but different k values; 2) the same k value, but different Eq D∞=11.79 x $(1-e^{(-0.47\,x\,age)})$ or 3) different k and eqD_∞ values.

Significant half-sib family effects were found within the four polyandry matings (Table 1). Comparing the half-sib families F616, F617 and F618, the male FF8 (sire of the F618 half-sib family) offers a growth advantage of 15% and 24% on average over families F617 and F616, respectively (Figure 2a). For the half-sibs F619, F620, F621 and F622 families, the male GV8 (sire of the F621 half-sib family) offers a growth advantage of 36, 17 and 19% on average in comparison to families F619, F620 and F622, respectively (Figure 2b). In contrast, the F702-F703 and F805-F806 pairs were not significantly different in growth performance (Figures 2c and 2d).

For the four half-sib combinations resulting from the polyandry mating design (a single female crossed with two to four different males), the result showed that the growth capacity could be determined by the male, resulting in a side effect, especially for the F618 and F621 families. This finding complements results showing half-sib family effect in an earlier study on spat growth performance [33], where only a maternal effect was suggested at this earlier life stage. The results obtained here with the family pairs F702-F703 and F805-F806, revealed no male effects, indicating that in some cases sire confer no growth benefit to their progenies. Selection of males remains an interesting prospect for genetic programs, as P. margaritifera is a protandric species and this therefore represents a means to save time.

Pearl weight and its relationship with shell growth performance of the families used as donors

After 14 months of culture, the 366 harvested pearls from the five randomly selected families used as donors (FO58, F612, F613, F615 and F622) showed an average weight of 1.16 ± 0.36 g, with a minimum–maximum range of 0.40 g to 2.72 g. A highly significant family effect was detected for pearl weight (F=12.00, p <0.0001). Tukey multiple comparisons showed that family O58 was significantly different from the other four, but that these latter were not significantly different from one another. A relationship was established between the k parameter of growth model of the five families (Table 1) and the weight of the corresponding pearls (Figure 3). The nonlinear relationship was significant (p<0.05, r=0.870, ddl=3).

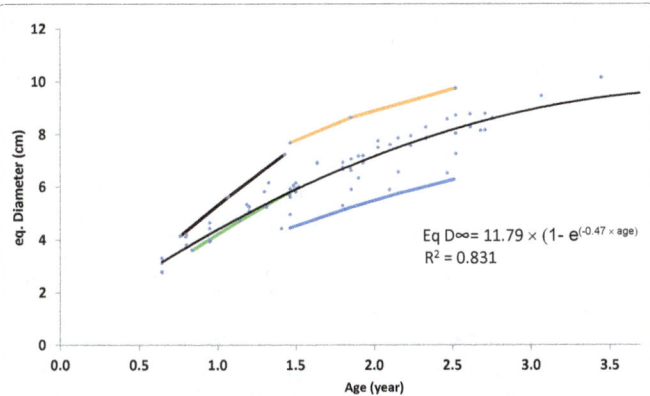

Figure 1: Establishment of a growth model for 22hatchery-produced families of *P. margaritifera*. The graph shows the relationship between their shell equivalent diameters (cm) and ages (years) all families combined (50 < N < 250). Families F616 (blue), F621 (orange), F733 (red) and F732 (green) shell growth capacities are highlighted. The Von Bertalanffy non-liner regression model is shown.

Eq D∞= 11.79 × (1- e$^{(-0.47 × age)}$)
R² = 0.831

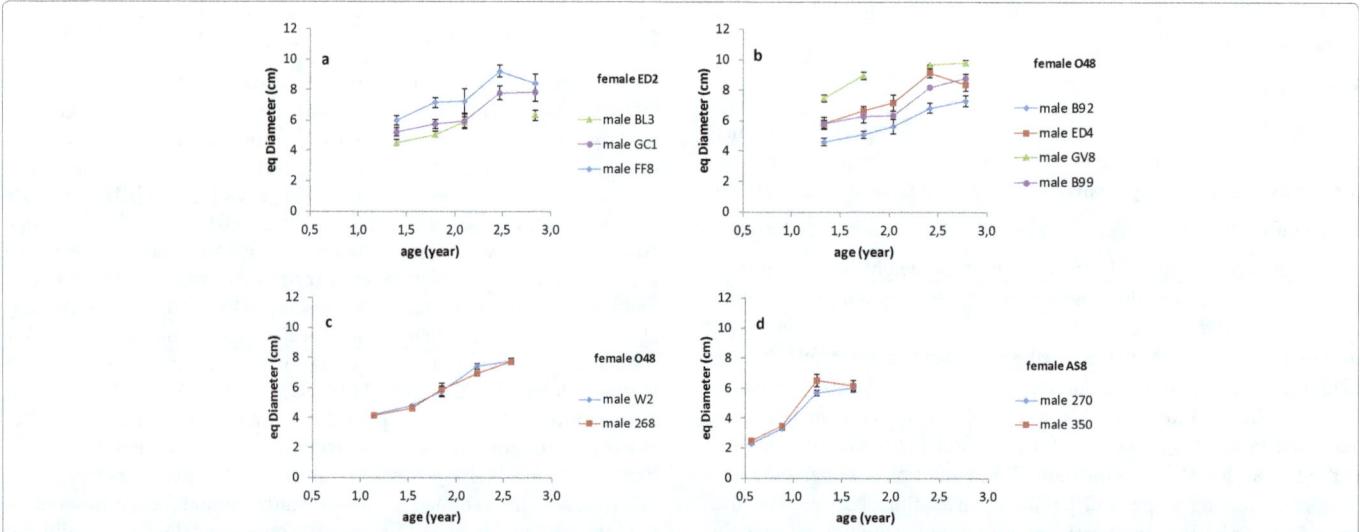

Figure 2: Growth of the 11 *P. margaritifera* half-sib families (corresponding to four polyandrous matings, refer Table 1); (a) families F616, F617 and F618 (b) families F619, F620, F621 and F622 (c) families F702 and F703 (d)families F805 and F806; the pearl oyster shell equivalent diameter (eq. Diameter) is expressed in cm and oyster age in years (means are presented with vertical bars representing 95% confidence intervals with 50 < N < 250).

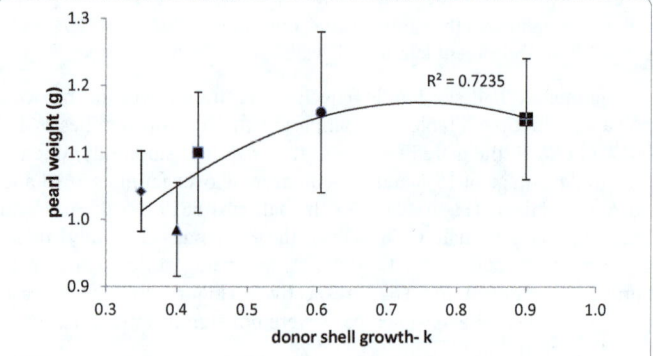

Figure 3: Correlations between the cultured pearl weight (in g) and donor oyster k values, following the Von Bertalanffy bivalve growth model (diamond: F613, triangle: F622, square: F615, circle: O58, cross: F612).

Pearl weight is an important commercial trait. Generally, the greater the nacre deposition rate (as measured by nacre weight) for oysters implanted with nuclei of the same size, the heavier (and more valuable) the resulting pearl should be, all other quality traits being equal. Consequently, there may be advantages to the industry in using pearl oyster donors promoting rapid nacreous deposition. Our results clearly show a correlation between the k growth parameter and the weight of the harvested pearls. Pearl weight is mainly dependent on calcium metabolism in the pearl sac, the epithelial tissue around the pearl epithelium derived from the mantle cells of the donor [34]. Although research on factors influencing the weight of pearls is limited, influence of donor oysters on pearl weight has been suggested [35]. Indeed, cells of donor origin appear to persist in the pearl sac. When genotyping the pearl sac and the corresponding recipient oysters with anonymous nuclear markers, alleles from the donor oyster are still detected in the pearl sac at pearl harvest [36]. The persistence of donor oyster cells and their activity in the pearl sac supports our result concerning the relationship between k values of donor pearl oysters and pearl weight obtained with these donors. Nevertheless, further studies should be conducted, particularly on the role of the recipient oyster, as its physiological condition and activity in interaction with the environment probably also strongly influence the potential for calcium metabolism in the pearl sac epithelium.

Care should be taken concerning possible relationships between rapid nacre deposition and other pearl quality traits, such as colour or calcite pearl formation. For example, Snow et al. [37] hypothesized that pearls with a smooth surface and brilliant lustre are produced when consistent and regular crystal formation occurs, which is not compatible with a high nacre deposition rate. Ky et al. [9] showed a strong family effect on pearl weight, but a significant effect of pearl colour was also detected, as well as a significant family·colour interaction. This indicates that potential relationships between weight and colour may vary among families. Selection for one trait may thus lead to indirect selection on others. As with any genetic improvement method, a better understanding of genetic correlations is necessary to avoid inadvertent selection for or against non-target traits.

Conclusion

Our study demonstrates that the potential exists to improve pearl growth rates in *P. margaritifera*. Selection of hatchery-produced *P. margaritifera* donors with high growth performances, and thus high potential for nacre deposition, could be used to increase cultured pearl weight, size and nacre deposition rate. Such family-based selection could be assisted by molecular tools such as those revealed in initial gene expression studies as potential biomarkers for pearl growth, e.g., *Pif*, *Aspein* and *Pearlin* gene families [38]. This genetic selection, using the routine digital method, could also be incorporated in hatchery-reared spat, which are now used for commercial production in *P. margartifera* in French Polynesia [39,40]. Overall, to improve cultured pearl size via selection, a multi-trait approach should be used, integrating several oyster lines, considering the key role played by recipients, and taking into account quantitative genetic control, environmental effects and associated correlations. In an effort to identify the genetic basis and molecular mechanisms underlying shell growth in *P. margaritifera* and provide fundamental information to assist selective breeding of superior pearl oyster lines with high performance growth potential, differentially expressed genes could be identified among the different spat shell size variants at the transcriptome level by RNA sequencing. Such integrative research should improve our understanding of the processes implicated in animal growth performance but also in the cultured pearl traits important for the *P. margaritifera* industry.

Acknowledgements

The authors are grateful to the team at Pearl Farm and support provided by the Direction des Ressources Marines et Minières (Mangareva Island, Gambier Archipelago, French Polynesia) for the experimental grafts and maintenance of the oyster cultures. The authors would also like to thank Mayalen Maihota, Jacques Moriceau, Manaarii Sham Koua, Roger Tetumu and Vincent Vanaa for their technical support with the biological material used in the present study that was produced between 2005 and 2008.

References

1. Gervis MH, Sims NA (1992) The biology and culture of pearl oysters (Bivalvia: Pterridae). ICLARM Stud 21: 49.

2. Pouvreau S, Jonquières G, Buestel D (2000) Filtration by the pearl oyster, Pinctada margaritifera, under conditions of low seston load and small particle size in a tropical lagoon habitat. Aquaculture 176: 295-314.

3. Yukihira H, Lucas JS, Klump DW (2000) Comparative effects of temperature on suspension feeding and energy budgets of the pearl oysters Pinctada margaritifera and P. maxima. Mar Ecol Prog Ser 195: 179-188.

4. Talvard C (2015) In: Points Forts de la Polynésie Française. Institut de la Statistique en Polynésie Française, Papeete, Polynésie Française.

5. Ellis S, Haws M (1999) Producing pearls using the black-lip pearl oyster. Center for Tropical and Sub-tropical Aquaculture Publications, Hawaii.

6. Demmer J, Cabral P, Ky CL (2016) Comparison of harvested rate and nacre deposition parameters between cultured pearls issued from initial graft and second nucleus insertion in P. margaritifera. Aquaculture Research 47: 3297-3306.

7. Ky CL, Demmer J, Sham Koua M, Cabral P (2015) Development of cultured pearl circles and shape after initial graft and second nucleus insertion in the black-lipped pearl oyster Pinctada margaritifera. Journal of Shellfish Research 34: 319-328.

8. Tayale A, Gueguen Y, Treguier C, Le Grand J, Cochennec-Laureau N, et al. (2012) Evidence of donor effect on cultured pearl quality from a duplicated grafting experiment on Pinctada margaritifera using wild donors. Aquatic Living Resource 25: 269-280.

9. Ky CL, Blay C, Sham Koua M, Vanaa V, Lo C, et al. (2013) Family effect on cultured pearl quality in black-lipped pearl oyster Pinctada margaritifera and insights for genetic improvement. Aquatic Living Resources 26: 133-145.

10. Ky CL, Blay C, Sham Koua M, Lo C, Cabral P (2014a) Indirect improvement of pearl grade and shape in farmed Pinctada margaritifera by donor "oyster" selection for green pearls. Aquaculture 432: 154-162.

11. Wada KT (1984) Breeding study of the pearl oyster Pinctada fucata. Bull Natl Res Inst Aquaculture 6: 79-157.

12. Velayudan TS, Chellam A, Dharmaraj S, Victor ACC, Kasim HM (1996) Comparison of growth and shell attributes for four generations of pearl oyster Pinctada fucata (Gould) produced in the hatchery. Indian Journal of Fisheries 43: 69-77.

13. He M, Guan Y, Yuan T, Zhang H (2008) Realized heritability and response to selection for shell height in the pearl oyster Pinctada fucata (Gould). Aquaculture Research 39: 801-805.

14. Wada KT, Komaru A (1996) Color and weight of pearls produced by grafting the mantle tissue from a selected population for white shell color of the Japanese pearl oyster Pinctada fucata martensii (Dunker). Aquaculture 142: 25-32.

15. Jin W, Bai Z, Fu L, Zhang, G, Li J (2012) Genetic analysis of early growth traits of the triangle shell mussel, Hyriopsis cumingii, as an insight for potential genetic improvement to pearl quality and yield. Aquaculture International 20: 927-933.

16. Jerry DR, Kvingedal R, Lind CE, Evans BS, Taylor JJU, et al. (2012) Donor-oyster derived heritability estimates and the effect of genotype x environment interaction on the production of pearl quality traits in the silver-lip pearl oyster, Pinctada maxima. Aquaculture 338: 66-71.

17. Ky CL, Blay C, Aiho V, Cabral P, Le Moullac G, et al. (2017) Macro-geographical differences influenced by family-based expression on cultured pearl grade, shape and colour in the black-lip pearl oyster Pinctada margaritifera: A preliminary case study in French Polynesia. Aquaculture Research 48: 270-282.

18. Le Pabic L, Parrad P, Sham KM, Nakasai S, Saulnier D, et al. (2016) Culture site dependence on pearl size realization in Pinctada margaritifera in relation to recipient oyster growth and mantle graft bio-mineralization gene expression using the same donor phenotype. Estuarine, Coastal and Shelf Science 182: 294-303.

19. Honkoop PJC, Beukema JJ (1997) Loss of body mass in winter in three intertidal bivalve species: An experimental and observational study of the interacting effects between water temperature, feeding time and feeding behaviour. J Exp Mar Biol Ecol 212: 277-297.

20. Pilditch CA, Grant J (1999) Effect of temperature fluctuations and food supply on the growth and metabolism of juvenile sea scallops (Placopecten magellanicus). Marine Biology 134: 235-248.

21. Marsden ID (2004) Laboratory and field experiments investigating the effects of reduced salinity and seston availability on growth of the little-neck clam Austrovenus stutchbury. Marine Ecology Progress Series 266: 157-171.

22. Pouvreau S, Tiapari J, Gangnery A, Lagarde F, Garnier M, et al. (2000) Growth of the black-lip pearl oyster, Pinctada margaritifera, in suspended culture under hydro-biological conditions of Takapoto lagoon (French Polynesia). Aquaculture 184: 133-154.

23. Pouvreau S, Prasil V (2001) Growth of the black-lip pearl oyster, Pinctada margaritifera, at nine culture sites of French Polynesia: Synthesis of several sampling designs conducted between 1994 and 1999. Aquatic Living Resources 14: 155-163.

24. Coeroli M, Mizuno K (1985) Study of different factors having an influence upon the pearl production of the black lip pearl oyster. Proceedings of The Fifth International Coral Reef Congress 5: 551-556.

25. Hui B, Vonau V, Moriceau J, Tetumu R, Vanaa V, et al. (2011) Hatchery-scale trials using cryopreserved spermatozoa of black-lip pearl oyster (Pinctada margaritifera). Aquatic Living Resources 24: 219-223.

26. Vakily JM (1992) Determination and comparison of bivalve growth, with emphasis on Thailand and other tropical areas. ICLARM Tech Rep 36: 125.

27. Ky CL, Lau C, Sham Koua M, Lo C (2015) Growth performance comparison of Pinctada margaritifera juveniles produced by thermal shock or gonad scarification spawning procedures. Journal of Shellfish Research 34: 811-817.

28. Ky CL, Nakasai S, Molinari N, Devaux D (2015) Influence of grafter skill and season on cultured pearl shape, circles and rejects in Pinctada margaritifera aquaculture in Mangareva lagoon. Aquaculture 435: 361-370.

29. Ky CL, Molinari N, Moe E, Pommier S (2014b) Impact of season and grafter skill on nucleus retention and pearl oyster mortality rate in Pinctada margaritifera aquaculture. Aquaculture International 22: 1689-1701.

30. Von-Bertalanffy L (1938) A quantitative theory of organic growth (Inquiries on growth laws II). Hum Biol 10: 181-213.

31. Nalluchinnappan I, Sudhandra Dev D, Irulandi M, Jeyabaskaran Y (1982) Growth of pearl oyster Pinctada fucata (Gould) in cage culture at Kundugal channel, Gulf of Mannar. Indian J Mar Sci 11: 193-194.

32. Numaguchi K (1994) Growth and physiological condition of the Japanese pearl oyster, Pinctada fucata martensii (Dunker, 1850) in Ohmura Bay, Japan. J Shellfish Res 13: 93-99.

33. Thompson RJ (1984) Production, reproductive effort, reproductive value and reproductive cost in a population of the blue mussel Mytilus edulis from a subartic environment. Mar Ecol Prog Ser 16: 249-257.

34. Ky CL, Blay C, Lo C (2015) Half-sib families' effect on cultured pearl quality traits in the black-lipped pearl oysters Pinctada margaritifera: testing for indirect benefits of polyandry and polygyny. Aquaculture International 24: 171-182.

35. Wada KT (1972) Relationship between calcium metabolism of pearl sac and pearl quality. Bull Natl Pearl Res Lab 16: 1949-2027.

36. McGinty EL, Evans BS, Taylor JUU, Jerry DR (2010) Xenografts and pearl production in two pearl oyster species, P. maxima and P. margaritifera: Effect on pearl quality and a key to understanding genetic contribution. Aquaculture 302: 175-181.

37. Arnaud-Haond S, Goyard E, Vonau V, Herbaut C, Prou J, et al. (2007) Pearl formation: persistence of the graft during the entire process of bio-mineralization. Marine Biotechnology 9: 113-116.

38. Snow MR, Pring A, Self P, Losic D, Shapter J (2004) The origin of the color of pearls in iridescence from nano-composite structures of the nacre. Am Mineral 89: 1353-1358.

36

Stock Identification of Critically Endangered Olive Barb, *Puntius sarana* (Hamilton, 1822) with Emphasis on Management Implications

Siddik M[1]*, Chaklader M[1], Hanif M[1], Islam M[2], Sharker M[1] and Rahman M[3]

[1]*Department of Fisheries Biology and Genetics, Patuakhali Science and Technology University, Bangladesh*
[2]*Department of Fisheries Management, Bangladesh Agricultural University, Bangladesh*
[3]*Department of Fisheries Biology and Genetics, Sheikh Fazilatunnesa Mujib Fisheries College, Bangladesh*

Abstract

The study was carried out to investigate the stock identification of the olive barb, *Puntius sarana* (Hamilton, 1822) through morphometric characters. A total of 110 sample ranging from 10.00-16.80 cm in total length (LT) and 13.94-63.46 g in body weight (BW) were examined to assess the morphometric variation of *Puntius sarana* from four mighty rivers; the Padma, Meghna, Jamuna and the Halda in Bangladesh. The univariate result showed significantly variation ($p<0.05$) in seven morphometric characters out of 23 characters among the populations. The discriminant analysis revealed a morphological segregation among the studied populations based on the characters of length of anal base (YZ) and pre-dorsal length (LM). Discriminant function analysis (DFA) showed 55.0% of the individuals were correctly classified into the four regions on the basis of morphological characters. The first principal component (PC I) analysis elucidated 51.56% of total variation whereas PC II and PC III were 10.72% and 8.28%, respectively. The dendrogram was drawn by using morphometric data showed that the Meghna and Halda population make one cluster and the Jamuna and the Padma population form another cluster and the distance between the Padma and Meghna river population were the highest. The canonical graph also revealed all population of the Meghna and Halda were highly overlapped compare to others. The results of the present study would help monitoring the species status in Bangladesh as a bid to take appropriate management measures for its wide geographical distribution.

Keywords: Stock structure; Critically endangered; Conservation; *Puntius sarana*

Introduction

Bangladesh is endowed with vast fisheries resources having ranked third in Asia after China and India [1,2]. However, the country's huge fisheries resources are dominated by 3 major river systems: the Padma, Jamuna, and Meghna. The Halda is another important river in Bangladesh which is geographically isolated from the 3 major river systems but globally recognized as a maiden breeding ground of freshwater fishes [3,4]. There are 260 freshwater fishes in Bangladesh under 145 genera and 55 families, of which 150 species (58%) have been categorised as small indigenous species (SIS) in Bangladesh [4]. *Puntius sarana* is an important component of SIS belongs to the family Cyprinidae commonly known as 'Olive barb' is categorized as critically endangered in Bangladesh [5-7] and vulnerable in India according to conservation status [8]. This species is a delicious and nutritious food item in South Asian countries including Bangladesh, Bhutan, India and Nepal due to rich lipoprotein content and soft bony structure [8,9].

Once *Puntius sarana* was available almost all around the year in ponds, lakes, ditches, floodplains, streams, coastal waters, estuaries, and rivers such as the Padma, Jamuna, Halda, Meghna, and also reported in the Gangetic river system of India and Bangladesh [10,11]. But this species is drastically declined in these water bodies over the years and currently it is on the verge of extinction [5,6,8]. There is an ever declining tendency in this fishery in recent years due to apparent deterioration of the habitat, over-exploitation and indeed lack of proper management [12-15]. The increasing water pollution and destruction of breeding grounds for various reasons also restricted the natural breeding of *Puntius sarana* [11]. In the present situation, it is very crucial to detect the reason of decline and understanding the ecology of the species with a view to managing the species efficiently. As a fishery management tool, morphological characters on fish are crucial from various points of view including evolution, ecology, behavior, conservation, water resource management, and stock assessment and identification [16]. The present study was therefore carried out to identify the stock structure of *Puntius sarana* in four rivers, the Padma, Jamuna, Meghna and Halda based on morphometric characters in Bangladesh.

Materials and Methods

Study area and sampling

A total of 110 samples were collected from the four important rivers in Bangladesh from November, 2013 to September, 2014 (Table 1 and Figure 1). The samples were caught by using the traditional fishing gears like cast net and conical trap. The specimens were moved to the Faculty of Fisheries, Patuakhali Science and Technology University, Bangladesh where all morphometric characteristics were observed following Froese and Pauly [17] method. Digital slide calipers (up to the nearest 0.1 cm)

	Population	Collection site (District)	No. of fish	Date of collection
01	Jamuna river	Sonatala (Bogra)	29	20.11.13
02	Padma river	Pangsha (Rajbari)	26	01.09.14
03	Meghna river	Araihazar (Narayangong)	30	10.09.14
04	Halda river	Hathazari (Chittagong)	25	20.09.14

Table 1: Sources, number of species and date of collection of *Puntius sarana* population from four rivers in Bangladesh.

*Corresponding author: Muhammad Abu Bakar Siddik, Department of Fisheries Biology and Genetics, Patuakhali Science and Technology University, Patuakhali-8602, Bangladesh, E-mail: siddik@pstu.ac.bd

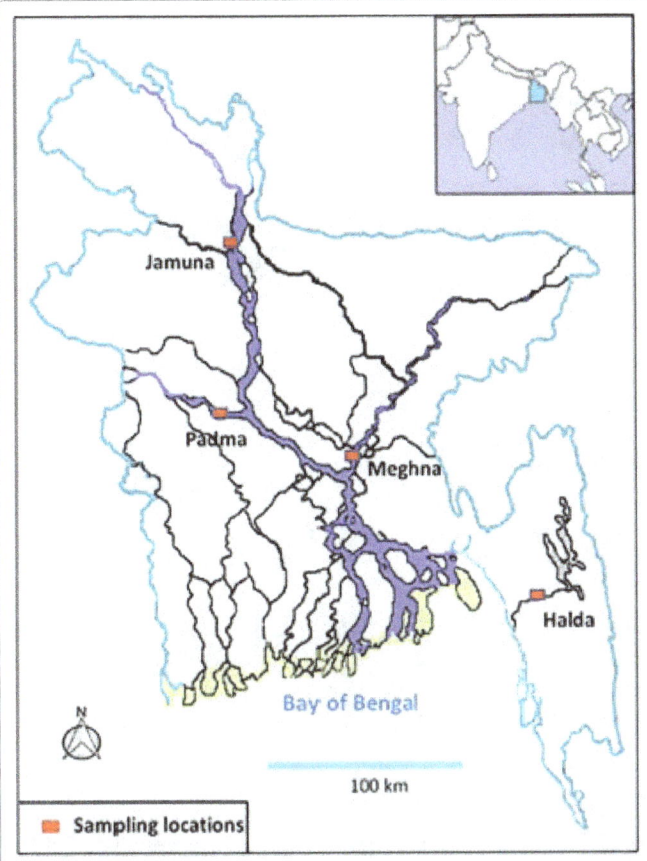

Figure 1: Map of Bangladesh showing the sampling site of four rivers; the Jamuna, Padma, Meghna and Halda.

were used to measure and an electric balance (up to the nearest 0.1 g) was used to weigh the specimen.

Statistical analysis

The morphometric characters of *Puntius sarana* were analyzed by different statistical methods. Univariate and multivariate statistics were used for analyzing character differences among different sources of samples. Most of the variability in a set of multivariate characters is due to size [18]. Thus, shape analysis should be free from the effect of size to avoid misinterpretation of the results [19]. In order to eliminate any variation resulting from allometric growth, all morphometric measurements were standardized according to Elliott et al. [20].

$$M_{adj} = M (L_s/L_o)^b$$

Where,

"M" is the original morphometric measurement,

"M_{adj}" is the size-adjusted measurement,

"L_o" is the total length of fish, and

"L_s" is the overall mean of total length for all fish from all samples for each variable.

"b" is the constant value of the equation

The parameter *b* was estimated for each character from the observed data as the slope of the regression of log M against log Lo, using all specimens. Multivariate post hoc Tukey tests were employed

to examine the statistical significant differences among mean value of morphometric characters. Discriminant analyses were then applied using stepwise insertion of variables to the size-adjusted traits to test for group membership and to identify discrimination among the populations within the different regions. Then Principal component analysis (PCA) was performed using the same morphological traits because discriminate analysis define the overall morphological variation within the regions [21]. However PCA helps in deduction data dimensions to a few principal components (PC) with identification of meaningful variables based on a combination of the original traits. Varimax rotation was selected for PCA as the rotation minimizes the number of variables with high loading factor. Kaiser [22] recommends that the value of KMO (Kaiser–Meyer–Olkin) less than 0·500 indicates the inadequacy statistic in a PCA for factor analysis. The Wilks' lamda test was performed to compare the differences between and among all groups. Cluster analysis were also applied to determine further the morphometric distances among the individuals of the two groups as a complement to discriminant analysis [23] by adopting the Euclidean distance as a measure of dissimilarity and the UPGMA (unweighted pair group method with arithmetical average) as the clustering algorithm. All statistical analyses were performed using SPSS v16 and SYSTAT v10.

Results

There were twenty three morphometric characters observed from the samples of four river population of *Puntius sarana* (Table 2 and Figure 2). Among which the LT, LF, B↓, LI, EH, IE, SnL, P↓, V↓, A↓, MN, WX, UZ, UL and UJ didn't show significant variation from each other while the head depth (K) and lowest body depth (B) of the Padma, Meghna and Halda river population revealed significant variation compare to the Jamuna river population. The head length (LH) of Jamuna, Padma

Characters	Jamuna	Padma	Meghna	Halda
LT	13.97 ± 0.31ᵃ	13.8 ± 0.51ᵃ	14.89 ± 0.28ᵃ	14.12 ± 0.39ᵃ
LS	10.99 ± 0.28ᵃᵇ	10.67 ± 0.40ᵇ	11.74 ± 0.20ᵃ	11.02 ± 0.38ᵃᵇ
LF	12.24 ± 0.26ᵃ	12.14 ± 0.44ᵃ	13.01 ± 0.23ᵃ	12.33 ± 0.35ᵃ
LH	2.32 ± 0.06ᵇ	2.47 ± 0.11ᵇ	2.83 ± 0.06ᵃ	2.55 ± 0.08ᵇ
K	2.88 ± 0.20ᵃ	2.27 ± 0.10ᵇ	2.47 ± 0.09ᵇ	2.27 ± 0.07ᵇ
M	4.21 ± 0.10ᵃ	4.01 ± 0.22ᵃ	4.31 ± 0.15ᵃ	4.10 ± 0.12ᵃ
B	1.68 ± 0.05ᵃᵇ	1.52 ± 0.06ᵇ	1.77 ± 0.08ᵃ	1.61 ± 0.06ᵃᵇ
LI	0.52 ± 0.02ᵃ	0.50 ± 0.01ᵃ	0.55 ± 0.02ᵃ	0.54 ± 0.01ᵃ
EH	1.25 ± 0.05ᵃ	1.22 ± 0.05ᵃ	1.32 ± 0.06ᵃ	1.23 ± 0.02ᵃ
IE	0.73 ± 0.02ᵃ	0.72 ± 0.04ᵃ	0.77 ± 0.06ᵃ	0.70 ± 0.01ᵃ
SnL	0.85 ± 0.04ᵃ	0.78 ± 0.03ᵃ	0.79 ± 0.03ᵃ	0.78 ± 0.02ᵃ
LM	5.32 ± 0.10ᵇ	5.25 ± 0.25ᵇ	5.85 ± 0.12ᵃ	5.39 ± 0.15ᵃᵇ
NS	3.87 ± 0.09ᵃ	3.45 ± 0.15ᵇ	3.44 ± 0.18ᵇ	3.31 ± 0.14ᵇ
D	2.31 ± 0.07ᵇ	2.43 ± 0.08ᵇ	2.69 ± 0.08ᵃ	2.54 ± 0.09ᵃᵇ
P	2.18 ± 0.08ᵃ	2.12 ± 0.08ᵃ	2.25 ± 0.06ᵃ	2.07 ± 0.08ᵃ
V	1.93 ± 0.06ᵃ	1.83 ± 0.07ᵃ	1.95 ± 0.05ᵃ	1.85 ± 0.07ᵃ
A	1.67 ± 0.06ᵃ	1.69 ± 0.07ᵃ	1.79 ± 0.09ᵃ	1.69 ± 0.05ᵃ
MN	1.88 ± 0.07ᵃ	1.77 ± 0.07ᵃ	1.96 ± 0.06ᵃ	1.76 ± 0.08ᵃ
QR	0.46 ± 0.04ᵇ	0.53 ± 0.02ᵃ	0.59 ± 0.02ᵃ	0.56 ± 0.02ᵃ
WX	0.59 ± 0.02ᵃ	0.61 ± 0.02ᵃ	0.61 ± 0.08ᵃ	0.59 ± 0.03ᵃ
YZ	1.17 ± 0.05ᵃ	1.07 ± 0.06a	1.21 ± 0.06ᵃ	1.18 ± 03ᵃ
JL	0.91 ± 0.02ᵃ	0.81 ± 0.04ᵃ	0.81 ± 0.04ᵃ	0.76 ± 0.03ᵃ
JU	0.77 ± 0.03ᵃ	0.73 ± 0.03ᵃ	0.70 ± 0.03ᵃ	0.66 ± 0.03ᵃ

Values are mean ± standard error. Mean values in each row bearing different superscripts are significantly different (P<0.05).

Table 2: Morphometric characters of *Puntius sarana* population observed in the Jamuna, Padma, Meghna and Halda rivers of Bangladesh.

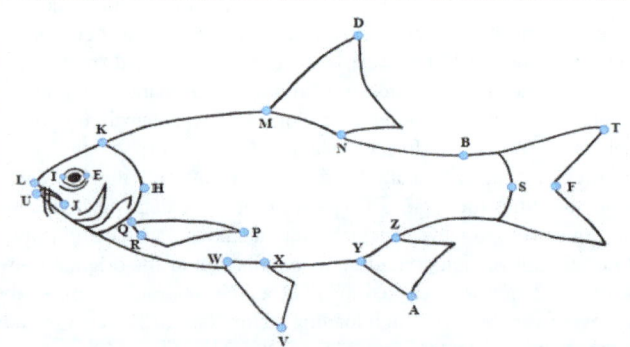

Figure 2: Morphometric measurements of *Puntius sarana*. Total length (LT); Fork length (LF); Standard length (LS); Head length (LH); Head depth ($K_\downarrow$); Highest body depth ($M_\downarrow$); Lowest body depth ($B_\downarrow$); Pre-orbital length (LI); Post-orbital length (EH); Eye diameter (IE); Pre-dorsal length (LM); Post-dorsal length (NS); Height of dorsal fin (D); Height of pectoral fin (P); Height of ventral fin (V); Height of anal fin (A); Length of dorsal base (MN); Length of pectoral base (QR); Length of ventral base (WX); Length of anal base (YZ); Upper jaw length (JL); Lower jaw length (JU).

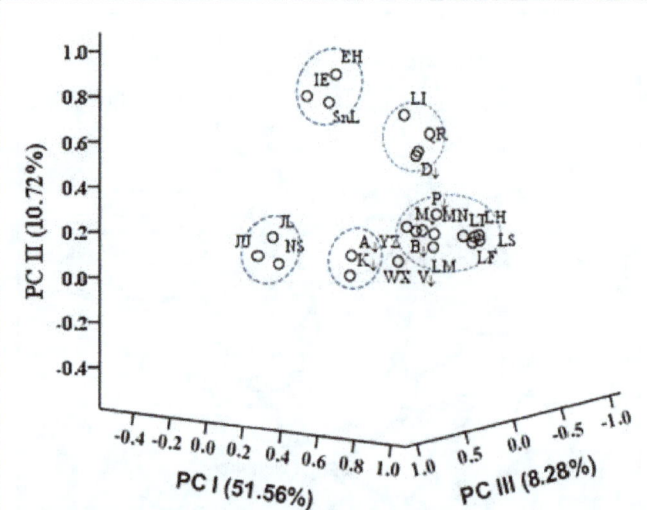

Figure 3: Principal component analysis (PCA) for 23 morphometric characters of *Puntius sarana* collected from four rivers, Bangladesh.

Group	LT:LS	$M_\downarrow$: $B_\downarrow$	LH:IE	LT: M	LT: B	TL:LH	LS:LH
Jamuna river	1.27ab	2.33a	3.23b	3.32a	8.37b	6.06a	4.76a
Padma river	1.29a	2.15a	3.51ab	3.52a	9.12a	5.62b	4.34b
Meghna river	1.28b	2.43a	3.89a	3.48a	8.58ab	5.27c	4.16b
Halda river	1.29a	2.32a	3.65ab	3.45a	8.83ab	5.56bc	4.33b

Vertically, letters a, b and c show statistically significant differences ($p < 0.05$) among the rivers.

Table 3: Different morphometric proportions of *Puntius sarana* population on four rivers of Bangladesh.

and Halda river population showed significant variation than Meghna river population. During multivariate analysis, an inadequate sample size is a common problem with many fish morphology studies [24]. Many authors performed theoretical works for decades on PCA and DFA recommended that the ratio of N: P (N-number of organisms measured and P-parameters included in the analysis) should be at least 3-3.5 [25]. In the present study, 7 characters were retained and the N:P ratio was 8.57 for all 7 morphometric measurements under these circumstance, suggesting adequate sample size for this study. The contributions of the variables to principle components (PC) were examined to determine the most effective morphometric measurement that discriminates the populations. Bartlett's Test of sphericity and the Kaiser-Meyer-Olkin (KMO) measure was executed to observe the fitness of the data for PCA. Generally the KMO statistics fluctuates between 0 and 1 [24] but Kaiser [22] recommends that values greater than 0.5 are acceptable. The value of KMO between 0.5 and 0.7 are mediocre, between 0.7 and 0.8 are good, and between 0.8 and 0.9 are superb [26]. In this study, the obtained value of KMO for the overall matrix is 0.83 and the Bartlett's Test of sphericity is significant (P<0.01). The results (KMO and Bartlett's) estimated from the present study suggest that the sampled data are appropriate to proceed with a factor analysis procedure.

Principal component analysis of 23 morphometric measurements extracted three factors with eigen-values > 1, explaining 70.57% of the variance (Figure 3). The first principal component (PC1) accounted for 51.56% of the variation, second (PC2), third (PC3) for 10.72% and 8.28% respectively. The most significant loadings on PC1 were LT, LF, LS, LH, $K_\downarrow$ $M_\downarrow$ $B_\downarrow$ LI, EH, IE, SnL, LM and NS JL were on PC II and EH, IE, SnL were on PC III, respectively. Nimalathasan [27] worthy mentioned that factor loading greater than 0.30 is considered significant, 0.40 more important, and 0.50 or greater is very significant. For parsimony, only those factors were considered significant in this study having loadings above 0.50. Visual examination of plots of PC I, PC II and PC III scores showed the most similar loadings were YZ, P↓, B↓ M↓, V↓ LT, LF, LH, LS, WX, and LM (Figure 3).

Different proportions of morphometric characteristics (LT:LS, M↓:B↓, LH:IE, LT:M↓, LT: B↓, LT:LH, LS:LH) of *Puntius sarana* are given in Table 3. There was no significant difference (*p*>0.05) observed in the ratio of M↓: B↓ and LT: M↓ through all four river populations

while the ratio of standard length and head length (LS:LH) of Jamuna population was significantly higher (*p*<0.05) than other three river populations (Table 3). Univariate statistics (ANOVA) revealed that 7 morphometric characters such as head length (LH), head depth (K↓), lowest body depth (B↓), post dorsal length (EH), height of dorsal fin (D↓), length of pectoral base (QR) and upper jaw length (UL) of 23 morphometric measurements significantly differed to varying degrees (*p* < 0.05, p < 0.01 or P< 0.001) among samples (Table 4).

A dendrogram was drawn for all population in four rivers based on morphometric characteristics (Figure 4). The Meghna river populations were nearest to Halda river population compare to two others (Jamuna and Halda) and the Jamuna river population were more adjacent to Padma river population than others (Meghna and Halda) according to the distance of Squared Euclidean Dissimilarity.

Three discriminate function (DF1, DF2 and DF3) produced based on morphometric characters during the analysis of discriminant function. The DF1 was accounted for 83.2%, the DF2 was accounted for 12% and the DF3 was accounted for 4.8% of among groups variability,

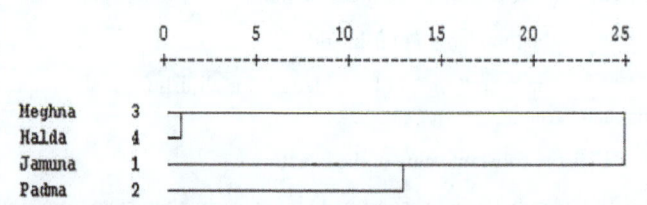

Figure 4: Dendrogram based on morphometric characters of *Puntius sarana* collected from four rivers, Bangladesh.

Characters	Wilks' Lambda	F	df1	df2	Sig.
LT	0.92	1.55	3	56	0.211
LS	0.91	1.95	3	56	0.131
LF	0.93	1.39	3	56	0.255
LH	0.73	7.00	3	56	0.000***
$K_\downarrow$	0.78	5.13	3	56	0.003**
$M_\downarrow$	0.96	0.73	3	56	0.537
$B_\downarrow$	0.87	2.83	3	56	0.047*
LI	0.91	1.78	3	56	0.162
EH	0.96	0.83	3	56	0.483
IE	0.97	0.64	3	56	0.592
SnL	0.94	1.15	3	56	0.335
LM	0.87	2.72	3	56	0.053
NS	0.87	2.86	3	56	0.045*
$D_\downarrow$	0.81	4.15	3	56	0.010*
$P_\downarrow$	0.94	1.12	3	56	0.350
$V_\downarrow$	0.95	0.99	3	56	0.401
$A_\downarrow$	0.97	0.57	3	56	0.637
MN	0.90	2.01	3	56	0.123
QR	0.78	5.18	3	56	0.003**
WX	0.99	0.28	3	56	0.844
YZ	0.92	1.62	3	56	0.195
JL	0.84	3.47	3	56	0.022*
JU	0.89	2.33	3	56	0.084

Table 4: Univariate statistics (ANOVA) of morphometric characters of *Puntius sarana* observed in four river of Bangladesh.

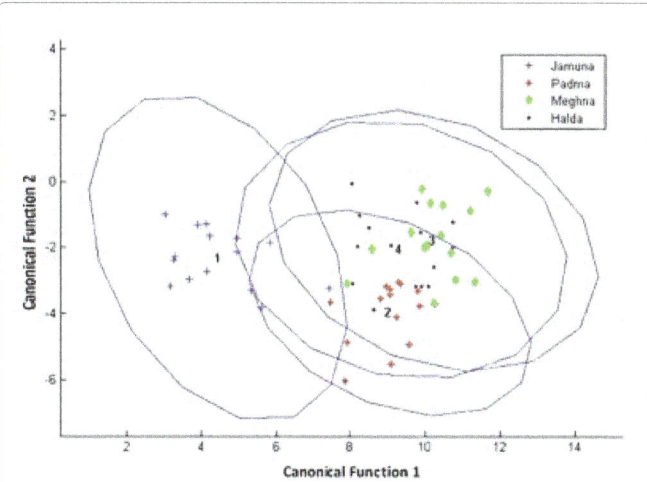

Figure 5: Coordinate plot of two discriminant functions from morphometric measurement of the four river populations of *Puntius sarana*. Samples referred to 1-Jamuna river, 2-Padma river, 3-Meghna river and 4-Halda river population.

explaining 100% of total among group variability as per morphometric measurements. The original grouped case classification in the discriminate function analysis have a high degree of correct classification with 55.0% of all fish being assigned to the correct population. Wilk's lamda tests of discriminant analysis indicated significant differences in first function (Wilk's lambda = 0.411, χ^2 = 49.77, d.f. = 6, P <0·001) of morphometric characters of all populations except the second function (Wilk's lambda = 0.897, χ^2 = 6.06, d.f. = 2, P > 0.05), which were non-significant. There was some intermingling relationship found among all populations and the populations were not separated. The diagram depicts that Meghna were highly dispersed in their morphological parameter compare to other groups of Jamuna, Padma and Halda. The Meghna and Halda river populations were highly overlapped than the Padma and Jamuna river population (Figure 5). The canonical graph for four river populations which was sequentially distributed in cluster form around their centroid value. This recommends that morphological growth trend of them more or less similar.

Discussion

The *Puntius sarana* is one of the most commercially important barb species having great potential for aquaculture not only in Bangladesh but all Southeast Asian countries as well [8,9]. In the recent years, there has been a growing concern about the conservation and sustainable management of this species. Therefore, study on basic biological aspects of *Puntius sarana* is crucial to impose adequate regulations for sustainable fishery management and conservation for this species. In the present study, the morphometric characters of all river population were somewhat similar but some significant variations were observed in HD, HBD, HL (Table 2). A similar study was conducted by Saroniya et al. [28] for different *Puntius* sp and found variation in HD, HBD among all populations of this species. These variation may have been due to the geography, ecology, human activities, genetic diversity and

experimental error of the population. There are also several studies have conducted on morphometric measurements and meristic counts on different fishes and found variations due to the geography, ecology and human activities [29-31]. However, morphometric variation of all river populations are expected because the specimens were collected from geographically separated location, may have originated from different ancestors and may be due to their adaptation capacity. Fish are very sensitive to environmental changes and adapt quickly by modifying their physiology and behavior to environmental changes [32]. Environmental changes (such as food abundance and temperature) directly affect the fish and the fish quickly change necessary morphometric in order to adapt themselves [19,24,32]. The total length and standard length of *Puntius sarana* populations were 1.27, 3.3-3.5, 8.3-9.1, 5.5-6.0 and 4.2-4.7 times respectively higher than SL, HBD, LBD and HL; and the highest body depth was 2.4-2.6 times higher than the lowest body depth. Again, the head length was 3.2-3.7 times higher than the eye diameter. A similar type of study was conducted by Hossain et al. [11] found that the total length was 5.5-6.0 and 4.0-5.0 times higher than head length and high body depth, respectively for *Puntius pangasius* populations. Moreover, the present findings agree with those of Schreck and Moyle [33] and Kurata [34] who reported that environmental factors (water temperature, pH etc.) considerably affect the morphology of fishes.

The dendrogram which are used in this study resulted in 2 clusters: the Meghna and Halda stocks in one and the Jamuna and Padma stock in another (Figure 4). The Meghna river population had high nearness with Meghna and the Jamuna river with Padma river. This similarity may be due to same genetic structure, environmental and geographical location. Considering DF1, the Jamuna population exhibited some similar characteristics among the Padma, Meghna and Halda river population. The Meghna and Halda river population are broadly overlapped, while the Payra river population clearly differed based on the DF2. In the present investigation, 55.0% of individuals were correctly classified into their respective groups by DFA, indicating intermingling among some the populations. DFA was applied by Turan et al. [35] on the anchovy (*Engraulis encrasicolus*) from different areas of the Mediterranean Sea, and found significant heterogeneity among different populations based on morphometric characters. Chaklader et al. [30] applied DFA for three populations of *Polynemus paradiseus* from the three coastal rivers and reported that environmental parameters and local migration of the fish were influenced the morphological discrimination among them.

The DFA segregation was partly confirmed by PCA, where the graphs of PCA scores for each sample revealed that, among four populations, some features showed overlapping and others were clearly distinct. Among five groups, one group revealed similar loadings of YZ, P↓, B↓, M↓, V LT, LF, LH, LS, WX, LM in four river population (Figure 3), indicated that these may be contributed by geographical location and environmental condition. Hossain et al. [36] applied PCA on *Labeo calbasu* collected from the Jamuna and Halda rivers as well as a hatchery, reported environmental factors and local migration of the fish attributing the morphological discrimination of fish.

Conclusion

The results indicated that *Puntius sarana* still has morphometric heterogeneity among the population from four different rivers of Bangladesh. The observation given in the present study can further be confirmed based on molecular and biochemical methods. The present study contributes baseline biological information that is expected to be helpful in facilitating the development of management strategies in relation to the fishery and conservation of *Puntius sarana* populations in selected rivers. In order to ameliorate the current situation and sustain the fishery stock of *Puntius sarana* in the selected four rivers, it is recommended to:

- Adopt separate management strategies to sustain the stocks from year after year.

- Reduce exploitation rates by implementing appropriate mesh size and gear selectivity to catch the optimum length which gives the maximum possible yield.

- Provide appropriate amenities to allow young individuals to reach the marketable size.

- Implement fisheries and conservation act, imposition ban on breeding time of fish species to sustain this resource for the future use.

References

1. Hussain M, Mazid M (2001) Genetic improvement and conservation of carp species in Bangladesh. Mymensingh, Bangladesh: BFRI and Penang, Malaysia.

2. Sharker M, Mahmud S, Siddik M, Alam M, Alam M, et al. (2015) Livelihood status of Hilsha fishers around Mohipur Fish Landing site, Bangladesh. World J Fish & Marine Sci 7: 77-81.

3. Sharker M, Siddik M, Nahar A, Shahjahan M, Faroque A, et al (2015) Genetic differentiation of wild and hatchery populations of Indian major carp Cirrhinus cirrhosis in Bangladesh. J Env Biol 36: 1223-1227.

4. Talwar P, Jhingran A (1991) Inland fishes of India and adjacent countries, India.

5. IUCN (2010) Red book of threatened fishes of Bangladesh, IUCN- The world conservation union 116.

6. Hanif M, Siddik M, Chaklader M, Mahmud S, Nahar A, et al. (2015) Biodiversity and conservation of threatened freshwater fishes in Sandha River, South West Bangladesh. World Applied Sciences Journal 33: 1497-1510.

7. Hanif M, Siddik M, Chaklader M (2015) Fish diversity in the southern coastal waters of Bangladesh: present status, threats and conservation perspectives. Croatian Journal of Fisheries 73: 251-274.

8. Siddik M, Nahar A, Ahamed F, Masood Z, Hossain M, et al. (2013) Conservation of Critically Endangered Olive Barb Puntius sarana (Hamilton, 1822) through Artificial Propagation. Our nature 11: 96-104.

9. Chakraborty B, Miah M, Mirza M, Habib M (2005) Growth, yield and returns to Puntius sarana (Hamilton, 1822) Sarpunti, in Bangladesh under Semi-intensive Aquaculture. Asian Fish Sci 18: 307-322.

10. Siddik M, Hanif M, Chaklader M, Nahar A, Mahmud S (2016) Fishery biology of gangetic whiting Sillaginopsis panijus (Hamilton, 1822) endemic to Ganges delta, Bangladesh. Egypt J Aquac Res 41: 307-313.

11. Hossain M, Ohtomi J, Ahmed Z (2009) Morphometric, meristic characteristics and conservation of the threatened fish, Puntius sarana (Hamilton, 1822) (Cyprinidae) in the Ganges River, northwestern Bangladesh. Turk J Fish Aquat Sci 9: 25-27.

12. Khan M, Miyan K, Khan S (2013) Morphometric variation of snakehead fish, Channa punctatus, populations from three Indian rivers. J Appl Ichthyol 29: 637- 642.

13. Mir F, Mir J, Chandra S (2013) Phenotypic variation in the Snow trout Schizothoraxri chardsonii (Gray, 1832) (Actinopterygii: Cypriniformes: Cyprinidae) from the Indian Himalayas. Contributions to Zool 82: 115-122.

14. Chaklader M, Nahar A, Siddik M, Sharker R (2014) Feeding habits and diet composition of Asian Catfish Mystus vittatus (Bloch, 1794) in shallow water of an impacted coastal habitat. World J Fish & Marine Sci 6: 551-556.

15. Nahar A, Siddik M, Alam M, Chaklader M (2015) Population genetic structure of paradise threadfin Polynemus paradiseus (Linnaeus, 1758) revealed by allozyme marker. Intl J Zool Res 11: 48-56.

16. AnvariFar H, Khyabani A, Farahmand H, Vatandoust S, Jahageerdar S et al (2011) Detection of morphometric differentiation between isolated up- and downstream populations of Siah Mahi (Capoeta capoeta gracilis) (Pisces: Cyprinidae) in the Tajan River (Iran). Hydrobiol 673: 41-52.

17. Froese R, Pauly D (2007) Fishbase 2007. World Wide Web electronic publication. Available at: Junguera S, Perez-Gandaras G (1993) Population diversity in Bay of Biscay anchovy, Engraulisen crasicolus (L., 1785) as revealed by multivariate analysis of morphometric and meristic characters. ICES J Mar Sci 50: 383-391.

18. Strauss R (1985) Evolutionary allometry and variation in body form in the South American catfish genus Corydoras (Callichthyidae). Systematic Biology 34: 381-396.

19. Elliott N, Haskard K, Koslow J (1995) Morphometric analysis of orange roughy, Hoplostethus atlanticus of the continental slope of southern Australia. Journal of Fish Biology 46: 202-220.

20. Solem O, Berg K (2011) Morphological differences in parr of Atlantic salmon Salmo salar from three regions in Norway. J Fish Biol 78: 1451-1469.

21. Kaiser H (1974) An Index of Factorial Simplicity. Psychometrika 39: 31-36.

22. Veasey E, Schammass A, Vencovsky R, Martins S, Bandel G, et al (2001) Germplasm characterization of Sesbania accessions based on multivariate analyses. Genet Resour Crop Evol 48: 79-90.

23. Mir I, Sarkar K, Dwivedi K, Gusain P, Jena K et al (2013) Stock structure analysis of Labeo rohita (Hamilton, 1822) across the Ganga basin (India) using a truss network system. J Appl Ichthyol 29: 1097-1103.

24. Kocovsky M, Adams V, Bronte C (2009) The effect of sample size on the stability of principal component analysis of truss-based fish morphometrics. Trans Am Fish. Soc 138: 487-496.

25. Field A (2000) Discovering statistics using SPSS for Windows. Sage, London.

26. Nimalathasan B (2009) Determinants of key performance indicators (KPIs) of private sector banks in Sri Lanka: an application of exploratory factor analysis. Ann. Stefan cel Mare Univ. Suceava Fac Econ Publ Admin 9: 9-17.

27. Saroniya K, Saksena N, Nagpure S (2013) The morphometric and meristic analysis of some puntius species from central India. Biolife 1: 144-154.

28. Pinheiro A, Teixeira M, Rego L, Marques F, Henrique N, et al (2005) Genetic and morphological variation of Solea lascaris (Risso, 1810) along the Portuguese coast. Fish Res 73: 67-78.

29. Chaklader R, Siddik M, Nahar A (2015) Taxonomic diversity of paradise threadfin Polynemus paradiseus (Linnaeus, 1758) inhabiting southern coastal rivers in Bangladesh. Sains Malaysiana 44: 1241-1248.

30. Vatandoust S, Mousavi-Sabet H, Razeghi-Mansour M, AnvariFar H, Heidari A, et al. (2015) Morphometric variation of the endangered Caspian lamprey, Caspiomyzon wagneri (Pisces: Petromyzontidae), from migrating stocks of two rivers along the southern Caspian Sea. Zool Stud 27: 54-56.

31. Swain P, Ridell B, Murray C (1991) Morphological differences between hatchery

and wild populations of coho salmon (Oncorhynchus kisutch): environmental versus genetic origin. Can J Fish Aquat Sci 48: 1783-1791

32. Schreck B, Moyle B (1990) Methods for Fish Biolog. American Fisheries Society, Bethesda.

33. Kurata H (1975) Environmental conditions, in "Feeding and growth of larval fish" (ed. By the Japanese Society of Fisheries), Koseishakoseikaku Tokyo.

34. Turan C, Erguden D, Gurlek M, Basusta N, Turan F, et al. (2004) Morphometric structuring of the anchovy (Engraulis encrasicolus L.) in the black, Aegean and Northeastern Mediterranean seas. Turk J Vet Anim Sci 28: 865-871.

35. Hossain M, Nahiduzzaman M, Saha D, Khanam M, Alam M (2010) Landmark-based morphometric and meristic variations of the endangered carp, Kalibaus Labeo calbasu, from stocks of two isolated rivers, the Jamuna and Halda and a hatchery. Zool Stud 49: 556-563.

37

Stock Structure Delineation of the African Catfish (*Clarius gariepinus*) in Selected Populations in Kenya Using Mitochondrial *DNA* (Dloop) Variability

Cynthia Nyunja[1]*, Joyce Maina[1], Joshua Amimo[1], Felix Kibegwa[1], David Harper[1,2] and Joseph Jung'a[1]

[1]*Department of Animal Production, University of Nairobi, Kenya*
[2]*University of Leicester, England*

Abstract

This study genetically characterized five populations of the African catfish (*Clarius gariepinus*) in Kenya. Samples were obtained from five sites in the country–Athi River hatchery, Kisii Fingerling Production Centre (FPC), Jewlett hatchery, Sagana Hatchery Station and Lake Baringo. DNA was extracted from tissue samples, followed by amplification and sequencing of the dloop region. Haplotype diversities, phylogenetic structure and variation at the dloop region of mitochondrial DNA were assessed.

Mitochondrial DNA analyses indicated that the sampled species showed genetic diversity between its populations. The genetic results were congruent indicating the differences in diversities and haplotype similarities of catfish samples from different sites. The Sagana, Kisii FPC, Jewlett and Baringo population cluster overlapped indicating possibly shared source of brood stock. The Athi river population was in a different cluster and its distinctiveness is attributed to imported brood stock. Both Athi River hatchery and Lake Baringo populations were highly variable and has great potential for production.

Keywords: *Clarius gariepinus*; Brood stock; Aquaculture

Introduction

The African catfish is an indigenous species in Kenya where it is widely cultured. Aquaculture in Kenya began with trout species mainly for recreational purposes [1], later tilapia species was cultured for food followed by the African catfish. *C. gariepinus* adapts well to artificial environments, and has rapidly gained status as an important aquaculture species [2]. The species have a rapid growth, a high reproductive potential and sturdy resistance to harsh environments as they have adaptive mechanisms. *C. gariepinus* as an aquaculture species has the advantage that it can be reared at a high stocking density without affecting fish survival [3]. The species now makes up a fifth of the total fish produced in the country [4]. This was a major result of the government's Economic Stimulus Programme in 2009 to encourage aquaculture production in Kenya [5]. There has been much interest in the culture of *C. gariepinus* to increase seed production and availability [5,6]. Many hatcheries have come up since 2009 and have been involved in artificial propagation of the species. Some of the hatcheries produce only fingerlings and some also produce table size catfish for consumption. However, there has been occasional losses by grow out farms due to poor quality seed supply and other stresses [7]. Hatcheries have faced many problems including identifying brood stock and structure of the populations.

Culture fish populations from hatcheries in Kenya with brood stock obtained from different sources; including natural water bodies; are heavily relied on by farmers for production. To increase the diversity of fish in fish farms [8] then more research needs to be undertaken to understand the population diversities. Molecular markers are used to characterize the population of catfish species proper identification [6,9]. Molecular tools such as mitochondrial DNA has been used to assess the phylogeny and haplotype variation of the African catfish in other studies. Mitochondrial variation assessments have enabled distinction of brood stock from wild populations for use by hatcheries in culturing populations and for phylogenetic trees [10]. Characterization of the populations was done using mitochondrial DNA in the current study. Mitochondrial DNA has some advantages over other tools for genetic analyses due to its maternal inheritance and fast mutation rate of the control region [11]. The objective was to ascertain genetic diversity and make recommendations for the sources of future farmed breeding stocks.

Materials and Methods

Sampled sites

African Catfish (*C. gariepinus*) samples were obtained from Lake Baringo and four hatcheries across the country as shown in Figure 1.

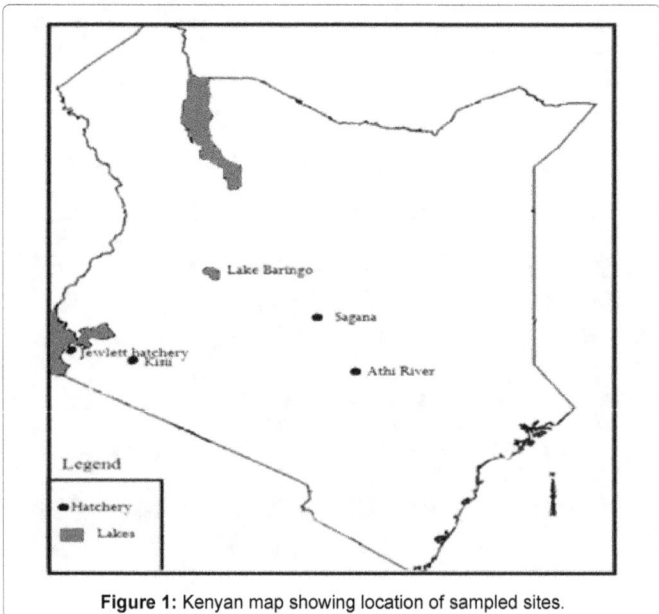

Figure 1: Kenyan map showing location of sampled sites.

*Corresponding author: Cynthia Nyunja, Department of Animal Production, University of Nairobi, Kenya, E-mail: cynyunja@yahoo.com

Lake Baringo is located in Kenya in the Great Rift Valley at an altitude of 975 m [12] with a maximum depth of 12 m. The lake has a surface area of approximately 130 km² located north of Lake Bogoria. The lake is fed by rivers Molo and Perkerra. The lake has a number of introduced fish species. The marbled lungfish (*Protopterus aethiopicus*) an introduced species provides the majority of fish output from the lake. Fish composition in the lake include *Clarius gariepinus* and others such as *Oreochromis niloticus*, *Protopterus aethiopicus*, *Barbus intermedius* and *Labeo cylindricus* [13].

The hatcheries were:

• Athi River hatchery near Nairobi was established in 2013.

• Jewlett hatchery to Western part of Kenya was established in 2011.

• The Kisii Fingerling Production Centre is a government institution.

• Sagana Centre was established in 1948 and is a governmental research station in Kenya.

The samples size was as follows: Athi river hatchery (22) Kisii FPC (20) Jewlett hatchery (20) Sagana (8) and Lake Baringo (23). The sampled fish were adults bought at the hatchery locations. Lake Baringo samples were bought from the commercial fisheries landings beach. The tissue samples were preserved in 96% ethanol in 1.5 ml tubes and transported to the University of Nairobi's Animal Production department genetics laboratory.

DNA extraction and Polymerase Chain Reaction (PCR)

The DNA extraction was done using the Qiagen extraction kit (Qiagen Valencia, CA USA) following the manufacturer's instructions with a few modifications. Twenty-five mg of catfish skeletal tissue was macerated, lysed and incubated for digestion at 56°C for 2 hours. Centrifugation for spinning down digested content was at 10000 rpm except for the final wash which was done at 14000 rpm, Elution was done with 50 μl of AE elution buffer for all the samples. Presence and quality of the extracted genomic DNA was assessed using 1% agarose gel electrophoresis.

Amplification of the targeted mitochondrial region in the extracted DNA, ~550bp, was by conventional polymerase chain reaction. The primer set used were forward primer L16473 (5'-CTAAAAGCATCGGTCTTGTAATCC-3') and reverse primer H355 (5'-CCTGAAATGAGGAGGAACCAGATG-3') [6,9].

The PCR reaction was with a master mix prepared in the laboratory at the Institute of Primate of Research, Nairobi, Kenya. To make a 20 μl reaction for each PCR reaction, 12.5 μl sterile deionised water, 2 μl of 5X PCR buffer, 2 μl of 10 μm dNTPs, 0.5 μl of 5 μm each of the forward and the reverse primers, 0.2 μl of Taq DNA polymerase (New England Biolabs, UK Ltd) and 2.5 μl of DNA template were used. The amplification was done in the DNA 480 Thermal cycler, Applied Biosystems USA.

The protocol for amplification of the D-loop region, adapted [6,9] as follows:

Initial denaturation was for 2 minutes at 94°C, 29 cycles of denaturation, annealing and extension for 94°C for 1 minute, 56°C for 1 minute 10 seconds and 72°C for 2 minutes respectively and the final extension at 72°C for 5 minutes. After PCR, 5 μl of each of the reaction was run on 1% agarose gel stained with ethidium bromide (5 mg/μl) to verify amplification. The samples were then purified using a Qiagen PCR purification kit (Qiagen Valencia, CA USA) following manufacturer's instructions. Ninety-six PCR products were selected with correct band size and good quality and were sequenced using an automated BigDye Terminator cycle chemistry (Sanger sequencing), by Genewiz˙ United Kingdom. The generated sequences were deposited in the GenBank with Accession Numbers: MF150204-MF150238.

Genetic analysis

The resulting sequences were edited and aligned in Bioedit version 7.1.9 software. The generated sequences were then compared with the nucleotide sequences in the GenBank using the Basic Local Alignment Search Tool (BLAST) to confirm species identity. MEGA V 7.0 [14] was used to construct the evolutionary phylogenetic trees. DNASP V5.10.01 [15] was used to calculate the haplotype diversities, nucleotide diversities and genetic differences. The Arlequin software version 3.5 [16] was used for Analysis of Molecular Variance (AMOVA) while the programme Network 5.0 version 8 was used to visualize the haplotypes in the populations using median joining tree.

Results

Phylogenetic relationships

The Neighbour-joining tree as shown in Figure 2 was constructed using maximum likelihood function based on the Tamura-Nei model [17]. The bootstrap census tree inferred from 1000 replicates is taken to represent the evolutionary history of the taxa analysed [18]. The percentage of replicate tress in which the associated taxa clustered together in the bootstrap test (1000 iterations) are shown next to the branches. Initial trees for the heuristic approach were obtained automatically by applying NeighborJoin and BioNJ algorithms to a matrix of pairwise differences estimated using the maximum likelihood function approach, and then selecting the topology with superior log likelihood value.

The phylogenetic tree indicated that the five fish populations formed one monophyletic assemblage and formed three clusters, AR3012 being very distant. *Clarius liocephalus* was used as an outgroup species [6]. This is confirmed to be another species as it forms a monophyletic line different from the other.

Genetic variation and haplotype analysis

A total of 33 haplotypes were detected with 60 polymorphic sites from the 433 nucleotide sites of the control region excluding *C. liocephalus*. The largest haplotype group consisted of twelve haplotypes from Baringo, followed by nine haplotypes from Athi River and Jewlett. Sagana had the lowest with four haplotypes.

The haplotype diversity (Hd) of all the samples was 0.988 ± 0.031 and the nucleotide diversity (π) was 0.02363 ± 0.02603 with 28 singleton variation sites and 32 parsimony informative sites.

The population at Athi River has the same number of haplotypes as Jewlett although of different types. Six haplotypes–2, 5, 7, 9, 13, and 16 occurred in more than one population as shown in Table 1. Haplotypes 5 and 9 were shared in three populations of Athi River, Kisii and Jewlett. Haplotype 2 occurred in Athi River and Sagana. Haplotype 7 occurred in Athi River and Jewlett. Haplotype 13 occurred in Jewlet and Kisii populations. Haplotype 16 occurred in Kisii and Sagana.

The haplotype distribution was drawn using median joining tree as shown in Figure 3.

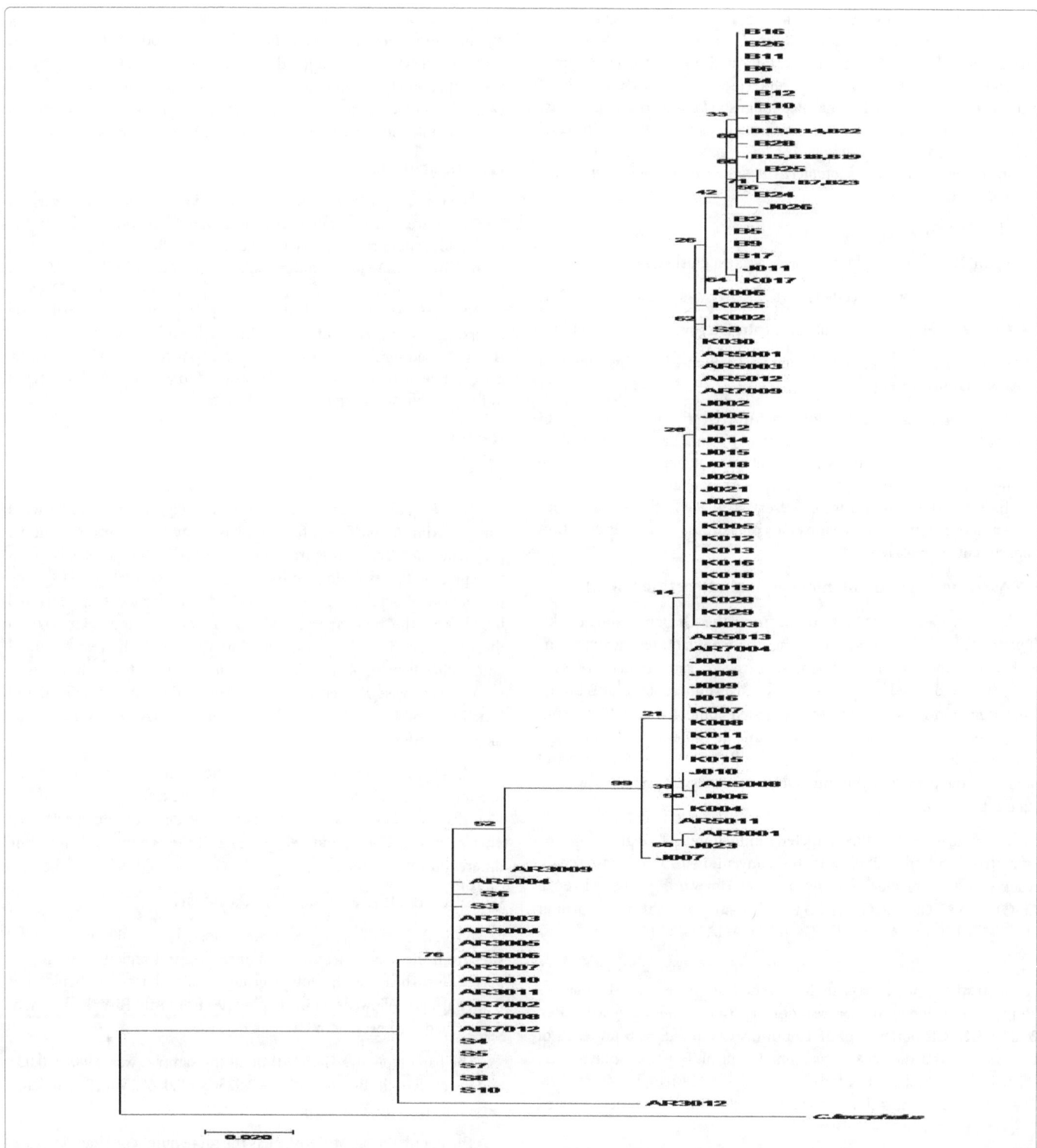

Figure 2: Molecular phylogenetic analysis of catfish samples from Kisii (K), Jewlett (J), Athi River (AR), Baringo, Sagana (S) with *C. liocephalus* as an outgroup constructed using maximum likelihood function based on the Tamura-Nei model.

Genetic differences and AMOVA

The overall nucleotide differences (Ks) and haplotype differences were according to Hudson et al. [19]. The pairwise differences based on haplotype and nucleotide statistics is shown in Table 2.

The AMOVA estimated 54.20% variation among populations and

45.80% variation to be from within populations (p<0.05; Fst 0.542) as shown in Table 3.

Population neutrality

The overall Tajima's D value [20] showed non-significant negative value (Table 4) (Tajima's D was 1.02 at P>0.05). The D value of each

Haplotype	No	Athi river	Jewlett	Kisii	Sagana	Baringo
Hap_1	1	AR3001	--	--	--	--
Hap_2	15	AR3003, AR3004, AR3005, AR3006, AR3007, AR3010, AR3011, AR7002, AR7008, AR7012	--	--	S4, S5, S7, S8, S10	--
Hap_3	1	AR3009	--	--	--	--
Hap_4	1	AR3012	--	--	--	--
Hap_5	23	AR5001, AR5003, AR5012, AR7009	J002, J005, J012 J014 J015, J018, J020, J021, J022	K003, K005, K012, K013, K016, K018, K019, K028 K029, K030	---	.
Hap_6	1	AR5004	--	--	--	--
Hap_7	2	AR5008,	J006	--	--	--
Hap_8	1	AR5011	--	--	--	--
Hap_9	11	AR5013, AR7004	J001, J008, J009, J016	K007, K008, K011, K014, K015	--	--
Hap_10	1	--	J003	--	--	--
Hap_11	1	--	J007	--	--	--
Hap_12	1	--	J010	--	--	--
Hap_13	2	--	J011	K017	--	--
Hap_14	1	--	J023	--	--	--
Hap_15	1	--	J026	--	--	---
Hap_16	2	--	--	K002	S9	--
Hap_17	1	--	--	K004	--	--
Hap_18	1	--	--	K006	--	--
Hap_19	1	--	--	K025	--	--
Hap_20-31		--	--	--	--	B2, B5, B9, B17, B3, B4, B6, B11, B16, B26, B7, B12, B13, B14, B22, B15, B18, B19, B23, B24, B25, B28
Hap_32	1	--	--	--	S3	--
Hap_33	1	--	--	--	S6	--

Table 1: The table shows haplotype distribution of the African catfish from Athi River, Jewlett, Sagana, Kisii and Baringo.

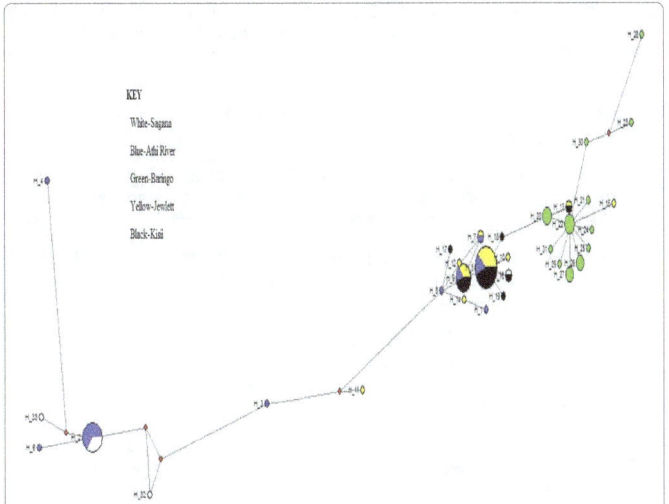

KEY

White-Sagana

Blue-Athi River

Green-Baringo

Yellow-Jewlett

Black-Kisii

Figure 3: Median network tree of African catfish mitochondrial DNA control region segments of Athi River, Jewlett, Sagana, Kisi and Baringo populations.

Discussion

Genetic variation at the mtDNA control region

The analysis of molecular variance demonstrated average levels of differentiation in the selected populations. The variation between populations in the current study was almost similar within populations. Although they are geographically isolated, a common origin of brood stock having been transported by humans for aquaculture purposes may have resulted in the almost equal between and within population variation.

Genetic diversity analysis revealed 33 haplotypes and 60 polymorphic sites. The number of haplotypes and polymorphic sites in the sampled

population considered individually were as shown in the Table 4 below. The Fu test revealed negative value of -4.45731 at an insignificant level. Kisii and Jewlett populations had relatively similar Tajima's D values, whilst Athi river was the only population that had a positive Tajima's value. The three populations Jewlett, Kisii and Baringo had no statistically significant values. Sagana had a negative significant value.

Population 1	Population 2	Hs	Ks
Athi River	Jewlett	0.7768	8.0222
Athi River	Kisii	0.7444	7.4658
Athi River	Sagana	0.7442	11.1048
Jewlett	Kisii	0.7447	1.9053
Jewlett	Sagana	0.7449	3.4211
Kisii	Sagana	0.6936	2.5865
Athi river	Baringo	0.8457	7.6698
Jewlett	Baringo	0.8512	2.5067
Kisii	Baringo	0.8196	1.8703
Baringo	Sagana	0.8530	3.3548

Table 2: Pairwise genetic differences of the five sampled populations: Athi River, Kisii, Jewlett, Sagana and Baringo.

Source of variation	Df	Percentage of variation	P value
Among populations	4	54.20	0.000
Within populations	88	45.80	--

Table 3: Hierarchical analyses of molecular variance showing amount of population genetic structure.

Statistics	Athi River	Jewlett	Kisii	Sagana	Baringo
Sample size	22	20	20	8	23
Polymorphic Sites	47	14	8	23	16
Pi	13.05	2.48	1.32	5.75	2.34
Tajimas D	0.048	-1.35	-1.39	-1.84	-1.64
P-value	>0.05	>0.05	>0.05	<0.001	>0.05

Table 4: Neutrality tests of populations.

Population	No.	Polymorphic sites	No. of haplotypes	Haplotype diversity (h) (± SD)	Nucleotide diversity (p) (± SD)
Athi river	22	47	9	0.775 ± 0.081	0.030 ± 0.005
Kisii FPC	20	8	7	0.711 ± 0.089	0.003 ± 0.001
Jewlett	20	14	9	0.779 ± 0.085	0.006 ± 0.002
Sagana	8	23	4	0.643 ± 0.184	0.013 ± 0.009
L. Baringo	23	16	12	0.913 ± 0.070	0.006 ± 0.003
L. Victoria	24	14	11	0.754 ± 0.093	0.008 ± 0.002
L. Kanyaboli	28	11	10	0.741 ± 0.064	0.005 ± 0.001

Table 5: Table showing haplotype diversity of the African catfish from four sampled hatcheries and one lake in the current study in comparison to two Kenyan lakes [6].

populations are shown in Table 5. The hatcheries had generally more polymorphic sites than the lake population. The population haplotype diversities ranged from 0.779-0.643 accommodating 0.754 and 0.741 of Lake Victoria and Lake Kanyaboli respectively [6].

There were many private haplotypes in Athi river hatchery and Lake Baringo populations. The common haplotypes were shared between Kisii, Jewlett and Athi river hatchery populations. Lake Baringo population had slightly more haplotypes and more polymorphic sites than Lake Kanyaboli and Lake Victoria and can be attributed to mixed brood stock introduced into the lake from Sagana station. The restocking of the lake was done by Sagana research centre. Lake populations often have higher diversity than cultured populations [21]. The high diversity in Lake Baringo could indicate that brood stock of catfish in the Sagana hatchery has been from different sources, giving potentially higher diversities.

The pairwise distances indicate how different the populations were from each other. The Kisii and the Sagana population were furthest from each other while Sagana and Baringo appear to be the closest. This further confirms that brood stock may have been introduced from Sagana into Lake Baringo. The shorter the genetic distance between populations, the more probable there has been some breeding between them and the less isolated they are from one another [22]. For the nucleotide differences, Athi River and Sagana were the most distant and this was evident in the number of high polymorphic sites in the two populations.

Phylogenetic structure and selection

The Kisii, Athi River and Jewlett samples clustered together, shown by the neighbour joining tree. This indicates the strong haplotype relatedness between the three populations. Some haplotypes were shared between the three populations indicating that there could be intermingling of individuals [6]. The mixed haplotypes support that some brooders in the hatcheries could have been obtained by human transportation from the same source such as Lake Victoria hence the paraphyletic groups of Kisii and Jewlett populations. In different cases [10] the aquarium trade was mainly responsible for having brought different populations together.

The selection test indicates how much a population has significantly deviated from neutral selection [23]. In our case, the deviations were present but insignificant. This cab attributed to different hatcheries having varying management and breeding practices [7], influencing overall genetic composition of brood stock. The differences are influenced by a variety of factors including source of brood stock [7].

Conclusion

From the current study, mitochondrial DNA revealed maternal linkage of the African catfish population in Kenya as in other studies of the African catfish from lakes [6] as well as in other livestock species such as goats [24]. The phylogenetic linkage attributed to transportation of brood stock by humans for catfish production to other sites.

Athi River had the highest diversities and number of polymorphic sites hence high potential for source of brood stock for farmers rearing catfish in Kenya. The exploitation of cultured *C. gariepinus* for baits can reduce the pressure on the wild populations. The levels of genetic diversity in the study can be used by the management of cultured populations to reduce chances of inbreeding in the populations and inform selection programmes. Lake Baringo also provides a potential source of diverse brood stock for use in brood stock selection pogrammes.

References

1. Vernon D, Someren V (1960) The inland fishery research station, Sagana Kenya. Nature.

2. Hecht T, Uys W, Britz PJ (1988) Culture of sharptooth catfish, *Clarias gariepinus* in Southern Africa. Foundation for Research Development: CSIR.

3. Shoko AP, Limbu SM, Mgaya YD (2016) Effect of stocking density on growth performance, survival, production and financial benefits of African sharptooth catfish (*Clarias gariepinus*) monoculture in earthen ponds. J Appl Aquaculture 28: 220-234.

4. Otieno MJ (2011) Fishery value chain analysis: Background Report-Kenya. FAO, Rome, Italy.

5. Musa S, Aura CM, Owiti G, Nyonje B, Orina P, et al. (2012) Fish farming enterprise productivity program (FFEPP) as an impetus to Oreochromisniloticus (L.) farming in Western Kenya: Lessons to learn. Afr J Agric Res 7: 1324-1330.

6. Barasa JE, Abila R, Grobler JP, Dangasuk OG, Njahira MN, et al. (2014) Genetic diversity and gene flow in *Clarias gariepinus* from Lakes Victoria and Kanyaboli, Kenya. Afr J Aquat Sci 39: 287-293.

7. Orina PS, Maina JG, Wangia SM, Karuri EG, Mbuthia PG, et al. (2014) Situational analysis of Nile tilapia and African catfish hatcheries management: A case study of Kisii and Kirinyaga counties in Kenya. Livestock Research for Rural Development 26: 5.

8. Munguti JM, Kim JD, Ogello EO (2014) An Overview of Kenyan aquaculture: Current Status, challenges, and opportunities for future development. Fish Aquat Sci 17: 1-11.

9. Nazia AK, Suzana M, Azhar H, Nguyen TTT, Azizah S (2010) No genetic differentiation between geographically isolated populations of *Clarias macrocephalus* Günther in Malaysia revealed by sequences of mtDNA cytochrome b and D-loop gene regions. J Appl Ichthyol 26: 568-570.

10. Wu LW, Liu CC, Lin SM (2011) Identification of exotic sailfin catfish species (*Pterygoplichthys Loricariidae*) in Taiwan Based on Morphology and mtDNA Sequences. Zool Stud 50: 235-246.

11. Liu ZJ, Cordes JF (2004) DNA marker technologies and their applications in aquaculture genetics. Aquaculture 238: 1-37.

12. Kallqvist T (1987) Primary production and phytoplankton in Lake Baringo and Naivasha, Kenya. Norwegian Institute for Water Research Report, Blinden, Oslo.

13. Britton JR, Harper DM (2005) Assessing the true status of the fish species *Labeo cylindricus* (Peters 1868) (Teleostei: Cyprinidae) in Lake Baringo Kenya African. J Aquat Sci 30: 203-205.

14. Kumar S, Stecher G, Tamura K (2016) MEGA7: Molecular Evolutionary genetics analysis version 7.0 for bigger datasets. Mol Biol Evol: msw054.

15. Librado P, Rozas J (2009) DnaSP v5: A software for comprehensive analysis of DNA polymorphism data. Bioinformatics 25: 1451-1452.

16. Excoffier L, Lischer HE (2010) Arlequin suite version 3.5: A new series of programs to perform population genetics analyses under Linux and Windows. Mol Ecol Res 10: 564-567.

17. Tamura K, Nei M (1993) Estimation of the number of nucleotide substitutions in the control region of mitochondrial DNA in humans and chimpanzees. Mol Biol Evol 10: 512-526.

18. Felsenstein J (1985) Confidence limits on phylogenies. An approach using the bootstrap. Evolution 39: 783-791.

19. Hudson RR, Slatkin M, Maddison WP (1992) Estimation of levels of gene flow from DNA sequence data. Genetics 132: 583-589.

20. Tajima F (1989) Statistical method for testing the neutral mutation hypothesis by DNA polymorphism. Genetics 123: 585-595.

21. Li Q, Park C, Endo T, Kijima A (2004) Loss of genetic variation at microsatellite loci in hatchery strains of the Pacific abalone ((Haliotis discus hannai). Aquaculture 235: 207-222.

22. Wright S (1943) Isolation by distance. Genetics 28: 139-156.

23. Maggio T, Andaloro MF, Arculeo M (2006) Genetic population structure of *Epinephelus marginatus* (Pisces, Serranidae) revealed by two molecular markers. Ital J Zool 73: 275-283.

24. Kibegwa FM, Githui KE, Jung'a JO, Badamana MS, Nyamu MN (2015) Mitochondrial DNA variation of indigenous goats in Narok and Isiolo counties of Kenya. J Anim Breed Genet 133: 238-247.

38

Role of Arachidonic Acid and COX Inhibitors in the Regulation of Reproduction in Freshwater Crab *Oziothelphusa senex senex*

K Prameswari[1], M Hemalatha[1], B Kishori[1]* and P Sreenivasula Reddy[2]

[1]*Department of Biotechnology, Sri Padmavati Mahila Visvavidyalayam (Women's University), Tirupati, Andhra Pradesh, India*
[2]*Department of Zoology, Sri Venkateswara University, Tirupati, Andhra Pradesh, India*

Abstract

Induced reproduction of cultured species helps to produce quantity and quality seed an important component for upright yield in crustacean aquaculture. The present study was aimed to investigate the role of arachidonic acid in the regulation of ovarian development in the freshwater crab, *Oziothelphusa senex senex*. Injection of AA significantly ($p<0.001$) increased the ovarian index, oocyte diameter and ovarian vitellogenin levels. Injection of COX inhibitors such as indomethacin and aspirin alone, and in combination with AA resulted in significant ($p<0.001$) reduction in ovarian index, oocyte diameter and ovarian vitellogenin levels in crabs. The results of the present study provide evidence that arachidonic acid and COX inhibitors involved in the regulation of female reproduction in the freshwater crab, *Oziothelphusa senex senex*.

Keywords: Arachidonic acid; COX inhibitors; Ovarian development; Crab

Introduction

Arachidonic acid (AA) is a polyunsaturated fatty acid and is the precursor for eicosanoids. It is metabolized by various enzymes and produces eicosanoids through the cyclooxygenase (COX) pathway and produce prostaglandin G_2 (PGG_2) and H_2 (PGH_2). Subsequently, prostaglandin synthase converts PGH_2 to prostaglandins. Eicosanoids have an important role in the regulation of essential functions such as reproduction, haemostasis, growth, and the immune system in vertebrates. The occurrence of eicosanoids and their precursors have been identified in various invertebrates [1,2]. The presence of AA has been also identified in different crustacean such as kuruma prawn *Marsupenaeus japonicus* [3,4], green tiger prawn *Penaeus semisulcatus* [5], shrimps *Penaeus monodon* [6,7], *Litopenaeus vannamei* [8] and *Penaeus merguiensis* [9]. Furthermore, the presence of prostaglandins PGE_2 and $PGF_{2\alpha}$ were reported in *Procambarus paeninsulanus* [10], *Marsupenaeus japonicus* [4] and *Penaeus monodon* [11], whereas, only PGE_2 was reported in *Carcinus maenas* [12]. Along with PGE_2 and $PGF_{2\alpha}$, PGD_2 also has been identified in the rice field crab *Oziothelphusa senex senex* [13]. The presence of PGE_2 has been detected in the muscle and hemolymph of shrimp *Penaeus monodon* [14].

Similar to vertebrates, eicosanoids may also regulate the different physiological functions in invertebrates [15]. The physiological role of prostaglandins in the regulation of female crustacean reproduction was reported in few studies. Involvement of PG in ovarian maturation was studied with penaeid prawn *Metapenaeus affinis* [16], PGE_2 and $PGF_{2\alpha}$ in the regulation of vitellogenesis and the induction of ovulation were reported in *Procambarus paeninsulanus* [10]. In the same species, PGE_2 administration induced significant elevation of cAMP in ovarian tissue [17]. Later, similar results were reported in the prawn *Macrobrachium rosenbergii* and in the crayfish *Cherax quadricarinatus* [18]. Similarly, ovarian synthesis of PGE_2, $PGF_{2\alpha}$ and PGD were reported in *Oziothelphusa senex senex in vitro* [13]. This study also reported that injection of PGE_2 and $PGF_{2\alpha}$ increased ovarian index and oocyte diameter in a dose-dependent manner and no changes were observed on ovarian growth with an injection of PGD in *Oziothelphusa senex senex*. Recently, the variation of PGE_2 levels at different vitellogenesis stages were demonstrated in freshwater prawn, *Macrobrachium rosenbergii* and administration of PGE_2 stimulated the ovarian maturation, increasing

ovarian somatic index, oocyte proliferation and vitellogenin (Vg) level in the hemolymph of prawn [19]. In contrast to these studies, it was reported that injection of PGE2 0.1 µg/g body weight did not stimulate the ovarian maturation in *Penaeus esculentus* [20].

The non-steroidal anti-inflammatory drugs (NSAID) may inhibit the biosynthesis of prostaglandins. NSAIDs exhibits competitive inhibition with arachidonic acid the substrate of COX enzyme thereby affects PGs biosynthesis and their physiological functions [21]. The negative role of NSAIDs in the synthesis of PGs also reported in crustaceans. Long-term exposure of ibuprofen (IBU) a NSAID, induced a dose-dependent reduction of reproduction in *Daphnia magna* [22]. In another study, it was reported that ibuprofen primarily affecting oogenesis rather than embryogenesis in daphnids, by interrupting eicosanoid (prostaglandin) metabolism [23]. In contrast, IBU affected reproduction negatively in *Daphnia magna* and *Moina macrocopa* [24]. Recently, Alfaro Montoya [25] reported the significant induced ovarian maturation with IBU treatment to unilateral eyestalk ablated of female *Litopaneus stylirostris*, whereas it had no effect on sperm counts and spermatophore weights in male *Litopenaeus* species. Though there are sporadic reports on the biosynthesis of PGs and their functions in crustaceans, the direct role of AA and COX inhibitors on reproduction in crustaceans are fragmentary. In view of this, the present study was tested the effect of AA and COX inhibitors, indomethacin and aspirin in the regulation of ovarian maturation in freshwater rice field crab *Oziothelphusa senex senex*.

Materials and Methods

Collection and maintenance of crabs

Female rice field crabs with a body weight 30 ± 5 g (intermolt,

---

*Corresponding author: Dr. B Kishori, Asst. Professor, Department of Biotechnology, Sri Padmavati Mahila Visvavidyalayam (Women's University), Tirupati-517 502, Andhra Pradesh, India, E-mail: kktinku@rediffmail.com

stage C4), uninjured were collected from in and around the Tirupati in Andhra Pradesh, India. The crabs were acclimated to laboratory conditions (temperature: 25-28°C; light/dark 12:12 h) for 3 days before being used for the experiments. Crabs were fed daily with sheep meat *ad libitum*.

Chemicals

Arachidonic Acid (AA) (99% Pure), Indomethacin (IM) and Aspirin (Asp) purchased from the Cayman Chemical company (USA), were used as test chemicals.

Experimental design

Randomly 140 crabs were divided into seven groups and 20 crabs each. Group 1 served as control and the crabs were sacrificed on the first day of the experiment. Crabs in the group 2 served as concurrent control and was received injections of crustacean saline [26] through the arthrodial membrane of the coxa of the third pair of walking legs. Crabs in the group 3 received injections of 1 µg/10 µl of ethanol AA. Crabs in the groups 4 and 5 injected with at a dose of 1 µg in 10 µl of ethanol indomethacin and aspirin respectively. Crabs in groups 6 and 7 injected with AA (as in group 3), IM and Asp (as in groups 4 and 5) on the 1st, 7th, 14th and 21st day and the crabs were sacrificed on day 30. No deaths were recorded in the concurrent controls or in the experimental groups.

Ovarian index

The crabs were weighed and the ovarian tissues were dissected out, blotted with paper towels and weighed wet using electronic balance and the gonad index was determined by using the following formula.

Ovarian index = Wet mass of the ovary
/Wet mass of the crab×100

Oocyte diameter

The diameters of twenty-five oocytes (µm) were measured from each freshly isolated ovary, using an ocular micrometer under the compound microscope. The average diameter of 25 oocytes in each ovary was considered as oocyte diameter.

Ovarian histology

The excised and clean ovaries were fixed in aqueous Bouin's fluid. After 24 h of fixation, the ovaries were dehydrated through an alcohol series, cleared in xylene and then embedded in paraffin (melting point 56-58°C). Serial sections (5 µm) were made and stained with haematoxylin and counter stained with eosin. The sections were photographed using phase contrast microscope.

Estimation of vitellogenin levels using ELISA

Vitellogenin was isolated from crab ovaries [27] and estimated the Vg content by Enzyme Linked Immuno Sorbent Assay (ELISA). Wells were coated with 20 µl of sample diluted in 1:10 ratio with coating buffer (carbonate buffer: 1.59 g sodium carbonate, 2.93 g sodium bicarbonate in 1000 ml distilled water, diethyl dithio carbonate (DIECA) 56 mg per 25 ml buffer). Blank values were obtained from wells coated with buffer alone. The plate is covered with a lid and placed in a humid chamber at 37°C for 2 h. After discarding the contents, the plate was washed five times with 0.1 M Phosphate Buffer Saline containing 0.05% Tween-20, pH 7.2 (PBST). Then, 200 µl of diluted (1:1000) primary antibody (antibodies raised in rabbits against crab vitellogenin isolated from vitellogenic stage III ovary from crab) in 0.1 M PBST with 2% polyvinyl

pyrrolidine, 0.2% ovalbumin (PBST-PO) is added to each well. The plate is covered with a lid and placed in a humid chamber at 37°C for 2 h. The plate was washed 5 times with PBST after discarding the contents. After that 200 µL of horseradish peroxidase (HRP) conjugated anti IgG antibody (1:1000 dilution with PBST-PO; purchased from Genei, Bangalore) was added to each plate and kept in dark for 1 h at 37°C. The plate was washed 5 times with PBST and 200 µL tetra methyl benzidene (TMB in 0.015% hydrogen peroxide) was added to each well. The plate was kept in dark for 1 h at 37°C. The reaction was stopped by adding 50 µL of 1 M phosphoric acid per well. Absorbance of each well was measured using an ELISA-plates reader (Bio-Rad Lab., Model 680) at 450 nm. All standard and sample measurements were performed in duplicate.

Statistical analysis

The significance of differences between the means was determined by performing one-way ANOVA (Tukey: Compare all pairs of columns) were carried out using SPSS 16.0 version software. Statistical significance was considered at p<0.001.

Results

The ovary of the initial controls is in immature stage with a mean ovarian index of 0.322 ± 0.024 and oocyte diameter of 26.65 ± 0.78. No change in the mean ovarian index (0.327 ± 0.021) and oocyte diameter (27.01 ± 0.63) were found in the concurrent control group (Table 1) and are similar to controls. Injection of AA significantly (p<0.001) increased the mean ovarian index (1.568 ± 0.11) and oocyte diameter (68.36 ± 1.19) with a maturation stage III (vitellogenic stage III) compared to control crabs (Table 1). Injection of COX-1 inhibitors indomethacin (IM) and aspirin (Asp) significantly (p<0.001) decreased the mean ovarian index (0.228 ± 0.023 and 0.231 ± 0.014 respectively) and mean oocyte diameter (22.12 ± 1.02 and 23.98 ± 1.06 respectively) compared to control crabs (Table 1). Co-injection of AA with either indomethacin or aspirin is also significantly decreased the ovarian index (0.226 ± 0.021 and 0.230 ± 0.024 respectively) and oocyte diameter (22.06 ± 1.01 and 23.66 ± 1.05 respectively) when compared with controls (Table 1). The ovarian vitellogenin levels of the initial control crabs were 0.158 ± 0.088 and similar (0.160 ± 0.055) in concurrent controls. The injection of Arachidonic acid significantly (p<0.001) increased the ovarian vitellogenin (0.251 ± 0.008) whereas

Groups	Ovarian index (g%)	Oocyte diameter (µm)	Color of the ovary
Control	0.322 ± 0.024	26.65 ± 0.78	White
Concurrent control	0.327 ^{NS} ± 0.021 (1.55)	27.01 ^{NS} ± 0.63 (1.35)	White
Arachidonic acid (AA)	1.568* ± 0.11 (386.95)	68.36* ± 1.19 (156.51)	Dark brown
Indomethacin (IM)	0.228* ± 0.023 (-29.19)	22.12* ± 1.02 (-16.69)	White
Aspirin (Asp)	0.231* ± 0.014 (-28.26)	23.98* ± 1.06 (-10.01)	White
AA+IM	0.226* ± 0.021 (-29.81)	22.06* ± 1.01 (-17.22)	White
AA+Asp	0.230* ± 0.024 (-28.57)	23.66* ± 1.05 (-11.21)	White
p value	<0.001	<0.001	
F value	2319.7	5882.4	

Values are mean ± S.D. of 20 individuals.
Values in parentheses are percent change from control.
Values with * are significantly different from controls; NS: Not Significant.

Table 1: Effect of arachidonic acid and COX-1 inhibitors, indomethacin and aspirin on ovarian in fresh water crab, *Oziothelphusa senex senex*.

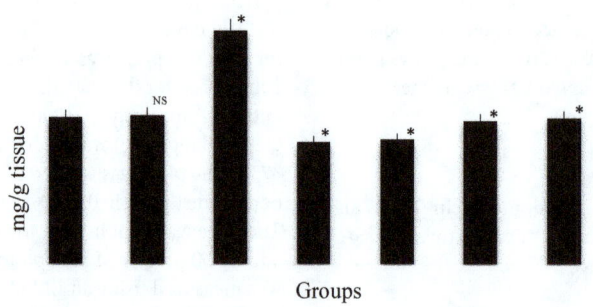

Bars with mean ± S.D. of 20 individuals.
Bars with * are significantly changed from controls at p<0.001; NS: Not significant.

Figure 1: Effect of arachidonic acid (AA), COX-1 inhibitors-indomethacin (IM) and aspirin (Asp) on ovarian vitellogenin levels in freshwater crab, *Oziothelphusa senex senex*.

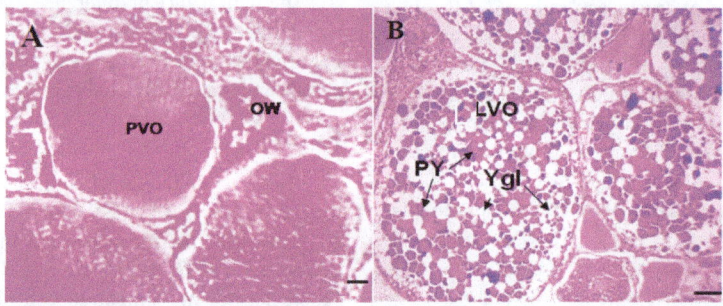

PVO: Pre-Vitellogenic Oocytes; OW: Thick Ovarian Wall; Ygl:Yolk Globules;
LVO: Late Vitellogenic Oocytes; PY: Protein Yolk;
Scale line=10 µm.

Figure 2: (A) Histological appearance of the immature stage ovary (control) and (B) Vitellogenic stage III ovary (arachidonic acid injected) of the freshwater crab *Oziothelphusa senex senex*.

injection of indomethacin or aspirin significantly (p<0.001) decreased the ovarian vitellogenin (0.131 ± 0.005 and 0.134 ± 0.006 respectively) (Figure 1). Co-injection of AA with either indomethacin or aspirin also significantly (p<0.001) decreased ovarian vitellogenin levels (0.154 ± 0.004 and 0.157 ± 0.002 respectively) compared to controls (Figure 1).

At the end of the experiment, only the ovaries of AA injected crabs were in vitellogenic stage III (dark brown in colour). Histological sections of with vitellogenic stage III contain a large accumulation of yolk globules occupies in the whole oocyte (Figure 2A) against immature ovary contain a thick ovarian wall with a centrally located germanium surrounded by no of oocytes (Figure 2B).

Discussion

In crustaceans, the ovaries undergo a series of developmental changes in color and increase in size during the reproductive cycle. Measurement of the ovarian index, oocyte diameter and histological changes of the ovary were the best indicators of measuring the female reproductive status in crustaceans [28,29]. Vitellogenesis is the synthesis and ovarian accumulation of yolk lipoprotein vitellogenin, a major nutritive component of embryo development after fertilization [30,31]. The vitellogenin concentration in various tissues correlates with the yolk accumulation in the oocyte [32,33]. The measurement of vitellogenin levels at different stages in the hemolymph and the ovary is used as additional criteria to determine the ovarian maturation in recent past [34].

In the present study, injection of AA resulted in a significant increase in the ovarian index, oocyte diameter, and ovarian vitellogenin levels with vitellogenic stage III ovary with enormous deposition of yolk

histologically indicating the involvement of AA on ovarian maturation in crab. Several investigators hypothesized that AA regulates female reproductive development in crustaceans and presence of AA have been identified in several crustaceans [11]. Injection of arachidonic acid significantly induces maturation, in shrimp to *Penaeus semisulcatus* at a dose of 5 µg/g body weight and 10 µg/g body weights [35]. Earlier it was reported that arachidonic acid supplementation through diet significantly promotes reproductive performance in tank domesticated *Penaeus monodon* [36], in *Daphnia magna* [37] and improved gonad somatic index, hepatosomatic index, egg clutch weight and fecundity in prawn, *Macrobrachium rosenbergii* [38]. In several fishes, also Arachidonic acid supplementation increases reproduction [39-42].

AA is a precursor of PGE2 and PGF2, the role of AA in shrimp maturation has been suggested as a precursor for prostaglandins [43]. Further, this was supported by several studies in different crustaceans where PGs, stimulates ovarian maturation in freshwater prawn *Metapenaeus affinis* [16], in crayfish [17], in freshwater crab, *Oziothelphusa senex senex* [13], in the shrimp *Penaeus monodon* [11] and in the freshwater giant prawn *Macrobrachium rosenbergii* [19]. Recently, it was demonstrated that administration of PGE2 increases the vitellogenin levels in hemolymph of *Macrobrachium rosenbergii* [19]. Similar to earlier studies, in the present study, the regulation of reproduction in crab, *Oziothelphusa senex senex* by AA may connect to its role as a precursor to prostaglandins.

The prostanoid pathway converts polyunsaturated fatty acids into PGs by COX enzymes from substrate AA. The presence of COX pathway in crustaceans is also available [18,44]. To date, COX enzymes isolated, sequenced and characterized in two crustacean species, *Gammarus pulex* and *Caprella mutica* [45]. COX has also been identified and sequenced in the shrimp *Penaeus monodon* [11]. Similar to mammals, NSAIDs affect the synthesis of PGs in invertebrates NSAIDs blocks the COX pathway [18,45], thereby the physiological roles of PGs in the present study the injection of COX inhibitors, indomethacin and aspirin individually and co-administration with AA resulted in a significant decrease in ovarian index, oocyte diameter and ovarian vitellogenin levels. There are no direct studies available to support the present results, but previous studies supported the COX-inhibiting role of non-steroidal anti-inflammatory drugs in crustaceans [21-25,19,46]. The exposure to IBU reduces the reproductive performance in *Daphina magna* and *Moina macrocopa* [24]. Similarly, the treatment of IBU (0.01 μg/g and 0.1 μg/g) reduced ovarian maturation in unilaterally eyestalk ablated female and male *Litopaneus* species. It is also reported that 0.1 μg/g of IBU significantly induced ovarian maturation in *Litopenaeus stylirostris*, but not in *Litopenaeus vannamei* whereas administration of IBU had no effect on sperm counts and spermatophore weights in *Litopenaeus* species [25]. COX inhibitor-indomethacin blocks the synthesis in *Macrobrachium rosenbergii* [47].

Conclusion

Administration of AA results in a significant induction in ovarian development, whereas administration of COX inhibitors, indomethacin and aspirin alone or in combination with AA inhibited the ovarian development in the crab. AA as a precursor induces the PGs synthesis, thereby maturation in crab, whereas the COX inhibitors block the PGs synthesis though AA is available it inhibits the maturation process in female crabs. This study provides potential evidence that AA and COX inhibitors indeed involved in the regulation of reproduction through the COX pathway in *Oziothelphusa senex senex*. However, the regulatory mechanism of AA and COX inhibitors in crustaceans is not known. Further emphasis is required to know the mechanistic action of the arachidonic acid or PGs in ovarian development in crustaceans.

References

1. Stanley D, Howard RW (1998) The biology of prostaglandins and related eicosanoids in invertebrates: Cellular, organismal and ecological actions. American Zoologist 38: 369-381.

2. Rowley AF, Vogan CL, Taylor GW, Clare AS (2005) Prostaglandins in non-insectan invertebrates: Recent insights and unsolved problems. Journal of experimental biology 208: 3-14.

3. Muriana FJ, Gutierrez V, Blaya JA, Bolufer J (1995) Phospholipid fatty acid composition of hepatopancreatic brush-border membrane vesicles from the prawn *Penaeus japonicus*. Biochimie 77: 190-193.

4. Tahara D, Yano I (2004) Maturation-related variations in prostaglandin and fatty acid content of ovary in the kuruma prawn *Marsupenaeus japonicus*. Comparative Biochemistry and Physiology Part A: Molecular & Integrative Physiology 137: 631-637.

5. Ravid T, Tietz A, Khayat M, Boehm E, Michelis R (1999) Lipid accumulation in the ovaries of a marine shrimp *Penaeus semisulcatus* (de haan). Indian Journal of Experimental Biology 202: 1819-1829.

6. Meunpol O, Meejing P, Piyatiratitivorakul S (2005) Maturation diet based on fatty acid content for male *Penaeus monodon* (Fabricius) brood stock. Aquaculture Research 36: 1216-1225.

7. Huang JH, Jiang SG, Lin H, Zhou FL, Ye L (2008) Effects of dietary highly unsaturated fatty acids and astaxanthin on the fecundity and lipid content of pond-reared *Penaeus monodon* (Fabricius) broodstock. Aquaculture Research 39: 240 251.

8. Felix GM, Velazquez PM, Alvarez QJ (2009) Effect of various dietary levels of docosahexanoic and arachidonic acids and different n-3/n-6 ratios on biological performance of Pacific white shrimp *Litopenaeus vannamei* raised in low salinity. Journal of the World Aquaculture Society 40: 194-206.

9. Chansela P, Goto-Inoue N, Zaima N, Hayasaka T, Sroyraya M (2012) Composition and localization of lipids in *Penaeus merguiensis* ovaries during the ovarian maturation cycle as revealed by imaging mass spectrometry. PLoS ONE 7: 1-12.

10. Spaziani EP, Hinsch GW, Edwards SC (1993) Changes in prostaglandin E2 and F2 alpha during vitellogenesis in the Florida crayfish *Procambarus paeninsulanus*. Journal of Comparative Physiology B 163: 541-545.

11. Wimuttisuk W, Tobwor P, Deenarn P, Danwisetkanjana K, Pinkaew D, et al. (2013) Insights into the prostanoid pathway in the ovary development of the penaeid shrimp *Penaeus monodon*. PLoS ONE 8: 1 15.

12. Hampson AJ, Rowley AF, Barrow SE, Steadman R (1992) Biosynthesis of eicosanoids by blood cells of the crab *Carcinus maenas*. Biochimica et Biophysica Acta 1124: 143-150.

13. Reddy PS, Reddy PR, Nagaraju GP (2004) The synthesis and effects of prostaglandins on the ovary of the crab *Oziothelphusa senex senex*. General and Comparative Endocrinology 135: 35-41.

14. Preechaphol R, Klinbunga S, Khamnamtong B, Menasveta P (2010) Isolation and characterization of genes functionally involved in ovarian development of the giant tiger shrimp *Penaeus monodon* by suppression subtractive hybridization (SSH). Genetics and Molecular Biology 33: 676-685.

15. Stanley D (2006) Prostaglandins and other eicosanoids in insects: Biological significance. Annual Review of Entomology 51: 25-44.

16. Sarojini R, Sambasiva RS, Jayalakshmi K, Nagabhushanam R (1989) Effect of PGF$_{1α}$ on ovarian maturation in a penaeid prawn *Metapenaeus affinis*. Journal of Zoological Research 2: 7-11.

17. Spaziani EP, Hinsch GW, Edwards SC (1995) The effect of prostaglandin E2 and prostaglandin F2α on ovarian tissue in the Florida crayfish *Procambarus paeninsulanus*. Prostate 50: 189-200.

18. Silkovsky J, Chayoth R, Sagi A (1998) Comparative study of effects of prostaglandin E$_2$ on cAMP levels in gonads of the prawn *Macrobrachium rosenbergii* and the crayfish *Cherax quadricarinatus*. Journal of Crustacean Biology 18: 643-649.

19. Sumpownon C, Engsusophon A, Siangcham T, Sugiyama E, Soonklang N, et al. (2015) Variation of prostaglandin E2 concentrations in ovaries and its effects on ovarian maturation and oocyte proliferation in the giant freshwater prawn, *Macrobrachium rosenbergii*. General and Comparative Endocrinology 1:129-138.

20. Koskela RW, Greenwood JG, Rothlisberg PC (1992) The influence of prostaglandin E$_2$ and the steroid hormones, 17α-hydroxyprogesterone and 17ß-estradiol on molting and ovarian development in the tiger prawn *Penaeus esculentus* Haswell, 1879 Crustacea: Decapoda. Comparative Biochemistry and Physiology 101: 295-299.

21. Charlier C, Michaux C (2003) Dual inhibition of cyclooxygenase-2 (COX-2) and 5-lipoxygenase (5-LOX) as a new strategy to provide safer non-steroidal anti-inflammatory drugs. European Journal of Medicinal Chemistry 38: 645-659.

22. Heckmann LH, Callaghan A, Hooper HL, Connon R, Hutchinson TH, et al. (2007) Chronic toxicity of ibuprofen to Daphnia magna: Effects on life history traits and population dynamics. Toxicology Letters 172: 137-145.

23. Heckmann LH, Sibly RM, Connon R, Hooper HL, Hutchinson TH, et al. (2008) Systems biology meets stress ecology: Linking molecular and organismal stress responses in Daphnia magna. Genome Biology 9: R40 1-15.

24. Han S, Choi K, Kim J, Ji K, Kim S, et al. (2010) Endocrine disruption and consequences of chronic exposure to ibuprofen in Japanese medaka *Oryzias latipes* and freshwater cladocerans Daphnia magna and Moina macrocopa. Aquatic Toxicology 98: 256-264.

25. Alfaro MJ (2015) The effect of ibuprofen on female and male reproduction of the open thelyca marine shrimp, *Litopenaeus*. Aquaculture research 46: 105-116.

26. Van Harreveld A (1936) A physiological solution for freshwater crustaceans. Proceedings of the Society for Experimental Biology and Medicine 34: 428-432.

27. Tsukimura B (2001) *Crustacean vitellogenesis*: Its role in oocyte development. American Zoologist 41: 465-476.

28. Pillai KK, Nair NB (1971) The annual reproductive cycle of *Uca annulipes*, *Portunus pelagicus* and *Metapenaeus affinis* (Decapoda: Crustacea) from the South West coast of India. Marine Biology 11: 152-166.

29. Charniaux-Cotton H, Payen G (1988) Endocrinology of selected invertebrate types: Crustacean Reproduction. Alan R Liss, New York.

30. Tseng DY, Chen YN, Kuo GH, Lo CF, Kuo CM (2001) Hepatopancreas is the extra-ovarian site of vitellogenin synthesis in the tiger shrimp, *Penaeus monodon*. Comparative Biochemistry and Physiology 129: 909-917.

31. Zapata V, Lopez GLS, Medesani D, Rodriguez EM (2003) Ovarian growth in the crab, *Chasmagnathus granulata* induced by hormones and neuroregulators throughout the year. *In vivo* and *in vitro* studies. Aquaculture 224: 339-353.

32. Quackenbush LS (1989) Yolk protein production in the marine shrimp *P. vannamei*. Journal of Crustacean Biology 9: 509-516.

33. Okamura T, Han CH, Suzuki Y, Aida K, Hanyu I (1992) Changes in hemolymph vitellogenin and ecdysteriod levels during the reproductive and non-reproductive molt cycles in the freshwater prawn *Macrobrachium nipponense*. Zoological Science 9: 37-45.

34. Tsukimura B, Bender JS, Linder CJ (2000) Development of an anti-vitellin ELISA for the assessment of reproduction in the ridgeback shrimp *Sicyonia ingentis*. Comparative Biochemistry Physiology 127: 215-224.

35. Maheswarudu G, Vineetha A (2013) Littoral Oligochaete *Pontodrilus bermudensis Beddard*: A potential source for Arachidonic acid that stimulates maturation in penaeid shrimp. The Journal of Veterinary Science. Photon 114: 290-300.

36. Coman GJ, Arnold SJ, Barclay M, Smith DM (2011) Effect of arachidonic acid supplementation on reproductive performance of tank domesticated *Penaeus monodon*. Aquaculture Nutrition 17:141-151.

37. Ginjupalli GK, Gerard PD, Baldwin WS (2015) Arachidonic acid enhances reproduction in *Daphnia magna* and mitigates changes in sex ratios induced by pyriproxyfen. Environmental Toxicology and Chemistry 34: 527-535.

38. Kangpanich C, Pratoomyot J, Siranonthana N, Senanan W (2016) Effects of arachidonic acid supplementation in maturation diet on female reproductive performance and larval quality of giant river prawn *Macrobrachium rosenbergii*. Peer J 4: e2735.

39. Bessonart M, Izquierdo MS, Salhi M, Hernandez-Cruz CM, Gonzalez MM, et al. (1999) Effect of dietary arachidonic acid levels on growth and fatty acid composition of gilthead sea bream *Sparus aurata L*. larvae. Aquaculture 179: 265-275.

40. Koven W, Barr Y, Lutzky S, Ben-Atia I, Weiss R, et al. (2001) The effect of dietary arachidonic acid (20:4n_6) on growth, survival and resistance to handling stress in gilthead sea bream *Sparus aurata* larvae. Aquaculture 193: 107-122.

41. Estevez A, Kaneko T, Seikai T, Tagawa M, Tanaka M (2001) ACTH and MSH production in Japanese flounder *Paralichthys olivaceus* larvae fed arachidonic acid-enriched live prey. Aquaculture 192: 309-319.

42. Bell GJ, Sargent JR (2002) Nutrition Group, Arachidonic acid in aquaculture feeds: Current status and future opportunities. Institute of Aquaculture, University of Stirling, Stirling FK9 4LA, Scotland, UK.

43. Middleditch BS, Missler SR, Hines HB, Brown A, Ward DC, et al. (1980) Metabolic profiles of penaeid shrimp: Dietary lipids and ovarian maturation. Journal of Chromatography 135: 359-368.

44. Valmsen K, Jarving I, Boeglin WE, Varvas K, Koljak R, et al. (2001) The origin of 15R-prostaglandins in the Caribbean coral *Plexaura homomalla*: Molecular cloning and expression of a novel cyclooxygenase. Proceedings of the National Academy of Sciences. USA 98: 7700-7705.

45. Varvas K, Kurg R, Hansen K, Jarving R, Jarving I (2009) Direct evidence of the cyclooxygenase pathway of prostaglandin synthesis in arthropods: Genetic and biochemical characterization of two crustacean cyclooxygenases. Insect Biochemistry and Molecular Biology 39: 851-860.

46. Hayashi Y, Heckmann LH, Callaghan A, Sibly RM (2008) Reproduction recovery of the crustacean Daphni magna after chronic exposure to ibuprofen. Ecotoxicology 17: 246-251.

47. Sagi A, Silkovsky J, Berkovich SF, Chayoth R (1996) Prostaglandin E2 in Previtellogeic Ovaries of the Prawn *Macrobrachium rosenbergii*: Synthesis and Effect on the Level of cAMP. General and Comparative Endocrinology 100: 308-313.

39

Vibrio Species Isolated from Farmed Fish in Basra City in Iraq

Asaad MR Al-Taee[1]*, Najem R Khamees[2] and Nadia AH Al-Shammari[2]

[1]*Marine Science Center, Basra University, Basra, Iraq*
[2]*College of Agriculture, Basra University, Basra, Iraq*

Abstract

Aim: This study was carried out to investigate the occurrence of potentially pathogenic species of *Vibrio* in seven types of fish sampled from fish farms located in different districts in Basra governorate, Iraq.

Methods and Results: A total of 153 live fishes was collected from fish farms during the period January-May 2016. Bacteria were isolated using selective medium thiosulfate citrate bile sucrose salt agar. Presumptive *Vibrio* colonies were identified using the VITEK 2 system and selected biochemical tests. In the present study *V. alginolyticus* (24 of 60) was the predominant species, followed by *V. cholerae* (10 of 60), *V. furnisii* (10 of 60), *V. diazotrophicus* (7 of 60), *V. gazogenes* (5 of 60) and *V. costicola* (4 of 60). The signs of vibriosis appeared in three types of fish, including *Cyprinus carpio*, *Coptodon zilli* and *Planiliza subviridius* in spite of the using Oxytetracycline in most fish farms.

Conclusion: The results of the present study demonstrated the presence of pathogenic *Vibrio* species nearly in all fish farms. So the farm owners should be concerned about the presence of these pathogenic bacteria which also contributes to human health risk and should adopt best management practices for responsible aquaculture to ensure the quality of fish.

Keywords: *Vibrio* spp; Vibriosis; *Cyprinus carpio; Coptodon zilli; Planiliza subviridius*

Introduction

The world fish production has grown recently- as a consequence of the decline production in capture fishery- with food fish supply increasing at an average annual rate of 3.2 percent, above the growth of the world population to 1.6 percent. World per capita apparent fish consumption increased from an average of 9.9 kg in the 1960s to 19.2 kg in 2012. According to the latest available statistics collected globally by FAO, world aquaculture production attained another all-time high of 90.4 million tons in 2012, including 66.6 million tons of food fish and 23.8 million tons of aquatic algae [1].

But this worldwide growth of aquaculture is overwhelmed by catastrophic fish diseases and spoilage caused by pathogenic bacteria, which are introduced to the fish farm through natural or artificial food sources, treated inlet water or through vertical transmission from brood stock [2-4]. The most diseases are caused by *Vibrio* spp., which are considered the well-known cause of a significant problem for the development of a sector with strong economic losses worldwide because of its high morbidity and mortality rates (mortality $\geq$ 50%) [5-8]. The Centers for Disease Control and Prevention [9] estimates that vibriosis causes 80,000 illnesses each year in the United States. About 52,000 of these illnesses are estimated to be the result of eating contaminated food and about 80% of infections occur between May and October when water temperatures are warmer.

Several factors have been proposed to influence the survival, persistence and ability of vibrios to cause infection. These include water temperature, UV or sunlight and salinity [10]. Many studies have been conducted on seasonal variation of pathogenic *Vibrio* species in natural environments [11-15].

However, there is a little information pertaining to vibriosis and the presence of *Vibrio* species in the fish farms especially in Iraq. Hence, this paper attempts to describe the presence of *Vibrio* spp in different types of fish.

Materials and Methods

Fish samples collection

A total of 153 live fish was collected from fish farms located in different districts in Basra governorate, Iraq (Figure 1), over a five month period (January- May 2016). The parameters of water have been measured. Fish Samples included common carp (*Cyprinus carpio*) (65), silver carp (*Hypophthalmichthys molitrix*) (15), sea bream (*Acanthopagrus arabicus*) (6), green mullet (*Planiliza subviridus*) (23), molly fish (*Poecilia latipinna*) (15), Bue tilapia (*Oreochromis aureus*) (15) and redbelly tilapia (*Coptodon zilli*) (14). Live fish samples were transported to the laboratory within a few hours, the measurements of the total length and body weight are recorded (Table 1).

Bacterial isolation and identification

The fish were killed by physical destruction of the brain, in order to prepare the samples for bacterial isolation. Initially a swab was taken from skin, fins and eyes while 1 gm of the gills and intestine were incised aseptically using a sterile scalpel. These samples were homogenized in 9 ml of sterile normal saline solution using a sterilized glass homogenizer (Brand- Germany). One milliliter aliquots of the homogenate solutions were serially diluted (10^{-1} to 10^{-7}). Aliquots of 0.1 ml of the serial dilutions were inoculated onto thiosulfate citrate bile sucrose salt agar (TCBS) (Hi media- India) in duplicate using the spread plate method and the plates were incubated at 30°C for 24-72 h.

*Corresponding author: Asaad MR Al-Taee, Marine Science Center, Basra University, Basra, Iraq, E-mail: amraltaee@yahoo.com

The presumptive *Vibrio* colonies, yellow- greenish- blue, on TCBS agar were picked and subjected to bacterial identification using VITEK 2 system (Biomerieux- USA) and biochemical tests [16] such as: oxidase test, H_2S production, urease, indole production, Voges- Proskauer, fermentation of: glucose, lactose, inositol, raffinose, mannitol, dextrose, adonitol, fructose, dolcitol, xylose, arabinose, trehalose, salicin, rahmnose, milibiose, galactose, sorbitol, sucrose, mannose and inuline.

Results and Discussion

Water quality parameters

The study was conducted during January- May, 2016 in which there is a fluctuation in water quality parameters in the aquaculture throughout the sampling sites (Table 2). The mean of temperature fluctuated from 24.5°C to 30.1°C. Meanwhile the mean of salinity was recorded to range from 1.23 to 6.22 ppt. The pH was relatively from 6.5 to 7.2.

The fish in a culture system always exposed to a variety of stressors which including high stocking density, handling, transportation and poor water quality [17]. On the other hand, fish immunity is reduced

Station	Temp (°C) Range (Mean)	Salinity ppt Range (Mean)	PH Range (Mean)
Hartha	18.5-32.4°C (24.5)	1.32-4.00 (1.34)	7.9-8.2 (7.1)
Mashab	13.0-33.0°C (29.8)	0.50-1.80 (1.23)	6.9-7.7 (6.8)
Basra University	19.0-32.0°C (26.4)	1.60-4.50 (2.1)	7.3-7.9 (7.1)
Muhaijran	16.0-30.0°C (26.2)	1.80-3.70 (1.54)	7.5-8.4 (6.5)
Seeba	12.0-30.0°C (28.2)	3.20-12.60 (6.22)	7.1-8.4 (7.2)
Marbad	18.0-34.0°C (30.1)	4.10-5.30 (4.5)	7.0-7.4 (7.1)

Table 2: The water parameters of studied stations.

Tests	V. gazogenes	V. alginolyticus	V. costicola	V. furnisii	V. diazotrophicus
Oxidase	-	+	+	+	+
Nitrate Reduction	-	+	+	+	+
Indole	-	+	-	-	+
V-P	-	+	+	-	-
H_2S	ND	-	-	-	-
Urease	ND	-	ND	-	ND
Fermentation of Glucose	-	-	+	+	-
Lactose	+	-	-	-	+
Inositol	-	-	-	-	-
Raffinose	-	-	-	-	-
Mannitol	+	+	+	+	+
Dextrose	+	+	+	-	+
Adonitol	-	-	-	-	-
Dolcitol	+	-	-	-	-
Xylose	+	-	-	-	+
Arabinose	+	-	-	+	+
Trehalose	-	+	+	+	+
Salicin	+	-	-	-	+
Rahmnose	-	-	-	-	-
Galactose	+	+	-	+	+
Sorbitol	+	-	-	-	-
Sucrose	+	+	+	+	+
Mannose	+	+	-	+	-
Inuline	-	-	-	-	-
Milibiose	-	-	-	-	-
+: Positive, -: Negative, ND: Not Determined					

Table 3: Biochemical profile of *Vibrio* sp.

during a stressful event which causes the fish to become susceptible to disease infection [18]

The growth of *Vibrio* in water is increased by high levels of organic matters, high salinity, high water temperature (25°C to 32°C) and pH (5-9). While the low salinity and high pH (>9.5) had shown to reduce the growth of this bacterium [19]. These favorable conditions for *Vibrio* were also observed in the present study.

In the present study, three types of aquaculture systems have been studied, the net cage aquaculture in Hartha station, which lies in the north of Shatt Al-Arab River. The second type is the terrestrial pond, which takes water either from Shatt Al-Arab River (Basra University, Muhaijran and Seeba stations) or from artesian wells as Marbad station. The third is Mashab station (net cage), which represent as a part of Hor Al-Hammar marsh and either take water from Shatt Al-Arab River and general downstream. There is an obvious effect of the temperature and salinity on the infected fish, especially in Seeba and Marbad stations. The present study agreed with Le Roux et al. [20] who reported that, *Vibrio* abound in the warm (>15°C) and saline aquatic environments.

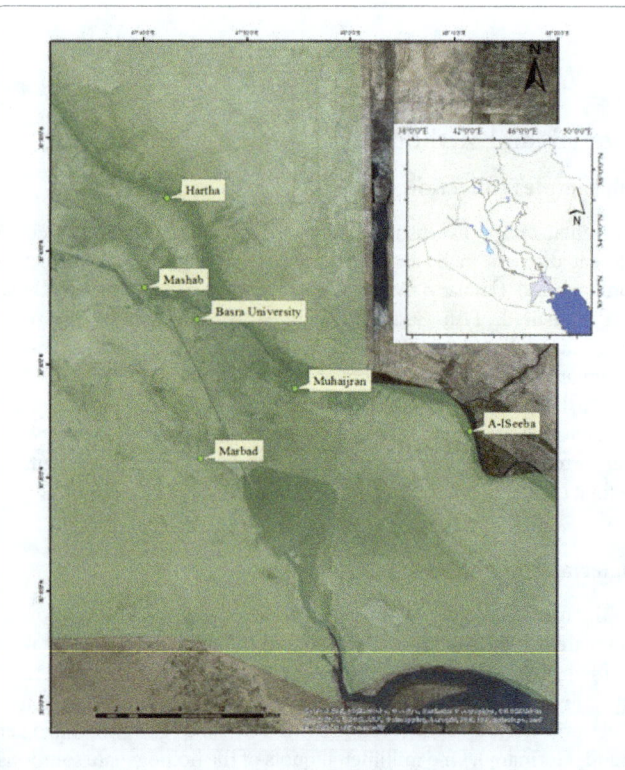

Figure 1: A map of sampling sites.

Fish Species	Total Length, cm Range (Mean)	Body Weight, gm Range (Mean)
Cyprinus carpio	16-49 (36.44)	13.0-2353.9 (890.4)
Hypophthalmichthys molitrix	32-54 (32.66)	65.0-1360 (653.3)
Acanthopagrus aerobics	15-29 (28.3)	13.0-123.8 (22.4)
Oreochromis aureus	12.3-14.2 (12.62)	10.0-62.0 (42)
Coptodon zilli	12-14 (12.23)	10.0-65.0 (40)
Planiliza subviridus	11.3-16.4 (14.23)	12.0-650 (42)
Poecilia latipinna	5.2-6.5 (4.21)	4.0-6.44 (3.88)

Table 1: Physical measurements taken at the time of collection.

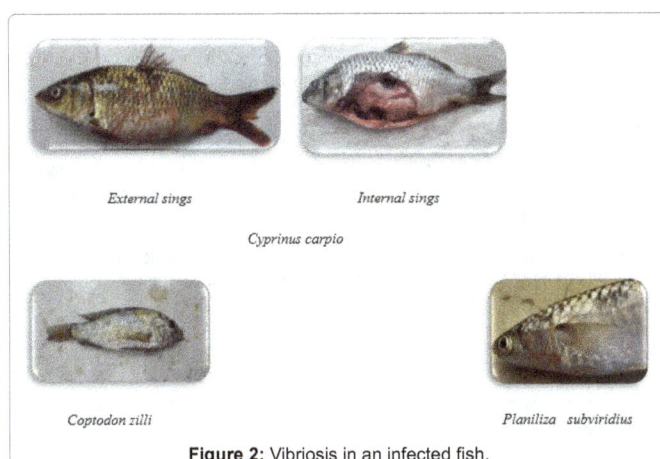

External sings   Internal sings

Cyprinus carpio

Coptodon zilli   Planiliza subviridius

Figure 2: Vibriosis in an infected fish.

Fish spp	Total Exam. Fish	Total Infected Fish	Disease
Cyprinus carpio	65	47	Vibriosis, Spring Viraemia of carp Virus (SVCV), Fin rot, Dropsy, Bacterial gill disease
Hypophthalmichthys molitrix	15	2	Viral hemorrhagic septicemia
Acanthopagrus arabicus	6	2	Pox disease
Oreochromis aureus	15	-	
Coptodon zilli	14	6	Vibriosis
Planiliza subviridius	23	5	Vibriosis
Poecilia latipinna	15	5	Red mouth

Table 4: The types of fish and their disease signs.

Finlay and Falkow [21] observed that, the occurrence of high total *Vibrio* count in the biofilm at high temperature concurred with the occurrence of disease outbreaks. Albert and Ransangan [17] revealed that, the water temperature has shown to play an important role in enhancing the growth of *Vibrio* spp, causing fish to stress and inducing severe vibriosis outbreak. Kaspart and Tamplin [22] noticed that, the optimal temperatures of survival of *V. vulnificus* 4965 was between 13 and 22°C in 10-ppt (noninhibitory salinity) sterile seawater, while the temperature outside this range reduced the time of survival.

Bacterial diagnosis

A total number of 153 fish were sampled during the study. Almost all the primary isolates from sampled organs (skin, round mouth, fin base, gill cover and intestine) showed green and yellow colonies on TCBS. From the samples of fish examined, the average rate of infection was 43.79%, including 15.03% of them as a bacterial infection and identified as *Vibrio* spp. Two methods were used for identification of *Vibrio spp.*, the VITEK 2 system which detect only *V. cholerae* (10 isolates) with probability 98% and confidence, excellent identification, while the other isolates were failed to identify with it, so its identified depending upon their biochemical profiles (Table 3) as *V. gazogenes* (5), *V. alginolyticus* (24), *V. costicola* (4), *V. furnisii* (10) *and V. diazotrophicus* (7).

The results of the present study indicated that, the infected fish display skin discoloration, red patches around the base of the fins and mouth and necrotic intestine. These signs have appeared in three types of fish, including *Cyprinus carpio, Coptodon zilli and Planiliza subviridius* (Figure 2 and Table 4). The infected fish were sampled from all stations (except Basra university station).

In the present study, although the managers of the farms stated that, they use Oxytetracycline (1%) in all stations. Oxytetracycline is widely used to treat bacterial infections in aquaculture farms, such as vibriosis and furunculosis [23,24]. In spite of that, many infected cases have been detected among fish, particularly common carp, and this is may be related to that, the an extensive use of antibiotics can cause the development of antibiotic-resistant pathogens which can infect both cultured animals as well as humans [25-27].

V. cholerae was the most species isolated from intestinal necrosis in common carp and the infections were distributed in all stations (except Basra University station), while other species of *Vibrio* were isolated from external infections.

In the present study *V. alginolyticus* (40%) was the predominant species, followed by *V. cholerae* (16.6), *V. furnisii* (16.6%), *V. diazotrophicus* (11.6), *V. gazogenes* (8.3%) and *V. costicola* (6.66%). The species of *Vibrio* were different between farms and this is may be related to the different source of larvae or different source of water. This agreed with Bhaskar et al. who reported the presence of *V. alginolyticus* as the most common, followed by *V. cholerae, V. parahemolyticus*, and *V. vulnificus* in *P. monodon* culture system. Sanjoy et al. [28] isolated five species of *Vibrio* from shrimp farm and found that, *V. cholerae* was the most common species.

Many researchers found that, the virulence level of Vibriois dependent on fish species, doses of infection, the time of exposure and age of host species and pathogenic factors of the bacterial strains [29-32].

Conclusion

The results of the present study demonstrated the presence of pathogenic *Vibrio* species as *V. cholerae, V. gazogenes, V. alginolyticus, V. costicola, V. furnisii* and *V. diazotrophicus* nearly in all fish farms. So the farm owners should be concerned about the presence of these pathogenic bacteria which also contributes to human health risk and should adopt best management practices for responsible aquaculture to ensure the quality of fish.

Acknowledgements

We thank the College of Agriculture and Marine Science Center, Iraq for assistance during working period. We would like to thank the managers of the fish farms who have made the sampling possible.

References

1. FAO (2014) The State of World Fisheries and Aquaculture 2014. Rome.

2. Sandaa RA, Magnesen T, Torkildsen L, Bergh O (2003) Characterization of the bacterial community associated with early stages of great scallop (*Pecten maximus*) using denaturing gradient gel electrophoresis (DGGE). Syst Appl Microbiol 26: 302-311.

3. Schulze AD, Alabi AO, Shaledrake AR, Miller KM (2006) Bacterial diversity in a marine hatchery: Balance between pathogenic and potentially probiotic bacterial strains. Aquaculture 256: 50-73.

4. Sahoo TK, Jena PK, Patel AK, Seshadri S (2014) Bacteriocins and their applications for the treatment of bacterial diseases in aquaculture: A review. Aquaculture Research 1-15.

5. Austin B, Austin DA (2007) Bacterial fish pathogens: disease of farmed and wild fish. (4th Edn) Springer Praxis Publishing, Chichester. UK.

6. Al- Sunaiher A, Abdelnasser SS, Ali AA (2010) Association of Vibrio species with disease incidence in some cultured fishes in the Kingdom of audi Arabia. World Appl Sci J 8: 653-660.

7. Frans I, Michiels CW, Bossier P, Willems KA, Lievens B, et al. (2011) Vibrio anguillarum as a fish pathogen: virulence factors, diagnosis and prevention. J Fish Dis 34: 643-661.

8. Chatterjee S, Haldar S (2012) Vibrio related diseases in aquaculture and development of rapid accurate methods. J Marine Sci Res Dev S:1.

9. Center for Disease Control and Prevention (2017) Vibrio species causing vibriosis.

10. Lipp EK, Huq A, Colwell RR (2002) Effects of global climate on infectious disease: The cholera model. Clin Microbiol Rev 15: 757-770.

11. Williams L, La Rock P (1985) Temporal occurrence of Vibrio Species and Aeromonas hydrophila in estuarine sediments. Appl Environ Microbiol 50: 1490-1495.

12. Venkateswaran K, Nakano H, Takayama OK, Matsuda O, Hashimoto H (1989) Occurrence and distribution of Vibrio spp., Listonella spp., and Clostridium botulinum in the Seto Inland Sea of Japan. Appl Environ Microbiol 55: 559-567.

13. Barbieri E, Falzano L, Fiorentini C, Pianetti A, Baffone W, et al. (1999) Occurrence, diversity, and pathogenicity of halophilic Vibrio spp. and non-O1 Vibrio cholera from estuarine waters along the Italian Adriatic coast. Appl Environ Microbiol 65: 2748-2753.

14. Pfeffer CS, Hite FM, Oliver JD (2003) Ecology of Vibrio vulnificus in Estuarine Waters of Eastern North Carolina. Appl Environ Microbiol 69: 3526-3531.

15. Hosseini HM, Cheraghali A, Yalfani R, Razavilar V (2004) Incidence of Vibrio spp. in shrimp caught off the south coast of Iran. Food Control 15: 187-190.

16. Bhaskar N, Setty TMR (1994) Incidence of vibrios of public health significance in the farming phase of tiger shrimp (Penaeus monodon). J Sci Food and Agricul 66: 225-231.

17. Albert V, Ransangan J (2013) Effect of water temperature on susceptibility of culture marine fish species to vibriosis. Inter. J Rese Pure and Appl Microbiol 3: 48-52.

18. Rijnsdorp AD, Peck MA, Engelhard GH, Mollmann C, Pinnegar JK (2009) Resolving the effect of climate change on fish populations. ICES J Mar Sci 66: 1570-1583.

19. Kiriratnikom S, Ruangsri J, Wanadet M, Songpradit A, Suanyuk N, et al. (2000) The abiotic factors influencing the growth of luminescent bacteria. Songklanakarin. J Sci Technol 22: 697-705.

20. Le RF, Wegner KM, Baker-Austin C, Vezzulli L, Osorio CR, et al. (2015) The emergence of Vibrio pathogens in Europe: ecology, evolution, and pathogenesis. Front Microbiol 6: 830.

21. Finlay BB, Falkow S (1997) Common themes in microbial pathogenicity revisited. Microbiol Mol Biol Rev 61: 136-169.

22. Kaspart CW, Tamplin ML (1993) effects of temperature and salinity on the survival of Vibrio vulnificus in Seawater and Shellfish. Appl Environ Microbiol 59: 2425-2429.

23. Capone GD, Weston PD, Miller V, Shoemaker C (1996) Antibacterial residues in marine sediments and invertebrates following chemotherapy in aquaculture. Aquaculture 145: 55-75.

24. Reed LA, Siewicki TC, Shah AC (2006) The bio-pharmaceutics and oral bioavailability of two forms of oxytetracycline to the white shrimp, Litopenaeus setiferus. Aquaculture 258: 42-54.

25. Khachatourians GG (1998) Agricultural use of antibiotics and the evolution and transfer of antibiotic-resistant bacteria. Can Med Ass J 159: 1129-1136.

26. Willis C (2000) Antibiotics in the food chain: their impact on the consumer. Rev Med Microbiol 11: 153-160.

27. Holmström K, Gräslund S, Wahlström A, Poungshompoo S, Bengtsson BE, et al. (2003) Antibiotic use in shrimp farming and implications for environmental impacts and human health. Inter J Food Sci Technol 38: 255-266.

28. Sanjoy B, Mei CO, Mohamed S, Helena K (2012) Antibiotic resistant Salmonella and Vibrio associated with farmed Litopenaeus vannamei. Sci World J 2012: 130-136.

29. Vera P, Navas JI, Quintero MC (1992) Experimental study of the virulence of three species of Vibrio bacteria in Penaeus japonicus (Bate 1881) juveniles. Aquaculture 107: 119-123.

30. Jun LI, Huai-Shu X (1998) Isolation and biological characteristics of Vibrio harveyi affecting hatchery reared Penaeus chinensis larvae. Chin J Oceanol Limnol 29: 353-361.

31. Gomez-Gill B, Herrera-Vega MA, Abreu GFA (1998) A Roque, Bioencapsulation of two different Vibrio species in Nauplii of the brine shrimp (Artemia franciscana). Appl Environ Microbiol 64: 2318-2322.

32. Ransangan J, Lal TM, Al-Harbi AH (2012) Characterization and experimental infection of Vibrio harveyi isolated from diseased Asian seabass (Lates calcarifer), Malay. J Microbiol 8: 104-115.

Permissions

List of Contributors

Sreenivasula Reddy P
Department of Zoology, Sri Venkateswara University, Tirupati-517 502, India

Srilatha M
Department of Biotechnology, Sri Venkateswara University, Tirupati-517 502, India

Istiaq Ahmad Chowdhury, Jewel Das and Nani Gopal Das
Institute of Marine Sciences and Fisheries, University of Chittagong, Bangladesh

Malik M Khalafalla
Department of Aquaculture, Faculty of Aquatic and Fisheries Sciences, Kafrelsheikh University, 33516 Kafr El-Sheikh, Egypt

EL-Sayed B
Agricultural Botany Department (Agric. Microbiology), Faculty of Agriculture, Kafrelsheikh University, 33516, Kafr El-Sheikh, Egypt

Md. Hasan Uddowla and Won-gyu Park
Department of Marine Biology, Pukyong National University, Busan 608-737, South Korea

Hyun-Woo Kim
Department of Marine Biology, Pukyong National University, Busan 608-737, South KoreaInterdisciplinary program of Biomedical Engineering, Pukyong National University, Busan, 608-737, South Korea

Ah Ran Kim
Interdisciplinary program of Biomedical Engineering, Pukyong National University, Busan, 608-737, South Korea

Agbabiaka LA
Department of Fisheries Technology, Federal Polytechnic Nekede Owerri, Imo State, Nigeria

Kuforiji OA
Department of Agricultural Technology, Federal Polytechnic Ekowe, Bayelsa State, Nigeria

Egobuike CC
Department of Fisheries Technology, Imo State Polytechnic, Umuagwo-Ohaji, Imo State, Nigeria

Mojekwu TO
Department of Biotechnology Unit, Nigerian Institute for Oceanography and Marine Research, Nigeria

Anumudu CI
Department of Zoology, University of Ibadan, Nigeria

Akewake Geremew
Department of Biology, Dilla University, P. O. Box 419, Dilla, Ethiopia

Abebe Getahun
Department of Zoological Sciences, Addis Ababa University, P. O. Box 1176, Addis Ababa, Ethiopia

Krishen Rana
Department of Sustainable Aquaculture, Stellenbosch University, Stellenbosch, South Africa

Doaa M Mokhtar
Department of Anatomy and Histology, Faculty of Veterinary Medicine, Assuit University, Egypt

Dinesh R, Chandra Prakash, Chadha NK and Sherry Abraham
Division of Aquaculture, Central Institute of Fisheries Education, Mumbai, India

Nalini Poojary
Division of Aquatic Environment and Health Management, Central Institute of Fisheries Education, Mumbai, India

Bamidele Oluwarotimi Omitoyin, Emmanuel Kolawole Ajani, Oluwabusayo Israel Okeleye and Benjamin Uzezi Akpoilih
Department of Aquaculture and Fisheries Management, University of Ibadan, Ibadan, Nigeria

Adeniyi Adewale Ogunjobi
Department of Microbiology, University of Ibadan, Ibadan, Nigeria

Valdez ASM and Castillo TR
Mindanao State University, General Santos City, Philippines

Azrita
Department of Biology Education, Faculty of Education of Bung Hatta University Padang, Indonesia

Yuneidi Basri and Hafrijal Syandri
Department of Aquaculture, Faculty of Fisheries and Marine Sciences of Bung Hatta University Padang, Indonesia

Elgendy MY and Kenawy AM
Department of Hydrobiology, Veterinary Research Division, National Research Centre, 12622 Dokki, Giza, Egypt

Abdelsalam M and Moustafa M
Department of Fish Diseases and Management, Faculty of Veterinary Medicine, Cairo University, 12211 Giza, Egypt

Seida A
a-Leibniz Research Institute for Environmental Medicine, Düsseldorf, Germany
b-Department of Microbiology and Immunology, Faculty of Veterinary Medicine, Cairo University, Giza Egypt

Yardimci RE and Gülşen Timur
Istanbul University, Department of Aquaculture, Ordu Street, No: 200, Laleli, Turkey

Pai IK and Maryem Shaikh Altaf
Department of Zoology, Goa University, Goa-403 206, India

Mohanta KN
Fishery field Laboratory, ICAR Complex, Old Goa, Goa, India

Mohammed Reshid, Yisehak Tsegaye Redda, Nesibu Awol and Awot Teklu
Mekelle University College of Veterinary Medicine, P.O.Box 231 Mekelle, Ethiopia

Marshet Adugna
Ethiopian Institute of Agricultural Research, National Fisheries and other Aquatic Life Research Center, P.O. Box 64, Sebeta, Ethiopia

Blessy G, Ajan C, Citarasu T and Michael Babu M
Planktology and Aquaculture Division, Centre for Marine Science and Technology, Manonmaniam Sundaranar University, Rajakkamangalam, Kanyakumari District, Tamil Nadu, India

Chaklader MR, Siddik MAB and Hanif MA
Department of Fisheries Biology and Genetics, Patuakhali Science and Technology University, Bangladesh

Ashfaqun Nahar
Department of Marine Fisheries and Oceanography, Patuakhali Science and Technology University, Bangladesh

Alam MJ
Department of Fisheries Management, Patuakhali Science and Technology University, Bangladesh

Sultan Mahmud
Department of Aquaculture, Patuakhali Science and Technology University, Bangladesh

Mohamed FA Abdel-Aziz and Mohammed A Ragab
Fish Rearing Lab, Aquaculture Division, National Institute of Oceanography and Fisheries (NIOF), Egypt

Yones AMM and Metwalli AA
National Institute of Oceanography and Fisheries, Shakshouk Fish Research Station, El-Fayoum, Egypt

Julinta RB, Abraham TJ, Anwesha Roy, Jasmine Singha and Gadadhar Dash
Department of Aquatic Animal Health, Faculty of Fishery Sciences, West Bengal University of Animal and Fishery Sciences, Chakgaria, Kolkata, West Bengal, India

NageshTS
Department of Fisheries Resource Management, Faculty of Fishery Sciences, West Bengal University of Animal and Fishery Sciences, Chakgaria, Kolkata, West Bengal, India

Patil PK
Central Institute of Brackishwater Aquaculture, Indian Council of Agricultural Research, Raja Annamalai Puram, Chennai, Tamil Nadu, India

Mzengereza K
Department of Fisheries Science, Mzuzu University, Private Bag 201, Mzuzu 2, Malawi

Kang'ombe J
Department of Aquaculture and Fisheries Science, Lilongwe University of Agriculture and Natural Resources, Bunda College, P.O. Box 219, Lilongwe, Malawi

Ibrahim M Aboyadak and Nadia GM Ali
Fish Disease Lab, National Institute of Oceanography and Fishery (NIOF), Egypt

Ashraf MAS Goda
Aquaculture Division, National Institute of Oceanography and Fishery (NIOF), Egypt

Walaa Saad
Central Diagnostic and Research Lab, Faculty of Veterinary Medicine, Kafrelsheikh University, Egypt

Asmaa ME Salam
Faculity of Aquatic and Fisheries Sciences, Kafrelsheikh University, Egypt

Eman A Abd El-Gawad and Ashraf M Abd El-latif
Department of Fish Diseases and Management, Faculty of Veterinary Medicine, Benha University, Egypt

Ramy M Shourbela
Department of Animal Husbandry and Animal Wealth Development, Faculty of Veterinary Medicine, Alexandria University, Egypt

Md. Anisur Rahman, Tayfa Ahmed, Md. Mehedi Hasan Pramanik, Flura, Md. Monjurul Hasan and Masud Hossain Khan
Bangladesh Fisheries Research Institute, Riverine Station, Chandpur, Bangladesh

M. G. S. Riar
Bangladesh Fisheries Research Institute, Freshwater Station, Mymensingh, Bangladesh

Khandaker Rashidul Hasan
Bangladesh Fisheries Research Institute, Freshwater Sub-station, Saidpur, Nilphamari, Bangladesh

Yahia Mahmud
Bangladesh Fisheries Research Institute, Headquarter, Mymensingh, Bangladesh

Tieliang Li, Zhihong Ma, Wei Xing, Na Jiang, Wentong Li, Xiangjun Sun and Lin Luo
Beijing Fisheries Research Institute, Beijing Key Laboratory of Fishery Biotechnology, Beijing, PR China

Chuan He
Extension Stations of Aquiculture Technology, Beijing Municipal Bureau of Agriculture, Beijing, PR China

George Ninan, Lalitha K.V, Zynudheen A.A and Jose Joseph
Central Institute of Fisheries Technology, Kochi 682 029, India

Alaa El-Din H. Sayed
Zoology Department, Faculty of Science, Assiut University, 71516 Assiut, Egypt

Nasser S. Abou Khalil
Medical Physiology Department, Faculty of Medicine, Assiut University, Assiut, Egypt

Weifeng Li, Xiaoyi Wu, Senda Lu, Shuntian Jiang, Yuan Luo, Mingjuan Wu and Jun Wang
Key Laboratory of Tropical Biological Resources of Ministry of Education, Hainan University, Haikou, China
Department of Aquaculture, Ocean College of Hainan University, Haikou, China

Ali SM
Microbiology Department, National Institute of Oceanography and Fisheries, Aswan Research Station, Egypt

Yones EM
Fish Diseases Lab, National Institute of Oceanography and Fisheries, Alexandria, Egypt

Kenawy AM, Ibrahim TB and Abbas WT
Hydrobiology Department, Veterinary Research Division, National Research Center, Dokki, Giza, Egypt

A. Hyder Ali, A. Jawahar Ali, M. Saiyad Musthafa, M.S. Arun Kumar, Mohamed Saquib Naveed, Mehrajuddin War and K. Altaff
Department of Zoology, The New College, Chennai, India

Tewodros Abate Alemayehu
Debre Berhan University, Department of Biology, Debre Berhan, Ethiopia

Ababe Getahun
Department of Zoological science, Faculty of Life Science, Addis Ababa University, Addis Ababa, Ethiopia

Ghada R Sallam and Hadir A Aly
Aquaculture Division, Fish Rearing laboratory, National Institute of Oceanography and Fisheries, Alexandria Branch, Alexandria, Egypt

Walied A Fayed
Department of Animal and Fish Production, Faculty of Agriculture (Saba Basha), Alexandria University, Alexandria, Egypt

Mohamed A El-Absawy and Zeinab A El-Greisy
Aquaculture Division, Fish Reproduction laboratory, National Institute of Oceanography and Fisheries, Alexandria Branch, Alexandria, Egypt

Fotedar R
Sustainable Aquatic Resources and Biotechnology, Department of Environment and Agriculture, Curtin University, Bentley 6102, Australia

Buller N
Department of Agriculture and Food Western Australia, Animal Health Laboratories, South Perth, WA, Australia

Ambas I
Sustainable Aquatic Resources and Biotechnology, Department of Environment and Agriculture, Curtin University, Bentley 6102, Australia

Faculty of Marine Science and Fishery, Department of Fishery, Hasanuddin University, Makassar, 90225, Indonesia

Chin-Long KY and Gilles LE MOULLAC
Ifremer, UMR 241 Ecosystèmes Insulaires Oceaniens (EIO), Labex Corail, Centre du Pacifique, BP 49, 98719 Taravao, Tahiti, French Polynesia

Siddik M, Chaklader M, Hanif M and Sharker M
Department of Fisheries Biology and Genetics, Patuakhali Science and Technology University, Bangladesh

Islam M
Department of Fisheries Management, Bangladesh Agricultural University, Bangladesh

Rahman M
Department of Fisheries Biology and Genetics, Sheikh Fazilatunnesa Mujib Fisheries College, Bangladesh

Cynthia Nyunja, Joyce Maina, Joshua Amimo, Felix Kibegwa and Joseph Jung'a
Department of Animal Production, University of Nairobi, Kenya

David Harper
Department of Animal Production, University of Nairobi, Kenya

University of Leicester, England

K Prameswari, M Hemalatha and B Kishori
Department of Biotechnology, Sri Padmavati Mahila Visvavidyalayam (Women's University), Tirupati, Andhra Pradesh, India

P Sreenivasula Reddy
Department of Zoology, Sri Venkateswara University, Tirupati, Andhra Pradesh, India

Asaad MR Al-Taee
Marine Science Center, Basra University, Basra, Iraq

Najem R Khamees and Nadia AH Al-Shammari
College of Agriculture, Basra University, Basra, Iraq

Index